HANDBOOK OF
COMPUTATIONAL FLUID MECHANICS

HANDBOOK OF
COMPUTATIONAL FLUID MECHANICS

edited by
Roger Peyret

CNRS
Laboratoire de Mathématiques J.-A. Dieudonné,
Université de Nice – Sophia Antipolis,
Parc Valrose,
06000 Nice, France

and

INRIA
2004, route des Lucioles,
06902 Sophia Antipolis, France

ACADEMIC PRESS

A Harcourt Science and Technology Company

San Diego   San Francisco   New York   Boston   London   Sydney   Tokyo

Academic Press Limited
24–28 Oval Road, London NW1 7DX, UK
http://www.hbuk.co.uk/ap/

Academic Press
A Harcourt Science and Technology Company
525 B Street, Suite 1900, San Diego, California 92101-4495, USA
http://www.apnet.com

ISBN 0-12-553010-2 (hdk)
ISBN 0-12-532200-3 (pbk)

A catalogue record for this book is available from the British Library

Printed in Great Britain by WBC Book Manufacturers Ltd, Bridgend, Mid Glamorgan

00 01 02 03 04 05 WB 9 8 7 6 5 4 3 2 1

Contents

List of contributors

G.B. Deng
CFD Group, LMF, URA 1217 CNRS, ECN, 1 Rue de Noe, 44072 Nantes Cedex, France

A. Dervieux
INRIA, 2004 route des Lucioles, BP 93, 06902 Sophia-Antipolis, France

T.B. Gatski
Mailstop 128, Aerodynamic and Acoustic Methods Branch, NASA Langley Research Center, Hampton, VA 23681-0001, USA

F. Grasso
Dip. Meccanica e Aeronautica, Università La Sapienza, Roma, Italy

M.D. Gunzburger
Department of Mathematics and Interdisciplinary Center for Applied Mathematics, Virginia Tech, Blacksburg, VA 24061-0531, USA

C. Härtel
Institute of Fluid Dynamics, Swiss Federal Institute of Technology, ETH Zentrum, CH 8092, Zurich, Switzerland

D.J. Mavriplis
Mailstop 132C, ICASE, NASA Langley Research Center, Hampton, VA 23681–0001, USA

C. Meola
Dip. Energetica Termofluidodinamica e Condizionamento Ambientale, Università di Napoli 'Federico II', Naples, Italy

J. Piquet
CFD Group, LMF, URA 1217 CNRS, ECN, 1 Rue de Noe, 44072 Nantes Cedex, France

P. Queutey
CFD Group, LMF, URA 1217 CNRS, ECN, 1 Rue de Noe, 44072 Nantes Cedex, France

M. Visonneau
CFD Group, LMF, URA 1217 CNRS, ECN, 1 Rue de Noe, 44072 Nantes Cedex, France

Preface

The purpose of this handbook is to provide graduate students, scientists and engineers with a well-documented critical survey of numerical methods for fluid mechanics. The rapid development of computational fluid mechanics, a result of the conjugate progress in numerical analysis, computer technology and visualization tools, has made it necessary to establish the state-of-the-art as it stands in the last years of the twentieth century. This has to be done while pointing out the basic foundations that have made possible the extraordinary spreading of the discipline, which nowadays plays an important role in all branches of fundamental and applied mechanics in both research and industry.

A complete description and thorough discussion of all the available efficient numerical methods would require more than one volume. Rather, the goal is to guide the reader towards a deeper understanding of the problems associated with the calculation of fluid motion in various situations: inviscid and viscous, incompressible and compressible, steady and unsteady, laminar and turbulent flows, simple and complex geometries. The best numerical calculations are those that combine an accurate discrete approximation, an efficient solution method and a well-adapted mesh. These various requirements are discussed in the course of this handbook.

The book is divided into seven chapters covering the main topics of computational fluid mechanics. Chapter 1, by A. Dervieux, is an introduction to numerical methods with a reflection on the basic principles. The next two chapters are devoted to the calculation of incompressible viscous flows based on the Navier–Stokes equations. In Chapter 2, G.B. Deng, J. Piquet, P. Queutey and M. Visonneau discuss mostly finite-volume and also finite-difference methods associated with general curvilinear coordinate systems, whose role is essential when dealing with complex geometries. In Chapter 3, M.D. Gunzburger presents the finite-element method and gives some important theoretical results. A well-established mathematical theory supports the finite-element method and this justifies the theoretical nature of this chapter compared to the others. Finite-volume methods for compressible inviscid and viscous flows are discussed by F. Grasso and C. Meola in Chapter 4, in which a number of fundamental questions concerning conservation laws are treated. The following two chapters are devoted to turbulent flow computations, each one considering an important aspect of them. In Chapter 5, C. Härtel discusses direct simulation and large-eddy simulation, while T.B. Gatski, in Chapter 6, presents a review of the turbulence models and associated numerical problems.

These two chapters clearly show the importance of the modelling (in nature and degree) in the numerical prediction of turbulent flows. Finally, Chapter 7, by D.J. Mavriplis, deals with mesh generation and adaption. These questions constitute an important aspect of computational fluid mechanics because the quality of the results strongly depends on the quality of the mesh.

The authors and the editor hope that the topics developed in the present handbook will be useful to anyone wishing to visit the fascinating world of the numerical representation of this natural yet so complex phenomenon: the motion of fluids.

R. Peyret
Nice, May 1995

1   About the basic numerical methods

Alain Dervieux

I INTRODUCTION

In the various chapters of this book, many sophisticated notions and methods of numerical analysis and engineering will be invoked. Introducing these notions is beyond the scope of this chapter; nor is this chapter written to give basic ideas to a neophyte. Instead, this chapter is intended for the reader who already has some knowledge of the topic; for a typical sample of numerical methods, we shall review briefly the existing mathematical presumptions that

HANDBOOK OF COMPUTATIONAL FLUID MECHANICS
ISBN 0-12-532200-3

make us believe that we shall obtain some numerical solutions and that these solutions are a good picture of the 'continuous solution' that is sought. A few standard references will help the reader to investigate further.

One striking feature is the relative weakness of the theoretical tools available for the numerical analysis of complex systems. The numerical analyst in computational fluid mechanics is forced to consider problems that are simpler than those encountered in practice. We shall therefore restrict our presentation to linear scalar partial differential equations and we concentrate on approximation questions.

The second striking feature is the difficulty in building bridges between the different numerical methods: if meshes are perfectly regular, truncation studies and Fourier stability analysis can be useful for most of the methods; if meshes are not regular, as occurs in real life, the user is somewhat deprived of ammunition; some specific analyses are possible for some particular approaches (e.g. finite-element theories).

- The accuracy of a method is the central quantitative question.
- The concept of stability is the central qualification condition (no solution, no accuracy).

These two basic considerations are in fact not independent. Accuracy theory assumes that mesh size can be made sufficiently small; furthermore, 'sufficiently' is not quantified. In practice, this assumption of the approximation theory is not clearly satisfied; applications are performed in a hazy land in which what is called 'spatial instability' may strike and hurt.

Spatial instability, although defined as a lack of stability in some special norm, is in fact an important point in approximation theory since a vast literature has been written for two main subtopics related to stabilization of flow approximations, namely primitive variable Stokes approximation and first-order hyperbolic systems.

This chapter will review the basic features of several numerical methodologies and discuss the answers they propose to the problem of spatial instability. This will be illustrated using a model problem of advection-diffusion, and basic ideas and properties of spatially stable advective schemes will be especially addressed.

The monographs and pedagogic works by Richtmyer and Morton (1967), Peyret and Taylor (1983), Dautray and Lions (1988), Fletcher (1988), Hirsch (1988) and Wendt (1992), may be very useful to the reader.

II CONTINUOUS MODELS

The purpose of this section is to define some basic models and to recall some important properties. For simplicity we consider only linear scalar models, which are sufficient for the introduction of several main features.

(i) *Unsteady advection-diffusion:*

$$\begin{array}{ll} 2D & u_t + au_x + bu_y - \nu\Delta u = f \\ 1D & u_t + au_x - \nu u_{xx} = f \end{array} \tag{1}$$

and associated initial and boundary conditions.

(ii) *Steady advection-diffusion:*

$$\begin{array}{ll} 2D & au_x + bu_y - \nu\Delta u = f \\ 1D & au_x - \nu u_{xx} = f \end{array} \tag{2}$$

and associated boundary conditions that can be Dirichlet or Neumann.

The relative weights of convective and diffusive terms are measured through the Peclet number (Reynolds number for Navier–Stokes) defined as follows:

$$Pe = |a|L/\nu,$$

where L is a typical length of the computational domain.

The main theories for the above models can be summarized as:

For the purely advective model ($\nu = 0$): the equation is hyperbolic; furthermore, the solution can be computed independently on one-dimensional curves, according to the theory of characteristics (see, for example, Pironneau, 1988). This theory has permitted the introduction of the maximum principle property: when f is zero, u is upper and lower bounded by the initial condition of the characteristic curve. This property extends to the diffusive case (Protter and Weinberger, 1967).

For advection-diffusion ($\nu \neq 0$): Hilbertian methods are much more developed around elliptic problems, which are coercive in the Sobolev space H^1.

(iii) *Linear conservation laws*:

$$\begin{array}{ll} 2D & u_t + (au)_x + (bu)_y - \nu\nabla \cdot (\nabla u) = f \\ 1D & u_t + (au)_x - (\nu u_x)_x = f \end{array} \tag{3}$$

Since (3) is equivalent to (1) with a right-hand-side depending on the divergence of the velocity $\mathbf{V} = (a, b)$ and not vanishing in general, then only positiveness of u is preserved.

Remark

The basic theories for the hyperbolic and elliptic models that we shall consider were defined a few decades ago. In the case of linear first-order hyperbolic systems, an abstract theory has been built around Friedrichs systems, in which coerciveness in L^2 of a bilinear form is the basic assumption (Friedrichs, 1958). In the case of non-linear systems, an abstract theory was first built (for scalar equations) by Kruskov (1970). For the elliptic case two complete theories can be found in Agmon *et al.* (1964) for Hölder spaces, and in Lions and Magenes (1968) for Sobolev spaces.

III DISCRETIZATIONS FOR STEADY EQUATIONS

This section will give an introduction to the main numerical methods for the research of 'continuous solutions'.

A Basic ideas in discretization

By modelling the behaviour of materials, the physicist has replaced the billions of atoms of matter by an infinitely complex continuous medium, described by a mathematical differential or integral equation:

$$Au = b$$

It was understood quite early on that closed formulas would not be obtained for the prediction u of the behaviour of the medium with complex data (geometry, etc.); an alternative was to compute an approximate behaviour, the so-called approximate solution u_h, an element of a finite-dimensional space, to be computed in a finite delay with a finite effort from an approximate finite-dimensional system:

$$A_h u_h = b_h$$

In the above expression, h is a small parameter assumed (in the dreams of numerical analysts) to tend towards zero while the dimension of the system tends to infinity. The approximation theory considers the question of the relation between u_h and u. Three families of tools are generally considered:

(i) Bridges between finite-dimensional and continuous spaces (interpolations, etc.) which allow A_h to be applied to u and finally the distance between u_h and u to be measured.
(ii) Consistency properties which state that A_h is closer to A when h tends to zero.
(iii) Stability properties which state that the discrete solutions u_h remain in a bounded/compact set when h tends to zero.

The fundamental question is the convergence one: Can we find a value h and a function u_h close enough to u ?

B Finite-Difference Methods (FDM)

The principle of the finite-difference method is to replace the differentiation operators by combinations of translation operators that can be derived from truncations of Taylor series, e.g.:

$$u(x + h) = u(x) + h\frac{du}{dx} + hR$$

$$\frac{du}{dx} = \frac{u(x + h) - u(x)}{h} + R$$

where R is a small remainder.

This technique is applied to the complete set of differential terms of the partial differential equation (PDE); the finite difference (FD) discrete system is then a function of linear combinations of translations of function u,

$$A_h u - f_h = \sum_k a_k u(x + kh) - \sum_k b_k f(x + kh)$$

and the difference between this expression and zero when u is the exact solution of the functional system $Au = f$ is called the FD *residual*.

The truncation order is the order p of the infinitely small quantity $A_h u - f_h = O(h^p)$ when h tends to zero. The scheme is said to be *consistent* if the truncation error tends towards zero when h tends to zero.

The consistency is not sufficient for obtaining an approximation of the exact solution u. One way to prove the *convergence* of u_h towards u is to observe that:

$$u_h - u = (A_h)^{-1}(A_h u - f_h)$$

The *stability* of the FD scheme can be defined as the uniform boundedness of $(A_h)^{-1}$ in the functional space L^2 of square-integrable functions. This property is generally deduced from 'simple boundedness':

$$\|(A_h)^{-1} u\| \le K(u)$$

by applying the Banach–Steinhaus theorem, the convergence in linear case is then deduced (following Kantorovitch; see Richtmyer and Morton, 1967).

In the case of unsteady problems that are time-marched, the system can be thought of as a (time-wise) triangular one and the stability condition can be analysed for a generic time step. The Lax equivalence theorem is the adaption of the Kantorovitch theorem to this important field. Fourier theory then yields a tool for analysing the unsteady FD schemes through this time-wise stability condition (Richtmyer and Morton, 1967), see Section IV.

The choice of L^2 as the functional space for stability analysis is adequate for most linear problems; one important exception concerns advection or advection-dominated advection-diffusion problems. Indeed, a central-differenced FD scheme can be L^2-stable but may provide very disappointing results owing to Gibbs-type oscillations. This phenomenon is generally referred to as *spatial instability*. Let us illustrate this with the following example:

$$u_x - \nu u_{xx} = 1; \quad u(0) = 0, \ u(1) = 0$$

The continuous or exact solution is given by

$$u = x - \frac{[1 - \exp{(x/\nu)}]}{1 - \exp{(1/\nu)}}$$

For ν positive, it is a concave solution with minima on the boundary; it tends to $u = x$ when ν tends to zero; for ϵ equal to zero, $u = x$ is one candidate solution, although it does not satisfy $u(1) = 0$.

Let us consider the usual central differenced finite-difference discretization for the steady advection-diffusion problem:

$$\frac{1}{2h}(u_{j+1} - u_{j-1}) - \nu\frac{1}{h^2}(u_{j+1} - 2u_j + u_{j-1}) = 1$$

where u_j is the approximation of u at point $x_j = jh$, $j = 0, \cdots, N$, $h = 1/N$.

It is easy to see that the corresponding discrete solution is given by:

$$u_j = jh - \frac{1 - q^j}{1 - q^N}, \quad q = \frac{2 + P}{2 - P}$$

where $P = h/\nu$ is the mesh Peclet number. We observe that the maximum-principle is satisfied only if the mesh Peclet number is less than 2.

Conversely, the solution for an upwind scheme

$$\frac{1}{h}(u_j - u_{j-1}) - \nu\frac{1}{h^2}(u_{j+1} - 2u_j + u_{j-1}) = 1$$

is given by the same formula but $q = 1 + 1/P$ and will fulfil this condition for any value of the Peclet number.

Remark

Any ν positive is enough to choose the right trend $u = x$ of the problem (this is true for the discrete case only if the mesh Peclet number is large enough). A somewhat equivalent approach is to consider that we are looking for the asymptotically steady solution of

$$u_t + u_x - \nu u_{xx} = 1$$

$$u(0, t) = 0, \quad u(1, t) = 0, \quad u(x, 0) = 0$$

In the case $\nu = 0$ the time-advancing scheme should involve some dissipation to ensure the convergence towards the $u = x$ solution (discarding the $u(1, t) = 0$ condition). This shows the relation between viscosity and sometime polarity that is seen more clearly for the case of non-linear systems with entropy.

C Finite-Volume Methods (FVM)

If the functions that are sought are not regular enough, then the remainder of the Taylor formula is not smaller than the first terms, so that, for example, many finite-difference schemes cannot be applied to the computation of discontinuous functions. Furthermore, most finite-difference methods apply through a mapping of the computational domain onto the physical one; this mapping also has to be regular enough to allow truncated Taylor formulas to be sufficiently accurate. Another manifestation of these difficulties is the necessity to conserve the integral of some important extensive quantities during an unsteady calculation. These complications in the design of FD schemes have motivated the introduction of integral formulations expressing conservation laws in a rather simple way (at least for second-order accuracy) in the physical domain, and that will allow the computation of discontinuous functions.

However, the first motivation in introducing finite-volumes was to express explicitly the discrete conservation laws by introducing the notion of *flux*. The physical domain is considered to be divided into cells. Between time t^n and time t^{n+1}, the increment of a physical (extensive) quantity, let us say the mass in cell C_j, denoted by:

$$mass_j = meas(C_j) \ density_j \tag{4}$$

is given by a sum of flux $flux_{jk}$ exchanged with each neighbouring cell C_j:

$$mass_j = mass_j + \sum_{k \in N(j)} flux_{jk} \tag{5}$$

Conservation of total mass is ensured by the equality:

$$flux_{jk} = -flux_{kj} \tag{6}$$

In the case of linear conservation laws, this may be written, for example, as

$$meas(C_j)u_j^{n+1} = meas(C_j)u_j^{n+1} + \sum_{k \in N(j)} (t^{n+1} - t^n)H_{\mathbf{V}}(u_j, u_k)$$

in which the elementary flux is assumed to satisfy a consistency equality and to approximate the normal flux of $u\mathbf{V}$ through the common interface of the two cells:

$$H_{\mathbf{V}}(u_j, u_j) = \int_{\partial C_j \cap \partial C_j} \mathbf{V} . \mathbf{n} \ u_j \ d\sigma \tag{7}$$

The Lax–Wendroff theorem (Lax and Wendroff, 1960) for the steady linear case can be summarized as:

If the scheme is conservative according to the flux definition (5),(6), and if the discrete solution converges to a limit, then the limit is a weak solution *(i.e. a distribution satisfying the equations).*

Theories establishing finite-volume convergence and accuracy further than the above theorem have not yet been developed, except through the ENO (Essentially Non-Oscillatory) developments (Harten and Osher, 1987) for higher order; see also for example Cockburn *et al.* (1994) for the unstructured case. However, the FDM theory is generally invoked for higher accuracy justification.

The choice of the form of the cell is also a rather important and controversial question. In cell-centred methods, primal meshes give the partition (triangular cells, for example). In vertex-centred methods, a dual partition (for example built from triangle medians) is used for flux derivation. Primal methods are better for capturing variation aligned with the mesh. Dual methods enjoy a larger number of direct neighbours, allowing more accurate accounting for irregular meshes.

Just as for FDM, the central FVM schemes may suffer instabilities for high Peclet computations, and upwinding or artificial viscosities are often introduced for stabilization.

D Finite-Element Methods (FEM)

The main idea of the finite-element method is to use finite-dimensional spaces of functions as basis and (test) functions in an integral formulation that allows a part of the differentiations to be transferred towards test functions.

For example, for solving problem (3), the following formulation is considered ($\vec{\mathbf{V}} = (a, b)$):

$$\int (\nu \nabla u . \nabla \varphi + \nabla . (u\mathbf{V})\varphi - f\varphi) \, \mathrm{d}x \, \mathrm{d}y = 0$$

where u is sought in a (shape) space $\mathcal{V}_h$ such that the above equation holds for any test function φ in a test space $\mathcal{W}_h$.

One important scheme is the Galerkin formulation, for which $\mathcal{V}_h$ and $\mathcal{W}_h$ are the same discrete subspace of the functional space. When the mesh is a regular one, and for some simple spaces $\mathcal{V}_h$, the standard Galerkin method can be interpreted as a central-differenced FDM and will suffer from advection-diffusion problems of spatial instability.

One way to circumvent these spatial instabilities in the FEM is to choose the *upwind Petrov–Galerkin approximation*. This approximation (Johnson *et al.*, 1984; Hughes and Mallet, 1986) consists of using a set of test functions that are not only different from the shape function, but also depend on the convection velocity:

$$\int (\nu \, \nabla u . \nabla \tilde{\varphi} + \nabla . (u\vec{\mathbf{V}})\tilde{\varphi} - f\tilde{\varphi}) \, \mathrm{d}x \, \mathrm{d}y = 0$$

with

$$\tilde{\varphi} = \varphi + h\nabla . (\varphi \mathbf{V})$$

where φ is the usual test function.

Remark

In several works (Hughes and Mallet, 1986), the Petrov–Galerkin denomination has evolved towards 'least-square Galerkin'. Indeed, the test function is adapted in such a manner that a second discretization of the equation is encapsulated in the test function. This discretization is somewhat equivalent to a least-square discretization; for a purely least-square approach we refer to Bruneau (1991).

Note that least-square discretizations are also equivalent to making the convection terms elliptic; another way to 'elliptize' convection was derived from the Lax–Wendroff scheme by Lerat et al. (1982).

E Compact Finite-Difference Methods (CFDM)

In the usual finite-difference methods the accuracy is increased by increasing the number of discretization points involved in the approximation of the derivatives at some given point. This leads to an extension of the stencil which is often undesirable (loss of the tridiagonal nature of the matrices, difficulties near a boundary, etc.). It is possible to increase the degree of accuracy while conserving a restricted stencil (for example a three-point stencil in a one-dimensional problem). One way to reach this goal is to use an hermitian method (Collatz, 1960) in which values of the derivatives are considered as dependent variables in complement to the values of the function. It is then necessary to define approximation formulas for connecting all these values: for example, for the first derivative

$$u'_{j+1} + 4u'_j + u'_{j-1} - \frac{3}{h}(u_{j+1} - u_{j-1}) = 0$$

and for the second derivative

$$u''_{j+1} + 10u''_j + u''_{j-1} - \frac{12}{h^2}(u_{j+1} - 2u_j + u_{j-1}) = 0$$

These two formulas are accurate to $O(h^4)$. When associating these formulas to the equation itself $Lu_j = f_j$ and adding boundary conditions as well as an additional boundary hermitian formula (Hirsh, 1975; Adam, 1977), we get a block tridiagonal system for the unknown (u_j, u'_j, u''_j). Note that we can easily eliminate the second-order derivatives and get a block tridiagonal system for the unknown (u_j, u'_j). More general three-point formulas are given by Rubin and Khosla (1976), Peyret and Taylor (1983), and higher-order, five-point formulas by Collatz (1960), and Lele (1992).

Other compact fourth-order schemes leading to simple tridiagonal system for the unknowns u_j have been proposed (Krause, 1971; Ciment et al., 1978; Dennis and Hudson, 1989).

F Spectral Methods (SM)

In spectral methods the approximation of the function makes use of a truncated
series of orthogonal basis functions. For example, Fourier series are used for
periodical problems. For boundary value problems Chebyshev or Legendre
polynomials are used as basis functions. Therefore, the approximation can be
written as:

$$u_N(x) = \sum_{k=1}^{N} \hat{u}_k \varphi_k(x) \tag{8}$$

where the $\hat{u}_k$ values have to be determined. The manner in which these
coefficients are determined characterizes the method. Let us consider the
boundary value problem

$$Lu = f, \quad a < x < b \tag{9}$$

$$u(a) = 0, \quad u(b) = 0 \tag{10}$$

The Galerkin method consists of annulating the residual $R_N = Lu_N - f$ in the
weak sense

$$(R_N, \varphi_i)_w = \int_a^b R_N \varphi_i w \, \mathrm{d}x = 0 \qquad i = 0, \cdots, N \tag{11}$$

where w is the weight associated with the orthogonality of the basis functions.
The Galerkin equations (11) assume that the values φ_k satisfy the homogeneous
boundary conditions (10). This is true with Fourier series, where boundary
conditions are replaced by periodicity. On the other hand, this cannot be done
for solving boundary value problems by means of Chebyshev or Legendre
polynomials, in which case, a modification of the above Galerkin method is
considered and leads to the so-called *tau* method.

The tau method simply consists of considering only the first $N - 1$ equations
of equations (11), that is for $i = 0, \cdots, N - 2$, and adding the boundary
conditions (10):

$$\sum_k \hat{u}_k \varphi_k(a) = 0, \quad \sum_k \hat{u}_k \varphi_k(b) = 0$$

Another method (*collocation*) consists of annulating the residual R_N on a given
set of points $x_j \in \,]a, b[$. Then the boundary conditions are added:

$$Lu_N(x_j) - f(x_j) = 0, \quad j = 1, \cdots, N - 1 \tag{12}$$

$$u_N(a) = 0, \quad u_N(b) = 0 \tag{13}$$

The collocation points x_j are generally the extrema of the Chebyshev (or
Legendre) polynomial of degree N. This choice is dictated mainly by reasons of

convergence of the approximation. In fact, the collocation method amounts to looking for the solution of the problem as the polynomial of degree N which will exactly satisfy the differential equation of some given points x_j where this polynomial will take the value $u_N(x_j)$. Therefore the truncated series expansion (8) can be reinterpreted as a Lagrangian interpolation polynomial

$$U_N(x) = \sum_{k=0}^{N} u_N(x_k)\Psi_k(x) \qquad (14)$$

where $\Psi_k(x_j) = \delta_{jk}$, and they can be easily determined (see Gottlieb *et al.*, 1984; Canuto *et al.*, 1988). By differentiation of (14), we express the derivatives at any collocation point in terms of the values u_N at all collocation points. So, bringing (14) as well as its derivatives into (12), we get an algebraic system for determining the values $u_N(x_j)$ rather than the coefficients $\hat{u}_k$. This strategy for the collocation method is now widely used in applications.

The main interest of spectral methods is their high degree of accuracy. It can be shown (Canuto *et al.*, 1988; Mercier, 1989) that the error between a given function $u(x)$ and its approximation $u_N(x)$ is such that

$$\|u - u_N\| \le \frac{c}{N^\alpha}$$

where α is related to the number of continuous derivatives of $u(x)$. Analogous results are also true for the error between the exact solution $u(x)$ of the differential problem and its approximation $u_N(x)$ (Canuto *et al.*, 1988). It is interesting to note that for a given (sufficiently large) number of points the degree of accuracy is governed by the regularity of the solution itself. In particular, for an infinitely derivable function the error is smaller than any power of $(1/N)$ (exponential or 'infinite' accuracy). This is completely different from FD or FE approximation where the error behaves as $1/N^p$, where p depends on the scheme and is essentially finite.

IV DISCRETIZATIONS FOR UNSTEADY EQUATIONS

In the following we restrict the presentation to cases where the mesh increment is constant.

A Stability, dispersion, dissipation

One remarkable property of the advection model is that the exact solution is well known and well behaved lending itself to an easy Fourier analysis for both stability and accuracy; we restrict the discussion here to advection with no diffusion.

1 Stability

The main result for stability in unsteady time-marched linear systems is the Lax equivalence theorem (Richtmyer and Morton, 1967):

> *Given a properly posed initial-value problem and a finite-difference approximation to it that satisfies the consistency condition, stability is the necessary and sufficient condition for convergence.*

The building of a complete L^2-stability theory for finite differences has yielded an efficient tool for the numerical analyst or the engineer. Indeed, the L^2-norm enjoys the following properties:

Fourier transforms are L^2-isometries and diagonalize both continuous and finite-difference operators when periodicity conditions and constant coefficients are assumed.

The Fourier transform of a partial differential operator:

$$\sum D_x^{\alpha_k} D_y^{\beta_k} u$$

is written:

$$\sum \xi_x^{\alpha_k} \eta_y^{\beta_k} \hat{u}$$

The Fourier transform of a finite-difference time-advancing scheme:

$$\sum a_k u_{j+k}^{n+1} = \sum b_k u_{j+k}^{n}$$

is written:

$$\sum a_k \exp\,(ik\theta)\hat{u}^{n+1} = \sum b_k \exp\,(ik\theta)\hat{u}^{n}$$

This means that passing from $\hat{u}^n$ to $\hat{u}^{n+1}$ is effected by multiplying by a complex number (the amplification factor) and therefore the transformed inverse matrix of the unsteady system is a triangular bidiagonal Jordan matrix in the time direction n. The stability problem of the unsteady discrete system is now reduced to the uniform boundedness of this approximate inverse matrix. It is then enough that the eigenvalue is bounded, which is expressed in a more general context by the standard Von Neumann criterion (Richtmyer and Morton, 1967):

$$\|u^{n+1}\| \leq (1 + k\Delta t)\|u^n\|$$

Remark

Note, however, that the L^2-stability has already shown its limitations; indeed, even for the simple advection model, the need for schemes that are L^∞-stable is evident since the positivity of the unknown or the maximum principle is often mandatory.

2 Accuracy

By Fourier transform, the exact solution of the advection equation is the rotation in $\mathbb{C}$ of a given (exact) angle of the previous state. An error on the angle (argument) is called a phase error. An error on the amplification modulus, i.e. the modulus of the above factor, represents a dissipation error if it is less than 1; if it is greater than 1, the time-advancing scheme is unstable.

The *equivalent equation* (see Hirsch, 1988), also referred to as the modified equation (Warming and Hyett, 1974), is a technique that casts the different parts of the error of finite-differences on further derivatives added to the continuous equation. Owing to the properties of Fourier for continuous functions, it is then easy to relate phase error with odd derivatives, and amplification error with even derivatives (or viscosities).

Remark
Fourier analysis is a powerful tool for stability analysis and for studying iterative algorithms: however, it cannot be applied to boundary conditions. Many works have attempted to build a half-space stability analysis; for one of the most elaborated, see Gustafsson *et al.* (1972) and Chapter 4.

B Method-of-lines

A large class of unsteady schemes can be considered as the result of a two-phase discretizing action, a spatial discretization and then a time discretization. This approach is often referred to as the method-of-lines.

Conversely, a discretization that cannot be derived in this way can be considered as a space-time discretization. The usual Lax–Wendroff (1960) schemes are space–time discretizations; a usual consequence for the unsteady research of steady solutions is the dependency of steady solutions on the time-step length used for reaching them.

In the case of method-of-lines schemes, the result of the first phase, spatial semi-discretization, is an ordinary differential system, to which the usual theory for ordinary differential equations (ODEs) can be applied and in particular the A-stability theory (see Gear, 1971).

Made even more popular by Jameson (1981), a particular A-stability analysis can be devised through the standard Fourier diagonalization. In this case, A-stability analysis is the comparison of two figures: the distribution of the eigenvalues of the space operator in the complex plane, and the contours of the time-amplification factor as a function of λ.

C Space–time approximations

This section is restricted to the description of some new trends; indeed, the golden age of such schemes is now passed.

The Lax–Wendroff scheme was introduced in 1960 as the first stable (dissipative in the sense of Kreiss) three-point scheme applicable to compressible flow problems. This allowed airfoil flows to be computed accurately for the first time. For the equation (1) with $v = f = 0$, the Lax–Wendroff scheme gives:

$$u_j^{n+1} = u_j^n - \frac{1}{2}\left(\frac{a\Delta t}{h}\right)(u_{j+1}^n - u_{j-1}^n) + \frac{1}{2}\left(\frac{a\Delta t}{h}\right)^2(u_{j+1}^n - 2u_j^n + u_{j-1}^n) \qquad (15)$$

The CFL (Courant, Friedrichs and Lewy, 1928) condition $\mid a \mid \Delta t \leq h$ ensures the stability of (15). Many versions of the Lax–Wendroff scheme were then developed, but the MacCormack (1969) version has been the most popular one and is still the basis of an important part of currently used finite-difference codes. For the advection-diffusion equation (1) with $f = 0$, one variant of the MacCormack scheme can be written as:

$$\bar{u}_j^{n+1} = u_j^n - \frac{a\Delta t}{h}(u_{j+1}^n - u_j^n) + \frac{\nu\Delta t}{h^2}(u_{j+1}^n - 2u_j^n + u_{j-1}^n) \qquad (16)$$

$$u_j^{n+1} = \frac{1}{2}(u_j^n + \bar{u}_j^{n+1}) - \frac{a\Delta t}{2h}(\bar{u}_j^{n+1} - \bar{u}_{j-1}^{n+1}) + \frac{\nu\Delta t}{2h^2}(\bar{u}_{j+1}^{n+1} - 2\bar{u}_j^{n+1} + \bar{u}_{j-1}^{n+1}) \qquad (17)$$

If $v = 0$, the MacCormack scheme reduces to the Lax–Wendroff scheme.

From the beginning of the 1980s, these codes have been progressively replaced by method-of-lines codes for which the steady solution does not depend on the time-step used. This dependence has also been turned to good advantage in some implicit versions by Lerat et al. (1982), who showed the possibility of stabilizing central differenced schemes with the Lax and Wendroff second-order derivative term while maintaining a particularly good accuracy.

V OSCILLATION CONTROL

In many physical problems, the admissible values of a variable are strictly limited to some interval; a classical constraint is to keep species concentration between 0 and 1. It is well known that Gibbs oscillations lead to violation of the corresponding maximum principle.

This section examines the ideas underlying attempts to improve a scheme or to design schemes that would enjoy one or both of the following two properties:

(i)   non-oscillatory schemes
(ii) schemes respecting maximum principles.

The control of oscillations can be obtained either by improving central differencing schemes by adding numerical viscosity or by applying staggered formulation. Numerical viscosity can be either added through a non-linear artificial term or it can be accounted for in an upwind formulation; note, however, that Petrov–Galerkin or least-squares Galerkin methods can be understood as staggering by upwinding the test function.

In most cases, non-oscillatory schemes disobey the maximum principle and produce over- or under-shoots. The maximum principle is respected by linear first-order accurate schemes (Godunov, 1959; Harten *et al.*, 1976) and by non-linear second-order accurate total variation diminishing (TVD) schemes.

A Godunov method

This section is a little alien to the present chapter since it addresses spatial stability for nonlinear problems. It is, however, an essential consideration in fluid dynamics.

It was soon noticed that the most accurate spatially stable (i.e. without oscillation) schemes for advection were upwind schemes: the optimum amount of numerical viscosity is that which annihilates the dependency of the solution on downward values. This derivation is quite straightforward for linear systems (Courant *et al.*, 1952).

Godunov (1959) introduced an enlightening idea for non-linear systems. The approach is to abandon the finite differences or the interpolation in finite volumes, since the solutions sought are singular ones. The focus is rather put on the discontinuities of a step function, which is constant on a cell partition of the interval considered. The evolution of the discontinuities can be considered as independent Riemann problems (unsteady problems with Heaviside initial conditions) provided the time is not great enough to allow the resulting waves to interact.

The resulting analytic solution is an exact one, but the process cannot continue another time step since the solution is no longer a step function.

The last phase of the time step is then the (conservative) projection of the analytic solution onto a step function that is constant in each cell of the above partition.

B Total Variation Diminishing property (TVD)

Specialists have long been searching for a scheme that would be more accurate than first-order monotone schemes like the upwind *donor cell* scheme and less oscillatory than second-order accurate schemes like the Lax–Wendroff schemes. Linear monotone schemes are non-oscillatory but cannot be second-order accurate. A less restrictive property has been introduced by Harten (1983).

Definition
A (scalar) two-level scheme is said to be TVD (total variation diminishing) if, for any mesh function $u^n = (u_j^n)_{j \in Z}$ of bounded variation,

$$TV(u^{n+1}) \leq TV(u^n) \tag{18}$$

where

$$TV(u^n) = \sum_j |u^n_{j+1} - u^n_j|. \tag{19}$$

Harten has proved that such schemes preserve monotonicity.

Furthermore, a sufficient condition for a scalar two-level scheme being a TVD scheme is to be a positive 'incremental scheme' according to the following definition:

$$u^{n+1}_j - u^n_j = C_{j+\frac{1}{2}}\delta u^n_{j+\frac{1}{2}} - D_{j-\frac{1}{2}}\delta u^n_{j-\frac{1}{2}} \tag{20}$$

with the notation:

$$\delta u^n_{j+\frac{1}{2}} = u^n_{j+1} - u^n_j \tag{21}$$

with:

$$C_{j+\frac{1}{2}} \geq 0, \quad D_{j+\frac{1}{2}} \geq 0 \quad \text{and} \quad C_{j+\frac{1}{2}} + D_{j+\frac{1}{2}} \leq 1 \tag{22}$$

in which the coefficients C and D may depend on u.

The above properties are easily verified for first-order accurate three-point monotone schemes like the usual upwind and Godunov schemes. We now examine a family of second-order advection schemes. In the following we consider the one-dimensional advection equation $u_t + au_x = 0$.

C Flux-Corrected Transport (FCT)

The first methodological throughput was provided by the Flux-Corrected Transport (FLT) scheme of Book *et al.* (1975); several further formalizations and extensions were introduced by Zalesak (1979). Starting from the Lax–Wendroff scheme applied to $u_t + au_x = 0$ (where $a > 0$):

$$\bar{u}^{n+1}_j = u^n_j - \frac{\sigma}{2}\left(\delta u^n_{j+\frac{1}{2}} + \delta u^n_{j-\frac{1}{2}}\right) + \left(\frac{\sigma^2}{2} + \frac{1}{8}\right)\left(\delta u^n_{j+\frac{1}{2}} - \delta u^n_{j-\frac{1}{2}}\right) \tag{23}$$

$$u^{n+1}_j = \bar{u}^{n+1}_j - \frac{1}{8}\left(l_{j+\frac{1}{2}}\delta W_{j+\frac{1}{2}} - l_{j-\frac{1}{2}}\delta W_{j-\frac{1}{2}}\right)$$

where σ is the Courant (or CFL) number defined by

$$\sigma = a\Delta t/h$$

and

$$W = \bar{u}^{n+1} \Rightarrow \text{Antidiffusion method}$$

or:

$$W = u^n \Rightarrow \text{Zalesak's method}$$

The $l_{j\pm\frac{1}{2}}$ coefficients must be built in terms of u in order to respect the maximum principle (see Zalesak, 1979). Note that for $l_{j\pm\frac{1}{2}} = 1$, we recover the Lax–Wendroff scheme; for $l_{j\pm\frac{1}{2}} = 0$, a monotone first–order scheme is obtained.

D Upwind and symmetric TVD principles

The FCT method as defined by Book *et al.* (1975) is a method for controlling *a posteriori* the monotonicity of the schemes. Its application relies in particular on the application of an explicit time advancing.

Conversely, the TVD methodology introduced by Harten allows an *a priori* control of the monotonicity.

To explain the TVD ideas, we start from a formulation by Sweby (1984):

$$u_j^{n+1} = \underbrace{u_j^n}_{1} - \underbrace{\sigma\{1 + \tfrac{1}{2}(1-\sigma)\left[\varphi(r_j^+)/r_j^+ - \varphi(r_{j-1}^+)\right]\}\delta u_{j-\frac{1}{2}}^n}_{2} \qquad (24)$$

with the following notation:

$$r_j^+ = \delta u_{j-\frac{1}{2}}^n / \delta u_{j+\frac{1}{2}}^n \qquad (25)$$

The function φ is discussed later.

We note that part 1 of (24) corresponds to the usual upwind scheme; part 2 of (24) is a term for counterbalancing the over-large diffusion ('antidiffusion term'):

$$[2] = \frac{\sigma(\sigma-1)}{2}\left[\varphi(r_j^+)/r_j^+ - \varphi(r_{j-1}^+)\right]\delta u_{j-\frac{1}{2}}^n$$

$$= \frac{\sigma(\sigma-1)}{2}\left[\varphi(r_j^+)\delta u_{j+\frac{1}{2}}^n - \varphi(r_{j-1}^+)\delta u_{j-\frac{1}{2}}^n\right]$$

Indeed, for $\varphi(r) = r$ we obtain an upwind scheme, and for $\varphi(r) = 1$ we obtain the usual Lax–Wendroff scheme.

This scheme can be written using the Harten formulation:

$$u_j^{n+1} = u_j^n - C_{j-\frac{1}{2}}\delta u_{j-\frac{1}{2}} + D_{j+\frac{1}{2}}\delta u_{j+\frac{1}{2}} \qquad (26)$$

with

$$\begin{cases} C_{j-\frac{1}{2}} = \sigma\{1 + \tfrac{1}{2}(1-\sigma)\left[\varphi(r_j^+)/r_j^+ - \varphi(r_{j-1}^+)\right]\} \\ D_{j+\frac{1}{2}} = 0 \end{cases} \qquad (27)$$

Then the above scheme is a TVD one via Harten's condition if:

$$C_{j-\frac{1}{2}} \in [0,1] \qquad (28)$$

and one sufficient condition for this is that:

$$0 \le \sigma \le 1 \qquad (29)$$

$$\varphi(r_j^+)/r_j^+ - \varphi(r_{j-1}^+) \in [-2, +2] \qquad (30)$$

Among the φ functions verifying (30), it is interesting to consider those providing antidiffusion:

$$\varphi(r) \geq 0 \qquad (31)$$

and vanishing at extrema:

$$\varphi(r) = 0 \quad \text{if} \quad r \leq 0 \qquad (32)$$

Finally, the conditions fulfilled by the function φ are:

$$\begin{cases} \varphi(r) = 0 \quad \text{if} \quad r \leq 0 \\ 0 \leq \varphi(r) \leq 2 \\ 0 \leq \varphi(r)/r \leq 2 \end{cases} \qquad (33)$$

We note that the above construction depends on the sign of the propagation velocity a; when a is negative, the scheme has to be written:

$$u_j^{n+1} = u_j^n + \sigma\left\{ -1 + \frac{1}{2}(1 + \sigma)\left[\varphi(r_{j+1}^-) - \varphi(r_j^-)/r_j^- \right] \right\} \delta u_{j+\frac{1}{2}}^n \qquad (34)$$

with

$$r_j^- = \delta u_{j+\frac{1}{2}}^n / \delta u_{j-\frac{1}{2}}^n, \quad \sigma = a\Delta t / h < 0 \qquad (35)$$

One of the disadvantages of (26), (34) is that it is necessary to know the sign of a. Sweby's approach (1984) consists of ignoring that sign, with a pessimistic majoration of the numerical viscosity; the scheme is reformulated as follows:

$$u_j^{n+1} = u_j^n - \frac{\sigma}{2}\left(\delta u_{j+\frac{1}{2}}^n + \delta u_{j-\frac{1}{2}}^n \right) + \left[\left(\frac{\sigma^2}{2} + K_{j+\frac{1}{2}} \right) \delta u_{j+\frac{1}{2}}^n - \left(\frac{\sigma^2}{2} + K_{j-\frac{1}{2}} \right) \delta u_{j-\frac{1}{2}}^n \right] \qquad (36)$$

Remark

We note that the above formula is similar to an FCT formulation, with the difference that:

$$K_{j\pm\frac{1}{2}} = \frac{1}{8}\left(1 - l_{j\pm\frac{1}{2}} \right) \qquad (37)$$

The Sweby approach gives:

$$K_{j+\frac{1}{2}}^{\text{Sweby}}\left(r_j^+, r_{j+1}^1, \sigma \right) = \begin{cases} K_{j+\frac{1}{2}}^+(r_j^+, \sigma) & \text{if} \quad a > 0 \\ K_{j+\frac{1}{2}}^-(r_{j+1}^-, \sigma) & \text{if} \quad a < 0 \end{cases} \qquad (38)$$

with

$$\begin{cases} K_{j+\frac{1}{2}}^+(r_j^+, \sigma) = \dfrac{\sigma}{2}(1 - \sigma)[1 - \varphi(r_j^+)] \\[2mm] K_{j+\frac{1}{2}}^-(r_{j+1}^-, \sigma) = \dfrac{|\sigma|}{2}(1 - |\sigma|)[1 - \varphi(r_{j+1}^-)] \end{cases} \qquad (39)$$

These methods have been tried with various steady and non-steady Euler flow computations. Davis (1984) and Yee (1987) have introduced a 'symmetric TVD' formulation that less explicitly dependent on the sign of a.

E The MUSCL method

The cause of oscillations in advection schemes was early interpreted as non-monotone reconstruction of the independent variables by Van Leer, who introduced a family of TVD referred to as MUSCL (Monotonic Upstream Scheme for Conservation Laws; Van Leer, 1983).

Let us return to the upwind TVD formulation of Sweby:

$$
K_{j+\frac{1}{2}} =
\begin{cases}
\dfrac{|\sigma|}{2}(1 - |\sigma|)(1 - \varphi(r_j^+)) & \text{if } \quad a > 0 \quad \text{with} \quad r_j^+ = \dfrac{\delta u_{j-\frac{1}{2}}}{\delta u_{j+\frac{1}{2}}} \\[3mm]
\dfrac{|\sigma|}{2}(1 - |\sigma|)(1 - \varphi(r_{j+1}^-)) & \text{if } \quad a < 0 \quad \text{with} \quad r_{j+1}^- = \dfrac{\delta u_{j+\frac{3}{2}}}{\delta u_{j+\frac{1}{2}}}
\end{cases}
\tag{40}
$$

We note that by choosing $\varphi(r) = r$ we get the (fully) upwind second-order scheme:

$$
u_j^{n+1} = u_j^n + \frac{\sigma}{2}\left(-3u_j^n + 4u_{j-1}^n - u_{j-2}^n\right) + \frac{\sigma^2}{2}\left(\delta u_{j-\frac{1}{2}}^n - \delta u_{j-\frac{3}{2}}^n\right) \quad \text{(if } a > 0)
$$

and for $\varphi(r) = \frac{1}{2}(r + 1)$, we obtain the half-upwind scheme of Fromm as an arithmetical average of (41) and Lax–Wendroff. This scheme presents some interesting symmetry properties which allow the MUSCL scheme of Van Leer to be built in an elegant manner. Indeed, if we consider the representation of the half-upwind scheme in the TVD region defined by (33), we observe that the limiting points, A and B, for the TVD region are symmetric to each other in the following way:

$$
\begin{aligned}
\varphi(r_A) = 2r_A \Rightarrow r_A = \tfrac{1}{3} \\
\varphi(r_B) = 2 \Rightarrow r_B = 3
\end{aligned}
\tag{41}
$$

The fact that the scheme belongs to the admissible interval can be interpreted geometrically as follows:

$$
\tfrac{1}{2}\left(|\delta u_{j-\frac{1}{2}}| + |\delta u_{j+\frac{1}{2}}|\right) \leq \min\left(|\delta u_{j+\frac{1}{2}}|, |\delta u_{j-\frac{1}{2}}|\right)
\tag{42}
$$

Let us consider, on the *dual interval* $](x_{j-1} + x_j)/2, \quad (x_j + x_{j+1})/2[$, a linear interpolation $\tilde{u}(x)$ of u that is *a priori* discontinuous at the limits of that interval, and which satisfies:

$$\begin{cases} \tilde{u}(x_j) = u_j \\ \tilde{u}\left(x_{j+\frac{1}{2}}\right) - \tilde{u}\left(x_{j-\frac{1}{2}}\right) = \frac{1}{2}\left(\delta u_{j+\frac{1}{2}} + \delta u_{j-\frac{1}{2}}\right) \end{cases} \tag{43}$$

then condition (42) means that the interpolation (43) is such that:

$$\tilde{u}(x) \in \left[\min\left(u_{j-1}, u_{j+1}\right), \max\left(u_{j-1}, u_{j+1}\right)\right] \tag{44}$$

Consequently, the TVD condition is related to the monotonicity of the interpolation (43); note that this link is a particular property of the half-upwind case.

F Essentially Non-Oscillatory method (ENO)

The reconstruction of the local interpolation of a function from a set of discrete values has early been recognized as a crucial issue for the design of advective schemes.

For example, the Lax–Wendroff scheme on a uniform 1D mesh can be interpreted as the following sequence:

- From a point j at time t^{n+1}, go backward along the trajectory, find x^* for time t^n.
- Interpolate the value u^* at x^* from values at x_j, x_{j-1}, x_{j+1}.
- Put $u_j^{n+1} = u^*$.

Of course, this parabolic interpolation does not preserve minima and maxima (or positivity) and one way to escape this disadvantage locally is to interpolate linearly in the places where the parabola makes over/under-shoots. However, this mixing with first-order interpolation produces a typical *clipping* of extrema, a phenomenon encountered frequently on TVD sophisticated schemes.

Van Leer (1983) long ago identified the importance of reconstructing functions for building non-oscillatory schemes, but within the TVD frame, involving clipping phenomena. Harten and Osher (1987) later identified a criterion that allowed non-TVD but non-oscillatory schemes to be built from reconstruction:

A reconstruction of degree k, $R(u, k)$ of a function u is said to be *essentially non oscillatory* if:

$$TV(R(u,k)) \leq TV(u) + O(h^r), \quad r \leq k$$

A typical ENO reconstruction is obtained by selecting the numerical molecule of a given cardinality that leads to the smallest divided difference (Harten, 1983) of maximal order:

$$|\Delta^k[x_j, x_{j+1}, ..., x_j + k]u| \leq |\Delta^k[x_l, x_{l+1}, ..., x_l + k]u|, \quad l = j+1 \text{ and } l = j - 1$$

Once the reconstruction is done, an exact finite volume integration is derived by putting:

$$\bar{u}_j(t) = \int_{x_{j-\frac{1}{2}}}^{x_{j+\frac{1}{2}}} u(x,t)\, dx$$

$$\bar{u}_j(t^{n+1}) - \bar{u}_j(t^n) + (x_{j+\frac{1}{2}} - x_{j-\frac{1}{2}})\left[\int_{t^n}^{t^{n+1}} h(u(x_{j+\frac{1}{2}}, s)\, ds - \int_{t^n}^{t^{n+1}} h(u(x_{j-\frac{1}{2}}, s)\, ds\right] = 0$$

It is remarkable that this method can choose molecules producing downward finite differences that are known to be unstable in time-advancing schemes. In fact the unstable modes that may appear will be non-linearly damped by the variation from one time level to the other of the molecules. This can result in a severe reduction in the accuracy observed *a posteriori*; this difficulty can be overcome by considering a set of more stable molecules.

Another disadvantage of the molecule adaption is that the non-linear scheme is *discontinuous*, and is therefore less able to reach steady states than the usual artificial viscosity or TVD schemes (TVD schemes still sometimes suffer this defect). Conversely, the bonus of ENO is the elimination of the clipping of extrema.

The link between reconstruction and mesh adaption is quite evident: a poor reconstruction (small molecule) indicates either a singularity or a small (possibly C^∞) detail. In both cases accuracy is locally less good, and mesh refinement should be applied. Conversely, where the molecules can be larger, a coarse mesh is sufficient. These observations have led to several mesh adaption theories such as multiresolution (Harten, 1994).

VI CONCLUDING REMARKS

In this short overview focusing on basic principles, we have left out the description of some new advective schemes involving low transverse numerical diffusion such as distribution schemes (Deconinck *et al.*, 1993).

Little emphasis has also been laid on the important role of the hyperbolic analysis of flow models, leading to flux splitting and fluctuation splitting (Deconinck *et al.*, 1993), that resulted in the introduction of a special treatment for each first-order derivative of a hyperbolic system. These methods are discussed in Chapter 4.

A striking feature is the considerable effort that has been devoted to obtaining the best second–order accurate schemes (instead, for example, of a fourth–order one). But, as has already demonstrated by the development of ENO methods, the door for even higher-order (more than three) accurate treatment of regular flow is now half opened.

REFERENCES

Adam, Y. (1977). *J. Comput. Phys.* **24**, 10–22.

Agmon, S., Douglis, S. and Nirenberg, L.(1964). *Comm. Pure Appl. Math.* **17**, 35–92.

Book, D.L., Boris, J.P. and Hain, K. (1975). *J. Comput. Phys.* **18**, 248–283.

Bruneau, C. H. (1991). *Computers and Fluids* **19**, 231–242

Canuto, C., Hussaini, N.Y., Quarteroni, A. and Zang, T.A. (1988). *Spectral Methods in Fluid Dynamics.* Springer-Verlag, Berlin.

Ciment, M., Leventhal, S.H. and Weinberg, B.C. (1978). *J. Comput. Phys.* **28**, 135–166.

Collatz, L. (1960). *The Numerical Treatment of Differential Equations.* Springer-Verlag, Berlin.

Cockburn, B., Coquel, F. and Le Floch, Ph. (1994). *Math. Comput.* **63**, 77–104, 207.

Courant, R., Friedrichs, K.O. and Lewy, M. (1928) *Math. Ann.* **100**, 32–76.

Courant, R., Isaacson, E. and Rees M. (1952). *Comm. Pure Appl. Math.* **5**, 243–255.

Dautray, R. and Lions, J.L. (1988). *Analyse mathématique et calcul numérique,* vol. 6. Masson, Paris.

Davis, S.F. (1984). *ICASE Report 84–20.*

Deconinck, H., Roe, P.L. and Struijs, R. (1993). *Computers and Fluids* **22**, 215–222.

Dennis, S.C.R. and Hudson, J.D. (1989). *J. Comput. Phys.* **85**, 390–416.

Fletcher, C.A.J. (1988). *Computational Techniques for Fluid Dynamics.* Springer-Verlag, Berlin.

Friedrichs, K.O. (1958). *Comm. Pure Appl. Math.* **2**, 333–418.

Gear, C.W. (1971). *Numerical Initial Value Problems in Ordinary Differential Equations.* Prentice Hall, Englewood Cliffs.

Godunov, S.K. (1959). *Mat. Sbornik* **47**, 271–295.

Gottlieb, D., Hussaini, M.Y. and Orszag S.A. (1984). In *Spectral Methods for Partial Differential Equations,* (R. G. Voigt, D. Gottlieb and M. Y. Hussaini, eds), pp. 1–54. SIAM, Philadelphia.

Gustafsson, B., Kreiss, H.O. and Sundström, A. (1972). *Math. Comput.* **26**, 649–686.

Harten, A. (1983). *J. Comput. Phys.* **49**, 357–393.

Harten, A., Hyman, J. M. and Lax, P. D. (1976). *Comm. Pure Appl. Math.* **29**, 197–322.

Harten, A. and Osher S. (1987). *SIAM J. Numer. Anal.* **24**, 279–309.

Harten, A. (1994). *J. Comput. Phys.* **115**, 319–338.

Hirsch, C. (1988). *Numerical Computation of Internal and External Flows.* J. Wiley & Sons, Chichester.

Hirsh, R.S. (1975). *J. Comput. Phys.* **19**, 90–109.

Hughes, T.J.R. and Mallet, M. (1986). *Comput. Methods Appl. Mech. Engrg.* **58**, 329–336.

Jameson, A., Schmidt, W. and Turkel, E. (1981). *AIAA Paper 81-1259.*

Johnson, C., Navert, U. and Pitkaranta, J. (1984). *Comput. Methods Appl. Mech. Engrg.* **45**, 285–312.

Krause, E. (1971). *DLR Mitt.* **71**, 109–138.

Krushkov, S.N. (1970). *Math. USSR Sb.* **10**, 217–243.

Lax, P.D. and Wendroff B. (1960). *Comm. Pure Appl. Math.* **13**, 217–237.

Lele, S.K. (1992). *J. Comput. Phys.,* **103**, 16–42.

Lerat, A., Sidès, J. and Daru, V. (1982).In *Proc. 8th International Conference on Numerical Methods in Fluid Dynamics,* Aachen, 1982 (E. Krause ed.), Lecture Notes in Physics, vol. 170, pp. 343–349. Springer-Verlag, Berlin.

Lions, J.L. and Magenes, E. (1968). *Problèmes aux limites non homogènes,* vol. 1. Dunod, Paris.

MacCormack, R.W. (1969). *AIAA Paper 69-354.*

Mercier, B. (1989). *An Introduction to the Numerical Analysis of Spectral Methods.* Springer-Verlag, Berlin.

Peyret, R. and Taylor, T.D. (1983). *Computational Methods for Fluid Flow.* Springer-Verlag, Berlin.

Pironneau, O. (1988). *Finite Element Methods for Fluids.* J. Wiley & Sons, Chichester.

Protter, M.H. and Weinberger, H.F. (1967). *Maximum Principles in Differential Equations.* Prentice Hall, Englewood Cliffs.

Richtmyer, R.D. and Morton, K.W. (1967). *Difference Methods for Initial-value Problems.* Interscience, J. Wiley & Sons, Chichester.

Rubin, S.G. and Khosla, P. K. (1976). *J. Comput. Phys.* **24**, 217–246.

Sweby, P.K. (1984). *SIAM J. Numer. Anal.* **21**, 995–1011.

Van Leer, B. (1983). In *Von Karman Institute, Computational Fluid Dynamics Lecture Series* 1983-04.

Warming, R.F. and Hyett, B.J. (1974). *J. Comput. Phys.* **14**, 159–179.

Wendt, J.F. (ed.) (1992). *Computational Fluid Dynamics, An Introduction.* Springer-Verlag, Berlin.

Yee, H. (1987). *NASA-TM 89464.*

Zalesak, S. (1979). *J. Comput. Phys.* **31**, 335–362.

2 Navier–Stokes equations for incompressible flows: finite-difference and finite-volume methods

G.B. Deng, J. Piquet, P. Queutey and M. Visonneau

I INTRODUCTION

The techniques developed in this chapter are applicable to problems where there is no obvious dominant flow direction. Also viscous–inviscid interaction is usually present, so that a separate calculation of wall regions and inviscid flow regions is not possible. Moreover regions of massive separation are usually present with the possibility of flow unsteadiness and turbulence. The restriction to incompressible flow introduces the computational difficulty that the continuity equation contains only velocity components, with no direct link with the pressure, as in the compressible flow case. There are several possible formulations, i.e. several choices for the dependent variables, to solve this difficulty. The most common formulation is the so-called primitive formulation in which the velocity field, $\mathbf{V}$, and the pressure field, ρP, are treated as unknowns in the fluid flow domain Ω.

$$\frac{\partial \mathbf{V}}{\partial t} + \mathbf{A}(\mathbf{V}) = -\nabla P + \nu \nabla^2 \mathbf{V} \tag{1}$$

$$\nabla \cdot \mathbf{V} = 0 \tag{2}$$

The non-linear convective term $\mathbf{A}(\mathbf{V})$ may be written in several alternative forms:

$$\mathbf{A}(\mathbf{V}) = (\mathbf{V} \cdot \nabla)\mathbf{V} \tag{3a}$$

$$\equiv \operatorname{div}(\mathbf{VV}) \tag{3b}$$

$$\equiv \operatorname{grad} \frac{V^2}{2} - \mathbf{V} \times \operatorname{curl} \mathbf{V} \tag{3c}$$

The form (3a) is the so-called convective form which states that the convective non-linear term is the sum of a curl-free and a solenoidal contribution:

$$(\mathbf{V} \cdot \nabla)\mathbf{V} \equiv \mathbf{A} = \operatorname{grad} P - \operatorname{curl} \mathbf{B}$$

where

$$\operatorname{curl} \mathbf{B} = \nu \nabla^2 \mathbf{V} - \frac{\partial \mathbf{V}}{\partial t}$$

The form (3b) is the conservative form, the latter form (3c) builds in the conservation of kinetic energy and involves the vector product of the velocity by the vorticity vector $\omega = \operatorname{curl} \mathbf{V}$. In both cases, the pressure gradient, which is the main source of momentum, adjusts itself to satisfy the continuity equation. If the divergence of the momentum equation is taken and if the velocity field is assumed to be known, the pressure is the solution of a Poisson equation which involves on its right-hand-side, besides the vorticity, the rate of strain tensor $\mathbf{D}$:

$$\nabla^2 P = -\nabla \cdot \mathbf{A} \equiv -\mathbf{D} : \mathbf{D} + \tfrac{1}{2}\omega^2 \quad \text{in } \Omega \tag{4}$$

or using the Stokes decomposition (Truesdell and Toupin, 1961):

$$P(\mathbf{x}) = \frac{1}{4\pi} \int_\Omega \frac{\nabla' \cdot \mathbf{A}'}{|\mathbf{x} - \mathbf{x}'|}\, \mathrm{d}v' + \frac{1}{4\pi} \int_\Sigma \frac{\mathbf{N} \cdot [|\mathbf{A}'|]}{|\mathbf{x} - \mathbf{x}'|}\, \mathrm{d}a' - \frac{1}{4\pi} \int_{\partial\Omega} \frac{\mathbf{n} \cdot \mathbf{A}}{|\mathbf{x} - \mathbf{x}'|}\, \mathrm{d}a' \tag{5}$$

$$\mathbf{B}(\mathbf{x}) = \frac{1}{4\pi} \int_\Omega \frac{\mathrm{curl}'\mathbf{A}'}{|\mathbf{x}-\mathbf{x}'|}\mathrm{d}v' + \frac{1}{4\pi}\int_\Sigma \frac{\mathbf{N}\times[|\mathbf{A}'|]}{|\mathbf{x}-\mathbf{x}'|}\mathrm{d}a' - \frac{1}{4\pi}\int_{\partial\Omega}\frac{\mathbf{n}\times\mathbf{A}}{|\mathbf{x}-\mathbf{x}'|}\mathrm{d}a' \quad (6)$$

where Σ is a discontinuity surface crossing the three-dimensional fluid volume Ω of outer boundary $\partial\Omega$; $\mathbf{N}$ is the unit normal to Σ, oriented from side 1 toward side 2 of the discontinuity, in agreement with the definition $[|\mathbf{A}|] \equiv \mathbf{A}_2 - \mathbf{A}_1$; and $\mathbf{n}$ is the outward normal to Ω on $\partial\Omega$. The prime indicates that the determination of P and $\mathbf{B}$ at any given point $\mathbf{x}$ within the domain Ω requires the integration to be carried out by summing elementary contributions at the generic points $\mathbf{x}'$ where all primed quantities are evaluated. Equation (6) justifies the fact that the velocity is the single solution of an integro-differential problem. Equation (5) indicates that given the velocity field (and hence the non-linear term), the value of the pressure within the fluid domain Ω can be calculated without any knowledge of the pressure over $\partial\Omega$. It indicates also that, given an initial divergence-free velocity field, the role of the pressure is to keep it divergence free. Moreover, for smooth advection terms the first surface integral in (5) vanishes and the boundary integral involves two contributions. The first of these is proportional to $q\partial q/\partial n$ (if q is the velocity modulus); the second is proportional to the product $(\mathbf{V}, \boldsymbol{\omega}, \mathbf{n})$. Although no pressure boundary condition is in principle required for the solution of the Navier–Stokes equations, the discrete problem does not, in general, see the pressure as a discrete form of (6) but involves it as a discrete form of (1), (2) and often introduces pressure values at the domain boundary. Since there is no primitive pressure equation, the pressure must be derived from the primitive equations (1), (2) by using the divergence-free constraint. Then the lack of natural boundary conditions for pressure presents particularly severe difficulties as reviewed in Orszag et al. (1986) and Gresho (1991a, b). Two cases must be distinguished and the discrete problem of pressure boundary conditions depends on whether a fractional-step or a one-step method is employed for the time discretization. In a fractional step or projection method, the appropriate discrete boundary condition for pressure is obtained quite naturally in the form of homogeneous or inhomogeneous Neumann conditions. In non-fractional-step methods, the problem is more complex, but it has been solved in an adequate way, using the so-called influence matrix technique (Glowinski and Pironneau, 1979; Kleiser and Schumann, 1980; Le Quéré and Alziary de Roquefort, 1985; Quartapelle and Napolitano, 1986; Dennis and Quartapelle, 1989). Unfortunately, the cost of solving the auxiliary problem (the influence matrix system) is prohibitively high in three dimensions.

Another interesting choice is the so-called velocity–vorticity formulation (Farouk and Fusegi, 1985; Speziale, 1987) in which the vorticity equation can be written as:

$$\frac{\partial\omega}{\partial t} = -\mathrm{div}\,(\mathbf{V}\omega - \omega\mathbf{V}) - \nu\,\mathrm{curl}\,(\mathrm{curl}\,\mathbf{V}) \quad (7a)$$

$$= -\mathrm{div}\,\mathbf{V}\omega + \boldsymbol{\omega}\cdot\mathbf{D} + \nu\,\nabla^2\boldsymbol{\omega} \quad (7b)$$

While the conservative form (7a) is the most suitable from a discrete point of view, the form (7b) emphasizes the fact that a vortex-stretching term $\omega \cdot \mathbf{D}$ supplies the eliminated pressure gradient in the transport equation. This term, which describes the interaction of the vorticity field with the rate of strain field $\mathbf{D}$, vanishes in two-dimensional flows. A time-marching scheme generally used for this equation yields a predicted vorticity field which is not divergence-free and must be corrected in agreement with $\omega = \text{curl } \mathbf{V}$. This equation points out the difficulty of getting updated $\mathbf{V}$ and ω as divergence-free fields. Also the determinations of the velocity cannot be uncoupled with that of the vorticity field because of the coupled boundary condition in the formulation. These difficulties can be solved in several different ways. We have the following choices:

(i)   The 'storage optimum' toroidal–poloidal representation of the velocity field (Clever and Busse, 1974).
(ii) Gatski *et al.* (1982) solve directly the discretized forms of div $\mathbf{V} = 0$ and $\omega = \text{curl } \mathbf{V}$ in a coupled way (see also Gatski, 1991). They report that the resulting block matrix equation is difficult to solve in a very efficient way. However, this method has been used successfully by Huang *et al.* (1992) for the lid-driven parallelopipedic cavity problem.
(iii) For the study of three-dimensional natural convection in a box, Aziz and Hellums (1967) and Mallinson and Van De Vahl Davis (1977) have used a vector potential Ψ such that $\mathbf{V} = \text{curl } \Psi$ so that $\nabla^2 \Psi = \omega$ (see also Richardson and Cornish, 1977; Yang and Camarero, 1986). For outer-flow problems, Hirasaki and Hellums (1968, 1970) have used the Stokes potential decomposition $\mathbf{V} = \text{curl } \Psi + \text{grad } \Phi$. They account for continuity through $\nabla^2 \Phi = 0$ with a Neuman-type boundary condition for the auxiliary potential Φ. Such a technique has been utilized for instance by Aregbasola and Burley (1977). However, Wong and Reizes (1984) have indicated difficulties in satisfying the (discrete) divergence-free constraint for $\mathbf{V}$ when using the discrete analogue of this method.
(iv) A fourth way to satisfy the solenoidality condition consists of solving the Poisson equation for the velocity:

$$\nabla^2 \mathbf{V} = -\text{curl } \omega$$

This method has been adopted by Dennis *et al.* (1979). Daube (1992) has used a matrix influence technique to satisfy this Poisson equation and $\omega = \text{curl } \mathbf{V}$ simultaneously in the two-dimensional case (see also Daube *et al.*, 1991; Ruas, 1991). Unfortunately, the influence technique used seems prohibitively expensive in the general (3D) case. Tromeur-Dervout and Ta Phuoc (1992) have however solved the lid-driven parallelopipedic cavity problem with a multigrid method for the Poisson equation. This method seems highly efficient, not only from a computational point of view, but also in that the reported residuals over div $\mathbf{V}$, div ω and curl $\mathbf{V} - \omega$ are of the order of 10^{-4}, 10^{-2} and 10^{-3}, respectively. This method has been further improved by Fontaine and Ta Phuoc (1994) using an influence technique for the vorticity–velocity coupling through the boundary condition $\omega \cdot \mathbf{n} = \mathbf{n} \cdot \text{curl } \mathbf{V}$. Finally, it is worth noting that the velocity–vorticity method has been also proposed by Guevremont *et al.* (1993) using a weak-Galerkin finite-element method followed by Newton linearization, with a direct solver for the simultaneous treatment of the velocity and vorticity components. Due to its high storage requirements, this method has been tested only on a coarse grid for the lid-driven cubic cavity problem.

In spite of the high potential of the velocity–vorticity formulation which bypasses the (cpu-consuming) pressure problem and involves directly the vorticity dynamics, the present work deals rather with the primitive variable formulation. It considers the use of this formulation for an arbitrary geometrical problem and includes the possible treatent of turbulent flows using an eddy-viscosity closure (see Chapter 6). The present chapter is outlined as follows. We first review the differential equations of the primitive variable formulation (Section II) in order to discuss the fundamental difficulties of the vector-form formulation, the possible ways to write down the possible component forms of the equations and the associated conservation problems. Section III is devoted to the discretization of the momentum equations. Attention is focused on schemes to be used in the so-called computational space. The discrete form of the continuity equation and the so-called coupling problem are discussed at length in Section IV. Strategies are discussed in Section V and some concluding remarks are put forward in Section VI.

II THE EQUATIONS

A Geometrical concepts

The Navier–Stokes equations are usually written in a cartesian coordinate system, with a basis of unit vectors $\mathbf{i}_a$ which do not depend on coordinates $\{\bar{x}^a\}$ and retain the same orientation relative to each other. In this cartesian system, the line element is defined by $d\mathbf{R} = d\bar{x}^a \mathbf{i}_a$ and can be integrated to $\mathbf{R} = \mathbf{0} + \bar{x}^a \mathbf{i}_a$, where $\mathbf{0}$ specifies the origin of the coordinate system. Here and throughout the chapter, repeated indices indicate summation over the three values of the index. Because turbulent viscous flows need to be studied past complex geometries, it seems suitable to write the master equations in a (body-fitted) curvilinear coordinate system $\{\xi^i\}$ in which the line element is defined by $d\mathbf{R} = d\xi^i \mathbf{R}_{;i} \equiv d\xi^i \mathbf{g}_i$, where the semicolon ';' denotes the partial derivative with respect to ξ^i. In contrast to the cartesian system, the basis vectors $\mathbf{g}_i$ are not of unit modulus, they are not dimensionless and their relative orientation changes with the point **R**.

A point of space **R** is therefore defined by its cartesian coordinates $\bar{x}^a$ in the basis $\mathbf{i}_a$ or by its curvilinear coordinates ξ^i in the basis $\mathbf{g}_i$. The transformation $\bar{x}^a(\xi^i)$ must of course be one-to-one and invertible (almost everywhere). A necessary and sufficient condition for these requirements is that the determinant of the matrix $g_i^a \equiv \bar{x}_{;i}^a$ of row vectors $\mathbf{g}_i$ does not vanish (we denote $g_i^a = \mathbf{g}_i \cdot \mathbf{i}_a$). This determinant is called the jacobian of the transformation $\bar{x}^a(\xi^i)$.

$$J = \frac{D(\bar{x}^1, \bar{x}^2, \bar{x}^3)}{D(\xi^1, \xi^2, \xi^3)} = \mathbf{g}_i \cdot (\mathbf{g}_j \times \mathbf{g}_k); \quad i, j, k \text{ in cyclic order} \tag{8}$$

The coordinate system $\{\xi^i\}$ is chosen so as to simplify the equation that defines the geometry around which the flow is to be predicted: instead of describing the body by the implicit equation $\mathcal{B}(\bar{x}^a) = 0$, the body surface is defined by, say, $\xi^3 = 0$. For this reason, the coordinate system $\{\xi^i\}$ is said to be a body-fitted coordinate (BFC) system. The simplicity gained for the specification of boundary conditions in the BFC system offsets the added complexity of the master equations with respect to their cartesian form.

Leaving aside the question of coordinate generation (see Chapter 7), we require some geometrical concepts before writing the master equations in the BFC system (see Thompson *et al.*, 1982; Vinokur, 1989). Apart from the covariant basis $\{\mathbf{g}_i\}$ of vectors tangent to the curves along which ξ^i varies, it is necessary to introduce the contravariant basis $\{\mathbf{g}^i\}$ of vectors $\mathbf{g}^i \equiv \text{grad } \xi^i$ normal to faces $\xi^i = \text{const}$. The relation between contravariant and covariant vectors is easily obtained from (9):

$$\frac{\partial \xi^i}{\partial \bar{x}^a} \cdot \bar{x}_{;j}^a = \delta_j^i \quad \text{or} \quad \mathbf{g}^i \cdot \mathbf{g}_j = \delta_j^i \tag{9}$$

and thus, because of (8):

$$\mathbf{g}^i = J^{-1} \mathbf{g}_j \times \mathbf{g}_k; \quad i, j, k \text{ in cyclic order} \tag{10}$$

Because $dV = d\bar{x}^1 \cdot d\bar{x}^2 \cdot d\bar{x}^3 = J \, d\xi^1 \cdot d\xi^2 \cdot d\xi^3$, the jacobian J is easily interpreted as the volume in the physical space $\{\bar{x}^a\}$ of a unit cube in the so-called computational space $\{\xi^i\}$ of unit sides $\Delta\xi^1 = \Delta\xi^2 = \Delta\xi^3 = 1$. Because the area element on a coordinate surface of constant ξ^i is: $dA = \|\mathbf{g}_j \times \mathbf{g}_k\| \, d\xi^j \cdot d\xi^k$, it is clear that the modulus of $\mathbf{S}^i = J\mathbf{g}^i$ is the surface area corresponding to a unit increment $\Delta\xi^j \Delta\xi^k$ (i, j, k cyclic) and that its direction is orthogonal to the surface $\xi^i = \text{const}$. An immediate consequence is that the flux associated to the vector **V** across the surface area of sides $\Delta\xi^j$, $\Delta\xi^k$ (i, j, k cyclic) is related to the contravariant component $V^i = \mathbf{V} \cdot \mathbf{g}^i$ by:

$$\text{Flux} \equiv JV^i = \mathbf{V} \cdot \mathbf{S}^i \tag{11}$$

The oriented area satisfies the so-called first fundamental metric identity (12):

$$\frac{\partial S_a^i}{\partial \xi^i} = 0 \quad \text{for all values of } a \tag{12}$$

This relationship must be fulfilled identically in a continuous as well as in a discrete sense. A second fundamental identity also has to be fulfilled when the curvilinear coordinate system $\{\xi^i\}$ moves with time, namely:

$$\frac{\partial J}{\partial t} = J\mathbf{g}^i \cdot \frac{\partial \mathbf{g}_i}{\partial t} \equiv J\mathbf{g}^i \frac{\partial \mathbf{w}}{\partial \xi^i} = \frac{\partial (Jw^i)}{\partial \xi^i} \tag{13}$$

where $\mathbf{w}$ is the velocity of the curvilinear coordinate sstem, i.e. the grid velocity whose cartesian components are:

$$\bar{w}^a = \left. \frac{\partial \bar{x}^a}{\partial t} \right|_{\xi=\text{const.}} \tag{14}$$

Only the component of the grid motion in the direction normal to the grid is significant; this explains why only the contravariant grid velocity component must enter (13). How such a relation can be satisfied in a discrete sense has been investigated by Demirdzic and Peric (1988) and Obayashi (1991) (see also Demirdzic and Peric, 1990; Rosenfeld and Kwak, 1991 for the use of the finite-volume method on moving grids).

The most important relationship to be used in the following restatement of the chain rule derivative formula which results from the definition of $\mathbf{g}^i$:

$$\frac{\partial \xi^k}{\partial \bar{x}^a} = J^{-1}\mathbf{S}^k \cdot \mathbf{i}_a \equiv J^{-1}S_a^k \tag{15}$$

Finally the covariant and contravariant metric tensors, $g_{ij} = \mathbf{g}_i \cdot \mathbf{g}_j$ and $g^{ij} = \mathbf{g}^i \cdot \mathbf{g}^j$ respectively, will be necessary. The contravariant metric tensor will be computed using the surface vectors, according to $g^{ij} = g^{-1}\mathbf{S}^i \cdot \mathbf{S}^j$ where $g = \det g_{ij} = J^2$.

B The different forms of the primitive formulation

The continuity equation is written:

$$0 = \nabla \cdot \mathbf{V} = \frac{\partial \bar{V}^a}{\partial \bar{x}^a} = \frac{\partial \xi^k}{\partial \bar{x}^a} \cdot \frac{\partial \bar{V}^a}{\partial \xi^k} \quad \text{(chain rule derivative)}$$

$$= J^{-1}\mathbf{S}^k \cdot \mathbf{i}_a \cdot \frac{\partial \bar{V}^a}{\partial \xi^k} \quad \text{(because of (13))}$$

$$= J^{-1}\frac{\partial (S_a^k \bar{V}^a)}{\partial \xi^k} \quad \text{(because of (12))} \qquad \text{PT form} \tag{16}$$

$$= J^{-1}\frac{\partial (JV^k)}{\partial \xi^k} \quad \text{(because of (11))} \qquad \text{TT form} \tag{17}$$

The partial transformation (PT) and total transformation (TT) forms (16) and (17) of the continuity equation share in common the fact that the independent variables are the time t and the transformed coordinates $\{\xi^i\}$ of the computational space. In the PT form, the dependent variables are the cartesian velocity components. In the TT form, the dependent variables are the contravariant velocity components in the curvilinear coordinate system. Other total transformation forms are possible using other dependent variables, such as the covariant velocity components $V_i = \mathbf{V} \cdot \mathbf{g}_i$. Contravariant or covariant physical components also may be used. They are respectively defined by $V^{(i)} = \mathbf{V} \cdot \mathbf{g}^i \sqrt{g_{ii}}$ or $V^{[i]} = \mathbf{V} \cdot \mathbf{g}^i / \sqrt{g^{ii}}$ and $V_{(i)} = \mathbf{V} \cdot \mathbf{g}_i / \sqrt{g_{ii}}$ or $V_{[i]} = \mathbf{V} \cdot \mathbf{g}_i \sqrt{g^{ii}}$ and their name stems from the fact that they retain the dimensions of velocity since the second and third ones are projections along unit vectors. Demirdzic *et al.* (1987) have successfully used the contravariant physical velocities.

Several partial transformation forms are also possible if, instead of working with the cartesian velocity components, one works with the velocity components in another reference – usually orthogonal – coordinate system (e.g. cylindrical or polar). However it can be shown that *only the cartesian velocity components make the momentum equation amenable to a strongly conservative form* like the PT or TT forms of the continuity equation, i.e. a form where no factor multiplies the divergence term. If a strongly conservative form of a partial differential equation is integrated over a finite number of control volumes and if it is properly discretized, the resulting fluxes from the strongly conservative form cancel in pairs at all inner control volume faces when summed, so that only boundary fluxes remain. This 'telescoping' property, which is analogous to a discrete divergence theorem, guarantees overall conservation. For this reason also, the total transformation is not retained since it produces in the momentum equation a source term that is not amenable to a divergence form. This is now shown in detail.

The cartesian form of the momentum equation is

$$\frac{\partial \bar{V}^a}{\partial t} + \frac{\partial}{\partial \bar{x}^b}(\bar{V}^a \bar{V}^b) + \frac{\partial}{\partial \bar{x}^b}\bar{P}^{ab} = 0 \tag{18}$$

where $\bar{P}^{ab}$ is the opposite of ρ times the total stress tensor:

$$\bar{P}^{ab} = P\delta^{ab} - \bar{\tau}^{ab} + \overline{v'^a v'^b} \tag{19}$$

Gathering the pressure term (δ^{ab} is the classical Kronecker symbol), the viscous transport term becomes:

$$\bar{\tau}^{ab} \equiv \nu \left[\frac{\partial \bar{V}^a}{\partial \bar{x}^b} + \frac{\partial \bar{V}^b}{\partial \bar{x}^a} \right]$$

(where the reader may notice that for a cartesian coordinate system, contravariant and covariant components are identical), and the turbulent Reynolds stress term is $-\rho \overline{v'^a v'^b}$ (see Chapter 6).

The time derivative for $\bar{x}^a = $ const. must be passed to a time derivative for $\xi = $ const. Again using the chain rule derivative for implicit functions gives:

$$\left.\frac{\partial \bar{V}^a}{\partial t}\right|_{\bar{x}^b=\text{const.}} \equiv \left.\frac{\partial \bar{V}^a}{\partial t}\right|_{\xi=\text{const.}} - \bar{w}^b g^i_{.b} \frac{\partial \bar{V}^a}{\partial \xi^i} \tag{20}$$

where $g^i_{.b} = \mathbf{g}^i \cdot \mathbf{i}_b$. Using (11) and the metric identities (12) and (13), it is easy to write (18) as:

$$J\frac{\partial \bar{V}^a}{\partial t} - J\bar{w}^b g^i_{.b} \frac{\partial \bar{V}^a}{\partial \xi^i} + \bar{V}^a\left[\frac{\partial J}{\partial t} - Jg^i_{.b}\frac{\partial \bar{V}^a}{\partial \xi^i}\right] + Jg^i_{.b}\left[\frac{\partial \bar{P}^{ab}}{\partial \xi^i} + \frac{\partial(\bar{V}^a \bar{V}^b)}{\partial \xi^i}\right] = 0$$

or under the so-called partial-transformation, strongly conservative form of the momentum equation:

$$\frac{\partial}{\partial t}(J\bar{V}^a) + \frac{\partial}{\partial \xi^i}[S^i_{.b}(\bar{V}^b - \bar{w}^b)\bar{V}^a] + \frac{\partial}{\partial \xi^i}[S^i_{.b}\bar{P}^{ab}] = 0 \tag{21}$$

where the mass flux with respect to the grid is defined by:

$$S^i_{.b}(\bar{V}^b - \bar{w}^b) \equiv J\mathcal{U}^i \tag{22}$$

The contravariant relative velocity component of the fluid with respect to the grid $\mathcal{U}^i$ satisfies the consequence (23) of the metric identity (13):

$$\frac{1}{J}\frac{\partial}{\partial \xi^i}[J\mathcal{U}^i] = \frac{\partial}{\partial t}[\log J] \tag{23}$$

which must also be fulfilled in a discrete sense.

Using now the fact that $\mathbf{P}$ is a tensor and $\mathbf{V}$ is a vector so that their components transform according to:

$$P^{ik} = \frac{\partial \xi^i}{\partial \bar{x}^a}\frac{\partial \xi^k}{\partial \bar{x}^b}\bar{P}^{ab}\,; \tag{24a}$$

$$V^i = \frac{\partial \xi^i}{\partial \bar{x}^a}\bar{V}^a \tag{24b}$$

(for instance), it is possible to recast the momentum equation (21) under the total transformation, strongly conservative form

$$\frac{\partial}{\partial t}\left[JV^i\frac{\partial \bar{x}^a}{\partial \xi^i}\right] + \frac{\partial}{\partial \xi^j}\left[J\mathcal{U}^j\frac{\partial \bar{x}^a}{\partial \xi^i}V^i\right] + \frac{\partial}{\partial \xi^i}\left[JP^{ki}\frac{\partial \bar{x}^a}{\partial \xi^k}\right] = 0 \tag{25}$$

The independent variables are now the contravariant velocity components V^i but the momentum equations are written along the orthonormal basis of the cartesian system. Each projection along $a = 1, 2, 3$ involves therefore the three fluxes $JV^i - J\mathcal{U}^i$ in (25) is specified by (22) – and a numerical solution of the momentum equation requires the simultaneous solution of the three equations

$a = 1, 2, 3$. It is noticed that (25) depends critically on the fact that $\mathbf{i}_a$ does not depend on the position. With another reference coordinate system $\{\mathbf{e}_\lambda\}$, a strongly conservative form would not be obtained along $\mathbf{e}_\lambda$.

Using $\mathbf{g}_i = \dfrac{\partial \bar{x}^a}{\partial \xi^i} \mathbf{i}_a$, the metric identity (13), and the definition (26) of Christoffel symbols:

$$\frac{\partial \mathbf{g}_i}{\partial \xi^j} = \Gamma^p_{ij} \mathbf{g}_p \tag{26}$$

it is possible to obtain the weakly conservative total transformation form of the momentum equation. However, the following manipulation is more expeditious:

$$\frac{\partial \bar{V}^i}{\partial t} + \frac{1}{J}\frac{\partial}{\partial \xi^k}[JV^i V^k] + V^k V^j \Gamma^i_{kj} + \frac{1}{J}\frac{\partial}{\partial \xi^k}[JP^{ik}] + P^{kj}\Gamma^i_{kj} = 0 \tag{27a}$$

Adding $V^i \left[\dfrac{1}{J}\dfrac{\partial J}{\partial t} - \dfrac{1}{J}\dfrac{\partial (Jw^k)}{\partial \xi^k} \right]$ and combining gives:

$$\frac{1}{J}\frac{\partial (JV^i)}{\partial t} + \frac{1}{J}\frac{\partial}{\partial \xi^k}[JU^k V^i] + \frac{1}{J}\frac{\partial}{\partial \xi^k}[JP^{ik}] + w^k \frac{\partial V^i}{\partial \xi^k} + \Gamma^i_{kj}[V^k V^j + P^{kj}] = 0 \tag{27b}$$

Equation (27a) involves explicit sources, such as $P^{kj}\Gamma^i_{kj}$, related to the geometric curvatures of the coordinate system $\{\xi^i\}$ as described by the Christoffel symbols. The way such terms are computed is critically important. First, the momentum source term involves the product of the momentum flux JP^{kj} by the Christoffel symbol. While the momentum flux is a property of the flow, the Christoffel symbol is a property of the curvilinear coordinate system. Because it is based on second-order derivatives of point coordinates, its evaluation is prone to serious inaccuracies and to a lack of smoothness. This is a very serious drawback which limits the practical use of this total transformation for numerical purposes. Moreover, such terms do not go into cell face fluxes, but instead are associated with cell centres so that there is no obvious discretization procedure which would guarantee overall conservation. Depending on how such terms are treated, dissipative properties of the system may not be transduced in a suitable way and instabilities are possible. Another technical difficulty is generated by the viscous contribution to P^{kj} which involves derivatives of Christoffel symbols. In spite of such drawbacks, there is a significant amount of research using the contravariant fluxes as unknowns (Demirdzic *et al.*, 1987; Rosenfeld and Kwak, 1989, 1992; Ikohagi and Chin, 1991; Mynett *et al.*, 1991; Rosenfeld *et al.*, 1991; Ikohagi *et al.*, 1992; Segal *et al.*, 1992; Wesseling *et al.*, 1992). In contrast, covariant velocity components have seldom been used (see however Karki, 1986; Davidson and Hedberg, 1989).

C The Reynolds–averaged Navier–Stokes equations

The stress variables P^{kj} or $\bar{P}^{ab}$ now need to be eliminated. This is easily done in the following way:

$$S^i_{.b}\bar{P}^{ab} = J\frac{\partial \xi^i}{\partial \bar{x}^b}\bar{P}^{ab} = Jpg^i_{.a} - \nu Jg^{ik}\frac{\partial \bar{V}^a}{\partial \xi^k} + J\overline{v'^a v'^b}\frac{\partial \xi^i}{\partial \bar{x}^b} \tag{28}$$

The turbulent contribution requires a closure assumption (see Chapter 6). In the following, the newtonian eddy-viscosity closure is invariably assumed.

$$\overline{v'^a v'^b} = \tfrac{2}{3}K\delta^{ab} - \nu_T\left[\frac{\partial \bar{V}^a}{\partial \bar{x}^b} + \frac{\partial \bar{V}^b}{\partial \bar{x}^a}\right] \tag{29}$$

The isotropic relationship (29) involves the turbulent kinetic energy K and the eddy-viscosity ν_T which is specified either by an algebraic relationship, as in the Baldwin and Lomax model, or given by a monomial relation involving turbulent parameters, like K and ε, the turbulent rate of dissipation, which are deduced from scalar transport equations (see Chapter 6). Substituting (29) into (28) yields:

$$\begin{aligned}
J\overline{v'^a v'^b}\frac{\partial \xi^i}{\partial \bar{x}^b} &= \tfrac{2}{3}JK\frac{\partial \xi^i}{\partial \bar{x}^b} - \nu_T Jg^{ik}\frac{\partial \bar{V}^a}{\partial \xi^k} - \nu_T J\frac{\partial \xi^i}{\partial \bar{x}^b}\frac{\partial \xi^k}{\partial \bar{x}^a}\frac{\partial \bar{V}^b}{\partial \xi^k} \\
&= \tfrac{2}{3}JK\mathbf{i}_a\mathbf{g}^i - \nu_T Jg^{ik}\frac{\partial \bar{V}^a}{\partial \xi^k} - \nu_T J[\mathbf{g}^i]_b[\mathbf{g}^k]_a\frac{\partial \bar{V}^b}{\partial \xi^k}
\end{aligned} \tag{30}$$

Substituting (29) into (25) gives the master equation (31)

$$\frac{\partial}{\partial t}[J\bar{V}^a] + \frac{\partial}{\partial \xi^i}[J\mathcal{U}^i\bar{V}^a + Jp\mathbf{i}_a\cdot\mathbf{g}^i] = \nu\frac{\partial}{\partial \xi^i}\left[Jg^{ik}\frac{\partial \bar{V}^a}{\partial \xi^k}\right] - \frac{\partial}{\partial \xi^i}\left[J\overline{v'^a v'^b}\frac{\partial \xi^i}{\partial \bar{x}^b}\right] \tag{31}$$

in which (30) can be substituted. Using the area vectors $\mathbf{S}^i = J\mathbf{g}^i$, the standard form for the momentum equations is then:

$$\begin{aligned}
\frac{\partial J\bar{V}^a}{\partial t} &+ \frac{\partial}{\partial \xi^i}[J\mathcal{U}^i\bar{V}^a] + \frac{\partial}{\partial \xi^i}[S^i_a(P + \tfrac{2}{3}K)] \\
&= \bar{T}^a \equiv \frac{\partial}{\partial \xi^i}\left[J(\nu + \nu_T)g^{ik}\frac{\partial \bar{V}^a}{\partial \xi^k}\right] + \frac{\partial}{\partial \xi^i}\left[J^{-1}\nu_T S^i_b S^m_a\frac{\partial \bar{V}^b}{\partial \xi^m}\right]
\end{aligned} \tag{32}$$

An alternative form of transport terms is given now:

$$\bar{T}^a \equiv \frac{\partial}{\partial \xi^i}\left[J(\nu + 2\nu_T)g^{ik}\frac{\partial \bar{V}^a}{\partial \xi^k}\right] + \frac{\partial}{\partial \xi^i}\left[J^{-1}\nu_T S^i_b S^m_a\frac{\partial \bar{V}^b}{\partial \xi^m}\right]_{a\neq b} \tag{33}$$

This form has been used for instance by Chen and Patel (1985); its main drawback is that it generates distinct equations for components $a = 1$, 2 or 3;

therefore it is not retained. The following momentum equation is usually preferred:

$$\frac{\partial J\bar{V}^a}{\partial t} + \frac{\partial (J\mathcal{U}^i\bar{V}^a)}{\partial \xi^i} + \frac{\partial P^{ia}}{\partial \xi^i} = 0 \tag{34a}$$

where

$$P^{ia} = -J(\nu + \nu_T)g^{ik}\frac{\partial \bar{V}^a}{\partial \xi^k}\bigg|_{i=k} - s^{ia} + S_a^i(P + \tfrac{2}{3}K) \tag{34b}$$

with the source term s^{ia} which vanishes if the grid is orthogonal and if the flow is laminar.

$$s^{ia} = \left[J(\nu + \nu_T)g^{ik}\frac{\partial \bar{V}^a}{\partial \xi^k}\right]_{i\neq k} + J^{-1}\nu_T S_b^i S_a^m \frac{\partial \bar{V}^b}{\partial \xi^m} \tag{35}$$

III DISCRETE TREATMENT OF THE MOMENTUM EQUATIONS

A The finite-volume technique: cell-centred treatment

The finite volume discretization starts from the integration of (34a) over a cubic control volume in the computational domain and applied the divergence theorem to derivatives with respect to ξ^i. Two types of integral quantities are therefore involved, namely:

$$\int_{\text{cell}} J\bar{V}^a \, dv \quad \text{and} \quad \int_{\text{face}} J\mathcal{U}^i\bar{V}^a \, d\xi^j \, d\xi^k \quad (i, j, k \text{ in cyclic order})$$

The cell mean value of $\bar{V}^a$ and the area mean value of the momentum flux are involved and pointwise approximations of such quantities are required to define the approximation problem.

To present the formulation of the approximate problem for the Navier–Stokes equations, we restrict ourselves to the two-dimensional case. A two-dimensional (structured) primary grid is introduced as a set of discrete points $\{\mathbf{R}(i,j), i = 0, \ldots, M, j = 0, \ldots, N\}$ where the centres of the mass control volumes are defined. The integers i and j are defined from the curvilinear coordinates ξ^1 and ξ^2 by:

$$i = M\frac{\xi^1 - \xi^1_{\min}}{\xi^1_{\max} - \xi^1_{\min}} \; ; \quad j = N\frac{\xi^2 - \xi^2_{\min}}{\xi^2_{\max} - \xi^2_{\min}} \; ; \quad \xi^i_{\min} \leq \xi^i \leq \xi^i_{\max} \tag{36}$$

together with a secondary grid (where the centres of the cell faces are defined) which involves the set of discrete points $\{\mathbf{R}(i + \tfrac{1}{2}, j) \text{ and } \mathbf{R}(i, j + \tfrac{1}{2}),$

$i = 0, \ldots, M, j = 0, \ldots N\}$. The primary grid is therefore regular in the computational plane where a unit square is the image of a quadrilateral in the physical plane. Points of the secondary grid are to be used to locate the computed face values. They can be localized half-way between regular grid points as in Figure 1. Finally, cell vertices A, B, C, D are taken at the intersection of medians of each quadrilateral defined from four primary grid points.

However, it is better to construct a grid twice as fine as the primary grid and to define in a first step the secondary grid points so that they define one colour of a checkerboard (Figures 2a or 2b), together with the control volumes. Primary grid points are then defined in a second step. They can be defined either from one half of the points of the other colour (remaining points being the cell vertices, like A, B, C, D), as in Figure 2a, or they can be taken at equal distances from neighbouring secondary points, as in Figure 2b (or in Figures 3 and 4). In the latter case, vertices are also taken at equal distances from neighbouring secondary points.

The foregoing choices have in common the fact that the grid spacing between primary grid points is fixed arbitrarily to unity in the computational domain. Hence the concept of accuracy has no sense in the computational domain and it is necessary to perform the Taylor analysis in the physical domain to know the accuracy of the numerical method used. This is of course tedious in the general case of Figures 1 and 2. Moreover, a one-dimensional example shows easily that norms specifying the order of accuracy of a method, while equivalent on a uniform grid, are not equivalent on a non-uniform grid. This is of course due to

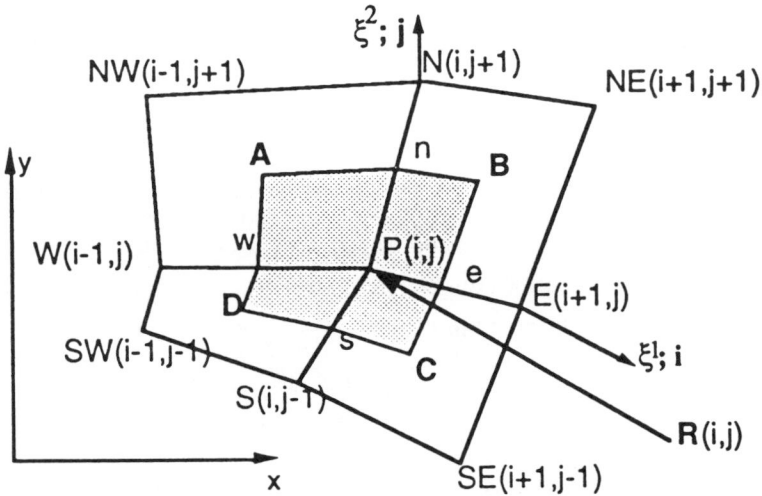

Figure 1 Presentation of the control volume ABCD (shaded) surrounding point P identifying the grid point $R(i,j)$. Points e, w, n, s are attached to faces of ABCD and are located as indicated in the figure.

(a)

(b)

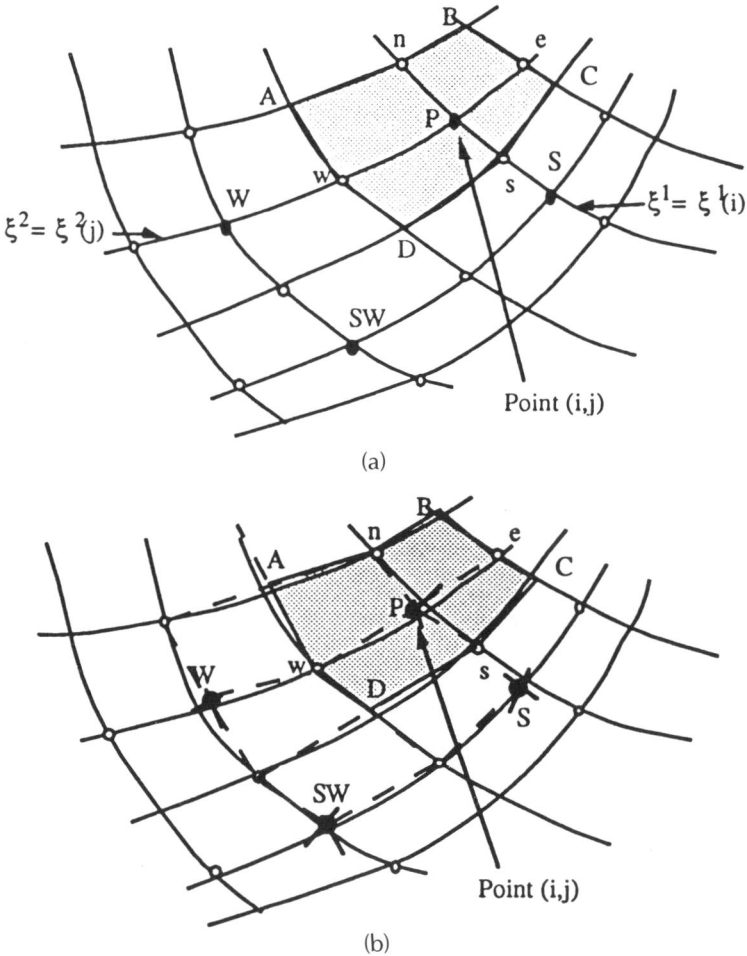

Figure 2 Generation of control volumes from the double grid. The shaded area represents the control volume ABCD. $\bigcirc$, secondary grid points; $\bullet$, primary grid points. The dashed line in part (b) connects primary and secondary points.

the existence of variable stretching factors and aspect ratios at the grid level. For such reasons, there is no guarantee that the accuracy proven for a constant-spacing analysis remains valid on a curvilinear grid.

If we denote the actual time level (at which the velocity and the pressure are still unknown) by the superscript $n + 1$ and the time increment by τ, a two- or three-level backward time-discrete formula is generally used:

$$\left(\frac{\partial V}{\partial t}\right)^{n+1} \approx \frac{1}{2\tau}\left(a_0 \mathbf{V}^{n+1} + a_1 \mathbf{V}^n + a_2 \mathbf{V}^{n-1}\right) \tag{37}$$

where $a_0 = -a_1 = 2$, $a_2 = 0$ for first-order accuracy (but only two time levels) and $a_0 = 3$, $a_1 = -4$, $a_2 = 1$ for second-order accuracy. Marx (1994) has investigated several time discretizations within the framework of artificial compressibility techniques and concluded that, for unsteady flow problems, second-order accuracy in time is required. However, for long time integrations, an improved time accuracy may be profitable as a means to retain a time step high enough, while compatible with the somewhat weakened stability constraints allowed by the fully implicit character of the momentum scheme: the stability is then controlled by the coupling procedure between solenoidality and momentum.

If ϕ denotes one of the cartesian velocity components, $\bar{V}^a$, the fully implicit time discrete version of (27) takes the following form:

$$\frac{a_0}{2\tau}[J\phi]^{n+1} + \frac{\partial}{\partial \xi^i}[J\mathcal{U}^i\phi]^{n+1} - \frac{\partial}{\partial \xi^i}\left[J(\nu + \nu_T)g^{ij}\frac{\partial \phi}{\partial \xi^j}\right]^{n+1} = \left[J^{-1}\nu_T S_b^i S_a^m \frac{\partial \bar{V}^b}{\partial \zeta^m}\right]^{n+1}$$

$$+ S_a^i[P + \tfrac{2}{3}K]^{n+1} - \frac{1}{2\tau}[a_1\phi^n + a_2\phi^{n-1}] \qquad (38a)$$

Anticipating the results in Section IV.F, it is assumed that a Picard linearization of convective terms has been performed so that $[J\mathcal{U}^i\phi]$ at time t^{n+1} is computed iteratively at any given iteration $(\nu + 1)$ from the previous iteration (ν) so that $[J\mathcal{U}^i\phi]^{n+1(\nu+1)} \approx [J\mathcal{U}^{i(\nu)}\phi^{(\nu+1)}]^{n+1}$, the index $(\nu + 1)$ being omitted from now on.

The problem of space discretization of equation (38a) is now viewed as that of the coordinate-invariant strong-conservation form (38b) of the linearized momentum equations:

$$\nabla \cdot \mathbf{P} = Q \quad \text{with} \quad \mathbf{P} = \rho\mathbf{V}\phi - \Gamma\nabla\phi \qquad (38b)$$

where Q gathers the right-hand-side of (38a) and the unsteady a_0-contribution. $\mathbf{P}$, not to be confused with point P, involves the second and third terms of (38a). The advection velocity $\mathbf{V}$ (with contravariant components u and v) in (38b) is simply $\mathcal{U}$, while $\Gamma = \mu + \mu_T$. The finite volume discretization of (38a) is shown in (38b). Integrating over a control volume ABCD surrounding point $\mathbf{R}(i,j)$ (Figure 1 or 2) yields in the two-dimensional case:

$$(\mathbf{P} \cdot \mathbf{S})_e - (\mathbf{P} \cdot \mathbf{S})_w + (\mathbf{P} \cdot \mathbf{S})_n - (\mathbf{P} \cdot \mathbf{S})_w = \bar{Q}_{i,j} \qquad (39)$$

The subscripts e, w, n, s specify the location of fluxes $(\mathbf{P} \cdot \mathbf{S})$ defined on the faces of the control volume ABCD: for example, e refers to the point $\mathbf{R}(i + \tfrac{1}{2}, j)$ and s refers to the point $\mathbf{R}(i, j - \tfrac{1}{2})$. In all the following, approximations of a quantity at point $\mathbf{R}(i, j)$ will be noted with i and j as subscripts. For instance, $\bar{Q}_{i,j}$ is the value of $\bar{Q}$ at $\mathbf{R}(i, j)$ or the approximation of Q averaged on ABCD. The

The expression for fluxes is given at points e or w by:

$$(\mathbf{P} \cdot \mathbf{S})_{e,w} = (\rho \mathbf{V} \cdot \mathbf{S}^1)_{e,w} \phi_{e,w} - (\Gamma \mathbf{S}^1 \cdot \nabla \phi)_{e,w}$$

$$= [\rho J u]_{e,w} \phi_{e,w} - \left[\Gamma \frac{\mathbf{S}^1 \cdot \mathbf{S}^1}{J}\right]_{e,w} \left[\frac{\partial \phi}{\partial \xi^1}\right]_{e,w} - \left[\Gamma \frac{\mathbf{S}^1 \cdot \mathbf{S}^2}{J}\right]_{e,w} \left[\frac{\partial \phi}{\partial \xi^2}\right]_{e,w}$$

$$= [\rho J u]_{e,w} \phi_{e,w} - [\Gamma J g^{11}]_{e,w} \left[\frac{\partial \phi}{\partial \xi^1}\right]_{e,w} - [\Gamma J g^{12}]_{e,w} \left[\frac{\partial \phi}{\partial \xi^2}\right]_{e,w} \tag{40a}$$

Similarly, at n or s:

$$(\mathbf{P} \cdot \mathbf{S})_{n,s} = [\rho J v]_{n,s} \phi_{e,w} - \left[\Gamma \frac{\mathbf{S}^2 \cdot \mathbf{S}^2}{J}\right]_{n,s} \left[\frac{\partial \phi}{\partial \xi^2}\right]_{n,s} - \left[\Gamma \frac{\mathbf{S}^2 \cdot \mathbf{S}^1}{J}\right]_{n,s} \left[\frac{\partial \phi}{\partial \xi^1}\right]_{n,s} \tag{40b}$$

We may distinguish the orthogonal flux associated with the first two terms and the non-orthogonal contribution associated with the last term which is to be included in the source term. The implicit treatment of the orthogonal flux yields in general a nine- or a nineteen-point molecule depending on two- or three-dimensional calculations. The resulting coefficient matrix has nine (or nineteen) diagonals, but the matrix is usually not diagonally dominant and the influence coefficients are not all positive. The non-orthogonal contribution is in general treated explicitly without significant stability penalties (see however Piquet and Visonneau, 1991, for an implicit treatment).

The source term is usually linearized as follows:

$$\bar{Q}_{i,j} = \int_{ABCD} Q \, dv = \bar{Q}_{1P} - \bar{Q}_{2P} \Phi_{i,j} \tag{41}$$

where $\Phi_{i,j}$ is the approximation of $\phi(\mathbf{R}_{i,j})$, while $\bar{Q}_{2P} \geq 0$ to ensure a numerically stable solution (the conditioning of the matrix for the Φ-unknowns is improved). Hence $\bar{Q}_{2P}$ incorporates the unsteady contribution $a_0 \rho J^{n+1}/2\tau$.

If (40) and (41) are substituted into (39), a general discretized solution is obtained for each cartesian velocity component that takes the form:

$$a_{Pi,j} \Phi_{i,j} = \sum_{nb} (a_{nb})_{i,j} \Phi_{nb} + b_{i,j} \tag{42a}$$

where

$$a_{Pi,j} = \sum_{nb} (a_{nb})_{i,j} + \bar{Q}_{2P} \tag{42b}$$

$$b_{i,j} = \bar{Q}_{1P} + [\Gamma J g^{12}]_e \left[\frac{\partial \phi}{\partial \xi^2}\right]_e - [\Gamma J g^{12}]_w \left[\frac{\partial \phi}{\partial \xi^2}\right]_w + [\Gamma J g^{21}]_n \left[\frac{\partial \phi}{\partial \xi^1}\right]_n$$
$$- [\Gamma J g^{12}]_s \left[\frac{\partial \phi}{\partial \xi^1}\right]_s \tag{42c}$$

The influence coefficients $(a_{nb})_{i,j}$ of the neighbouring unknowns depend on the selected scheme and on the retained unknowns. In general nb refers to the eight neighbours of point $\mathbf{R}_{i,j}$. In the following we consider the numerical approximation of fluxes in terms of nodal values. For this purpose, the momentum equation is treated as an independent advection-diffusion equation.

The fluxes are often approximated as follows:

$$P_e^O \equiv [\rho Ju]_e \phi_e - [\Gamma Jg^{11}]_e \left[\frac{\partial \phi}{\partial \xi^1}\right]_e \approx [\rho Ju]_e \Phi_{i,j} - \left[\frac{\Gamma Jg^{11}}{\Delta \xi^1}\right]_e A(R_e)(\Phi_{i+1,j} - \Phi_{i,j})$$

$$(43a)$$

$$P_w^O \equiv [\rho Ju]_w \phi_w - [\Gamma Jg^{11}]_w \left[\frac{\partial \phi}{\partial \xi^1}\right]_w$$

$$\approx [\rho Ju]_w \Phi_{i,j} - \left[\frac{\Gamma Jg^{11}}{\Delta \xi^1}\right]_w A(-R_w)(\Phi_{i,j} - \Phi_{i-1,j})$$

$$(43b)$$

where $A(R)$ is a function of the cell Reynolds number along $\xi^1 : R = \rho Ju\Delta\xi^1/[\Gamma Jg^{11}]$. For all values of R, positive and negative, $A(R) = A(|R|) + \max[-R, 0]$. The centred scheme is given by:

$$\phi_e = \sigma_{Pi}\phi_E + (1 - \sigma_{Pi})\phi_P \quad \text{with} \quad \sigma_{Pi} = \overline{Pe}/\overline{PE} \quad \text{(Figure 3)}$$

so that $A(|R|) = 1 - (1 - \sigma)|R|$. The pure upwind scheme is given by: $\sigma_{Pi} = [1 + |\rho Ju_e| \rho Ju_e]/2$ so that $A(|R|) = 1$. Hence the hybrid scheme is $A(|R|) = \max[0, 1 - (1 - \sigma)|R|]$. The construction of ϕ_w proceeds in the same way.

The multiexponential scheme (Deng et al., 1991a) uses $A(|R|) = |R|/(e^{|R|} - 1)$. For constant coefficient data and a uniform grid of spacing h, the multiexponential scheme yields the exact one-dimensional solution of $\phi'' - 2A\phi' = 0$, $\phi(0) = 0$, $\phi(1) = 1$, even at high cell Reynolds numbers Ah, in the boundary layer close to $x = 0$ (Spalding, 1972). The resulting uniform accuracy of the scheme (in the one-dimensional case) results from the use of the

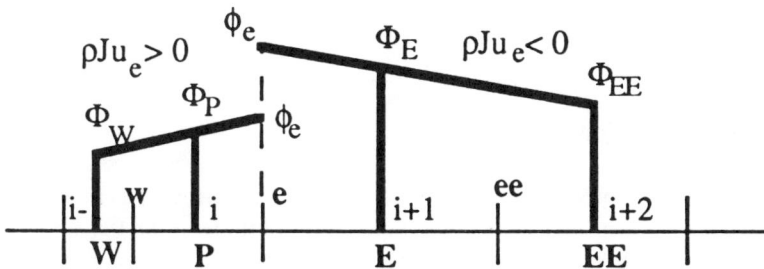

Figure 3 Principle of the second-order upwind scheme.

exponential solution (of the model problem) as a component of the shape function.

The power-law scheme of Patankar (1980), which is a convenient polynomial fit to the multiexponential scheme, is defined by:

$$A(|R|) = \max\left[0, (1 - 0.1\,|R|)^5\right]$$

Substituting such expressions of $A(|R|)$ into $A(R)$ and the result into (43) yields the influence coefficients:

$$(a_E)_{i,j} = \left[\frac{\Gamma Jg^{11}}{\Delta\xi^1}\right]_e A(R_e); \quad (a_W)_{i,j} = \left[\frac{\Gamma Jg^{11}}{\Delta\xi^1}\right]_w A(-R_w) \qquad (44)$$

The second-order upwind scheme uses two upstream neighbouring nodes for the estimation of the cell face value, as shown in Figure 3. This yields:

$$\phi_e = \Phi_E + \sigma_{Pi+1}(\Phi_E - \Phi_{EE}) \quad \text{if} \quad \rho Ju_e < 0,$$
$$\phi_e = \Phi_P + \zeta_{Pi}(\Phi_P - \Phi_W) \quad \text{if} \quad \rho Ju_e > 0$$

The interpolation coefficients are defined in the general case by:

$$\sigma_{Pi+1} = \overline{eE}/\overline{E.EE}; \quad \zeta_{Pi} = \overline{Pe}/\overline{WP}$$

In the following, we particularize to the case where primary grid points are localized half-way between secondary grid points. Then, we may work only with the following σ coefficients:

$$\sigma_{Pi-1} = \overline{Ww}/\overline{WP} = 1 - \zeta_{Pi}; \quad \sigma_{Pi} = \overline{Pe}/\overline{PE}; \quad \sigma_{Pi+1} = \overline{E.ee}/\overline{E.EE}$$

Again using central differencing for the diffusive flux, the influence coefficients in (42a) take the following form:

$$(a_E)_{i,j} = \left[\frac{\Gamma Jg^{11}}{\Delta\xi^1}\right]_e + (1 + \sigma_{Pi+1})\max\left[0, -\rho Ju_e\right] + \sigma_{Pi}\max\left[0, -\rho Ju_w\right]$$

$$(a_W)_{i,j} = \left[\frac{\Gamma Jg^{11}}{\Delta\xi^1}\right]_w + (2 - \sigma_{Pi-2})\max\left[0, \rho Ju_w\right] + (1 - \sigma_{Pi-1})\max\left[0, \rho Ju_e\right]$$

$$(a_{EE})_{i,j} = -\sigma_{Pi+1}\max\left[0, -\rho Ju_e\right]; \quad (a_{WW})_{i,j} = -(1 - \sigma_{Pi-2})\max\left[0, \rho Ju_w\right]$$

$$(45)$$

and similar expressions are obtained in the ξ^2 direction. Hence, *nb* in (42) is equal to *E, N, S, W, EE, WW, NN, SS* and the molecule uses nine points, instead of five, for the orthogonal contribution (13 in the 3D case). The second-order upwind scheme is not positive since the coefficients of the most distant nodes can be negative. To improve the robustness of the solving algorithm, it is advisable to incorporate the coefficients of the most distant nodes into the source term and treat them explicitly. The resulting extra contribution to add is:

$$\Delta b_{i,j} = a_{EE}(\Phi_{EE} - \Phi_P) + a_{WW}(\Phi_{WW} - \Phi_P) + a_{NN}(\Phi_{NN} - \Phi_P) \\ + a_{SS}(\Phi_{SS} - \Phi_P) \tag{46}$$

Now nb in (42) simply consists of E, W, N, S.

One of the most commonly used schemes is the QUICK scheme due to Leonard (1979). This scheme, which has been used frequently (for practical implementation problems, see Han and Humphrey, 1981; Pollard and Siu, 1982; Shyy et $al.$, 1985; Hayase et $al.$, 1992), is an attempt to combine the accuracy of central differencing with the stability inherent in upwinding by using in each direction separately a quadratic upstream interpolation. A convenient way to obtain the QUICK reconstructed value for, say, ϕ_e (Figure 4) is to use the adequate second-order Lagrange interpolation polynomial between values Φ_E, Φ_P, Φ_W if $(\rho Ju)_e > 0$, or between Φ_E, Φ_P, Φ_{EE} if $(\rho Ju)_e < 0$, along the line $j = $ const.

If the interpolation is written in the physical plane (which is preferable) then ϕ_e, which is a function of the curvilinear abscissa, s, along $j = $ const., can be written down from the divided difference formula:

$$\phi(S) = \Phi_P + (s - s_P)\frac{\Phi_E - \Phi_P}{\overline{PE}} + \tfrac{1}{2}(s - s_P)(s - s_E) \, \text{CURV} \tag{47a}$$

where CURV is twice the second-order divided difference:

$$\text{CURV} = 2\left[\frac{\Phi_W - \Phi_E}{\overline{EW}\,\overline{PW}} - \frac{\Phi_E - \Phi_P}{\overline{PE}\,\overline{PW}}\right] = -\frac{2}{\overline{PW}\,\overline{PE}}\left[\frac{\overline{EP}}{\overline{EW}}\Phi_W + \frac{\overline{WP}}{\overline{WE}}\Phi_E - \Phi_P\right] \tag{47b}$$

This approximation allows QUICK to be interpreted as the pure upwind scheme augmented by gradient and curvature-type correction terms. Hence:

$$\phi_e \equiv \phi(s_e) = \sigma_{Pi}\Phi_E + (1 - \sigma_{Pi})\Phi_P \\ + \sigma_{Pi}(1 - \sigma_{Pi})\frac{\overline{PE}}{\overline{PW}}\left[\frac{\overline{EP}}{\overline{EW}}\Phi_W + \frac{\overline{WP}}{\overline{WE}}\Phi_E - \Phi_P\right] \tag{48a}$$

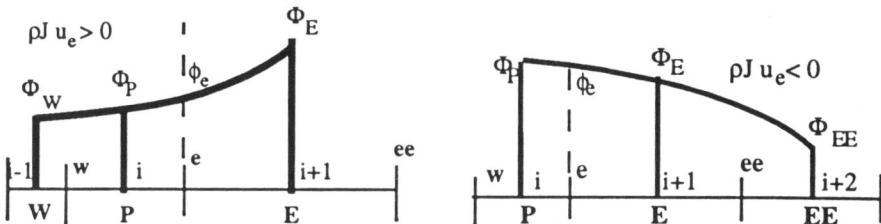

Figure 4 Quadratic upstream interpolation in the case where $\rho Ju_e > 0$ (left) and $\rho Ju_e < 0$ (right).

The first-order derivative may be also reconstructed:

$$\frac{\partial \phi}{\partial s}\Big|_e = \frac{\Phi_E - \Phi_P}{\overline{PE}} + (1 - 2\sigma_{Pi})\frac{1}{\overline{PW}}\left[\frac{\overline{EP}}{\overline{EW}}\Phi_W + \frac{\overline{WP}}{\overline{WE}}\Phi_E - \Phi_P\right] \quad \text{while}$$

$$\frac{\partial^2 \phi}{\partial s^2} = \text{CURV}$$

(48b)

If the reconstruction is performed in the computational plane, s is supplied by ξ^1 and the grid space is uniform and equal to one. Then:

$$\phi_e \equiv \phi(\tfrac{1}{2}) = \tfrac{1}{2}[\Phi_E + \Phi_P] - \tfrac{1}{8}[\Phi_E + \Phi_W - 2\Phi_P] \tag{49}$$

Because of the one-dimensional form of the reconstruction, the QUICK scheme is appropriate to steady or quasisteady flows in which the cell Reynolds number is large no more than in one direction. Leonard (1979) has extended this approach to the unsteady case (QUICKEST scheme) – see e.g. Davis and Moore (1982) for a significant application of QUICKEST to the computation of a vortex street behind a rectangle. To overcome stability problems, the contribution:

$$\frac{\overline{EP}}{\overline{EW}}(\Phi_W - \Phi_P)$$

is treated as a source term, and only the remaining contribution in (47) is taken to contribute to a_E and a_P (Han and Humphrey, 1981; Huang *et al.*, 1985; see also Pollard and Siu, 1982 for attempts to solve the convergence problems induced by the source term contributions).

B Finite-difference techniques and the closure stencil concept

1 Finite-difference techniques

An alternative way to provide a discrete aproximation uses a finite-difference technique, starting from the convective form of the equations:

$$[\rho Ju - \Gamma Jg^{11}]\frac{\partial \phi}{\partial \xi^1} - \Gamma Jg^{11}\frac{\partial^2 \phi}{\partial(\xi^1)^2} = S^1_{i,j} \equiv \frac{\partial}{\partial \xi^1}\left[\Gamma Jg^{12}\frac{\partial \phi}{\partial \xi^2}\right] + \bar{Q}^1_{i,j} \tag{50}$$

Then, and this defines the finite difference character of the approximation, (50) is discretized at point $P(i,j)$, in the computational plane. $A_\phi = \rho Ju - \Gamma Jg^{11}$ and $B_\phi = \Gamma Jg^{11}$ are assumed to be locally constant along the line $j = \text{const.}$, in the vicinity of P, the index i referring to the ξ^1-direction. We then write:

$$A_\phi\frac{\partial \phi}{\partial \xi^1} - B_\phi\frac{\partial^2 \phi}{\partial(\xi^1)^2}\Big|_{i,j} \approx C_0\Phi_{i,j} + C_1\Phi_{i+1,j} + C_2\Phi_{i+2,j} + C_3\Phi_{i-1,j} + C_4\Phi_{i-2,j}$$

(51)

The coefficients C_i, $i = 0$, ..., 4 may be determined from consistency arguments (using the Taylor formula), or in such a way that the approximation is exact for several shape functions, for example $\phi = \text{const.}$, $\phi = \xi^1$, $\phi = (\xi^1)^2$ and $\phi = \exp\left(A_\phi \xi^1 / B_\phi\right)$. The result, which involves the cell Reynolds number $R_{i,j}$ at point P, is:

$$R_{i,j} \mid \tfrac{1}{2}(\Phi_{i+1,j} - \Phi_{i-1,j}) \ - A(|R_{i,j}|)(\Phi_{i+1,j} - 3\Phi_{i,j} + 3\Phi_{i-1,j} - \Phi_{i-2,j}) \mid$$
$$- (\Phi_{i+1,j} + \Phi_{i-1,j} - 2\Phi_{i,j}) = 0 \tag{52a}$$

or:

$$R_{i,j} \mid \tfrac{1}{2}(\Phi_{i+1,j} - \Phi_{i-1,j}) \ + A(|R_{i,j}|)(\Phi_{i-1,j} - 3\Phi_{i,j} + 3\Phi_{i+1,j} - \Phi_{i+2,j}) \mid$$
$$- (\Phi_{i+1,j} + \Phi_{i-1,j} - 2\Phi_{i,j}) = 0$$
$$\tag{52b}$$

depending on whether $A > 0$ or $A < 0$, with:

$$A(|R|) = \frac{1}{2}\frac{1 + \exp(-|R|)}{(1 + \exp(-|R|))^2} - \frac{1}{[R|(1 - \exp(-|R|))]} \quad \text{such that}$$

$$A(0) = \frac{1}{12} \le A(|R|) \le \frac{1}{2} - \frac{1}{|R|}$$

This implies that C_1 and C_3 are negative while C_2 and C_4 are positive. The equations (52) indicate that such schemes introduce a formally 'third-order' dispersion effect:

$$BR_{i,j}\Delta\xi^1 A(|R_{i,j}|)\frac{\Delta^3\phi_{i-1/2}}{\Delta(\xi^1)^3} \quad \text{where}$$
$$\Delta^3\phi_{i-\frac{1}{2}} \equiv \Phi_{i+1,j} - 3\Phi_{i,j} + 3\Phi_{i-1,j} - \Phi_{i-2,j} \tag{53}$$

From the fact that:

$$\phi_e = \tfrac{1}{2}(\Phi_E - \Phi_P) - A(|R_{i,j}|)(\Phi_E - 2\Phi_P - \Phi_W)$$

it follows that the centred scheme results from $A(|R|) = 0$, the QUICK scheme from $A(|R|) = \tfrac{1}{8}$, while the values of $\tfrac{1}{6}$ and $\tfrac{1}{2}$ have been proposed by Agarwal (1981) and Atias et al. (1977), respectively. The scheme of Gushkin and Schennikov (1974) also belongs to this category. Extended stencils have also been proposed in the form of third-order upwinding (Barret, 1982; Kawamura and Kuwahara, 1984) which yields a formally 'fourth-order' dissipation effect:

$$aR_{i,j}(\Delta\xi^1)^2\frac{\Delta^4\phi_i}{\Delta(\xi^1)^4} \quad \text{with} \quad a = \tfrac{1}{3} \quad \text{(Barrett)}$$
$$\text{or} \quad a = \tfrac{1}{4} \quad \text{(Kawamura and Kuwahara)} \tag{54}$$

where:

$$\Delta^4\phi_i \equiv \Phi_{i+2,j} - 4\Phi_{i+1,j} + 6\Phi_{i,j} - 4\Phi_{i-1,j} + \Phi_{i-2,j}$$

The aforementioned finite-difference and finite-volume discretizations share in common the fact that the reconstruction of the orthogonal momentum fluxes $\mathbf{P}^O$ at interfaces are one-dimensional. As a result, only the truncation error part of the numerical diffusion can be controlled while the cross-flow diffusion part is not.

2 The closure stencil

To deal with this problem, we introduce the concept of *closure stencil*. The closure stencil is defined as the set of connected nodal points which are used to reconstruct the fluxes where necessary, i.e. at e, w, n, s, prior to their elimination. The volume covered by the closure stencil is the closure volume. As a result of the elimination, the neighbours involved in the discrete equation (42) will either define the nine-point (compact) molecule of Figure 1, or involve extra points. In the former case, the reconstruction will be compact, while it will be non-compact in the latter. Most of the proposed attempts tend to minimize diffusion using compact flux reconstructions yielding a nine-point molecule for (42) (eight neighbours).

The most often used scheme is the so-called skew-upwind difference scheme (SUDS) (Raithby, 1976; see also Raithby and Torrance, 1974; Galpin *et al.*, 1986; De Henau *et al.*, 1989). The line which supports the velocity vector issuing from the point (say w) where the reconstruction for ϕ is required intersects the contour of the closure volume upstream of w between two nodal points which are used to fix the constants C_1' and C_2' of the assumed linear shape function of Φ within the closure volume (Figure 5). Using local coordinates $\mathbf{r}' = \mathbf{R} - \mathbf{R}_w$ (with components x', y') from the origin w such that $\mathbf{R}_w = \sigma_{Pi-1}\mathbf{R}_P +$

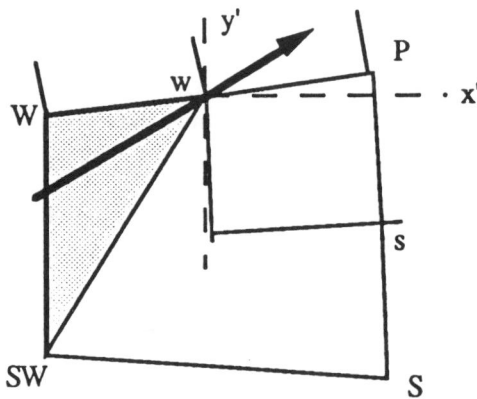

Figure 5 The closure volume for point w (shaded). Dashed lines indicate local coordinate axes x' and y'. The mass control volume surrounding point P is as in Figure 1. The local streamline at w is indicated in bold.

$(1 - \sigma_{Pi-1})\mathbf{R}_w$, the linear shape function is:

$$\phi = C_1' + C_2'\left(y'\frac{\bar{U}_w}{Q_w} - x'\frac{\bar{V}_w}{Q_w}\right); \quad Q_w = \sqrt{\bar{U}_w^2 + \bar{V}_w^2} \tag{55}$$

where $\bar{U}_w$ and $\bar{V}_w$ are cartesian velocity components at point w. The required 'boundary conditions' to determine C_1' and C_2' are $\phi = \Phi_W$ if $\mathbf{R} = \mathbf{R}_W$ and $\phi = \Phi_{SW}$ if $\mathbf{R} = \mathbf{R}_{SW}$. The required flux value is then simply $\phi_w = C_1'$. All other face values are aproximated in the same manner. SUDS, while removing flow-skewness problems, shares with one-dimensional reconstructions the disadvantage of not responding to cross-stream diffusion and sources. SUDS has often been evaluated in the past and compared with other upwind schemes (Leschziner, 1980; Smith and Hutton, 1982; Syed et al., 1985; Syed and Chiapetta, 1985; Shyy, 1985; Patel et al., 1985; Huang et al., 1985; Sharif and Busnaina, 1988a, b, 1993; Zurigat and Ghajar, 1990). Another problem is that the scheme is in error for cell Reynolds numbers close to 1, in particular when the convection velocity is along the WP direction. Such problems, coupled with the high computational expense associated with the need to account for six distinct interpolation ranges for every face, diminish somewhat the scheme's attractiveness.

Rather than being linear in the physical plane according to (55), the shape function can instead be taken as being two-dimensional in the computational plane. This is more in the spirit of finite-difference techniques. As an example, a piecewise quadratic shape function can be retained.

$$\phi = C_1 + C_2\xi + C_3\xi^2 + C_4\eta + C_5\eta^2 + C_6\xi_\eta \quad \text{where}$$
$$\xi = \xi^1 - \xi_P^1; \quad \eta = \xi^2 - \xi_P^2 \tag{56}$$

This is what is accomplished in another version of the QUICK scheme where the coefficients in (56) are estimated by use of data at the upstream-weighted group of grid points defining the closure stencil for point w: e.g. Φ_{NW}, Φ_W, Φ_{WW}, Φ_P, Φ_{SW}, Φ_S, if $\bar{U}$, $\bar{V} \geq 0$. This yields along the west face (Figure 6):

$$\phi_w = \tfrac{1}{2}[\Phi_W + \Phi_P] - \tfrac{1}{8}[\Phi_P + \Phi_{WW} - 2\Phi_W] + \tfrac{1}{24}[\Phi_{NW} + \Phi_{SW} - 2\Phi_W] \tag{57a}$$

Compared with (49), the last term in (57a) is an upstream-weighted transverse curvature effect which does not appear in the one-dimensional case. Similarly, if $\bar{U} < 0$,

$$\phi_w = \tfrac{1}{2}[\phi_W + \phi_P] - \tfrac{1}{8}[\phi_E + \phi_W - 2\phi_P] + \tfrac{1}{24}[\phi_N + \phi_S - 2\phi_W] \tag{57b}$$

The QUICK scheme has also been applied to unsteady problems (see in particular Chen and Falconer, 1992).

In order to put some more physics in the interpolation practice, it is possible to constraint the shape function to be itself the solution of an advection-diffusion problem in the nine-point neighbourhood of point P. Then the starting

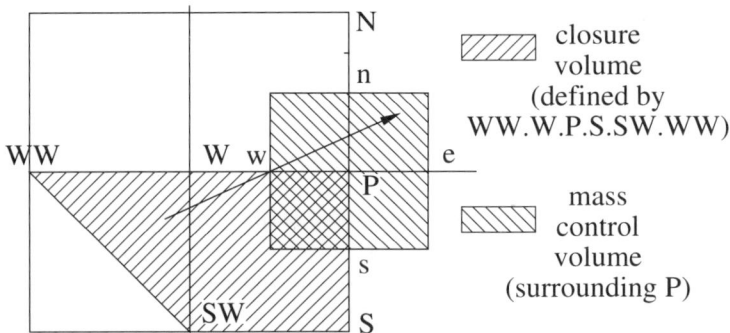

Figure 6 Control and closure volumes for the QUICK scheme in the computational plane. The streamline issuing from *w* is shown in bold.

equation is the following normalized (convective) transport equation:

$$\frac{\partial^2 \phi}{\partial \xi^{*2}} + \frac{\partial^2 \phi}{\partial \eta^{*2}} - \left[2A \frac{\partial \phi}{\partial \xi^*} + 2B \frac{\partial \phi}{\partial \eta^*} \right] = R_\phi$$

where

$$\xi^* = \xi^1/\sqrt{g^{11}}, \quad \eta^* = \xi^2/\sqrt{g^{22}}, \quad A = (A_\phi)_P/\sqrt{g_P^{11}}, \quad B = (B_\phi)_P/\sqrt{g_P^{22}}$$

while R_ϕ involves crossed second-order derivatives, as well as the contribution of the time derivative. A, B and R_ϕ are assumed to be constant in the vicinity of P. A and B define an advection velocity field in the computational plane wth a modulus $V = (A^2 + B^2)^{1/2}$. Since we work in the computational plane, the discrete molecule involves nine points defining a rectangular volume whose sides are given by $\Delta \xi^* = h = 1/\sqrt{g^{11}}$ and $\Delta \eta^* = k = 1/\sqrt{g^{22}}$.

Raithby (1976) has proposed a solution of the following type:

$$\phi = C_1' + C_2' \left(\eta^* \frac{A}{V} - \xi^* \frac{A}{V} \right) + C_3' \exp \left(2A\xi^* + 2B\eta^* \right) \tag{59}$$

in the cartesian case with $R_\phi = 0$. The three constants C_i', are then determined from $\phi = \Phi_P$ if $\mathbf{R} = \mathbf{R}_P$, $\phi = \Phi_W$ if $\mathbf{R} = \mathbf{R}_W$ and $\phi = \Phi_{SW}$ if $\mathbf{R} = \mathbf{R}_{SW}$. The resulting skew-upstream weighted difference scheme appears to be determined from diffusive and convective mechanisms alone. Hence it is difficult to justify since source terms, like the pressure force, are neglected.

In order to define their so-called finite-analytic scheme, Chen and Chen (1984) assume that a shape function, similar to (59), and solution of (58) in the nine-point neighbourhood of P, admits the following boundary:

$$\phi(\xi^1 = \xi_w^1, \xi^2) = \phi_W(\xi^2) = a_0 + a_1\xi^2 + a_2[\exp(2B\xi^2) - 1] \tag{60}$$

where the constants a_0, a_1, a_2 can be specified by the three nodal values of ϕ on the boundary, namely:

$$a_0 = \Phi_W, \quad a_1 = \frac{1}{2k}[\Phi_{NW} - \Phi_{SW} - (\Phi_{NW} + \Phi_{SW} - 2\Phi_W)\coth Bk],$$

$$a_2 = (\Phi_{NW} + \Phi_{SW} - 2\Phi_W)/(4\sinh^2 Bk)]$$

The boundary conditions for the north, south and east (N, S, E) sides can be approximated similarly. Equation (58) with boundary conditions ϕ_W, ϕ_S, ϕ_E, ϕ_N, specified by the eight nodal values, can be solved analytically by the method of separation of variables, while the inhomogeneous term is treated as a known constant. When the analytical solution is evaluated at the interior node P, we have equation (42a) with the finite analytic coefficients a_E, etc. specified in terms of the cell Reynolds numbers Ah and Bk and of multipliers P_A and P_{AB}

$$a_S = \frac{\exp(Bk)}{2\cosh Bk}P_A; \quad a_W = \frac{\exp(Ah)}{2\cosh Ah}P_B; \quad a_N = \exp(-2Bk)a_S;$$

$$a_E = \exp(-2Ah)a_W; \quad a_{SW} = \frac{\exp(Ah + Bk)}{4\cosh Ah \cosh Bk}[1 - P_A - P_B];$$

$$a_{SE} = \exp(-2Ah)a_{SW}; \quad a_{NW} = \exp(-2Bk)a_{SW}; \tag{61a}$$

$$a_{NE} = \exp(-2Ah - 2Bk)a_{SW};$$

$$a_P = \frac{h\tanh Ah}{2A}(1 - P_A) = \frac{k\tanh Bk}{2B}(1 - P_B)$$

with

$$P_A = 4E_2 Ah \cosh Bk \coth Ah; \quad P_B = 1 + \frac{Bh\coth Bk}{Ak\coth Ah}(P_A - 1) \tag{61b}$$

The multipliers P_A and P_B involve the evaluation of an infinite series:

$$E_2 = \sum_{m=1}^{\infty} \frac{(-1)^m(\lambda_m h)}{[(Ah)^2 + (\lambda_m h)^2]\cosh\sqrt{A^2 + B^2 + \lambda_m^2 k}}; \quad \lambda_m h = (m - \tfrac{1}{2})\pi \tag{61c}$$

The convergence of this series slows down when at least one of the cell Reynolds numbers Ah or Bk increases. In this case, Chen and Patel (1985) have proposed alternative values for P_A and P_B which avoid the use of E_3.

$$Ak\coth Ah \geq Bh\coth Bk; \quad P_A = 0, \quad P_B = 1 - Bh\coth Bk/Ak\coth Ah$$

and

$$Ak\coth Ah < Bh\coth Bk; \quad P_B = 0, \quad P_A = 1 - Ak\coth Ah/Bh\coth Bk \tag{61d}$$

Unfortunately, such expressions result from the hypothesis that, when one velocity component dominates, say A, the solution is equivalent to that provided by a one-dimensional problem along ξ^1. This assumption is not satisfied since the diffusive term $\partial^2\phi/\partial(\xi^2)^2$ has no reason to vanish, especially in the boundary layers which develop along the lines $\xi^2 = \xi_S^2$ and $\xi^2 = \xi_N^2$, and which are responsible for the slowing down of convergence of E_2. This single important defect is enough to explain the poor performance of the finite-analytic scheme, when compared, as in Sotiropoulos et al. (1994), to other more classical treatments. Another drawback of the finite-analytic method lies in its high cost in the three-dimensional case where the influence coefficients require the summation of three series. To avoid this problem, it is possible to combine the finite-analytic scheme in flow cross-planes with a one-dimensional scheme in the third direction, as was done by Chen et al. (1988). A similar, but formally fourth-order accurate, method has been proposed for steady flows by Chen et al. (1993) following the idea of the finite-analytic method; however, it has not been tested on curvilinear grids.

Deng et al. (1991a) have proposed a uniexponential scheme which is easier to use on curvilinear grids. The idea is briefly outlined below. We start from the normalized transport equation alternatively written:

$$\left[\frac{\partial^2\phi}{\partial s^2} - 2V\frac{\partial\phi}{\partial s}\right] + \left[\frac{\partial^2\phi}{\partial\xi^{*2}} + \frac{\partial^2\phi}{\partial\eta^{*2}} - \frac{\partial^2\phi}{\partial s^2}\right] = S_\phi \tag{62}$$

where s is the local advection direction in the computational plane (Figure 7) and $V = \sqrt{A^2 + B^2}$. The first bracket in (62) can be expressed using the exponential discretization of Spalding (1972), which yields:

$$\left[\frac{\partial^2\phi}{\partial s^2} - 2V\frac{\partial\phi}{\partial s}\right] \approx (C_U\Phi_U - \Phi_P + C_D\Phi_D)/C_P,$$

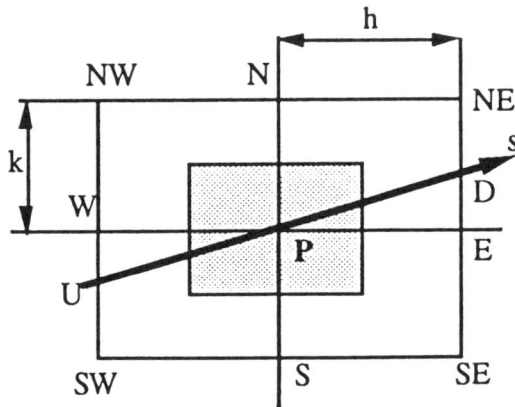

Figure 7 Schematic sketch for the uniexponential scheme. s is the direction of the streamline issuing from P.

with

$$C_U = \frac{\exp(V)}{\exp(V) + \exp(-V)}; \quad C_D = \frac{\exp(-V)}{\exp(V) + \exp(-V)}; \quad C_P = \frac{\tanh V}{2V} \tag{63}$$

The second bracket in (62) is discretized using standard second-order finite differences. A parabolic interpolation function is used to express the intermediate values Φ_U and Φ_D in terms of dependent variables; for instance:

$$\Phi_U = \Phi_{NW} \frac{\eta_U^*(k + \eta_U^*)}{2k^2} - \Phi_W \frac{(\eta_U^* - k)(k + \eta_U^*)}{2k^2} + \Phi_{SW} \frac{\eta_U^*(\eta_U^* - k)}{2k^2} \tag{64}$$

which results in a nine-point formula. Despite the fact that the uniexponential scheme is not positive, no serious difficulties have been encountered with it or with one of its blended variants using a centred scheme (Deng and Piquet, 1992).

The aforementioned nine-point schemes share in common their formally first-order accuracy on a cartesian uniform grid. Moreover, they involve only the single evaluation of S_ϕ at point P. An important property of the matrix of influence coefficients is its strong diagonal dominance which is given by:

$$|(a_P)_{i,j}| \ge \sum_{nb} |(a_{nb})_{i,j}| \quad \text{with the strict inequality for at least one value of } i,j \tag{65}$$

Because of the consistency of the schemes, the equality is realized at inner points of the domain in the absence of source terms. Diagonal dominance is then a sufficient condition for a discrete maximum principle to be satisfied: in other words, Φ_P does not lie outside the range of its neighbours, but is a weighted mean of the values of Φ at surrounding nodes nb and, thus, the solution does not contain overshoots, undershoots or undesirable oscillations. To achieve this so-called monotonicity property, it is sufficient that $(a_{nb})_{i,j} > 0$, because of the consistency of the scheme. Furthermore, if a scheme is not diagonally dominant or if the positive property is not satisfied, then Φ_P is no longer a weighted mean of its neighbours and there is a possibility of oscillations. Failure of diagonal dominance or of the positive property is often related to the failure of a mesh Reynolds number condition. Hence the magnitude of oscillations may depend on the Reynolds number range, but other causes possibly occur, such as the treatment of boundary conditions and of imposed source profiles (Lillington, 1981). From an algebraic point of view, upwinding is a means of generating a system of velocity unknowns defining a so-called M-matrix (Varga, 1962) which makes the efficient use of classical iterative methods possible.

Several negative results are important. They are already valid on a cartesian grid of uniform spacing h.

(i) For a steady problem, there is no diagonally dominant uniformly accurate second-order nine-point scheme (Van Veldhuizen, 1979; Berger et al.,

1981). This result, verified on the pure convection model $U\partial\phi/\partial x + V\partial\phi/\partial y = 0$, is a consequence of the Taylor formula which yields the equality (66) in which a_P has been eliminated using the consistency condition:

$$
\frac{1}{2}\left(\frac{V_P}{U_P} + \frac{U_P}{V_P}\right)(a_N + a_S + a_{NE} + a_{NW}) + \frac{1}{2}\frac{V_P}{U_P}(a_E + a_W)
$$

$$
+ \frac{1}{2}\frac{U_P}{V_P}(a_N + a_S) = a_{NE} - a_{NW} + a_{SW} - a_{NW} \tag{66}
$$

(ii) For a steady advection-diffusion equation with $R_\phi = 0$, there is no nine-point diagonally dominant scheme free of numerical diffusion and oscillations for cell Reynolds numbers higher than 2. The mesh Reynolds number is actually the ratio of $|(\partial\phi/\partial x)_P|$ and $|(\partial^2\phi/\partial x^2)_P|$ coefficients:

$$
Rx_P = 2\frac{a_E - a_W + a_{NE} - a_{NW} + a_{SE} - a_{SW}}{a_E + a_W + a_{NE} + a_{NW} + a_{SE} + a_{SW}}
$$

(iii) For a steady advection-diffusion equation with $R_\phi = 0$, there is no nine-point diagonally dominant scheme which contains diffusion terms acting only in the streamwise direction (Lillington, 1981). This is because the coefficient of the second-order derivative in the direction n normal to the flow is:

$$
\mathrm{cosec}^2\,\theta[a_E + a_W] + \sec^2\,\theta[a_N + a_S] + [\sec\theta - \mathrm{cosec}\,\theta]^2[a_{NE} + a_{SW}]
$$

$$
+ [\sec\theta + \mathrm{cosec}\,\theta]^2[a_{NW} + a_{SE}]
$$

where θ is the angle between the streamwise direction of the flow and the x-coordinate line. This coefficient cannot vanish for all values of θ. Hence the objective is rather to find a compromise between accuracy and diagonal dominancy.

(iv) Any nine-point scheme of the type (42a) is an approximation to $U\partial\phi/\partial x + V\partial\phi/\partial y$ which is at most second-order accurate (Shay, 1981). The truncation error analysis shows that it is not possible to make coefficients of $\partial^2\phi/\partial x^2$, $\partial^2\phi/\partial y^2$, $\partial^2\phi/\partial x\partial y$, $\partial^3\phi/\partial x^3$, $\partial^3\phi/\partial y^3$ vanish in the truncation error expansion. Shay was able to construct a second-order accurate solution by dropping $\partial^2\phi/\partial x^2$, $\partial^2\phi/\partial y^2$, $\partial^2\phi/\partial x\partial y$, $\partial^3\phi/\partial x\partial y^2$, $\partial^3\phi/\partial x^2\partial y$ so that three degrees of freedom remain which can be used to satisfy a monotonicity property. As a result, the leading terms in the truncation error at point P are:

$$
\frac{Rh^2}{6}\left[U\frac{\partial^3\phi}{\partial x^3} + V\frac{\partial^3\phi}{\partial x^3}\right]_P + h^3[a_{NW} - a_{SW}]\frac{\partial^3\phi}{\partial x^2\partial y}\bigg|_P
$$

$$
+ h^3[a_{SE} - a_{NW}]\frac{\partial^3\phi}{\partial x\partial y^2}\bigg|_P
$$

The presence of the mesh Reynolds number again indicates the non-uniform accuracy of the scheme.

While the main part of the momentum equations based on cartesian velocity components is treated like the scalar equation, the pressure term is usually considered as a part of the source term. It can be treated in the physical plane as:

$$(PS^1)_e - (PS^1)_w + (PS^2)_n - (PS^2)_s \qquad (67)$$

The required closure assumption for the face values P_e, P_n, P_s, P_w is made as simple as possible: a linear interpolation between neighbouring nodes yields classically (Figure 3):

$$P_e = \sigma_{Pi}P_E + (1 - \sigma_{Pi})P_P \quad \text{with} \quad \sigma_{Pi} = \overline{Pe}/\overline{PE} \qquad (68)$$

Alternatively, the pressure terms can be treated in the computational plane using:

$$\left.\frac{\partial P}{\partial x}\right|_P = J^{-1}\left[S_1^1\frac{\partial P}{\partial \xi^1} + S_1^2\frac{\partial P}{\partial \xi^2}\right]_P \qquad (69)$$

with centred approximations of pressure derivatives in the computational plane.

Up to now, we have been mainly concerned with the momentum equation and we have treated its discretization as an independent procedure. The discretization processes described (of finite-volume or finite-difference type) could be viewed in all cases as a means to express the (momentum) fluxes through a particular interpolation in terms of nodal unknowns. In all cases, a face velocity value was expressed in terms of nodal (cell-centred) velocity unknowns; a face pressure value was expressed in terms of nodal (cell-centred) pressure unknowns. It remains now to consider the continuity equation and to solve the so-called *coupling problem*, i.e. to find a solution of the system of mass and momentum equations.

IV THE MASS EQUATION AND COUPLING ALGORITHMS

A The discrete form of the mass equation

Since mass conservation is of vital importance in incompressible flows, the continuity equation should be treated in a conservative form to avoid spurious mass losses or generation. The discrete TT form of the continuity equation (17), which involves the contravariant velocity components u and v, may be written as:

$$(Ju)_{AB} - (Ju)_{CD} + (Jv)_{BC} - (Jv)_{DA} = 0 \qquad (70)$$

If $Ju \equiv JV^1$ and $Jv \equiv JV^2$ are defined as the averages of the mass flux across the faces of the control volume ABCD (Figure 1) surrounding point $\mathbf{R}_{i,j}$, the form (70) is exact. If u and v are defined as the discrete approximations of the contravariant velocity components V^1 and V^2 at some point lying on the faces of the control volume ABCD,

$$(Ju)_{i+\frac{1}{2},j} - (Ju)_{i-\frac{1}{2},j} + (Jv)_{i,j+\frac{1}{2}} - (Jv)_{i,j-\frac{1}{2}} = 0 \tag{71}$$

is the discrete aproximation of the continuity equation in the computational plane.

We may use instead the approximations $\overline{U}$, $\overline{V}$ for the cartesian velocity components $\overline{V}^1$, $\overline{V}^2$. They are connected to u and v by:

$$Ju = S_1^1 \overline{U} + S_2^1 \overline{V}; \quad Jv = S_1^2 \overline{U} + S_2^2 \overline{V} \tag{72}$$

along each face of ABCD (Figure 1). However, $\overline{U}$ and $\overline{V}$ are available, not along cell faces, but at the centres of the finite volumes. This is the flux reconstruction problem: if (71) is to be used, the unknown fluxes Ju, Jv at cell faces must be reconstructed from the available unknown nodal values of the velocity components. We leave aside this problem and substitute (72) into (71) in order to work only with velocity unknowns; when this is done the discrete equation for the PT form (16) of the continuity equation:

$$[S_1^1 \overline{U} + S_2^1 \overline{V}]_{i+\frac{1}{2},j} - [S_1^1 \overline{U} + S_2^1 \overline{V}]_{i+\frac{1}{2},j} + [S_1^2 \overline{U} + S_2^2 \overline{V}]_{i,j+\frac{1}{2}}$$
$$- [S_1^2 \overline{U} + S_2^2 \overline{V}]_{i,j-\frac{1}{2}} = 0 \tag{73a}$$

where projected surface values of $\mathbf{S}$ are known exactly from the locations of the face vertices, while $\overline{U}$ and $\overline{V}$ (or, equivalently, Ju and Jv) will be ultimately specified from the solution of the flux reconstruction problem, as linear combinations of nodal unknowns. Hence the continuity equation can be viewed as an algebraic equation of the form (73b):

$$\mathbf{DV} = \mathbf{g} \tag{73b}$$

where $\mathbf{D}$ (not to be confused with the rate of strain tensor) is the product of a discrete divergence operator ∇ and an interpolation matrix $\mathcal{M}_u$ operating on the set of velocity nodal unknowns, $\mathbf{V}$, and $\mathbf{g}$ is a (known) right-hand-side depending on (velocity) boundary conditions. The choice of $\mathcal{M}_u$ results from the selected method for the reconstruction $\mathbf{v} = \mathcal{M}_u \mathbf{V}$ of mass fluxes Ju and Jv (defining the vector $\mathbf{v}$ (such that $\nabla \mathbf{v} = \mathbf{g}$) of auxiliary unknowns) from the available nodal unknowns $\mathbf{V}$; this will be examined later.

B The general form of the discrete momentum equation

Following equation (37), we assume (i) a one-step backward time-discrete (Euler) scheme of time step τ, and (ii) that additional defining sources are known from the previous iteration at the computed time step momentum equation and can be written under the following 'rescaled form' (74) of (42):

$$\left[1 + K_P \frac{Re}{\tau}\right]\overline{U}_P^{n+1} = \sum_{nb} K_{nb}\overline{U}_{nb}^{n+1} + K_P\left[\frac{Re}{\tau}\overline{U}_P^n + 2g^{12}\frac{\delta^2\overline{U}}{\delta\xi\delta\eta}\bigg|_P\right]$$
$$+ ReK_P\left[S_1^1\frac{\delta P}{\delta\xi}\bigg|_P + S_1^2\frac{\delta P}{\delta\eta}\bigg|_P\right] \tag{74a}$$

$$\left[1 + K_P \frac{Re}{\tau}\right]\overline{V}_P^{n+1} = \sum_{nb} K_{nb}\overline{V}_{nb}^{n+1} + K_P\left[\frac{Re}{\tau}\overline{U}_P^n + 2g^{12}\frac{\delta^2\overline{V}}{\delta\xi\delta\eta}\bigg|_P\right]$$
$$+ ReK_P\left[S_2^1\frac{\delta P}{\delta\xi}\bigg|_P + S_2^2\frac{\delta P}{\delta\eta}\bigg|_P\right] \tag{74b}$$

where K_P and K_{nb} are influence coefficients determined by the discretization method, such that $K_P > 0$, while $\sum_{nb} K_{nb} = 1$, each coefficient K_{nb} being positive if the discrete scheme is positive.

The influence coefficients K_{nb} and K_P depend in principle of the velocity field itself since convective terms are non-linear, so that they involve the velocity field $\mathbf{V}^{n+1}$ at the previous iteration (ν). The matrix form corresponding to the two discretized momentum equations written at all cell centres P is:

$$(\mathbf{E} - \mathbf{A})\mathbf{V} + \mathbf{GP} = \mathbf{f} \tag{75}$$

where $\mathbf{E}$ stems for the diagonal matrix of the coefficient $1 + K_P Re/\tau$ for the unknowns $\overline{U}^{n+1}$ and $\overline{V}^{n+1}$ at the cell centres P_{ij}. $\mathbf{A}$ is the matrix involving the influence coefficients K_{nb}, $\mathbf{G}$ is the matrix involving the discrete pressure gradient terms (last square bracketed term on the right-hand-side of (74)), i.e. the product of ∇^T with an adequate interpolation $\mathcal{M}_p$ of pressure unknowns. $\mathbf{f}$ groups together the explicit terms (source terms) and the boundary conditions. The matrix $\mathbf{E} - \mathbf{A}$ is invertible (and often diagonally dominant with a positive discrete scheme, for all time steps). Hence, if the pressure field is known, the velocity field results from:

$$\mathbf{U} = (\mathbf{E} - \mathbf{A})^{-1}(\mathbf{f} - \mathbf{GP}) \tag{76}$$

To determine the pressure field, the following pressure equation is derived by applying the operator $\mathbf{D}$ to (76). Then:

$$\mathbf{D}(\mathbf{E} - \mathbf{A})^{-1}\mathbf{GP} = \mathbf{D}(\mathbf{E} - \mathbf{A})^{-1}\mathbf{f} - \mathbf{g} \tag{77}$$

Unfortunately, such an approach is almost impossible to use since $\mathbf{D}(\mathbf{E} - \mathbf{A})^{-1}\mathbf{G}$ is a full matrix and inversion of $(\mathbf{E} - \mathbf{A})^{-1}$ cannot easily be achieved for large three-dimensional computations. Iterative methods are thus required to avoid the direct inversion of $\mathbf{E} - \mathbf{A}$.

C Iterative algorithms for the coupled systems

An approximate inverse $\mathbf{C}$ of $\mathbf{E} - \mathbf{A}$ is constructed. Equation (75) can be recast as:

$$\mathbf{V} = [\mathbf{I} - \mathbf{C}(\mathbf{E} - \mathbf{A})]\mathbf{V} + \mathbf{C}(\mathbf{f} - \mathbf{G}\mathbf{P})$$

where the right-hand-side factor of $\mathbf{V}$ is small in some sense. Then (77) becomes:

$$\mathbf{DCG\,P} = \mathbf{D}\hat{\mathbf{V}} - \mathbf{g}, \tag{78}$$

with the definition of the pseudo-velocity $\hat{\mathbf{V}} = [\mathbf{I} - \mathbf{C}(\mathbf{E} - \mathbf{A})]\mathbf{V} + \mathbf{C}\mathbf{f}$ so that the momentum equation simplifies to:

$$\mathbf{V} = \hat{\mathbf{V}} - \mathbf{CGP} \tag{79}$$

If the previous velocity field is used to evaluate the pseudo-velocity, Eq. (78) yields a solution for the pressure. Equation (79) is then used to update the velocity field. Hence, the calculation of pressure and velocity is handled using separate equations; this justifies referring to such methods as 'segregated methods'. The coupling algorithm between velocity and pressure is as follows:

(i)   Initialization of velocity and pressure fields $\{\mathbf{U}^{(\nu)}, \mathbf{P}^{(\nu)}\}$.

(ii) Solution of (78) to yield $\mathbf{P}^{(\nu+1)}$ from

$$\mathbf{DCGP}^{(\nu+1)} = \mathbf{D}\{[\mathbf{I} - \mathbf{C}(\mathbf{E} - \mathbf{A})]\mathbf{U}^{(\nu)} + \mathbf{Cf}\} - g.$$

(iii) Solution of the momentum equations $(\mathbf{E} - \mathbf{A})\mathbf{U}^{(\nu+1)} = \mathbf{f} - \mathbf{GP}^{(\nu+1)}$ in the form:

$$\mathbf{U}^{(\nu+1)} = [\mathbf{I} - \mathbf{C}(\mathbf{E} - \mathbf{A})]\mathbf{U}^{(\nu)} + \mathbf{C}[\mathbf{f} - \mathbf{GP}^{(\nu+1)}], \quad \text{so that} \quad \mathbf{DU}^{(\nu+1)} = \mathbf{g}.$$

(iv) Steps (ii) and (iii) are repeated until convergence.

This coupling algorithm is the basis algorithm for all segregated methods. When $\mathbf{I} - \mathbf{E}^{-1}\mathbf{A}$ is convergent, the spectral radius of $\mathbf{E}^{-1}\mathbf{A}$ is less than one and $(\mathbf{E} - \mathbf{A})^{-1} = \mathbf{E}^{-1}[\mathbf{I} + \mathbf{E}^{-1}\mathbf{A} + (\mathbf{E}^{-1}\mathbf{A})^2 + ...]$ defines a convergent series. An approximate inverse of $\mathbf{E} - \mathbf{A}$ is obtained from a truncation of this series at the mth power. The most common approximate inverse is obtained by the choice $m = 0$, $\mathbf{C} = \mathbf{E}^{-1}$, from which the pressure equation to solve becomes simply:

$$\mathbf{DE}^{-1}\mathbf{GP} = \mathbf{DE}^{-1}(\mathbf{AV} + \mathbf{f}) - \mathbf{g} \tag{80}$$

For $m \neq 0$, the sparsity of the truncated expansion increases with m without necessarily yielding an improved approximation (Connell and Stow, 1986). Interative procedures, like PISO (Issa, 1985), SIMPLE, SIMPLER (Patankar, 1980) or other variants share in common the choice $m = 0$, and implement the coupling procedure in an incremental way, solving for $\mathbf{P}^{(\nu+1)} - \mathbf{P}^{(\nu)}$ and $\mathbf{V}^{(\nu+1)} - \mathbf{V}^{(\nu)}$ rather than for $\mathbf{P}^{(\nu+1)}$ and $\mathbf{V}^{(\nu+1)}$.

In SIMPLE (Patankar, 1980), the guessed pressure field $\mathbf{P}^{(\nu)}$ is used to compute $\mathbf{V}^*$ from the momentum equation:

$$(\mathbf{E} - \mathbf{A})\mathbf{V}^* = \mathbf{f} - \mathbf{GP}^{(\nu)}$$

A pressure correction is then computed from the solution of

$$\mathbf{DE}^{-1}\mathbf{GP}' = \mathbf{DV}^* - \mathbf{g}$$

so that $\mathbf{P}^{(\nu+1)} = \mathbf{P}^{(\nu)} + \mathbf{P}'$. The velocity is finally corrected using $\mathbf{V}^{(\nu+1)} = \mathbf{V}^* - \mathbf{E}^{-1}\mathbf{GP}'$. These three steps are then repeated until convergence. Strong underrelaxation (equivalent to a diminution of the time step) of the procedure is required (Patankar, 1980; Wachpress, 1981; Van Doormaal and Raithby, 1984; Miller and Schmidt, 1988). Like all other procedures, the SIMPLE algorithm may be accelerated (Wen and Ingham, 1993), for instance by means of a multigrid procedure; SIMPLE then becomes a smoother whose behaviour has been analysed by Shaw and Sivaloganathan (1988) and Sivaloganathan and Shaw (1988).

In PISO (Issa, 1985), the starting velocity field $\mathbf{V}^* \equiv \mathbf{V}^{(\nu)}$ is obtained from the given pressure field $\mathbf{P}^{(\nu)}$ using

$$(\mathbf{E} - \mathbf{A})\mathbf{V}^* = \mathbf{f} - \mathbf{GP}^{(\nu)}$$

The predicted pressure field $\mathbf{P}^*$ results then from (80) with $\mathbf{V} = \mathbf{V}^*$ and it is used to compute the corrected velocity field $\mathbf{V}^{**} = \mathbf{E}^{-1}(\mathbf{AV}^* + \mathbf{f} - \mathbf{GP}^*)$. The last two operations are usually repeated to provide a second correction $\{\mathbf{P}^{**}, \mathbf{V}^{***}\}$ and $\mathbf{P}^{**} = \mathbf{P}^{(\nu+1)}$, $\mathbf{V}^{***} = \mathbf{V}^{(\nu+1)}$. The first correction of PISO is algebraically equivalent to one iteration of SIMPLE, although the right-hand-sides of the solved systems are different.

In SIMPLER (Raithby and Schneider, 1979; Patankar, 1980), one starts from a guessed velocity field $\mathbf{V}^{(\nu)}$ and computes $\hat{\mathbf{V}} = \mathbf{E}^{-1}(\mathbf{AV}^{(\nu)} + \mathbf{f})$. According to phase (ii), once the pressure equation (80) has yielded a pressure update $\mathbf{P}^*$ from $\mathbf{V}^{(\nu)}$, according to:

$$\mathbf{DE}^{-1}\mathbf{GP}^* = \mathbf{D}\hat{\mathbf{V}} - \mathbf{g}$$

the corresponding velocity prediction of phase (iii) is obtained from

$$(\mathbf{E} - \mathbf{A})\mathbf{V}^* = \mathbf{f} - \mathbf{GP}^*$$

The pressure field $\mathbf{P}^*$ is correct if the velocity field $\mathbf{V}^{(\nu)}$ is correct. Hence, there is no need to alter it and instead of updating in turn the pressure field, as in the second correction of PISO, a pressure correction field $\mathbf{P}' = \mathbf{P}^{**} - \mathbf{P}^*$ is computed as the solution of

$$\mathbf{DE}^{-1}\mathbf{GP}' = \mathbf{DV}^* - \mathbf{g}$$

and it is used *only* to correct the velocity, according to $\mathbf{V}^{**} = \mathbf{V}^* - \mathbf{E}^{-1}\mathbf{GP}'$. The updated iteration is finally $\mathbf{V}^{(\nu+1)} = \mathbf{V}^{**}$, $\mathbf{P}^{(\nu+1)} = \mathbf{P}^*$.

In spite of the improved convergence behaviour of SIMPLER and PISO over that of SIMPLE (Jang *et al.*, 1986; Wanik and Schnell, 1989; Kim and Benson, 1992; McGuirk and Palma, 1993), these procedures do not appear to be as free from difficulties: underrelaxation is usually found to be necessary, especially on fine grids and for complex geometries. The use of the pressure correction equation in SIMPLER seems to destabilize the convergence process in some cases and it is not adequate for unsteady problems. For this reason it may be preferable to use PISO, which makes use of a true pressure equation, although the required underrelaxation factors may be very low (current values of 0.1 are often encountered on complex three-dimensional fine grids) and do not necessarily prevent the non-linear residuals from showing limit-cycle behaviour.

D Improving convergence of segregated methods

Most incompressible flow solution algorithms adopts a particular form of the segregated solution procedures discussed in Section IV.C. The most important contribution to the total cost of solving the flow problem lies in the pressure solver. As a result of the required interpolations, the pressure matrix is often non-symmetric so that special techniques and suitable acceleration procedures are currently required. The available techniques are reviewed briefly.

1 Approximate factorization techniques

The most commonly adopted technique for solving linearized algebraic equations is the alternating direction line Gauss–Seidel solver. However, uncertain convergence rates are encountered for the pressure solver, especially when the grid skewness generates strongly anisotropic metric coefficients, or when mixed boundary conditions are used. A particular class of more implicit solvers is the so-called incomplete LU decomposition (ILU). The exact LU decomposition of the pressure matrix generates a full lower, $\mathbf{L}'$, and upper, $\mathbf{U}'$, triangular matrix. Such matrices are approximated by some triangular matrices retaining a sparsity similar to that of the original unfactorized pressure system ($\mathbf{DE}^{-1}\mathbf{A} \equiv \mathbf{L}'\mathbf{U}'$ is usually a nanodiagonal matrix for two-dimensional problems, and it has 19 or 27 diagonals in three-dimensional problems). The coefficients of the approximate triangular matrices, $\mathbf{L}$ and $\mathbf{U}$, for $\mathbf{L}'$ and $\mathbf{U}'$, respectively, are determined from constraints imposed by the product $\mathbf{LU}$ of the approximate decomposition matrices. The constraints are such that $\mathbf{LU}$ must be identical to the product $\mathbf{L}'\mathbf{U}'$ for all its non-zero entries. Unfortunately, $\mathbf{LU}$ contains additional non-zero entries (with respect to $\mathbf{DE}^{-1}\mathbf{A}$), the fill-ins, where none should appear; in other words, $\mathbf{DE}^{-1}\mathbf{A} \equiv \mathbf{LU} + \mathbf{R}$, where $\mathbf{R}$ is the residual. Among the ILU decompositions, one of the most useful is the so-called strongly implicit procedure, SIP (Stone, 1968; Zedan and Schneider, 1979). In SIP an attempt is made to partially cancel the effect of non-zero entries of $\mathbf{LU}$. A parameter, α, is introduced to control the degree of partial cancellation. When this parameter is zero, the resulting method is algebraically equivalent to the so-called ILU(0) method. Another possibility is to introduce approximate triangular matrices with entries along extradiagonals where the matrix $\mathbf{DE}^{-1}\mathbf{A}$ is sparse. Depending on the number n, of intrduced extradiagonals in L and U, the resulting method is called an ILU(n) decomposition (Meijerink and Van der Vorst, 1977, 1981). In spite of their theoretical interest, the performance of ILU(n) methods is seldom better (at least for pressure solvers of segregated methods) than more classical ILU(0) methods. Fortunately the rate of convergence of ILU decompositions can be enhanced by means of acceleration techniques.

2 Conjugate-gradient type accelerations

There are a large number of available methods for solving systems $\mathbf{Ax} = \mathbf{b}$ where $\mathbf{A}$ is a non-symmetric matrix of $R^{N,N}$. Such methods are designed to be adequate for a suitable acceleration effect, when the preconditioning method may be one of the ILU decompositions introduced in Section IV.D.1. Conjugate-gradient-type methods (CG) seek the kth approximate solution $\mathbf{x}^k$ to equation $\mathbf{Ax} = \mathbf{b}$ from the subspace $\mathbf{x}^0 + K_k(\mathbf{r}^0, \mathbf{A})$, where $\mathbf{x}^0$ is the initial iterate and $K_k(\mathbf{r}^0, \mathbf{A}) = \mathrm{span}\,\{\mathbf{r}^0, \mathbf{Ar}^0, ..., \mathbf{A}^{k-1}\mathbf{r}^0\}$ is the kth Krylov subspace generated by the coefficient matrix $\mathbf{A}$ and the initial residual vector $\mathbf{r}^0 = \mathbf{b} - \mathbf{Ax}^0$. The residual vector of the kth step $\mathbf{r}^k = \mathbf{b} - \mathbf{Ax}^k$ can be written as:

$$\mathbf{r}^k = \mathbf{r}^0 - \mathbf{A}q_{k-1}(\mathbf{A})\mathbf{r}^0 = p_k(\mathbf{A})\mathbf{r}^0$$

where $q_{k-1} \in \Pi_{k-1}$ is a polynomial of degree at most $k-1$ and $p_k(z) = 1 - zq_{k-1}(z) \in \Pi_k$ with $p_k(0) = 1$. Π_k denotes the set of all polynomials of degree at most k. Voevodin (1983) and Faber and Manteuffel (1984) have demonstrated that, in general, CG-like schemes cannot satisfy the following properties simultaneously: (i) construct iterates $\mathbf{x}^k$ that minimize the residual norm in the sense that, for instance, $\|\mathbf{r}^k\|_2 \equiv \sqrt{\mathbf{r}^{k\mathrm{T}}\mathbf{r}^k}$ is minimal for all $\mathbf{x}^k \in \mathbf{x}^0 + K_k(\mathbf{r}^0, \mathbf{A})$; (ii) find $\mathbf{x}^k$ by a short recurrence such that the work and storage requirements remain small and constant while the iterative process moves forward. A notable exception is the classical CG method for symmetric positive definite matrices.

We may subdivide the CG-type methods into four groups, according to their convergence properties.

(i) *Generalized minimal residual methods.* Generalized minimal residual methods, like GMRES (Saad and Schultz, 1986; Saad, 1989), construct iterates that satisfy the residual minimization property:

$$\|\mathbf{r}^k\|_2 = \min_{p\in\Pi_k, p(0)=1} \|p(\mathbf{A})\mathbf{r}^0\|_2$$

The approximate solution $\mathbf{x}^k$ is computed by first creating an orthonormal basis for the Krylov subspace $K_k(\mathbf{r}^0, \mathbf{A})$ and then solving a least-square problem to determine the residual that satisfies the mentioned minimization property. Saad and Schultz demonstrated that the least-square problem has a unique solution so that GMRES cannot breakdown. Moreover, since the residual norm is minimized in each step, it cannot increase as the iteration proceeds and the convergence is monotomic. The most important drawback of the iteration stems from the fact that the Arnoldi orthogonalization procedure requires the storage of k vectors and the execution of $O(k)$ vector operations in each step. This makes GMRES very expensive when k increases. If we assume that $\mathbf{A}$ is diagonalizable, i.e. that there exists an invertible matrix $\mathbf{T}$ such that $\mathbf{T}^{-1}\mathbf{A}\mathbf{T} = \mathbf{Y} \equiv \mathrm{diag}\,(\lambda_1, \ldots, \lambda_N)$, the convergence rate of GMRES is bounded above by

$$\frac{\|\mathbf{r}^k\|_2}{\|\mathbf{r}^0\|_2} \le \kappa(\mathbf{T}) \min_{p \in \Pi_k, p(0)=1} \max_{\lambda_i \in \Lambda} |\, p(\lambda_i)\, |$$

where $\Lambda = \{\lambda_1, \ldots, \lambda_N\}$ is the spectrum of $\mathbf{A}$ and $\kappa(\mathbf{T}) = \|\mathbf{T}\|_2 \|\mathbf{T}^{-1}\|_2$ is the condition number of any matrix $\mathbf{T}$ of eigenvectors of $\mathbf{A}$. Unfortunately $\kappa(\mathbf{T})$ is huge in practical applications, i.e. eigenvalues of $\mathbf{A}$ may be very sensitive to small perturbations in the matrix entries. Nachtigal *et al.* (1992) have shown that, in this case, $\mathbf{A}$ is far from normality and the convergence of GMRES depends on polynomial approximation problems defined on the so-called pseudospectrum of $\mathbf{A}$ rather than on the spectrum of $\mathbf{A}$. Moreover, it is necessary to restart GMRES after some number, m, of steps (between 5 and 20) in order to keep storage requirements under control. Hence the algorithm works in practice with an m-dimensional Krylov space, spending a large amount of time in relearning information obtained during previous cycles. This problem is important since the iretative process may begin to converge superlinearly only after m has reached a certain value, as has been shown by Vuik (1993) for the pressure problem. For all these reasons, GMRES should be used in a hybrid method (Nachtigal *et al.*, 1992).

(ii) *Lanczos-type schemes.* The bi-conjugate gradient method (Bi-CG) (Fletcher, 1976) and the quasi-minimal residual (QMR) method (Freund and Nachtigal, 1991) seek the kth approximate solution $\mathbf{x}^k$ to $\mathbf{A}\mathbf{x} = \mathbf{b}$ from the same Krylov subspace $\mathbf{x}^0 + K_k(\mathbf{r}^0, \mathbf{A})$ as GMRES. However, the 2-norm of the residual computed by Bi-CG and QMR does not satisfy the minimization property as in GMRES. Instead of using the Arnoldi orthogonalization procedures as does GMRES, Bi-CG and QMR use the non-symmetric Lanczos process to generate basis vectors from the Krylov subspace. Instead of building an orthogonal basis of K_k, it builds a pair of bi-orthogonal bases $\{\mathbf{v}_i\}$ and $\{\mathbf{w}_i\}$, for the two subspaces $K_k(\mathbf{v}_0, \mathbf{A})$ and $K_k(\mathbf{w}_0, \mathbf{A}^{\mathrm{T}})$. The basis vectors are computed from a pair of three-term recurrences (for details, see Golub and Van Loan, 1983) so that the Lanczos process involves little work and storage per iteration. The coefficients of the recurrence relations are obtained by factorizing a tridiagonal matrix without pivoting. This is an important source of instability which partially explains the erratic behaviour of the convergence curves of Bi-CG (see e.g. Piquet *et al.*, 1987). In contrast with the Arnoldi method (where breakdown is favourable since exact eigenvalues of the invariant subspace are caught), the Lanczos algorithm breaks down with the existence in $K_k(v_0, A)$ of a vector orthogonal to $K_k(w_0, A^T)$, so that no oblique projection results (for more details, see e.g. Saad, 1989). Such unwanted terminations may be avoided using 'look-head' steps (Freund *et al.*, 1993). Bi-CG generates the same sequences of bi-orthogonal vectors as the Lanczos process. Bi-CG also has the finite termination property (it finds x in at most N steps in exact precision arithmetics) but there is no minimization property for the intermediate steps, so that the (non-monotonic) convergence properties of Bi-CG are less well understood than those of GMRES. QMR updates recurrence relations by applying a least-square procedure 'quasi-minimization' to only one part of the residual expression. QMR can be viewed as a more stable implementation of Bi-CG since the least-square procedure has a unique solution. This eliminates, with the problem of the LU-factorization of Bi-CG, one of the potential sources for breakdown and instability.

(iii) *Transpose-free Lanczos-type algorithms (CGS, Bi-CGSTAB, TFQMR).* Sonneveld (1989) was the first to recognize that matrix-vector products with A^T could be avoided, permitting the development of transpose-free algorithms. In the Bi-CG algorithm, it is possible to replace the residual vector $r^k = p_k(A)r^0$ by $r^k = p_k^2(A)r^0$, for the same polynomial p_k, with no increase in the amount of work per step, and with a convergence (or divergence) faster than that of Bi-CG by a factor of between 1 and 2. The residual vector constructed with $p_k^2(A)$ defines the so-called conjugate-gradient square method (CGS) method. Although breakdown is rarely encountered for the pressure problem, a modification that avoids exact breakdowns of CGS is available (Brezinski and Sadok, 1991). To smooth the erratic behaviour of CGS, Van der Vorst (1992) replaced $p_k^2(A)$ by a product of two polynomials $p_k(A)s_k(A)$, where $s_k \in \Pi_k$ is a polynomial of degree k, whose coefficients are updated recursively, using a local steepest descent procedure. The convergence curves of the resulting algorithm, Bi-CGSTAB, are smoother than those of CGS, although the new algorithm is not necessarily more stable than CGS. Freund (1993) proposed another variant of CGS, the so-called transpose-free quasi-minimal residual TFQMR algorithm. The convergence of these transpose-free methods is in general not well understood, in spite of their remarkable efficiency.

(iv) *CG applied to normal equations (CGN).* CGN is the classical CG algorithm applied to the system of equations $A^T A x = A^T b$. Therefore CGN retains the nice properties of CG methods, namely the exact arithmetic finite termination property, the minimization property (and thus monotonic convergence) and the existence of a three-term recurrence. Unfortunately, the following convergence formula is found:

$$\frac{\|\mathbf{r}^k\|_2}{\|\mathbf{r}^0\|_2} \leq \min_{p \in \Pi_k, p(0)=1} \max_{\lambda_i \in \Lambda(\mathbf{A}^T\mathbf{A})} |p(\lambda_i)|$$

This inequality indicates that the convergence behaviour of CGN is determined solely by the singular values of $\mathbf{A}$, i.e. by the roots of the eigenvalues of $\mathbf{A}^T\mathbf{A}$. This implies that if $\mathbf{A}$ is not normal, the moduli of its eigenvalues are not equal to the singular values, so that the convergence of GCN does not result from a study of the eigenvalues of $\mathbf{A}$ alone. If $\mathbf{A}$ is normal, the condition number of $\mathbf{A}^T\mathbf{A}$ is much greater than that of $\mathbf{A}$ so that an accelerated procedure for solving the normal system is as slow as an unaccelerated procedure for solving the non-symmetric system. Several polynomials may be selected for p_k, for instance the classical Chebyshev polynomials transplanted over the interval of extreme singular values of $\mathbf{A}$, or the product of two polynomials of lower degree p_n and p_{k-n}. Owing to the observation relative to the high condition number of $\mathbf{A}^T\mathbf{A}$, the GCN method is not recommended for solving a non-symmetric problem, except in some special cases.

To conclude this section, it is considered that preference should be given to methods which introduce a three-term recursion between generated iterates, so as to minimize the extra storage penalty. Among the possible candidates, transpose-free methods should be selected, like the CGS method, the Bi-CGSTAB method, or the TFQMR method, though all these methods suffer from a lack of monotonic convergence and may require a significant number of iterations before superlinear convergence occurs. Significant comparisons between the aforementioned methods have been developed for the pressure solver by Vuik (1993) and for an advection-diffusion equation by Peters (1993). A detailed analysis is also available in Haroutunian *et al.* (1993), but in the context of segregated finite-element methods.

3 Block-correction accelerations

Based on the procedure of Settari and Aziz (1973), the use of block-correction accelerations has been discussed in some detail by Hutchinson and Raithby (1986). The solution is adjusted in blocks, normally along rows or columns of the computational domain, by an additive constant such that the residuals in each block sum to zero. A particular mode of application of block corrections (BC) could be along both rows and columns with one ILU iteration between, with the following sequence of operations: (i) application of BC to ensure that the residuals in each column sum to zero; (ii) application of one ILU iteration; (iii) application of BC to ensure that the residuals in each row sum to zero; (iv) application of one ILU iteration. A tridiagonal matrix algorithm can be used to solve the block-correction equation. Such a method seems particularly useful to accelerate the convergence of a procedure when the solution for pressure varies in a single direction.

4 Multigrid acceleration

The foregoing iterative methods (Gauss–Seidel, ILU) may reduce the high-frequency error modes rather efficiently, although the low-frequency error modes are not affected. The multigrid technique (MG) solves this problem by employing a sequence of grids ranging from fine to very coarse to eliminate the error modes in a particular frequency band on a given grid structure. After smoothing the residuals on the finest grid using a preconditioner of the basis iterative method as the smoother, the residuals are transferred to a coarser grid using a restriction operator. Then a coarse grid smoother is applied. This procedure is repeated until the coarsest grid is reached, where the exact solution of the remaining algebraic system is reached. The coarse grid corrections or solutions are then transferred back to finer grids, using a prolongation operator. On each grid, the smoother is applied again. There are several possibilities for cycling through the fine and coarse grids, called multigrid cycles. As expected, better coarse grid corrections require more visits of coarse grids. A well-designed MG solution method has a convergence rate bounded by a constant much smaller than one and independent of grid size, so that problems with m unknowns can be in principle solved in $O(m)$ operations. Comprehensive overviews of MG methods are provided by Brandt (1977, 1982), Hackbusch (1985) and Wesseling (1991, 1992).

The two-grid algorithm for $\mathbf{A}_h\mathbf{x}_h = \mathbf{b}_h$, where $\mathbf{x}_h$ is the fine grid solution of the (possibly non-) linear problem and $\mathbf{b}_h$ is the known fine grid right-hand-side, consists of seven steps which are presented below. Apart from the subscript h which defines fine grid operators and variables, we require the subscript H for coarse grid operators and variables as well as the restriction operators $\hat{I}_H^h$ and I_H^h for the solution and residuals, respectively, and the prolongation operator I_h^H.

(i)   Perform ν_1 pre-smothing iterations to $\mathbf{A}_h\mathbf{x}_h = \mathbf{b}_h$.
(ii) Compute the fine grid residuals $\mathbf{r}_h = \mathbf{b}_h - \mathbf{A}_h\mathbf{x}_h$.
(iii) Choose the coarse grid solution $\tilde{\mathbf{x}}_H$.
(iv) Apply the restriction to the fine grid residual $\mathbf{r}_H = I_H^h\mathbf{r}_h$.
(v)   Solve the coarse grid equation for $\mathbf{x}_H$, using $\mathbf{A}_H\mathbf{x}_H = \mathbf{b}_H \equiv s_H\mathbf{r}_H + \mathbf{A}_H\tilde{\mathbf{x}}_H$.
(vi) Apply the prolongation operator to the coarse grid solution $\mathbf{x}_H$:
      $\mathbf{x}_h = \mathbf{x}_h + I_h^H(\mathbf{x}_H - \tilde{\mathbf{x}}_H)$.
(vii) Apply ν_2 post-smoothing iterations.

The parameter s_H is introduced to keep the right-hand-side of the coarse grid equation in the range of the coarse grid operator (Hackbusch, 1985). The foregoing non-linear MG method has been devised by Brandt (1982) and Hackbusch (1985); it slightly generalizes the so-called full approximation storage (FAS) method which correspond to $s_H = 1$, $\tilde{\mathbf{x}}_H = \hat{I}_H^h\mathbf{x}_H$. It is drawn up

streamwise pressure gradient which allows departure-free behaviour. Such methods have been developed in the framework of the so-called partially parabolic assumption, for instance by Hoekstra (1989) or Rosenfeld *et al.* (1991) (see also Rubin and Reddy, 1983; Liu and Pletcher, 1986; Pouagare and Lakshminarayana, 1986; Saint Victor, 1986). Unfortunately, in such methods, the coupling is performed only in subdomains (a cell volume with the given longitudinal station and girth) so that the resulting matrices to solve are easy to handle, but poor convergence rates are sustained, especially on fine grids (mainly because of the weak coupling between subdomains). The situation is in some respects improved with multigrid methods (Vanka, 1986a, b, c; Bruneau and Jouron, 1990; Sockol, 1993) although the fully coupled method, in itself, does not appear to bring sufficient improvements with respect to standard Poisson-based methods (Arakawa *et al.*, 1988; Linden *et al.*, 1989). Such methods, like the one used by Vanka, use a smoother, called the symmetric Gauss–Seidel method, which couples on the level of the mass control volume, the fluxes on its faces (i.e. the four corresponding velocity unknowns of the so-called staggered grid arrangement, see Section V) and the cell-centred pressure unknown. A coupled line version of this smoother has been presented by Thompson and Ferziger (1989) for the cartesian form of the Navier–Stokes equations. All the unknowns on a line of cells are updated simultaneously with alternate zebra-sweeping and the band matrix to solve along each line uses a direct band solver from the LINPACK library. Comparisons between cell and line-cell methods have been performed by Oosterlee (1993), while extensions to a collocative layout of variables have been realized by Barcus *et al.* (1987).

It appears, therefore, that the coupling between the solenoidal velocity field and the pressure has to be performed over the whole subdomain, in spite of the increased complexity of the algebraic system. Velocity and pressure are then simultaneously updated in a linear sense, iterations being performed only to solve for the non-linearity. In contrast to the methods of Vanka (1986a, b), Bruneau and Jouron (1990) and McArthur and Patankar (1990) where the continuity equation is retained in its primitive form, it appears better to follow the Schneider and Zedan (1984) practice in the use of a pressure equation and many possibilities remain in the choice of coupled dependent variables and iterative techniques. Other possibilities include the use of a penalty relation, as in Braaten and Shyy (1986), following a classical practice of finite-element methods.

The starting point of the procedure used by Deng *et al.* (1991c) is provided by the continuity equation (73) and the momentum equations (74) which are easily written as:

$$
\bar{U}_P = \hat{U}_P - C_P J^{-1} Re \left[S_1^1 \frac{\delta P}{\delta \xi} \bigg|_P + S_1^2 \frac{\delta P}{\delta \eta} \bigg|_P \right];
$$
$$
\bar{V}_P = \hat{V}_P - C_P J^{-1} Re \left[S_2^1 \frac{\delta P}{\delta \xi} \bigg|_P + S_2^2 \frac{\delta P}{\delta \eta} \bigg|_P \right]
$$

(81)

where the pseudo-velocities are defined by:

$$\bar{U}_P = \sum_{nb} C_{nb} \bar{U}_{nb}^{n+1} + K_P[e_1(\bar{U}_P^n - \bar{U}_P^{n+1}) - S_U] \tag{82a}$$

$$\bar{V}_P = \sum_{nb} C_{nb} \bar{V}_{nb}^{n+1} + K_P[e_1(\bar{V}_P^n - \bar{V}_P^{n+1}) - S_V] \tag{82b}$$

In the continuity equation (73), written as:

$$J_e V_e^1 - J_w V_w^1 + J_n V_n^2 - J_s V_s^2 = 0 \tag{83}$$

it is necessary to specify the mass fluxes with respect to the nodal unknowns. The most common approach is to write a momentum equation at face points e, w, n, s. The Rhie and Chow (1983) method reconstructs the mass fluxes from the cartesian velocities at interfaces:

$$\bar{U}_w = \hat{U}_w - C_w Re \frac{\delta P}{\delta x}\bigg|_w \quad ; \quad \bar{V}_e = \hat{V}_e - C_e Re \frac{\delta P}{\delta y}\bigg|_e \quad ; \quad \text{etc.} \ldots \tag{84}$$

Hence:

$$V_e^1 = \hat{V}_e^1 - C_e Re \left[S_1^1 \frac{\delta P}{\delta x}\bigg|_e + S_2^1 \frac{\delta P}{\delta y}\bigg|_e \right];$$
$$V_w^1 = \hat{V}_w^1 - C_w Re \left[S_1^1 \frac{\delta P}{\delta x}\bigg|_w + S_2^1 \frac{\delta P}{\delta y}\bigg|_w \right] \tag{85a}$$

$$V_n^2 = \hat{V}_n^2 - C_n Re \left[S_1^2 \frac{\delta P}{\delta x}\bigg|_n + S_2^2 \frac{\delta P}{\delta y}\bigg|_n \right];$$
$$V_s^2 = \hat{V}_s^2 - C_s Re \left[S_1^2 \frac{\delta P}{\delta x}\bigg|_s + S_2^2 \frac{\delta P}{\delta y}\bigg|_s \right] \tag{85b}$$

Unknown variables in (85), like $\hat{V}_e^1$, $\hat{V}_w^1$, $\hat{V}_n^2$, $\hat{V}_s^2$, and coefficients C_e, C_w, C_n, C_s, are linearly interpolated from their known values at P, N, S, W, and E, while pressure gradients are discretized at interfaces in the computational plane using (86):

$$\frac{\delta P}{\delta x}\bigg|_{e,w,n,s} = \frac{1}{J} \left[S_1^1 \frac{\delta P}{\delta \xi}\bigg|_{e,w,n,s} + S_1^2 \frac{\delta P}{\delta \eta}\bigg|_{e,w,n,s} \right];$$
$$\frac{\delta P}{\delta y}\bigg|_{e,w,n,s} = \frac{1}{J} \left[S_2^1 \frac{\delta P}{\delta \xi}\bigg|_{e,w,n,s} + S_2^2 \frac{\delta P}{\delta \eta}\bigg|_{e,w,n,s} \right] \tag{86}$$

Substituting the discrete form (86) into (85) and the result into (83) yields a pressure equation which takes the following form:

$$\left[C_e ReJ\left(g^{11}\frac{\delta P}{\delta \xi}+g^{12}\frac{\delta P}{\delta \eta}\right)\right]_e - \left[C_w ReJ\left(g^{11}\frac{\delta P}{\delta \xi}+g^{12}\frac{\delta P}{\delta \eta}\right)\right]_w$$
$$+\left[C_n ReJ\left(g^{21}\frac{\delta P}{\delta \xi}+g^{22}\frac{\delta P}{\delta \eta}\right)\right]_n - \left[C_s ReJ\left(g^{21}\frac{\delta P}{\delta \xi}+g^{22}\frac{\delta P}{\delta \eta}\right)\right]_s$$
$$= [S_1^1 \hat{U}+S_2^1 \hat{V}]_e - [S_1^1 \hat{U}+S_2^1 \hat{V}]_w - [S_1^2 \hat{U}+S_2^2 \hat{V}]_n - [S_1^2 \hat{U}+S_2^2 \hat{V}]_s$$

$$(87)$$

The left-hand-side is a discrete approximation of J div $[CRe$ grad $P]$:

$$\sum_{i=1}^{2}\sum_{j=1}^{2}\frac{\partial}{\partial \xi^i}\left[CReJg^{ij}\frac{\partial}{\partial \xi^i}\right]$$

and the right-hand-side is a discrete approximation of J div $\hat{U} \equiv \dfrac{\partial}{\partial \xi^i}[J\hat{U}^i]$.

Treating crossed second-order pressure derivatives as source terms and evaluating them from the previous non-linear iteration yields the resulting linear coupled system:

$$\hat{U}_P - \sum_{nb} C_{nb}\bar{U}_{nb}^{n+1} + e_1 K_P \bar{U}_P^{n+1} = e_1 K_P \bar{U}_P^{n} - S_U \qquad (82a)$$

$$\hat{V}_P - \sum_{nb} C_{nb}\bar{V}_{nb}^{n+1} + e_1 K_P \bar{V}_P^{n+1} = e_1 K_P \bar{V}_P^{n} - S_V \qquad (82b)$$

$$\bar{U}_P^{n+1} - \hat{U}_P - C_P J^{-1} Re\left[S_1^1 \frac{\delta P}{\delta \xi}\bigg|_P + S_1^2 \frac{\delta P}{\delta \eta}\bigg|_P\right] = 0 \qquad (81a)$$

$$V_P^{n+1} - \hat{V}_P - C_P J^{-1} Re\left[S_2^1 \frac{\delta P}{\delta \xi}\bigg|_P + S_2^2 \frac{\delta P}{\delta \eta}\bigg|_P\right] = 0 \qquad (81b)$$

$$\left[C_e ReJ\left(g^{11}\frac{\delta P}{\delta \xi}+g^{12}\frac{\delta P}{\delta \eta}\right)\right]_e - \left[C_w ReJ\left(g^{11}\frac{\delta P}{\delta \xi}+g^{12}\frac{\delta P}{\delta \eta}\right)\right]_w$$
$$+\left[C_n ReJ\left(g^{21}\frac{\delta P}{\delta \xi}+g^{22}\frac{\delta P}{\delta \eta}\right)\right]_n - \left[C_s ReJ\left(g^{21}\frac{\delta P}{\delta \xi}+g^{22}\frac{\delta P}{\delta \eta}\right)\right]_s$$
$$= [S_1^1 \hat{U}+S_2^1 \hat{V}]_e + [S_1^1 \hat{U}+S_2^1 \hat{V}]_w - [S_1^2 \hat{U}+S_2^2 \hat{V}]_n + [S_1^2 \hat{U}+S_2^2 \hat{V}]_s$$

$$(87)$$

The system (82, 81, 87) can be organized in different ways depending on the ordering of unknowns. For instance, it can be:

$$\mathbf{A}_P \mathbf{X}_P + \sum_{nb}\mathbf{A}_{nb}\mathbf{X}_{nb} = \mathbf{S}_P \qquad (88)$$

where $\mathbf{X}_P = \|\bar{U}_P, \bar{V}_P, \hat{U}_P, \hat{V}_P, P_P\|^T$ and $\mathbf{X}_{nb} = \|\bar{U}_{nb}, \bar{V}_{nb}, \hat{U}_{nb}, \hat{V}_{nb}, P_{nb}\|^T$, with $nb = N, S, W, E, NW, SW, NE, SE$. For a two-dimensional problem, it is easily seen that a 5*5 block nanodiagonal system remains to be solved for each block with the following structure:

$$\mathbf{A}_P = \begin{pmatrix} -1 & 0 & 1 & 0 & -C_P^{up} \\ 0 & -1 & 0 & 1 & -C_P^{up} \\ 1 & 0 & e_1 C_P & 0 & 0 \\ 0 & 1 & 0 & e_1 C_P & 0 \\ C_P^{pu} & C_P^{pu} & 0 & 0 & 1 \end{pmatrix}$$

$$\mathbf{A}_{nb} = \begin{pmatrix} 0 & 0 & 0 & 0 & -C_{nb}^{up} \\ 0 & 0 & 0 & 0 & -C_{nb}^{vp} \\ 0 & 0 & -C_{nb} & 0 & 0 \\ 0 & 0 & 0 & -C_{nb} & 0 \\ C_{nb}^{pu} & C_{nb}^{pv} & 0 & 0 & C_{nb}^{pp} \end{pmatrix} \; ; \quad \mathbf{S}_P = \begin{pmatrix} 0 \\ 0 \\ C_P(e_1 \bar{U}_P^n - S_U) \\ C_P(e_1 \bar{U}_P^n - S_V) \\ S_P \end{pmatrix}$$

In this formulation, pseudo-velocities are treated as auxiliary unknowns. The explicit inverse of each block is required and care has to be taken with its sparsity in order to minimize the storage requirements (Deng et al., 1991c). It is also possible to express the pseudo-velocities in terms of velocities, but the nanodiagonal property of the system is then lost, as indicated by the Figures 8 and 9 which specify respectively the nine-point stencil for the present choice of 5*5 blocks and the 21-point stencil for the resulting choice of 3*3 blocks.

The coupled system (88) is solved by preconditioned conjugate-gradient-squared algorithms (the Bi-CGSTAB version of Van der Vorst (1992) appears at present optimal, as indicated by Deng et al., 1991b). Ad hoc preconditionings are therefore necessary.

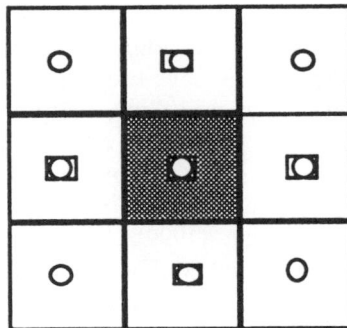

Figure 8 Rhie and Chow's method. Nine-point stencil for pressure, pseudovelocities and velocities. $\bigcirc$, pressure nodes; $\square$, pseudo-velocity nodes.

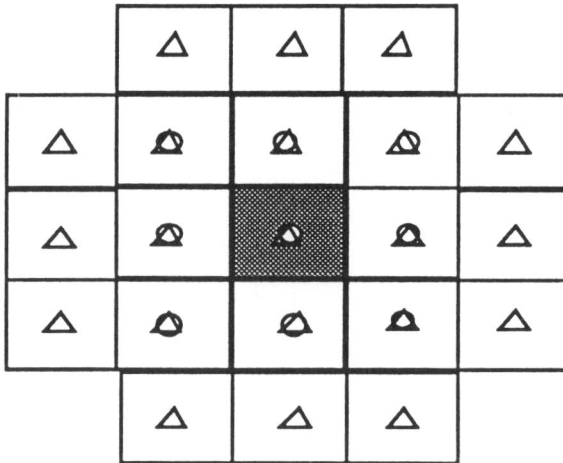

Figure 9 Rhie and Chow's method. Pressure and velocity stencil for 3*3 blocks. ○, pressure nodes; △, velocity nodes.

Several possibilities have been tested for the method using 5*5 blocks; among them, the simplest is Jacobi preconditioning, which has been proved to be very effective for problems with cartesian grids. On curvilinear grids, ILU preconditionings have been found to be more robust (Deng *et al.*, 1991c). Such a formulation allows a systematic study of segregated methods which can be interpreted as a particular preconditioning of the fully coupled system. The main drawback of preconditioned CG-type methods is that the global storage requirements of the method are increased by a factor slightly less than two (Deng *et al.*, 1991c). This is partly due to the large amount of velocity nodes involved in Rhie and Chow's reconstruction. This fact, alone, justifies the search for more economical reconstructions, like the CPI reconstruction (Deng *et al.*, 1991b, 1994a), which yields a fully coupled block 3*3 nanodiagonal system and a less severe storage penalty (about 30% extra storage, see Deng *et al.*, 1994b).

Instead of looking for a special preconditioner for the matrix arising from the fully coupled system, some authors have tried to use an efficient direct solver, like gaussian elimination (Braaten and Shyy, 1986) or YSMP (Eisenstat *et al.*, 1977), which incorporates a Dyakonov iteration procedure or a direct factorization without pivoting as in contributions of Vanka and Leaf (1983, 1984) and Vanka (1985), or an efficient LU decomposition, as in Braaten and Shyy (1986), Karki and Mongia (1990), McArthur and Patankar (1990), Habashi *et al.* (1993) or Knoll and McHugh (1993). However, direct solvers lead to strong storage limitations which forbid their use in three-dimensional problems.

Mixed methods represent an attempt to reduce significantly the storage of direct methods, while still retaining the high rates of convergence resulting from the fully coupled treatment of flow equations. The solution domain is divided up into a number m of subdomains. M is selected as a trade-off between required computer storage and desired rates of convergence for the procedure. An example of this method is provided by Vanka and Leaf (1984), who developed a telescoping subdomain analysis in which subdomains are stacked axially to reconstruct the global domain. They are visited in a regular sequence which is repeated until convergence. For some axisymmetric expansion flows, the convergence rate was reported to be roughly the same as for the full direct approach, while the computer storage was greatly reduced. A similar idea was put forward by Dietrich *et al.* (1975), who developed a block-implicit relaxation method in which the domain decomposition was performed in one direction.

F Other flux reconstruction techniques

The discretization of the momentum equations has been studied in Section III on a cell-centred collocated grid using closure methods for interface values based on Taylor series expansions. An integration point value was expressed only in terms of dependent variables of the same family: a velocity face value was expressed only in terms of velocity nodal unknowns, a pressure face value in terms of pressure nodal unknowns. The Rhie and Chow method was used to reconstruct mass fluxes at interfaces arising in (83) by means of a discretized momentum equation, like (84) or (85), which mimics the discretized momentum equation (81) written as point P: the pressure gradient is discretized at the interface while the remaining terms of the momentum equation are linearly interpolated from their known counterparts in the momentum discrete equations (81). The resulting fluxes are used only in the continuity equation and this prevents the occurrence of spurious pressure modes. However, although the approach provides a satisfactory solution for the discretization of the continuity equation, the momentum equation is not concerned.

In order to reconstruct the fluxes in a more systematic way, Schneider and Raw (1987a, b) deduce the flux reconstruction formulas from discretized momentum equations written on staggered grids and use the reconstructed fluxes in the discrete continuity equation as well as in the momentum equations. This idea is the key to the development of control-volume finite-element methods which belong to the class of null-space-free methods (Dvinsky and Dukowicz, 1993) in that the interpolation operator $\mathcal{M}_u$ between the nodal velocity field $\mathbf{U}$ and the flux field $\mathbf{u} = \mathcal{M}_u \mathbf{U}$ at cell faces can be seen as generating an approximate projection method in which the velocity field itself is necessarily non-solenoidal cell-by-cell (Abdallah, 1987) but only in some approximate sense, while the interpolated flux field $\mathbf{u}$ is exactly cell-by-cell solenoidal. The method of Schneider and Raw (1987a, b) was further developed by Deng *et al.* (1991b, 1992, 1994a), who corrected several of its drawbacks and constructed a second-order accurate consistent physical interpolation of minimal compacity, the so-called CPI method (see also Ferry and Piquet, 1991, in which the reconstruction of fluxes satisfies the momentum equations as well as the continuity equation). The CPI closure transforms the continuity equation (83) into a pressure equation similar to (87). It builds in an automatic upwinding and hence correct limit behaviours for large and small values of the Reynolds number. This automatic upwinding applies in the momentum equations not only to the convective terms but also to the pressure gradient terms through the physical flux interpolation. It is also worth noticing that this pressure equation is not obtained as a discrete divergence of the momentum equations. The CPI closure generates a linear system which can be solved either with a segregated method or with a fully coupled method. In the latter case, a 3*3 block tridiagonal system has to be solved for two-dimensional problems. Again a preconditioned CG method is used for this purpose (Deng *et al.*, 1992, 1994b).

G The non-linear problem

When we consider for instance the system (82, 81, 87), it is clear that convective terms have been already linearized, as indicated in Section III.A. It is necessary now to discuss again briefly this point since it affects the global behaviour of the method. Moreover this problem is rarely mentioned (see however Galpin and Raithby, 1986). If (ν) is the iteration index, the Picard substitution method treats the non-linear equation $\mathbf{F}(\mathbf{X}) \equiv \mathbf{A}(\mathbf{X})\mathbf{X} - \mathbf{b} = 0$ as:

$$\mathbf{F}^{(\nu+1)}(\mathbf{X}^{(\nu+1)}) = \mathbf{A}(\mathbf{X}^{(\nu)})\mathbf{X}^{(\nu+1)} - \mathbf{b} = 0$$

In other words, the convective term $[(\mathbf{V} \cdot \nabla)\mathbf{V}]^{(\nu+1)}$ is treated as $(\mathbf{V}^{(\nu)} \cdot \nabla)\mathbf{V}^{(\nu+1)}$, with the consequence that the linearization procedure does not couple the momentum equations; this explains why the Picard method has been used here, as usual.

The Newton method can be written:

$$\mathbf{X}^{(\nu+1)} = \mathbf{X}^{(\nu)} - [\partial \mathbf{F}/\partial \mathbf{X}]^{(\nu)} \mathbf{X}^{(\nu)}$$

where the jacobian matrix $\partial \mathbf{F}/\partial \mathbf{X}$ is large, sparse and unsymmetric. In this case, the convective term $[(\mathbf{V} \cdot \nabla)\mathbf{V}]^{(\nu+1)}$ is treated as $(\mathbf{V}^{(\nu)} \cdot \nabla)\mathbf{V}^{(\nu+1)} + (\mathbf{V}^{(\nu+1)} \cdot \nabla)\mathbf{V}^{(\nu)} - (\mathbf{V}^{(\nu)} \cdot \nabla)\mathbf{V}^{(\nu)}$. The second contribution couples the three momentum equations, introducing additional terms in the linear system with respect to the Picard method. This technical difficulty explains the infrequent use of this method in spite of its quadratic convergence. The Newton method has a radius of convergence which is sometimes less than that of the Picard method, implying that the initial guess must be correctly chosen. A recent example of the application of the Newton method for direct coupled solvers is given by Knoll and McHugh (1993).

Non-linear iterations have to be performed in order to decrease non-linear residuals by a significant amount for each solution of the linear (fully coupled) system. This is necessary to maintain the global time accuracy to the formally second-order accuracy of the two-level (time) backward Euler scheme.

H Perturbed continuity equation and false transient methods

The pseudo-compressibility approach, which was introduced by Chorin (1967), consists of a modification of the continuity equation through the addition of a temporal pressure derivative (see also Yanenko, 1971). The continuity equation then becomes an evolution equation for pressure. Thus numerical methods developed for the study of compressible flows can be adapted for the study of incompressible flows, using the pseudo-compressibility approach. The method introduces a compressibility parameter β (homogeneous to a velocity scale squared), which influences the convergence of the method, the incompressibility constraint being supplied by:

$$\frac{1}{\beta}\frac{\partial P}{\partial t} + \operatorname{div} \mathbf{V} = 0 \tag{89}$$

The equation (89) has no physical sense, except when the steady state is reached. For unsteady problems, the pseudo-compressibility approach can still be used; in this case t in (89) is a fictitious time scale and the convergence of (89) has to be reached at every real time step. With (89) in place of strict incompressibility, the equations for the inviscid motion become hyperbolic and, therefore, waves of finite speed are introduced. Their magnitude depends on β and these numerical waves may interact with the boundaries of the domain. Chang and Kwak (1984) performed an analysis of the one-dimensional inviscid model which yielded the following equations:

$$\frac{\partial u}{\partial t} + \frac{1}{u \pm c}\frac{\partial P}{\partial t} + [u \pm c]\left[\frac{\partial u}{\partial t} + \frac{1}{u \pm c}\frac{\partial P}{\partial x}\right] = 0 \tag{90}$$

where $c = \sqrt{u^2 + \beta}$ is the artificial sound speed of numerical waves. The Mach number u/c is always subsonic and there are two Riemann invariants:

$$Q^{\pm} = u + \int \frac{\mathrm{d}P}{u \pm c}$$

For downstream-propagating waves, $Q^+ = $ const. along the characteristic such that $\mathrm{d}x^+/\mathrm{d}t = u + c$, while for upstream propagation waves, $Q^- = $ const. along the characteristic $\mathrm{d}x^-/\mathrm{d}t = u - c$. The interaction of such waves with viscous regions is studied using an acoustic approximation for (90). The inviscid motion equations are linearized around a steady state. If the pressure, velocity and friction increments are written with the same symbols as their primitive counterparts, the resulting acoustic approximation is found to be:

$$\left[\frac{\partial}{\partial t} + (u+c)\frac{\partial}{\partial x}\right]\left[\frac{\partial}{\partial t} + (u-c)\frac{\partial}{\partial x}\right]\binom{P}{u} = \binom{\beta\partial\tau_w/\partial x}{-\partial\tau_w/\partial t} \tag{91}$$

Since the shear stress depends on the velocity, coupling between pressure waves and the vorticity spreading depends on their respective length scales. If the characteristic time for the slower upstream-propagating wave to travel a distance L, namely $L/(c-u)$, is small with respect to the characteristic time of diffusion Re^{-1}, i.e. if $L/(u-c)Re \ll 1$, waves and vorticity effects are decoupled and the slow fluctuations of the separation at the walls, provoked by the high-frequency wave-induced adverse pressure gradients, are avoided. Hence:

$$\beta \gg [A + L/Re]^2 - 1 \tag{92a}$$

Now the time needed for the waves to travel downstream and back upstream, in order to distribute pressure, is $L/(c+u) + L/(c-u)$, and the minimum time step necessary to allow convergence needs N computational time steps where:

$$N > \sqrt{(1+\beta)2L/\beta\tau} \tag{92b}$$

With such conditions, the pseudo-compressibility method is effective in propagating long- and short-wavelength errors on $\nabla \cdot \mathbf{V}$ from one location to another; however, it is unable to damp out quickly the artificial sound waves after pressure equilibration. Inequalities (92) indicate that the parameter β should be neither too small nor too big. This can also be seen for a Von Neumann analysis which is usually performed on two different sets of linear models of the Navier–Stokes equations (Peyret and Taylor, 1983). The first set consists in neglecting the pressure in the momentum equation so that the divergence equation is disregarded and each momentum equation, of the advection-diffusion type, is decoupled. The second set is obtained by neglecting the convective non-linear terms while retaining pressure. A Stokes-type analysis is accomplished: the pressure velocity coupling is treated, but the velocity is sufficiently low to make viscous effects predominant. Such an analysis is most often performed on the following explicit algorithm:

$$\frac{\mathbf{V}^{n+1} - \mathbf{V}^n}{\tau} + (\mathbf{C} - \Lambda)\mathbf{V}^n + \mathbf{GP}^n = \mathbf{h}; \qquad \frac{\mathbf{P}^{n+1} - \mathbf{P}^n}{\tau} + \beta(\mathbf{DV}^n - \mathbf{g}) = 0 \quad (93a, b)$$

where $\mathbf{C}$ is the discrete convection operator and Λ the discrete diffusion operator ($\tau^{-1}\mathbf{I} + \mathbf{C} - \Lambda \equiv \mathbf{E} - \mathbf{A}$, $\mathbf{f} = \mathbf{h} + \tau^{-1}\mathbf{V}^n$). The pressure-decoupled Von Neumann analysis for a spatially centred scheme of constant space steps h on a two-dimensional cartesian grid yields (Peyret and Taylor, 1983; corrected 2nd printing):

$$[u^2 + v^2]\frac{Re\tau}{2} \le 1; \qquad \frac{4\tau}{Reh^2} \le 1$$

where u and v are the maximum advection velocities in the two cartesian directions and Re is the Reynolds number. The Stokes stability analysis gives:

$$\frac{4\tau}{h^2}\left[\frac{1}{Re}+\frac{\tau\beta}{2}\right]\le 1$$

The artificial compressibility method may also be viewed as an iterative technique to solve a Poisson equation for the pressure. Taking the discrete divergence of (93a) and using (93b) yields:

$$\mathbf{P}^{n+1}-\mathbf{P}^n-\tau^2\beta\,\mathbf{DG}\,\mathbf{P}^n=\tau\beta[\mathbf{g}-\tau\mathbf{Dh}+\mathbf{D}(1+\mathbf{C}-\Lambda)\mathbf{V}^n] \tag{94}$$

This implies that at convergence, when $\mathbf{P}^{n+1}=\mathbf{P}^n$, what is solved is an aproximation of a Poisson equation for pressure.

For steady flows, the artificial compressibility approach is very popular since it is usually implemented as an extension of methods available for compressible flows. Another reason for its popularity is that, for the hyperbolic system of equations, the treatment of the boundary conditions is relatively easy, using available arguments relative to the characteristic directions and developed for compressible flow studies. The hyperbolic system of equations is generally solved with an implicit approximate factorization algorithm, such as the one given by Beam and Warming (1976) or Briley and McDonald (1980), the divergence-free velocity field being obtained as the steady state solution is reached, once artificial pressure waves of finite speed have been removed. Recent improvements of the method include the use of a similarity transform which diagonalizes the jacobian matrices and uncouples the system of equations. The equations can then be solved using standard scalar tridiagonal algorithms (Rogers *et al.*, 1987).

Turkel (1987) has considered another model, namely:

$$\beta^{-1}\frac{\partial p}{\partial t}+\nabla\cdot\mathbf{V}=0 \tag{95a}$$

$$\frac{\alpha+1}{\beta}\frac{\partial p}{\partial t}+\frac{\partial\mathbf{V}}{\partial t}+\mathbf{A}(\mathbf{V})+\nabla p=Re^{-1}\nabla^2\mathbf{V} \tag{95b}$$

In cartesian coordinates, the two-dimensional system (95) can be written for its inviscid part:

$$\frac{\partial\mathbf{V}}{\partial t}+\mathbf{A}\frac{\partial\mathbf{V}}{\partial x}+\mathbf{B}\frac{\partial\mathbf{V}}{\partial y}=0 \text{ with } \mathbf{V}=[p\ \ u\ \ v]^{\mathrm{T}} \tag{96a}$$

where the Jacobian matrices are:

$$\mathbf{A}=\begin{pmatrix}0 & \beta & 0\\ 1 & (1-\alpha)u & 0\\ 0 & -\alpha v & u\end{pmatrix} \tag{96b}$$

$$\mathbf{B}=\begin{pmatrix}0 & 0 & \beta\\ 0 & v & -\alpha u\\ 1 & 0 & (1-\alpha)v\end{pmatrix} \tag{96c}$$

Wave speeds of (96) are the eigenvalues of the matrix $\omega_1 \mathbf{A} + \omega_2 \mathbf{B}$ where ω_1 and ω_2 are the x and y wave numbers along x and y, respectively, in the two-dimensional Fourier transform. If

$$q = u\omega_1 + v\omega_2$$

the eigenvalues of $\omega_1 \mathbf{A} + \omega_2 \mathbf{B}$ are real and given by:

$$d_0 = q; \quad d_\pm = \tfrac{1}{2}[(1 - \alpha)q \pm \sqrt{(1 - \alpha)^2 q^2 + 4\beta^2}] \tag{97}$$

β is now chosen so as to minimize the largest possible ratio of waves speeds: $\max (| d_i/d_j |)$ where i,j are 0, $+$ or $-$ and it is found that:

$$\frac{\beta}{q^2} = \begin{cases} 2 - \alpha \text{ if } \alpha < 1 & \text{the condition number is } |2 - \alpha| \tag{97a} \\ \alpha \quad \text{ if } \alpha \geq 1 & \text{the condition number is } \alpha \tag{97b} \end{cases}$$

It follows from (97) that the optimal value of α is $\alpha = 1$, from which β is found to depend on the local speed $\sqrt{u^2 + v^2}$. The system (95) can be symmetrized if and only if:

$$\beta > \alpha(u^2 + v^2) \tag{98}$$

Equation (98) implies that in order for the system (95) to be close to optimal and symmetrizable, β has to be chosen slightly larger than the velocity modulus. However, it does not seem that such preconditionings improve the behaviour of the artificial compressibility method. Such methods can also be viewed as a particular false transient method leading to a steady-state solution through the time marching solution of a non-physical unsteady equation. Viviand (1980) and Peyret and Viviand (1985), (see also de Jouette et al., 1991) have considered generalized preconditioning procedures for hyperbolic systems representing the Euler equations and given specific rules that ensure that the preconditioned problem is well posed. These works were further confirmed by Storti et al. (1990) from the point of view of eigenvalue control.

Other perturbative approaches are possible; for instance the penalty approach, first proposed by Temam (1968), has received considerable attention in the finite-element community. In this approach, the continuity equation is supplied by:

$$\varepsilon P = -\text{div } \mathbf{V} \quad \text{with } \varepsilon \ll 1 \tag{99}$$

The resulting evolution equation for div $\mathbf{V}$ is an advection-diffusion equation in which the diffusivity is $\nu + \varepsilon^{-1}$ (with a source term proportional to the second invariant of the velocity gradient tensor) so that the level of div $\mathbf{V}$ is simply smoothed out away from non-zero values that exist or arise. Accurate results are therefore possible only if the characteristic time for diffusion of div $\mathbf{V}$ remains short compared with the physical time scales of the problem. This requires $\nu\varepsilon \ll 1$ and the Reynolds number based on ε^{-1} (namely $uL\varepsilon$) is small with respect to 1. The method is effective in damping short-wavelength components of div **V** and tends to produce smooth solutions. However, long wavelengths tend to persist. It is possible to combine penalty and pseudo-compressibility approaches in order to gain from their different benefits. Ramshaw and Mesina (1991) have proposed to compute the pressure using:

$$\frac{\partial P}{\partial t} = -\beta \, \text{div } \mathbf{V} - \sigma \frac{\partial}{\partial t} (\text{div } \mathbf{V}) \tag{100}$$

where β and σ are supposed to be large. Brooks and Hughes (1982) have proposed a similar idea, letting the velocity divergence be a linear combination of P and $\partial P/\partial t$; however, unlike (100), such a choice does not allow div **V** to be driven to zero in a steady state. Ramshaw and Mousseau (1990) have solved (100) using an explicit (MAC-type, see Harlow and Welsh, 1965; Amsden and Harlow, 1970) conditionally stable scheme for (100) and for momentum equations. Tests of the method are focused on the square driven cavity with very coarse grids.

V STRATEGIES FOR LAYOUTS OF DEPENDENT VARIABLES

A Methods without interpolation

The simplest way to discretize the continuity equation is to localize the dependent variables on the surfaces of the control volumes so that fluxes can be evaluated without interpolation. A first possibility is indicated in Figure 10.

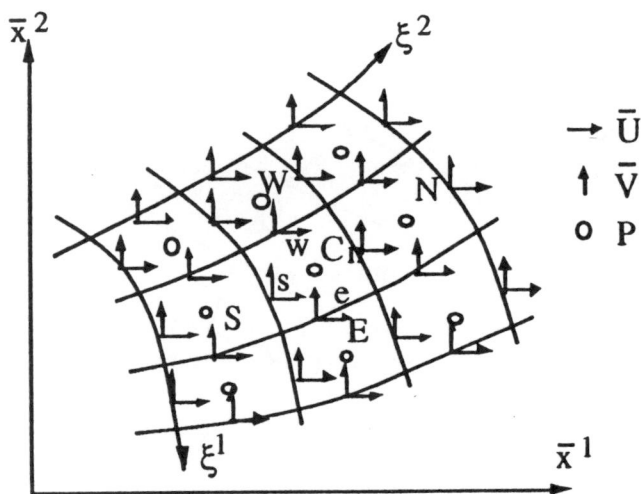

Figure 10 Layout of independent variables.

In this case, the cartesian components of the velocity are located at the centres of the faces of the control volume while the pressure is located at the centre of the control volume. This case is called a *partial staggering of unknowns*; it allows the continuity equation to be easily handled since the fluxes on the faces of the control volume are immediately available. For the momentum equation, two possibilities arise. The discretization is performed either at velocity nodes (Figure 11a) or at nodes and centres of the grid volumes (Figure 11b).

In the former case (Figure 11a), the variables located on surfaces ξ^1 = const. are connected to those located at ξ^2 = const. surfaces only through the pressure field; the resulting discrete system may therefore be ill conditioned. The

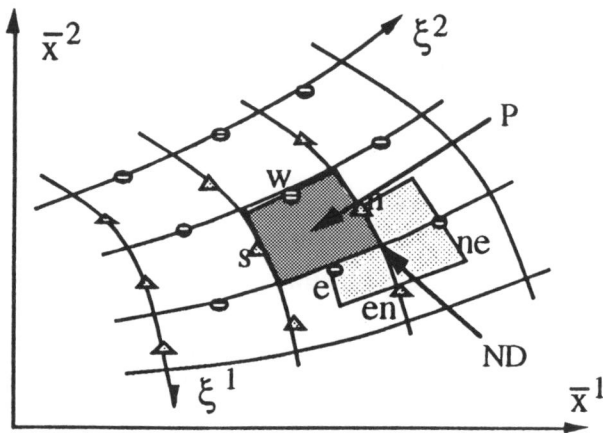

(a)

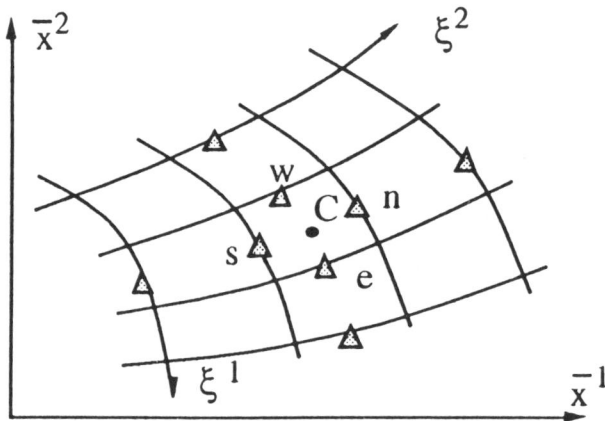

(b)

Figure 11 Discretization at nodes of velocity components (a) or at grid centres (b).

(continuity) primary control volume (strongly shaded) is limited by faces w, e, n, s and it is seen that there is no real reason to use it rather than the (weakly shaded) control volume limited by non-specified faces e, ne, n, en. This other continuity control volume involves a non-represented pressure field based on unknowns located at the vertices of the primary control volume: conservation of mass, while imposed on fluxes at w, e, s, n is not imposed on the same family of fluxes (namely n, e, ne, en) through the weakly shadowed control volume. Hence there is a weak coupling between the pressure field located at centres of the primary control volumes and the pressure field located at vertices of the primary control volumes. In the latter case (Figure 11b), the molecule that is obtained is not compact in the sense that, even for an orthogonal grid, it involves more than three points per direction. For such a reason, this layout is not retained.

If the contravariant components of the velocity field are taken as independent variables (Kwak *et al.*, 1986), see Figure 12, these problems can be avoided to some extent. The continuity equation is still easy to handle without interpolations, but the momentum equation raises other problems. With the contravariant components of the velocity, the momentum equation introduces, in its convective form, Christoffel symbols which are difficult to estimate accurately. Karki (1986) has shown that this drawback can be avoided if the independent variables are estimated in a local cartesian frame $\{X', Y'\}$ in which the momentum equations are written down (Figure 13). Unfortunately, the reconstruction of the cartesian components at every face is very costly since a 3*3 matrix is inverted on each face. Also, the strong conservation form of the equations cannot be retained.

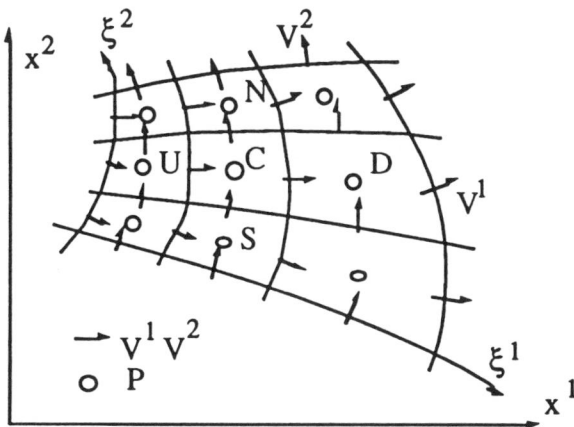

Figure 12 Staggered grid approach: localization of contravariant components and of the pressure field.

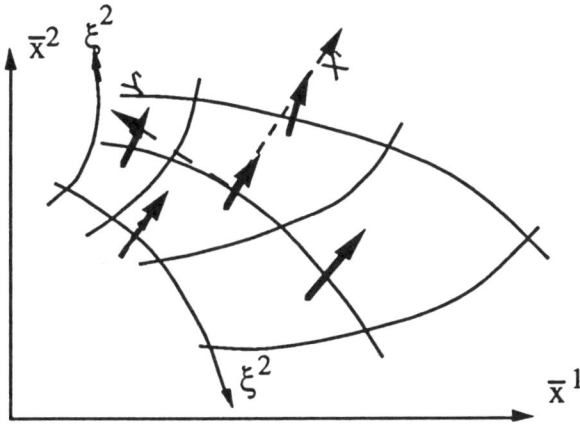

Figure 13 Introduction of a local cartesian frame. The dotted lines show the local cartesian frame $\{X', Y'\}$.

The case of the partial transformation is not considered here since interpolations then become necessary.

B Methods with interpolation

1 The chequerboard problem in the collocated case

We now consider methods which need interpolations to discretize the continuity equation. Again, the staggering mode of dependent variables and the interpolation practice determine the discretization methods. The most natural choice is to locate the unknowns at the same point in the centre of the grid volumes (Figure 14). If a linear interpolation is used for the variables and with first-order central derivatives, this grid layout generates chequerboard oscillations for the pressure field. If the pressure gradient is discretized with centred derivatives, the momentum equation for U_C uses an approximation for $\partial P/\partial \xi^1$ which involves the difference $P_E - P_W$, while no approximation of $\partial P/\partial \xi^2$ is needed in the simple case where the grid is aligned with velocity components; similarly the momentum equation for U_E involves P_{EE} and P_C. Then the velocity field is insensitive to an arbitrary constant added to pressure unknowns at points '⬤' or at points '△'. This implies the singularity of the pressure system in the case of alignment or, in the general case (because of the variation of the geometric coefficients), at least an ill-conditioned character of the pressure system. To solve this well-known problem, staggering the unknowns or modifying the interpolation practice is necessary. Suitable

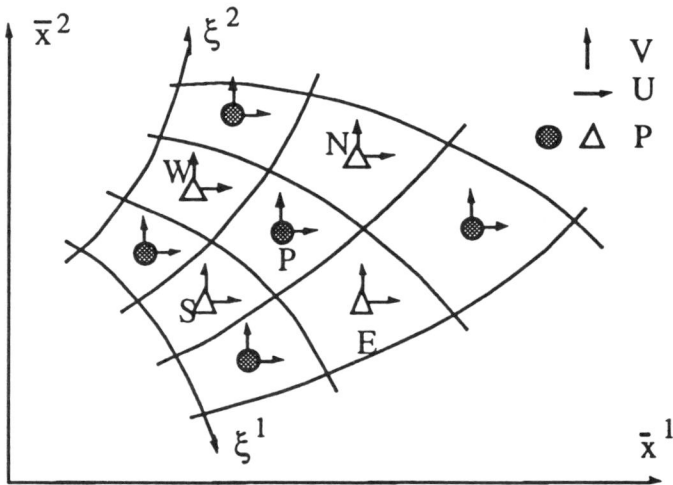

Figure 14 Collocated grid, showing grid-centred location of unknowns.

modifications of the interpolation practice have been put forward by Rhie and Chow (1983); their success has contributed to the increasingly popular character of the collocative arrangement which is nowadays the most commonly used (Dick, 1988; Majumdar, 1988; Peric *et al.*, 1988; Soh and Goodrich, 1988; Acharya and Moukalled, 1989; Goodrich and Soh, 1989; Deng *et al.*, 1991a; Kobayashi and Pereira, 1991; Sotiropoulos and Abdallah, 1991, 1992; Aksoy and Chen, 1992; Melaeen, 1992a, b; Piquet and Queutey, 1992). Even on a cartesian grid such a layout has been proved to be as accurate as the classical staggered layout discussed in Section V.B.3 (Peric *et al.*, 1988; Armfield, 1991; Aksoy and Chen, 1992).

2 The partially staggered grid: the ICED-ALE practice

Fortin *et al.* (1971), Hirt *et al.* (1974), Kuznetsov (1974), Pracht (1975) and Vanka *et al.* (1980) use a partially staggered grid in which the pressure is cell centred while both velocity components are located at the vertices of the volume (Figure 15). This is the so-called ICED-ALE arrangement. To evaluate the pressure gradient $\partial P/\partial \xi^1$ which drives U at the corner *ne*, an expression like

$$\frac{P_N + P_P}{2} - \frac{P_{NE} + P_E}{2}$$

is used. Mass conservation for the (shadowed) control volume centred at point P is checked using a velocity for each face obtained through an average of the velocities at the corners. The two averaging processes lead to the result that mass conservation for the P control volume is satisfied if P_{NE}, P_{SE}, P_{SW}, P_{NW}

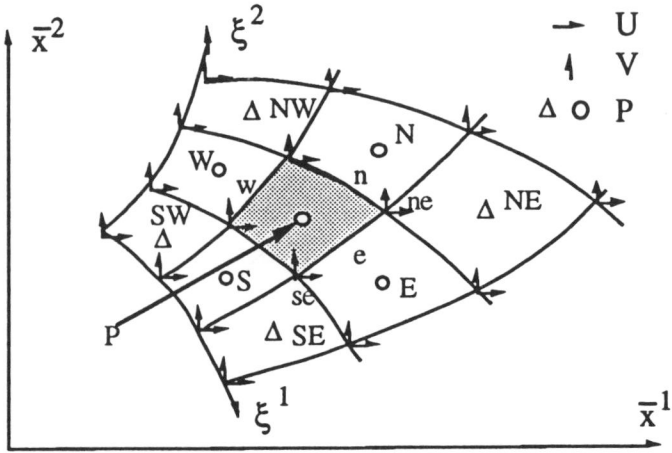

Figure 15 Partial staggering in the ICED-ALE technique.

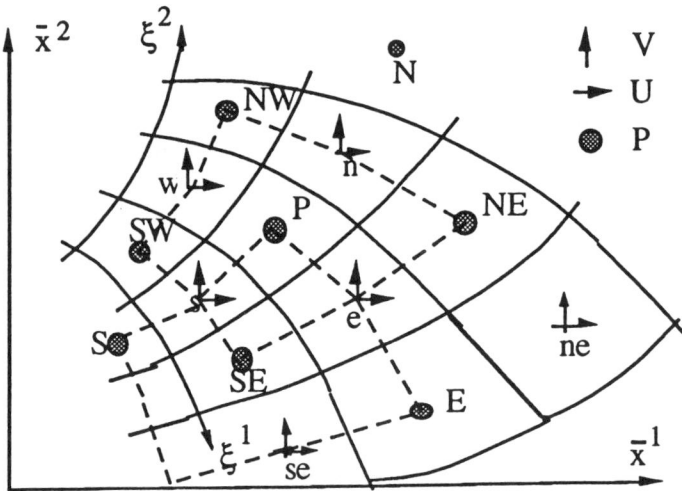

Figure 16 Elimination of half of the variables over a collocated grid.

and P_C are correct independently of P_E, P_N, P_S, P_W. Mass conservation at E provides similarly a constraint for P_N, P_{NEE}, P_{SEE}, P_E, P_S. Again, chequerboard oscillations are possible and a treatment similar to that of Rhie and Chow is necessary.

Coming back to Figure 14, if the pressure at non-shaded points '$\triangle$' and the velocity at shaded points '$\oplus$' are discarded, the grid layout of Figure 16 is obtained. The pressure and velocities are now suitably staggered but, at least if centred differences are used, there is still weak coupling between the pressures at P_P, P_E, P_N, P_W, P_S and P_{NE}, P_{SE}, P_{SW}, P_{NW} in the case of alignment. Conservation of mass, for the control volume centred at P, depends only on the velocities normal to the faces, e, s, w, n; these, in turn, depend only, through momentum equations, of P_C, P_E, P_N, P_W, P_S, in the case of alignment. If these pressures are correct, mass conservation for the control volume centred at P is satisfied even if P_{NE}, P_{SE}, P_{SW}, P_{NW} are wrong. The latter pressures are checked by the mass conservation for the control volume centred at point SE. Again two pressure distributions are present which are only weakly coupled through boundary conditions.

3 The fully staggered grid: the MAC practice

If one of the two pressure fields of Figure 12 is eliminated together with the velocity components driven by the discarded pressure field in the case of alignment of U, V, W with ξ^1, ξ^2, ξ^3 coordinate directions, the grid layout of Figure 17 results: this is the staggered grid of Harlow and Welch (1965) – see also Amsden and Harlow, 1968, 1970; Hirt *et al.*, 1975; Cloutman *et al.*, 1976 – which has been used in curvilinear coordinates more recently by Shih *et al.* (1989), Oosterlee (1993) and Oosterlee *et al.* (1993).

On each face of the control volume, the vertices of which are the grid nodes, one velocity component is known; others need to be interpolated from the dependent variables located on other faces in order to establish the mass balance over the control volume. *Such a method is proved to be highly efficient when grid*

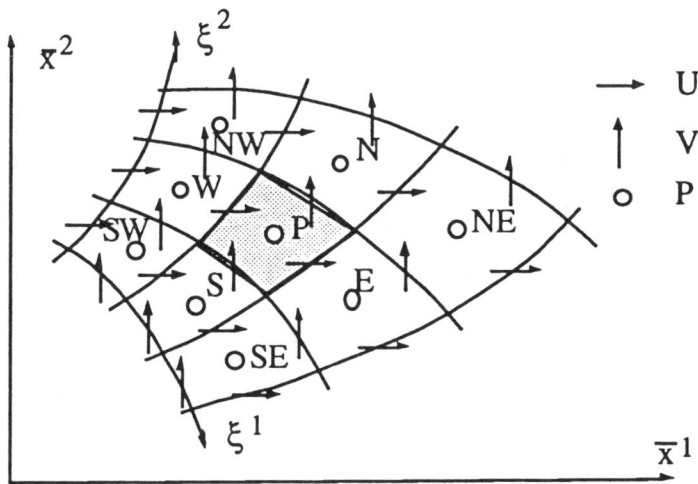

Figure 17 Staggered grid of Harlow and Welch (1985). The mass control volume at point P is shaded.

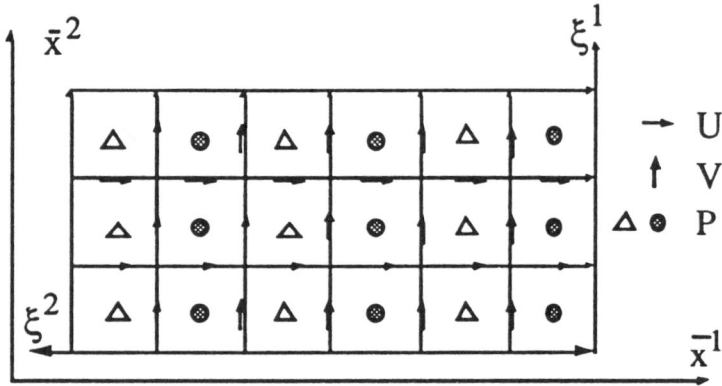

Figure 18 Decoupling phenomenon for the pressure field on a staggered grid as a result of misalignment of grid lines and cartesian directions.

lines are parallel with the directions of cartesian velocity components so that fluxes are computed mainly from the velocity components that are not interpolated. When the interpolated part of the flux becomes dominant, the system becomes ill conditioned and even singular in the extreme case of Figure 18 where the velocity components that are not interpolated are parallel to the faces of the mass control volume and thus do not contribute to the fluxes required in the continuity equation.

The velocity field is insensitive to a constant added to the pressure field at nodes '△'. An answer to this apparent difficulty is to use the computed cartesian components (U, V) only to calculate the contravariant velocity components that enter the mass conservation constraint. When other contravariant components are required, they should be obtained rather by interpolation from contravariant components constrained by mass conservation. The previously mentioned drawback of the staggered grid which appears as soon as there is no more grid alignment has prompted the attempt of Wachpress (1979).

4 The TURF practice

Wachpress (1979) stores both velocity components at cell faces since they are necessary to compute the fluxes in the case of non-alignment. To match this arragement of velocity components, additional pressure nodes are introduced at the vertices of the mass control volume (Figure 19). The consequence of this so-called 'TURF' practice is that there are now two sets (for two-dimensional problems) of momentum equations and two associated (overlapping) control volumes for the velocity components. The coupling between the equations for the two sets is again very weak and results, as usual, from the non-alignment.

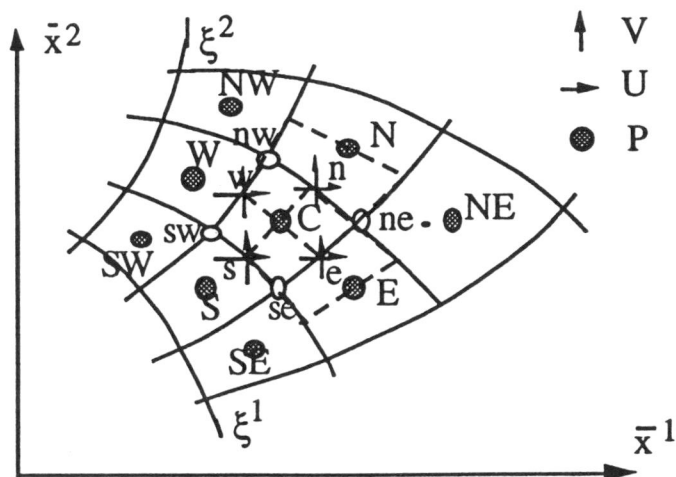

Figure 19 TURF layout.

This is easily seen from the momentum equations written at locations e and n. U_e and V_n are linked with P_C, P_D and P_N, as in the staggered grid case, whereas V_e and U_n are linked with P_{ne}, P_{nw} and P_{se}. Two overlapping staggered solutions result: for instance two distinct uniform pressure fields, arranged as in Figure 19, are compatible with the momentum equations. This serious drawback is probably the reason for the lack of published results with this arrangement.

To summarize: *If the cartesian velocity components are retained as dependent variables and if the grid is not aligned with velocity components, and therefore not cartesian, no staggering practice is able to provide, alone, a good pressure conditioning.*

The conditioning of the discrete system can be of course improved using artificial dissipation. This amounts to solving a modified continuity equation like:

$$\nabla \cdot \mathbf{V} = h^2 \left[\omega_1 \nabla^2 P + \omega_2 \frac{\partial^2}{\partial x \partial y} \left(\frac{\partial u}{\partial y} + \frac{\partial v}{\partial x} \right) \right] \tag{101}$$

where h is a characteristic grid size and ω_1 and $\omega_2 = 0$ are small parameters (Linden *et al.*, 1989). Such methods weaken somewhat the quality of the discretization (it is difficult to control the amount of added dissipation) while reintroducing in an artificial way a coupling between pressure and velocity. This method has been discussed in detail by Abdallah (1987); it may be noticed also that (101) is similar to the first differential equation (Shokin, 1989) generated by the CPI method which produces on a uniform grid $\omega_2 = -\omega_1/3 = -1/24$ (Deng *et al.*, 1994b).

The most efficient way to improve the conditioning of the discrete system is to modify the interpolation procedure. A first possibility is to use upwind derivatives instead of centred derivatives in the continuity as well as in the momentum equation (otherwise chequerboard oscillations would not be avoided). Dick (1988) has applied the so-called Steger–Warming flux splitting to the solution of the steady problem. The accuracy of the results then depends on the order of accuracy of the upwinded derivatives. A second possibility is to use the so-called Rhie and Chow interpolation, as explained in Section IV. In this method, the pseudo-velocity field and the influence coefficient of the pressure gradient are interpolated but the pressure gradient itself is computed directly from the nodal values of the pressure. Although the momentum equations are still insensitive to a chequerboard oscillation, the oscillations are filtered out when the interpolation is used and the discrete system becomes well conditioned. For curvilinear grids, the Rhie and Chow interpolation can be used on a staggered as well as on a collocated grid. *The collocated grid should be preferred since the discretization of the three components of the momentum equations* (and of the other advection-diffusion equations for turbulence quantities) *generates only one linear system.* Also, the influence coefficients for the velocity unknowns as well as for the pressure gradients are evaluated at the same points. Therefore the resulting pressure equation will more easily take a symmetric form. As already mentioned, the Rhie and Chow interpolation produces a velocity stencil of 21 points for two-dimensional problems and this makes the algebraic solution of the problem rather complex. Although correct, it raises some problems relative to the accuracy of the linear interpolation of mass fluxes; also it embodies the implicit assumption that the main driving influence for the mass fluxes results from local pressure gradients. Hence, this procedure should introduce a dependence of volumic forces if buoyancy or rotation effects become significant. The interpolation of fluxes needed for a pressure equation coincides with the closure problem for convective terms, as already discussed, therefore it appears advisable to use a flux closure which (i) involves the local pressure gradients to be used directly in the continuity equation and (ii) guarantees the compatibility between shape assumptions in the pressure equation and in the momentum equation (Prakash and Patankar, 1985; Schneider and Raw, 1987a; Deng *et al.*, 1991b, 1994a). There are still several possibilities: independent variables can be located either at the grid nodes or at the cell centres. Such choices mainly influence the implementation of boundary conditions.

C Boundary conditions

We first look at the node centred grid in which (shaded) the mass control volumes surround the nodes of the grid (Figure 20). If the velocity is imposed at the boundary, the Navier–Stokes equations being well determined, no boundary condition is in principle necessary for the pressure. However, pressure nodes

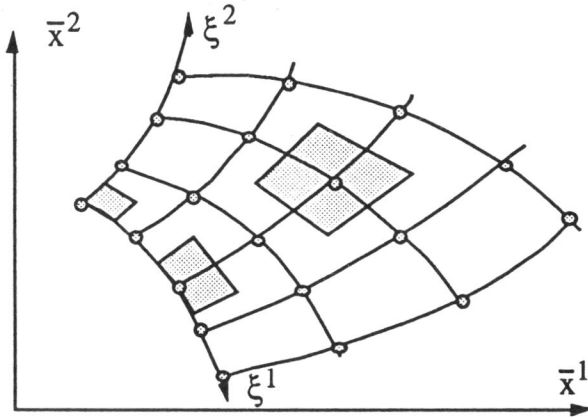

Figure 20 Node-centred collocated grids and mass control volumes.

appear at the boundary so that closure assumptions have to be introduced on boundary mass control volumes. Two methods can be used to specify the necessary closures.

First, it is possible to stress numerical boundary conditions over the pressure field. Such conditions can be deduced from the velocity boundary conditions using the continuity and momentum equations. This problem is difficult to solve correctly. In most cases, the normal pressure gradient is set equal to zero over a wall boundary, but this condition is hardly valid in an inlet section or when the Reynolds number is low (then the normal component of the momentum equation, or one of its parabolized versions, should be preferred). The second possibility uses imposed boundary conditions over the velocity field to deduce the pressure field without resorting to supplementary boundary conditions. The pressure field then results from the continuity equation and as many discrete continuity equations as pressure unknowns are necessary. Figure 16 indicates that irregular control volumes are then required close to the wall. Unfortunately the accuracy of the scheme is often of a lower order close to the boundary. Moreover, this method is not suitable for conservative forms since nodes where independent variables are defined are not at cell centres; also interpolated variable nodes are neither on the surfaces of the control volumes nor at the centres of these surfaces.

The layout presented in Figure 21 (Peric, 1985) avoids such drawbacks. The grid nodes correspond to the vertices of the mass control volumes so that the dependent variables are located at the centres of the control volumes while interpolated variables are at the centres of the faces of the (shaded) mass control volumes. Boundary conditions can now be specified easily. Pressure unknowns inside the domain can be determined without losing accuracy and without

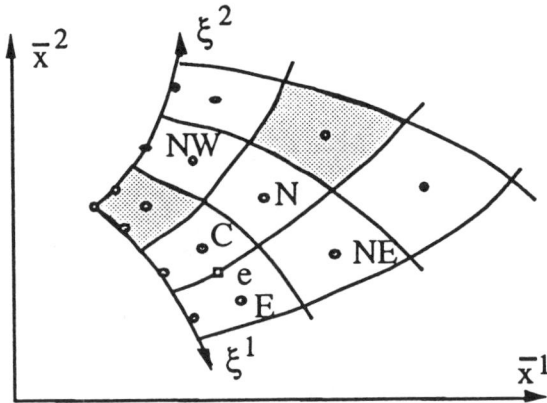

Figure 21 Cell-centred collocated grid; treatment of boundary conditions.

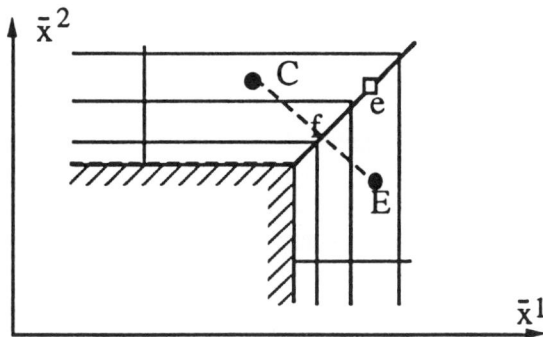

Figure 22 Interpolation problem at a singular grid point or along a grid line issuing from an angle point.

pressure numerical boundary conditions constructed from velocity-imposed boundary conditions. This is because the pressure unknown is inside the control volume. However, face values are introduced on the boundaries, the axes ξ^1 and ξ^2 in Figure 21, where mass fluxes can be in general specified: on each face of the control volume, the contravariant velocity component is evaluated when forming the pressure equation. The sum of pressure derivatives estimated on each face of the control volume gives the pressure gradient, in the direction normal to the face, which drives the corresponding contravariant velocity component. For instance, if the grid is orthogonal to the wall (Figure 21), the direction of the contravariant velocity component at e is along the covariant

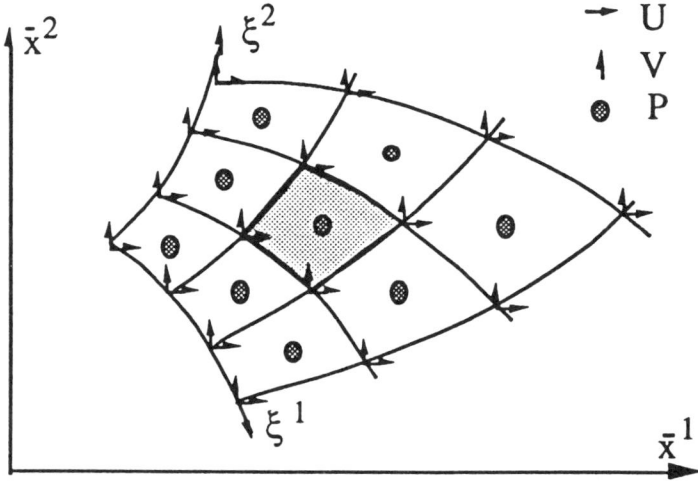

Figure 23 Boundary conditions for the ICED-ALE mode.

vector $\mathbf{g}_1$ so that the pressure gradient along ξ^1 does not involve points along the direction ξ^2. Usually the coefficient multiplying $\partial P/\partial \xi^2$, although non-zero, is small. An upwind derivative may be used to estimate it for instance:

$$\frac{\partial P}{\partial \xi^2}\bigg|_d \approx \frac{P_N + P_{ND} - P_C - P_D}{2} \tag{102}$$

Such a measure appears to be sufficient since the grid is usually fine close to the boundaries, unless strong departure from orthogonality is present. However, the aforementioned layout still presents several drawbacks. First, stencils for the discrete momentum equation are irregular close to the boundary. Also, the nodes for interpolated variables are not located halfway between the nodes of dependent variables: Figure 22 indicates that a variable to interpolate at location e, identified by a square, will involve dependent variables at locations C and E, identified by black circles. Hence the information retained at point e is in fact pertinent to point f instead. In spite of such problems, the cell-centred collocated grid is considered to be well suited for finite volume techniques.

A third alternative is provided by the mixed mode of the ICED-ALE practice which is recalled in Figure 23. The velocity boundary conditions are easily imposed while the pressure does not appear on the boundaries. Moreover, nodes of interpolated variables are halfway between nodes of dependent variables so that this solution seems to offer a nice compromise, although it still requires flux interpolation.

VI CLOSURE

The numerical aspects of the discrete solutions of Reynolds-averaged Navier–Stokes equations have been reviewed in this chapter. The various performances of the techniques considered have been discussed. However, the literature indicates that discussion about the performance of the various techniques is rather inhomogeneous. As regards the choice of the formulation, it appears that methods which depart from the primitive formulation have not been investigated in general coordinates, so that their present degree of validation is not as high as that for the pressure–velocity formulation.

Concerning numerical schemes for the advection-diffusion equation, it appears that the quest for accuracy is still a real problem, given the difficulty of defining this accuracy in the case of a problem involving complex geometries. A possible idea could be to combine higher-order interpolation in conjunction with a low-order scheme via a deferred correction approach. Moreover, the performance of the techniques considered should be evaluated on a broader scope of test problems that emphasize relevant numerical aspects, before more general conclusions can be reached regarding their practical utility.

If we consider the question of robustness of methods, it appears that the availability of strongly coupled solvers is at least a useful tool for diagnostics: the availability of a method which solves the pressure–velocity coupling problem exactly allows difficulties in the convergence of the global procedure to be attributed to the linearization technique which is used. It is considered that the question of linearization of the equations deserves further attention. This is also true whenever we consider the convergence of the turbulence characteristics. For instance, the global convergence of the numerical procedure is mainly controlled by the convergence of the equations describing the (positive-definite) turbulence characteristics, and more precisely, that of the ε equation, if a $K - \varepsilon$ model is used, as studied in Chapter 6.

Most of the numerical methods discussed lead to a compact stencil for the discrete momentum equations. Owing to the fact that more accurate discrete schemes should involve non-compact stencils, the convergence and stability of the resulting solvers is not strictly assured. To circumvent potential problems that could arise wth these alternative discretization schemes, development and use of appropriate solvers should be considered.

The significant reduction in cpu costs provided by acceleration techniques should orientate development of 'grid-insensitive' methods, perhaps by a combination of MG- and CG-type methods. Also, estimations of the truncation error should be possible, through a MG tau-type technique. This could provide an accuracy estimator to be used to adapt the grid to the flow (either through straining or local enrichment, see Chapter 7) and thus to reduce the truncation error of the method. This could be a plausible economical means of obtaining grid-independent solutions.

REFERENCES

Abdallah, S. (1987). *J. Comput. Phys.* **70**, 182–202.

Acharya, S. and Moukalled, F.H. (1989). *Numer. Heat Transfer B* **15**, 131–152.

Agarwal, R.K. (1981). In *Computers in Flow Prediction and Fluid Dynamics Experiments* (K.N. Ghia, T.L. Müller and B.R. Patel, eds), Proc. ASME Winter Annual Meeting, Washington, pp. 73–82.

Aksoy, H. and Chen, C.J. (1992). *Numer. Heat Transfer B* **21**, 287–306.

Amsden, A.A. and Harlow, F.H. (1968). *J. Comput. Phys.* **3**, 94–110.

Amsden, A.A. and Harlow, F.H. (1970). *J. Comput. Phys.* **6**, 322–325.

Arakawa, C., Demuren, A.O., Rodi, W. and Schönung, B. (1988). In *Notes in Numerical Fluid Mechanics* (M. Deville, ed.), vol. 20, pp. 1–8. Vieweg-Verlag, Braunschweig.

Aregbasola, Y.A.S. and Burley, D.M. (1977). *J. Comput. Phys.* **24**, 398–415.

Armfield, S.W. (1991). *Computers and Fluids* **20**, 1–17.

Atias, M., Wolfshtein, M. and Israeli, M. (1977). *AIAA J.* **15**, 263–266.

Aziz, K. and Hellums, J.D. (1967). *Phys. Fluids* **10**, 314–324.

Barcus, M., Paric, M. and Scheuerer, G. (1987). In *Notes in Numerical Fluid Mechanics* (M. Deville, ed.), vol. **20**, pp. 9–16. Vieweg-Verlag, Braunschweig.

Barrett, K. (1982). In *Numerical Modelling in Diffusion Convection* (J. Caldwell and A.O. Moscardini, eds), pp. 33–48. Pentech Press, London.

Beam, R.M. and Warming, R.F. (1976). *J. Comput. Phys.* **22**, 87–110.

Berger, A.E., Solomon, J.M., Ciment, M., Leventhal, S.H. and Weinberg, B.C. (1981). *Math. Comput.* **35**, 695–731.

Braaten, M.E. and Shyy, W. (1986). *Int. J. Numer. Methods Fluids* **6**, 325–349.

Braaten, M.E. and Shyy, W. (1987). *Numer. Heat Transfer* **11**, 417–442.

Brandt, A. (1977). *Math. Comput.* **31**, 333–390.

Brandt, A. (1982). In *Multigrid Methods* (W. Hackbusch and U. Trottenberg, eds), Lecture Notes in Mathematics, vol. 960, pp. 270–312. Springer-Verlag, Berlin.

Brandt, A. and Dinar, N. (1984). In *Numerical Methods for PDE* (S.V. Parter, ed.). Academic Press, New York.

Brandt, A., Dendy, J.E. and Ruppel, H. (1980). *J. Comput. Phys.* **34**, 348–370.

Brezinski, C. and Sadok, H. (1991). *Numer. Alg.* **1**, 199–206.

Briley, W.R. and MacDonald, H. (1980). *J. Comput. Phys.* **34**, 54–77.

Brooks, A.N. and Hughes, T.J.R. (1982). *Comput. Methods Appl. Mech. Engrg* **32**, 199.

Bruneau, C.H. and Jouron, C. (1990). *J. Comput. Phys.* **89**, 389–413.

Caretto, L.S., Curr, R.M. and Spalding, D.B. (1972). *Int. J. Heat Mass Transfer* **15**, 1878–1806.

Chang, J.L.C. and Kwak, D.C. (1984). *AIAA Paper 84–252*, AIAA 22nd Aerospace Sciences Meeting, Reno, NV.

Chen, C.J. and Chen, H.C. (1984). *J. Comput. Phys.* **53**, 203–226.

Chen, G.Q., Gao, Z. and Yang, Z.F. (1993). *J. Comput. Phys.* **104**, 129–139.

Chen, H.C. and Patel, V.C. (1985). *IIHR Report No. 285*, Iowa City.

Chen, H.C., Patel, V.C. and Ju, S. (1988). *IIHR Report No. 323*, Iowa City.

Chen, Y. and Falconer, R.A. (1992). *Int. J. Numer. Methods Fluids* **15**, 1171–1196.

Chorin, A.J. (1976). *J. Comput. Phys.* **2**, 12–26.

Clever, R.M. and Busse, F.H. (1974). *J. Fluid Mech.* **65**, 625–646.

Cloutman, L.D., Hirt, C.W. and Romero, N.C. (1976). *Los Alamos Scientific Lab. Report LA-6236 Report.*

Connell, S. and Stow, P. (1986). *Computers and Fluids* **14**, 1–10.

Daube, O. (1992). *J. Comput. Phys.* **103**, 404–414.

Daube, O., Guermond, J.L. and Sellier, A. (1991). *CRAS Paris, Sér. II* **313**, 377–382.

Davidson, L. and Hedberg, P. (1989). *Int. J. Numer. Methods Fluids* **9**, 531–540.

Davis, R.W. and Moore, E.F. (1982). *J. Fluid Mech.* **116**, 475–506.

de Jouette, C., Viviand, H., Wornom, S. and Le Gouez, J.M. (1991). *Proc. 4th Symposium on Computational Fluid Dynamics,* Davis, California, pp. 270–275.

De Henau, V., Raithby, G.D. and Thompson, B.E. (1989). *Int. J. Numer. Methods Fluids* **9**, 855–864.

Demirdzic, I. and Peric, M. (1988). *Int. J. Numer. Methods Fluids* **8**, 1037–1050.

Demirdzic, I. and Peric, M. (1990). *Int. J. Numer. Methods Fluids* **10**, 771–790.

Demirdzic, I., Gosman, A.D., Issa, R. and Peric, M. (1987). *Computers and Fluids* **15**, 251–273.

Deng, G.B. and Piquet, J. (1992). *Int. J. Numer. Methods Fluids* **15**, 99–124.

Deng, G.B., Piquet, J., Queutey, P. and Visonneau, M. (1991a). *Int. J. Numer. Methods Engrg* **31**, 1427–451.

Deng, G.B., Ferry, M., Piquet, J., Queutey, P. and Visonneau, M. (1991b). In *Notes on Numerical Fluid Mechanics* (J.B. Vos, A. Rizzi and I.L. Ryhming, eds), vol. 35, pp. 191–200. Vieweg-Verlag, Braunschweig.

Deng, G.B., Piquet, J. and Visonneau, M. (1991c). *Proc. 2nd International Colloquium on Viscous Fluid Dynamics in Ship and Ocean Technology*, Osaka, pp. 186–202.

Deng, G.B., Piquet, J., Queutey, P. and Visonneau, M. (1992). In *Notes on Numerical Fluid Mechanics* (M. Deville, T.H. Le and Y. Morchoisne, eds), vol. 36, pp. 34–45. Vieweg-Verlag, Braunschweig.

Deng, G.B., Piquet, J., Queutey, P. and Visonneau, M. (1994a). *Computers and Fluids* **23**, 1029–1047.

Deng, G.B., Piquet, J., Queutey, P. and Visonneau, M. (1994b). *Int. J. Numer. Methods Fluids* **19**, 605–640.

Dennis, S.C.R. and Quartapelle, L. (1989). *Int. J. Numer. Methods Fluids* **9**, 871–890.

Dennis, S.C.R., Ingham, D.B. and Cook, R.N. (1979). *J. Comput. Phys.* **33**, 325–339.

Dick, E. (1988). *Int. J. Numer. Methods Fluids* **8**, 317–326.

Dietrich, D., McDonald, B.E. and Warn-Varnas, (1975). *J. Comput. Phys.* **18**, 421–434.

Dvinsky, A.S. and Dukowicz, J.K. (1993). *Computers and Fluids* **22**, 685–696.

Eisenstat, S.C., Gursky, M.C., Schultz, M.H. and Sherman, A.H. (1977). *Res. Rep.* **114**, Department of Mechanical Engineering, Yale University.

Faber, V. and Manteuffel, T.A. (1984). *SIAM J. Numer. Anal.* **21**, 352–262.

Farouk, B. and Fusegi, T. (1985). *Int. J. Numer. Methods Fluids* **5**, 1017–1034.

Ferry, M. and Piquet, J. (1991). *GMD-Studien Nr* **189**, GMD Sankt Augustin.

Fletcher, R. (1976). In *Lecture Notes in Mathematics*, vol. 506. Springer-Verlag, Berlin.

Fontaine, J. and Ta Phuoc, L. (1994). In *Notes on Numerical Fluid Mechanics* (F.K. Ebeker, R. Rannacher and G. Wittum, eds) vol. 47, pp. 79–88, Vieweg–Verlag, Braunschweig.

Fortin, M., Peyret, R. and Temam, R. (1971). *J. Méca* **10**, 357–390.

Freund, R.W. (1993). *SIAM J. Statist. Comput.* **14**, 470–482.

Freund, R.W. and Nachtigal, N. (1991). *Numer. Math.* **60**, 315–339 (1991).

Freund, R.W., Gutnecht, M.H. and Nachtigal, N. (1993). *SIAM J. Statist. Comput.* **14**, 137–158 (1993).

Fuchs, L. and Zhao, H.S. (1984). *Int. J. Numer. Methods Fluids* **4**, 539–555.

Galpin, P.F. and Raithby, G.D. (1986). *Int. J. Numer. Methods Fluids* **6**, 409–426.

Galpin, P.F., Van Doormaal, J.P. and Raithby, G.D. (1985). *Int. J. Numer. Methods Fluids* **5**, 615–625.

Galpin, P.F., Raithby, G.D. and Van Doormaal, J.P. (1986). *Numer. Heat Transfer*, **9**, 241–246.

Gatski, T. (1991). *Appl. Numer. Math.* **7**, 227–240.

Gatski, T., Grosch, C.E. and Rose, M.E. (1982). *J. Comput. Phys.* **48**, 1–22.

Glowinski, R. and Pironneau, O. (1979). *SIAM Rev.* **21**, 167–212.

Golub, G.H. and Van Loan, C.F. (1983). *Matrix Computations.* Johns Hopkins University Press, Baltimore, MD.

Goodrich, J.W. and Soh, W.Y. (1989). *J. Comput. Phys.* **84**, 207–241.

Gresho, P.M. (1991a). In *Advances in Applied Mechanics*, vol. 28. Academic Press, London.

Gresho, P.M. (1991b). In *Annual Reviews of Fluid Mechanics*, vol. 23 pp. 413–453. Annual Reviews Inc., Palo Alto, CA.

Guevremont, G., Habashi, W.G., Kotiuga, P.L. and Hafez, M.M. (1993). *J. Comput. Phys.* **107**, 176–187.

Gushkin, V.A. and Schennikov, V.V. (1974). *USSR Comput. Maths. Math. Phys.* **14**, 252–256.

Habashi, W.G., Peeters, M.F., Robichaud, M.P. and Nguyen, V.N. (1993). In *Incompressible Computational Fluid Dynamics, Trends and Advances* (M.D. Gunzburger and R.A. Nicolaides, eds), pp. 151–182. Cambridge Univ. Press, Cambridge.

Hackbusch, W. (1985). *Multigrid Methods and Applications*, Springer Series in Comp. Maths, vol. 4. Springer-Verlag, Berlin.

Han, T. and Humphrey, J.A.C. (1981). *Comput. Methods Appl. Mech. Engrg.* **29**, 81–95.

Harlow, F.H. and Welch, J.R. (1965). *Phys. Fluids* **8**, 2182–2189.

Haroutunian, V., Engelman, M.S. and Hasbani, I. (1993). *Int. J. Numer. Methods Fluids* **17**, 323–348.

Hayase, T., Humphrey, J.A.C. and Grief, R. (1992). *J. Comput. Phys.* **98**, 108–118.

Hirasaki, G.J. and Hellums, J.D. (1968). *Quart. Appl. Maths.* **26**, 331–342.

Hirasaki, G.J. and Hellums, J.D. (1970). *Quart. Appl. Maths.* **28**, 293–296.

Hirt, C.W., Amsden, A.A. and Cook, J.L. (1974). *J. Comput. Phys.* **14**, 227–254.

Hirt, C.W., Nichols, B.D. and Romero, N.C. (1975). *Los Alamos Scientific Lab. Report LA-5852.*

Hoekstra, M. (1989). In *Proc. 5th International Conference on Numerical Ship Hydrodynamics*, Hiroshima (K.H. Mori, ed.), pp. 87–100.

Huang, P.G., Launder, B.E. and Leschziner, M.A. (1985). *Comput. Methods Appl. Mech. Engrg.* **48**, 1–24.

Hutchinson, B.R. and Raithby, G.D. (1986). *Numer. Heat Transfer* **9**, 511–537.

Ikohagi, T. and Chin, B.R. (1991). *Computers and Fluids* **19**, 479.

Ikohagi, T., Chin, B.R. and Daiguji, H. (1992). *Computers and Fluids* **21**, 163–175.

Issa, R.I. (1985). *J. Comput. Phys.* **62**, 40–65.

Jang, D.S., Jetli, R. and Acharya, S. (1986). *Numer. Heat Transfer* **10**, 209–228.

Karki, K.C. (1986). *A Calculation Procedure for Viscous Flows at All Speeds in Complex Geometries*, PhD Thesis, Univ. Minnesota.

Karki, K.C. and Mongia, H.C. (1990). *Int. J. Numer. Methods Fluids* **11**, 1–20.

Karki, K.C., Vanka, S.P. and Mongia, H.C. (1989). *AIAA Paper 89–0483*, AIAA 27th Aerospace Sciences Meeting, Reno, NV.

Kawamura, T. and Kuwahara, K. (1984). *AIAA Paper 84-0340*, AIAA 22nd Aerospace Sciences Meeting, Reno, NV.

Kim, S.W. and Benson, T.J. (1992). *Computers and Fluids* **21**, 435–454.

Kleiser, L. and Schumann, U. (1980). In *Notes on Numerical Fluid Mechanics* (E.H. Hirschel, ed.), vol. 2, pp. 165–173. Vieweg-Verlag, Braunschweig.

Knoll, D.A. and McHugh, P.R. (193). *Int. J. Numer. Methods Fluids* **17**, 449–461.

Kobayashi, M.H. and Pereira, J.C.F. (1991). *Numer. Heat Transfer* **19**, 243–262.

Kuznetsov, B.G. (1974). *Fluid Dyn. Trans.* **4**, 367–396.

Kwak, D.C., Chang, J.L., Shanks, S.P. and Chakravarthy, S. (1986). *AIAA J.* **24**, 390–396.

Leonard, B.P. (1979). *Comput. Methods Appl. Mech. Engrg.* **19**, 59–98.

Le Quéré, P. and Alziary de Roquefort, T. (1985). *J. Comput. Phys.* **57**, 210–228.

Leschziner, M.A. (1980). *Comput. Methods Appl. Mech. Engrg.* **23**, 293–312.

Lillington, J.N. (1981). *Int. J. Numer. Methods Fluids* **1**, 3–16.

Linden, J., Lonsdale, G., Steckel, B. and Stüben, K. (1989). In *Lecture Notes in Physics* vol. 323, pp. 57–68. Springer-Verlag, Berlin.

Liu, X. and Pletcher, R.H. (1986). *Numer. Heat Transfer* **8**, 539–556.

Lonsdale, R.D. (1993). *Int. J. Numer. Heat Fluid Flow* **3**, 3–14.

Lonsdale, G. and Walsh, J.E. (1988). *Int. J. Numer. Methods Fluids* **8**, 671–686.

Majumdar, M. (1988). *Numer. Heat Transfer* **13**, 125–132.

Mallinson, G.D. and Van de Vahl Davis, G. (1977). *J. Fluid Mech.* **83**, 1–31.

Markatos, N.C. and Pericleous, K.A. (1984). *Int. J. Heat Mass Transfer* **27**, 755–767.

Marx, Y.P. (1994). *J. Comput. Phys.* **112**, 182–309.

McArthur, J.W. and Patankar, S.V. (1990). *Int. J. Numer. Methods Fluids* **9**, 325–340.

McGuirk, J.J. and Palma, J.M.L.M. (1993). *Computers and Fluids* **22**, 77–88.

Meijerink, J.A. and Van der Vorst, H.A. (1977). *Math. Comput.* **31**, 148–162.

Meijerink, J.A. and Van der Vorst, H.A. (1981). *J. Comput. Phys.* **44**, 134–155.

Melaaen, M.C. (1992a). *Numer. Heat Transfer* **21**, 1–19.

Melaaen, M.C. (1992b). *Numer. Heat Transfer* **21**, 21–39.

Miller, T.F. and Schmidt, F.W. (1988). *Numer. Heat Transfer* **14**, 213–233.

Mynett, A.E., Wesseling, P., Segal, A. and Kassels, C.G.M. (1991). *Appl. Sci. Res.* **48**, 157–191.

Nachtigal, N.M., Reichel, L. and Trefethen, L.N. (1992). *SIAM J. Matrix Anal. Appl.* **13**, 796–825.

Obayashi, S. (1991). *NASA CR 177572*.

Oosterlee, C.W. (1993). *Robust Multigrid Methods for the Steady and Unsteady Incompressible Navier–Stokes Equations in General Coordinates*, PhD Thesis, Techn. Univ. Delft.

Oosterlee, C.W., Wesseling, P., Segal, A. and Brakkee, E. (1993). *Int. J. Numer. Methods Fluids* **17**, 301–321.

Orszag, S.A., Israeli, M. and Deville, M.O. (1986). *J. Sci. Comput.* **1**, 75–116.

Patankar, S.V. (1980). *Numerical Heat Transfer and Fluid Flow*. Hemisphere Publishing Corp., New York.

Patel, M.K., Markatos, N.C. and Cross, M. (1985). *Int. J. Numer. Methods Fluids* **5**, 225–244.

Peric, M. (1985). *A Finite Volume Method for the Prediction of Three-Dimensional Fluid Flow in Complex Ducts*, PhD Thesis, Imperial College, Univ. London.

Peric, M., Kessler, R. and Scheuerer, G. (1988). *Computers and Fluids* **16**, 389–403.

Peters, A. (1993). *Int. J. Numer. Methods Fluids* **17**, 955–974.

Peyret, R. and Taylor, T.D. (1983). *Computational Methods for Fluid Flow*. Springer-Verlag, New York.

Peyret, R. and Viviand, H. (1985). In *Recent Advances in the Aerospace Sciences* (C. Casci, ed.), pp. 41–71. Plenum Press, New York.

Piquet, J. and Queutey, P. (1992). *Computers and Fluids* **21**, 599–625.

Piquet, J. and Visonneau, M. (1991). *Computers and Fluids* **19**, 183–215.

Piquet, J., Queutey, P. and Visonneau, M. (1987). In *Lecture Notes in Physics*, vol. 323, pp. 489–493. Springer-Verlag, Berlin.

Pollard, A. and Siu, A.L.W. (1982). *Comput. Methods Appl. Mech. Engrg.* **35**, 293–313.

Pouagare, M. and Lakshminarayana, G. (1986). *J. Comput. Phys.* **64**, 389–415.

Pracht, W.E. (1975). *J. Comput. Phys.* **17**, 132–159.

Prakash, C. and Patankar, S.V. (1985). *Numer. Heat Transfer* **8**, 259–280.

Quartapelle, L. and Napolitano, M. (1986). *J. Comput. Phys.* **62**, 340–348.

Raithby, G.D. (1976). *Comput. Methods Appl. Mech. Engrg* **9**, 153–164.

Raithby, G.D. and Torrance, K.E. (1974). *Computers and Fluids* **2**, 191–206.

Raithby, G.D. and Schneider, G.E. (1979). *Numer. Heat Transfer* **2**, 417–440.

Ramshaw, J.D. and Mesina, G.L. (1991). *Computers and Fluids* **20**, 165–175.

Ramshaw, J.D. and Mousseau, V.A. (1990). *Computers and Fluids* **18**, 361–368.

Rayner, F. (1991). *Int. J. Numer. Methods Fluids* **13**, 507–518.

Rhie, C.M. and Chow, W.L. (1983). *AIAA J.* **17**, 1525–1532.

Richardson, S.M. and Cornish, A.R.H. (1977). *J. Fluid Mech.* **82**, 309–319.

Rogers, S.E., Chang, J.L.C. and Kwak, D. (1987). *J. Comput. Phys.* **73**, 364–379.

Rosenfeld, M. and Kwak, D. (1989). *AIAA Paper 89-0466*, AIAA 30th Aerospace Sciences Meeting, Reno, NV.

Rosenfeld, M. and Kwak, D. (1991). *Int. J. Numer. Methods Fluids* **13**, 1311–1321.

Rosenfeld, M. and Kwak, D. (1992). *AIAA Paper 92-0185*, AIAA 30th Aerospace Sciences Meeting, Reno, NV.

Rosenfeld, M., Kwak, D. and Vinokur, M. (1991). *J. Comput. Phys.* **94**, 102–137.

Ruas, V. (1991). *CRAS. Paris, Sér. I* **313**, 639–644.

Rubin, S.C. and Reddy, D.R. (1983). *Computers and Fluids* **11**, 281–306.

Saad, Y. (1989). *SIAM J. Sci. Stat. Comput.* **10**, 1200–1232.

Saad, Y. and Schultz, M.H. (1986) *SIAM J. Sci. Stat. Comput.* **7**, 856–869.

Saint Victor, X. (1986). *Résolution des Equations de Navier–Stokes Bi ou Tridimensionnelles par méthodes de Marche, Application au calcul de Mélanges d'Ecoulements Cisaillés*, PhD Thesis, Univ. Toulouse.

Schneider, G.E. and Raw, M.J. (1987a). *Numer. Heat Transfer* **11**, 363–390.

Schneider, G.E. and Raw, M.J. (1987b). *Numer. Heat Transfer* **11**, 391–400.

Schneider, G.E. and Zedan, M. (1984). *AIAA Paper 84-1743*, AIAA 19th Thermophysics Conference, Snowmass.

Segal, A., Wesseling, P., Van Kan, J., Oosterlee, C.W. and Kassels, K. (1992). *Int. J. Numer. Methods Fluids* **15**, 411–426.

Settari, A. and Aziz, K. (1973). *SIAM J. Numer. Anal.* **10**, 506.

Sharif, M.A.R. and Busnaina, A.A. (1988a). *J. Comput. Phys.* **74**, 143–176.

Sharif, M.A.R. and Busnaina, A.A. (1988b). *Appl. Math. Modelling* **12**, 98–108.

Sharif, M.A.R. and Busnaina, A.A. (1993). *ASME J. Fluids Engrg.* **115**, 33–40.

Shaw, G.J. and Sivaloganathan, S. (1988). *Int. J. Numer. Methods Fluids* **8**, 441–461.

Shay, W.A. (1981). *Computers and Fluids* **9**, 279–299.

3 Navier–Stokes equations for incompressible flows: finite-element methods

Max D. Gunzburger

HANDBOOK OF COMPUTATIONAL FLUID MECHANICS
ISBN 0-12-532200-3

I INTRODUCTION

This chapter is concerned with the finite-element approximation of viscous, incompressible, laminar flows. The governing partial differential equations are the continuity equation and the Navier–Stokes equations. The first of these is the mathematical realization of the incompressibility of the flow. The Navier–Stokes equations are the mathematical realization of Newton's second law of motion along with a linear constitutive law relating stresses to rates of strains. Derivations of these equations may be found in numerous texts on fluid mechanics. The specific form of the governing equations depends on the choice of dependent variables. At first, we concentrate on the 'primitive variable' formulation; subsequently, we consider other commonly used formulations of the equations of viscous, incompressible, laminar flow. It should be noted that we only consider *homogeneous* incompressible flows, i.e. incompressible flows such that the density is constant throughout the flow. Also, for the most part, we concentrate on issues that are specific to finite-element methods, and do not dwell too much on material that is common to all discretization schemes for the simulation of incompressible flows.

There are now many books solely or partly devoted to the subject of this chapter; among these are Baker (1985), Canuto *et al.* (1988), Cuvelier *et al.* (1986), Fletcher (1988), Girault and Raviart (1979, 1986), Glowinski (1984), Gunzburger (1989), Gunzburger and Nicolaides (1993), Peyret and Taylor (1983), Pironneau (1989), Quartapelle (1993), Temam (1979) and Thomasset (1981).

II THE PRIMITIVE VARIABLE FORMULATION

A The mathematical model

1 Differential equations

The derivation of the governing equations from first principles is in terms of the velocity field $\mathbf{u}$ and the pressure field p; for this reason, the equations written in terms of $\mathbf{u}$ and p are referred to as the *primitive variable formulation* and are given by the *Navier–Stokes equation*

$$\mathbf{u}_t - \nabla \cdot \left(\nu\big((\nabla\mathbf{u}) + (\nabla\mathbf{u})^{\mathrm{T}}\big) \right) + (\mathbf{u} \cdot \nabla)\mathbf{u} + \nabla p = \mathbf{f} \tag{1}$$

and the *continuity equation or incompressibility constraint*

$$\nabla \cdot \mathbf{u} = 0 \tag{2}$$

Here, ν denotes the kinematic viscosity of the fluid, $\mathbf{f}$ a prescribed body force per unit mass, $(\cdot)_t$ the time derivative, and $(\nabla\mathbf{u})^{\mathrm{T}}$ the transpose of the tensor $\nabla\mathbf{u}$.

Note that the pressure and body force have been scaled by the constant density of the flow. One should also recall that for incompressible flows, the pressure is a conveniently defined mechanical variable and is not a thermodynamic variable. For details concerning the derivation of (1) and (2) one may consult any of the many books on fluid mechanics; a particularly lucid account is given in Serrin (1959).

In general, the viscosity coefficient ν is a function of temperature; however, in many situations it is nearly constant. In such cases, using (2) and well-known vector identities, the viscous term in (1) may be written in the alternate forms

$$
\begin{aligned}
&\text{(i)} \quad \nabla \cdot \left(\nu \left((\nabla \mathbf{u}) + (\nabla \mathbf{u})^{\mathrm{T}} \right) \right) = \\
&\text{(ii)} \quad \nu \Delta \mathbf{u} = \\
&\text{(iii)} \quad -\nu \nabla \times (\nabla \times \mathbf{u}) = \\
&\text{(iv)} \quad \nu \left(\nabla (\nabla \cdot \mathbf{u}) - \nabla \times (\nabla \times \mathbf{u}) \right).
\end{aligned}
\tag{3}
$$

Again, it should be emphasized that if $\nu \neq$ constant, then only the first choice can be used; however, if $\nu =$ constant throughout the flow, then any of the four alternatives given in (3) may be used. The particular choice is influenced by the specific boundary conditions one wishes to impose; this issue will be discussed below.

2 Initial conditions

For unsteady, i.e. time-dependent problems, the velocity field at some instant of time, usually chosen to be $t = 0$, is prescribed. It is customary, and indeed, many time-stepping schemes require that the initial condition on the velocity be solenoidal. Thus, the initial condition for the primitive variable formulation is given by

$$
\mathbf{u}(0, \mathbf{x}) = \mathbf{u}_0(\mathbf{x}) \qquad \text{with} \qquad \nabla \cdot \mathbf{u}_0 = 0 \,,
\tag{4}
$$

where $\mathbf{u}_0(\mathbf{x})$ is a given solenoidal function of position $\mathbf{x}$.

3 Boundary conditions

Boundary conditions arise from two sources. In the first place, there are *physical boundary conditions* that should be imposed at the boundaries of the flow, e.g. at solid walls or at interfaces with other fluids or at infinity. Second, due to either the methods used for the discretization of the mathematical model or the methods used for the solution of the discrete equations, numerical boundary conditions are sometimes needed in order to effect a computational simulation. One example is problems such that the flow field is of infinite extent, e.g. flows in infinitely long channels or flows exterior to airfoils, in which case *outflow boundary conditions* are needed at the artificial boundaries created when an infinite domain is truncated to a finite computational domain. The different forms for the viscous term in the Navier–Stokes equations given in (3) can be used to facilitate the implementation of different boundary conditions within finite-element methods.

The fluid adjacent to a *solid wall* moving with prescribed velocity $\mathbf{g}(t, \mathbf{x})$ adheres to the wall so that we have the boundary condition

$$\mathbf{u}(t, \mathbf{x}) = \mathbf{g}(t, \mathbf{x}) \qquad \text{for } \mathbf{x} \text{ on a solid wall} \tag{5}$$

The most interesting case is that of a *fixed solid wall* for which $\mathbf{g} \equiv \mathbf{0}$ so that

$$\mathbf{u}(t, \mathbf{x}) = \mathbf{0} \qquad \text{for } \mathbf{x} \text{ on a fixed solid wall} \tag{6}$$

Either (5) or (6) may be directly applied as a boundary condition for any of the choices given in (3) for the viscous term.

Other possible choices of boundary conditions include specification of the *stress vector*

$$-p\mathbf{n} + \nu \mathbf{n} \cdot \left((\nabla \mathbf{u}) + (\nabla \mathbf{u})^{\mathrm{T}} \right) = \tau(t, \mathbf{x}) \qquad \text{for } \mathbf{x} \text{ on the boundary} \tag{7}$$

or of some component of the stress vector, where $\tau(t, \mathbf{x})$ is a given function defined on the boundary. One may also mix different components of (5) and (7); however, not all possible combinations lead to physically interesting or mathematically well-posed problems; see Verfurth (1985). The boundary condition of (7) can be of use at outflows and at interfaces, e.g. free boundaries, between two fluids.

We defer the discussion of other boundary conditions that are useful in numerical simulations until we consider weak formulations of the Navier–Stokes system.

B Weak formulation of the steady-state Navier–Stokes system

The mathematical theory for the estimation of the errors in finite element approximations of viscous, incompressible flows requires the introduction of some notions and results from functional analysis and the theory of partial differential equations. In particular, Sobolev spaces play a central role. In addition, the development of finite element algorithms for the approximate solution of the Navier–Stokes equations is based on recasting the Navier–Stokes system into 'weak form'. Throughout, we let $L^2(\mathcal{D})$ denote the space of functions that are square integrable in the Lebesque sense with respect to the domain $\mathcal{D} \subset R^n$, $n = 2$ or 3.

We consider the *steady-state* primitive variable formulation with the *homogeneous Dirichlet boundary condition* $\mathbf{u}(t, \mathbf{x}) = \mathbf{0}$ on Γ, where Γ denotes the boundary of the flow domain Ω. This simple setting is introduced to discuss the basic theory of mixed finite element-methods applied to the Navier–Stokes equations.

1 Sobolev spaces

Given a non-negative integer s, we define the Sobolev space

$$H^s(\mathcal{D}) = \left\{ \phi \in L^2(\mathcal{D}) \mid D^m\phi \in L^2(\mathcal{D}),\ m = 1,\ldots,s \right\}$$

where $D^m\phi$ denotes any and all partial derivatives of order m. Note that $L^2(\mathcal{D}) = H^0(\mathcal{D})$. We also have the associated norm

$$\|\phi\|_{s,\mathcal{D}} = \left(\|\phi\|_{0,\mathcal{D}}^2 + \sum_{m=1}^{s} \|D^m\phi\|_{0,\mathcal{D}}^2 \right)^{1/2}$$

where $\|\cdot\|_{0,\mathcal{D}}$ denotes the $L^2(\mathcal{D})$-norm defined by

$$\|\phi\|_{0,\mathcal{D}} = \left(\int_{\mathcal{D}} \phi^2\,d\mathbf{x} \right)^{1/2}$$

We will also use the subspaces

$$L_0^2(\mathcal{D}) = \left\{ \phi \in L^2(\mathcal{D}) \mid \int_{\mathcal{D}} \phi\,d\mathbf{x} = 0 \right\}$$

and

$$H_0^1(\mathcal{D}) = \left\{ \phi \in H^1(\mathcal{D}) \mid \phi = 0 \text{ on } \partial\mathcal{D} \right\},$$

where $\partial\mathcal{D}$ denotes the boundary of $\mathcal{D}$. Note that

$$|\phi|_{1,\mathcal{D}} = \left(\sum_{i=1}^{n} \|\partial\phi/\partial x_i\|_{0,\mathcal{D}}^2 \right)^{1/2}$$

defines a norm on $H_0^1(\Omega)$. We also have the dual space $H^{-1}(\mathcal{D})$ of $H_0^1(\mathcal{D})$ with norm

$$\|\phi\|_{-1,\mathcal{D}} = \sup_{v \in H_0^1(\mathcal{D}),\, v \neq 0} \frac{\int_{\mathcal{D}} \phi v\,d\mathbf{x}}{\|v\|_{1,\mathcal{D}}}$$

For vector-valued functions $\mathbf{v}$ having components v_i, $i = 1,\ldots,n$, we have the corresponding spaces

$$\mathbf{H}^s(\mathcal{D}) = [H^s(\mathcal{D})]^n = \left\{ \mathbf{v} \mid v_i \in H^s(\mathcal{D}),\ i = 1,\ldots,n \right\}$$

and norm

$$\|\mathbf{v}\|_{s,\mathcal{D}} = \left(\sum_{i=1}^{n} \|v_i\|_{s,\mathcal{D}}^2 \right)^{2}$$

Similarly, we have that $\mathbf{H}_0^1(\mathcal{D}) = [H_0^1(\mathcal{D})]^n$ and $\mathbf{H}^{-1}(\mathcal{D}) = [H^{-1}(\mathcal{D})]^n$. Often, whenever there is no chance for ambiguity, we will omit the domain from the norm designator, e.g. we will simply use $\|\cdot\|_s$ in place of $\|\cdot\|_{s,\mathcal{D}}$. The $\mathbf{L}^2(\mathcal{D})$ inner product

$$(\mathbf{f}, \mathbf{v}) = \int_{\mathcal{D}} \mathbf{f} \cdot \mathbf{v} \, d\mathbf{x} \quad \forall \mathbf{v} \in \mathbf{H}_0^1(\mathcal{D})$$

also denotes the duality pairing between functions $\mathbf{f}$ and $\mathbf{v}$ belonging to $\mathbf{H}^{-1}(\mathcal{D})$ and $\mathbf{H}_0^1(\mathcal{D})$, respectively.

Finally, we define the *subspace of weakly solenoidal functions*

$$\mathbf{Z} = \{\mathbf{v} \in \mathbf{H}_0^1(\Omega) \mid b(\mathbf{v}, q) = 0 \quad \forall q \in L_0^2(\Omega)\},$$

where $b(.,.)$ is defined below. Thus, functions belonging to $\mathbf{Z}$ have zero divergence almost everywhere in Ω.

Details concerning Sobolev spaces in general may be found in Adams (1975) and concerning Sobolev spaces of particular use in the Navier–Stokes settings in Girault and Raviart (1986) and Temam (1979).

2 Weak formulation

Weak formulations are defined in terms of the bilinear forms

$$a(\mathbf{u}, \mathbf{v}) = \nu \int_{\Omega} \nabla \mathbf{u} : \nabla \mathbf{v} \, d\mathbf{x} \quad \forall \mathbf{u}, \mathbf{v} \in \mathbf{H}^1(\Omega)$$

and

$$b(\mathbf{v}, q) = \int_{\Omega} q \nabla \cdot \mathbf{v} \, d\mathbf{x} \quad \forall \mathbf{v} \in \mathbf{H}^1(\Omega), \, q \in L^2(\Omega)$$

and the trilinear form

$$c(\mathbf{w}, \mathbf{u}, \mathbf{v}) = \int_{\Omega} \mathbf{w} \cdot \nabla \mathbf{u} \cdot \mathbf{v} \, d\mathbf{x} \quad \forall \mathbf{u}, \mathbf{v}, \mathbf{w} \in \mathbf{H}^1(\Omega)$$

For the steady-state primitive variable formulation with homogeneous Dirichlet boundary conditions we have the weak formulation: given $\mathbf{f} \in \mathbf{H}^{-1}(\Omega)$, seek $\mathbf{u} \in \mathbf{H}_0^1(\Omega)$ and $p \in L_0^2(\Omega)$ such that

$$a(\mathbf{u}, \mathbf{v}) + c(\mathbf{u}, \mathbf{u}, \mathbf{v}) + b(\mathbf{v}, p) = (\mathbf{f}, \mathbf{v}) \quad \forall \mathbf{v} \in \mathbf{H}_0^1(\Omega) \tag{8}$$

and

$$b(\mathbf{u}, q) = 0 \quad \forall q \in L_0^2(\Omega) \tag{9}$$

Details concerning weak formulations of the Navier–Stokes system may be found in Ladyzhenskaya (1969), Temam (1979) and Girault and Raviart (1986).

C Finite-element theory for the primitive variable formulation

One of the great successes of finite-element theory in the realm of non-linear-problems is that for the steady-state Navier–Stokes system. Among the many books on the subject of finite-element theory for the Navier–Stokes equations, Girault and Raviart (1986) remains to this day the definitive reference; other books of interest in this regard are Temam (1979), Glowinski (1984), Cuvelier *et al.* (1986), Gunzburger (1989), Pironneau (1989), Gunzburger and Nicolaides (1993), and Quartapelle (1993).

1 The discrete problem

Finite-element discretizations are defined as follows. Choose finite-element spaces $\mathbf{V}_0^h$ for the velocity approximations and S_0^h for the pressure approximations. These spaces are parameterized by the parameter h which is usually chosen to be some measure of the grid size. Then, the *approximate problem* is given by: seek $\mathbf{u}^h \in \mathbf{V}_0^h$ and $p^h \in S_0^h$ such that

$$a(\mathbf{u}^h, \mathbf{v}^h) + c(\mathbf{u}^h, \mathbf{u}^h, \mathbf{v}^h) + b(\mathbf{v}^h, p^h) = (\mathbf{f}, \mathbf{v}^h) \quad \forall \mathbf{v}^h \in \mathbf{V}_0^h \tag{10}$$

and

$$b(\mathbf{u}^h, q^h) = 0 \quad \forall q^h \in S_0^h \tag{11}$$

If $\mathbf{V}_0^h \subset \mathbf{H}_0^1(\Omega)$ and $S_0^h \subset L_0^2(\Omega)$, then the finite-element method is called a *conforming method*; if $\mathbf{V}_0^h \not\subset \mathbf{H}_0^1(\Omega)$ or $S_0^h \not\subset L_0^2(\Omega)$, the method is called a *non-conforming method*.

The discrete finite-element problem is equivalent to a system of non-linear algebraic equations. To see this, let $\{ q_j(\mathbf{x}) \}_{j=1}^J$ and $\{ \mathbf{v}_k(\mathbf{x}) \}_{k=1}^K$ denote bases for S_0^h and $\mathbf{V}_0^h$, respectively. Then,

$$p^h = \sum_{j=1}^J \alpha_j q_j(\mathbf{x}) \quad \text{and} \quad \mathbf{u}^h = \sum_{k=1}^K \beta_k \mathbf{v}_k(\mathbf{x})$$

for some constants α_j and β_k. If $\vec{P}$ and $\vec{U}$ denote the vector of unknowns having components α_j and β_k, respectively, then the approximate problem is equivalent to the non-linear algebraic system

$$\begin{pmatrix} A(\vec{U}) & B^{\mathrm{T}} \\ B & 0 \end{pmatrix} \begin{pmatrix} \vec{U} \\ \vec{P} \end{pmatrix} = \begin{pmatrix} \vec{F} \\ 0 \end{pmatrix} \tag{12}$$

where A and B are the matrices having entries

$$A_{\ell k} = a(\mathbf{v}_k, \mathbf{v}_\ell) + \sum_{m=1}^K c(\mathbf{v}_m, \mathbf{v}_k, \mathbf{v}_\ell)\beta_m = a(\mathbf{v}_k, \mathbf{v}_\ell) + c(\mathbf{u}^h, \mathbf{v}_k, \mathbf{v}_\ell)$$

$$= \nu \int_\Omega \nabla \mathbf{v}_k : \nabla \mathbf{v}_\ell \, d\mathbf{x} + \int_\Omega \left(\sum_{m=1}^K \beta_m \mathbf{v}_m \right) \cdot \nabla \mathbf{v}_k \cdot \mathbf{v}_\ell \, d\mathbf{x} \text{ for } k, \ell = 1, \dots, K$$

and

$$B_{ik} = b(\mathbf{v}_k, q_i) = \int_\Omega q_i \nabla \cdot \mathbf{v}_k \, d\mathbf{x} \text{ for } i = 1, \ldots, J \text{ and } k = 1, \ldots, K$$

respectively, and $\vec{F}$ is the vector having components

$$F_\ell = (\mathbf{f}, \mathbf{v}_\ell) = \int_\Omega \mathbf{f} \cdot \mathbf{v}_\ell \, d\mathbf{x} \text{ for } \ell = 1, \ldots, K$$

Note that the entries of A depend on the components of $\vec{U}$.

Before giving details about specific choices for the finite-element spaces $\mathbf{V}_0^h$ and S_0^h, we discuss in general terms some requirements on these spaces that will guarantee that the resulting approximating schemes defined by (10) and (11) are stable and that approximate solutions obtained through these schemes are optimally accurate.

2 The LBB condition

Unlike positive definite problems, i.e. the Poisson equation, the inclusions $\mathbf{V}_0^h \subset \mathbf{H}_0^1(\Omega)$ and $S_0^h \subset L_0^2(\Omega)$ are not sufficient to guarantee the stability and optimal accuracy of the discrete approximations to the Navier–Stokes system obtained from (10) and (11). These can be guaranteed by five conditions, four of which follow are usually easily verified. The first three of these are the *continuity conditions*

$$|a(\mathbf{u}^h, \mathbf{v}^h)| \leq \kappa_a \|\mathbf{u}^h\|_1 \|\mathbf{v}^h\|_1 \quad \forall \mathbf{u}^h, \mathbf{v}^h \in \mathbf{V}_0^h \tag{13}$$

$$|b(\mathbf{v}^h, q^h)| \leq \kappa_b \|\mathbf{v}^h\|_1 \|q^h\|_0 \quad \forall \mathbf{v}^h \in \mathbf{V}_0^h, q^h \in S_0^h \tag{14}$$

and

$$|c(\mathbf{w}^h, \mathbf{u}^h, \mathbf{v}^h)| \leq \kappa_c \|\mathbf{u}^h\|_1 \|\mathbf{v}^h\|_1 \|\mathbf{w}^h\|_1 \quad \forall \mathbf{u}^h, \mathbf{v}^h, \mathbf{w}^h \in \mathbf{V}_0^h \tag{15}$$

which follow from the inclusions $\mathbf{V}_0^h \subset \mathbf{H}_0^1(\Omega)$ and $S_0^h \subset L_0^2(\Omega)$ and the similar continuity conditions that hold over the entire underlying spaces $\mathbf{H}^1(\Omega)$ and $L^2(\Omega)$. Here, κ_a, κ_b, and κ_c are constants whose values are independent of h. See Babuska and Aziz (1972) or Ciarlet (1978) for the derivation of (13) and (14) and Girault and Raviart (1986) and Temam (1979, 1983) for the derivation of (15).

Analogous to the subspace $\mathbf{Z}$ we may define the subspace of *discretely-solenoidal functions*:

$$\mathbf{Z}^h = \{ \mathbf{v}^h \in \mathbf{V}_0^h \mid b(\mathbf{v}^h, q^h) = 0 \quad \forall q^h \in S_0^h \}$$

Note that, in general, even when $\mathbf{V}_0^h \subset \mathbf{H}_0^1(\Omega)$ and $S_0^h \subset L_0^2(\Omega)$, $\mathbf{Z}^h \not\subset \mathbf{Z}$. A measure of the 'angle' between $\mathbf{Z}^h$ and $\mathbf{Z}$ is given by

$$\Theta = \sup_{\mathbf{z}^h \in \mathbf{Z}^h, \|\mathbf{z}^h\|_1 = 1} \inf_{\mathbf{z} \in \mathbf{Z}} \|\mathbf{z} - \mathbf{z}^h\|_1$$

Note that $0 \leq \Theta \leq 1$. Also note that $\mathbf{u}^h \in \mathbf{Z}^h$ since $b(\mathbf{u}^h, q^h) = 0$ for all $q^h \in S_0^h$. The fourth condition is the *coercivity condition*

$$a(\mathbf{z}^h, \mathbf{z}^h) \geq \gamma_a |\mathbf{z}^h|_1^2 \quad \forall \mathbf{z}^h \in \mathbf{Z}^h \tag{16}$$

where γ_a is a constant whose value is independent of h. In most cases this condition also follows from the inclusions $\mathbf{V}_0^h \subset \mathbf{H}_0^1(\Omega)$ and $S_0^h \subset L_0^2(\Omega)$ and the analogous condition that holds on the entire space $\mathbf{H}_0^1(\Omega)$; see, e.g. Babuska and Aziz (1972), Ciarlet (1978) and Girault and Raviart (1986).

The fifth condition does not follow from the inclusions and its verification is not usually a simple matter. It is known as the *div-stability* or *Ladyzhenskaya–Babuska–Brezzi* or *LBB* or *inf-sup condition* and is given by

$$\sup_{\mathbf{0} \neq \mathbf{v}^h \in \mathbf{V}_0^h} \frac{b(\mathbf{v}^h, q^h)}{|\mathbf{v}^h|_1} \geq \gamma_b \|q^h\|_0 \quad \forall q^h \in S_0^h \tag{17}$$

or, recalling the definition of $b(\cdot, \cdot)$,

$$\sup_{\mathbf{0} \neq \mathbf{v}^h \in \mathbf{V}_0^h} \frac{\int_\Omega q^h \nabla \cdot \mathbf{v}^h \, d\mathbf{x}}{|\mathbf{v}^h|_1} \geq \gamma_b \|q^h\|_0 \quad \forall q^h \in S_0^h$$

where γ_b is a constant whose value is independent of h. This condition is thoroughly discussed in Brezzi (1974), Girault and Raviart (1986), Gunzburger (1989), and Brezzi and Fortin (1991); see also Ladyzhenskaya (1969) and Babuska (1973).

Before discussing some means for verifying this condition, we consider some of its consequences.

Assuming the chosen finite-element spaces satisfy the LBB condition, one can then prove various results about the discrete system given by (10) and (11). For example, we have the *existence* of solutions, i.e. given $\mathbf{f} \in \mathbf{H}^{-1}(\Omega)$, there exist $\mathbf{u}^h \in \mathbf{V}_0^h$ and $p^h \in S_0^h$ satisfying (10) and (11). Moreover, for sufficiently small data $\mathbf{f}$ or sufficiently small $Re = 1/\nu$ one can prove the *uniqueness* of solutions of (10) and (11). Details concerning these issues may be found in, e.g. Brezzi (1974), Brezzi and Fortin (1991), Girault and Raviart (1986), Glowinski (1984), Gunzburger (1989) and Temam (1979).

3 Error estimates

We introduce the notion of a *branch of regular solutions*. Let I denote a compact interval of $\mathbb{R}^+$. Then, a branch of solutions $\{\mathbf{u}(Re), p(Re) ; Re \in I\}$ of the Navier–Stokes system is called a regular branch of solutions if the linear problem

$$-\nu \Delta \mathbf{w} + \mathbf{u} \cdot \nabla \mathbf{w} + \mathbf{w} \cdot \nabla \mathbf{u} + \nabla r = \mathbf{f} \text{ in } \Omega$$

$$\nabla \cdot \mathbf{w} = 0 \text{ in } \Omega$$

and

$$\mathbf{w} = \mathbf{0} \quad \text{on } \Gamma$$

has a unique solution $(\mathbf{w}, r)$ for each $Re = 1/\nu \in I$. Note that this system is the Navier–Stokes system linearized about $\mathbf{u}(Re)$. Basically, a solution belongs to a regular branch if it does not correspond to a singular point such as a bifurcation point or turning point; for details, see Brezzi *et al.* (1980) and Girault and Raviart (1986).

Assuming that the LBB condition given by (17) holds, one can also prove that the discrete solution *converges* to the exact solution whenever the uniqueness condition holds or when the exact solution is known to belong to a regular branch of solutions. Indeed, in these cases we have the *optimal error estimates*

$$|\mathbf{u} - \mathbf{u}^h|_1 \leq C_1 \inf_{\mathbf{v}^h \in \mathbf{V}_0^h} \|\mathbf{u} - \mathbf{v}^h\|_1 + C_2 \Theta \inf_{q^h \in S_0^h} \|p - q^h\|_0 \tag{18}$$

and

$$\|p - p^h\|_0 \leq C_3 \inf_{\mathbf{v}^h \in \mathbf{V}_0^h} \|\mathbf{u} - \mathbf{v}^h\|_1 + C_4 \inf_{q^h \in S_0^h} \|p - q^h\|_0 \tag{19}$$

If the solution of a linearized adjoint problem is sufficiently regular, then one can use a duality argument to prove that

$$\|\mathbf{u} - \mathbf{u}^h\|_0 \leq C_5 h |\mathbf{u} - \mathbf{u}^h|_1 \tag{20}$$

where h is a parameter measuring the size of the finite-element grid. In (18)–(20), the constants C_i, $i = 1, \ldots, 5$, are independent of h. For a detailed derivation of (18)–(20), one should consult Girault and Raviart (1986) and the references cited therein.

If one examines the two terms on the right-hand-sides of (18) and (19), we see that in order to equilibrate these terms, one would usually use polynomials of one degree higher for $\mathbf{u}^h$ than for p^h. Also, if $\mathbf{Z}^h \subset \mathbf{Z}$, i.e. if discretely solenoidal functions are solenoidal, then $\Theta = 0$ and the velocity error uncouples from the pressure error. Moreover, the constants in the error estimates are proportional to $1/\gamma_b$, where γ_b is the constant in the LBB condition of (17).

We now explore the role of the LBB condition in deriving the error estimates of (18) and (19). The first four conditions given in (13)–(16) 'easily' imply that

$$|\mathbf{u} - \mathbf{u}^h|_1 \leq \tilde{C}_1 \inf_{\mathbf{z}^h \in \mathbf{Z}^h} |\mathbf{u} - \mathbf{z}^h|_1 + C_2 \Theta \inf_{q^h \in S^h} \|p - q^h\|_0$$

The first term on the right-hand-side requires knowledge of how well solenoidal functions, i.e. functions in $\mathbf{Z}$, are approximated by discretely solenoidal functions, i.e. functions in $\mathbf{Z}^h$. The LBB condition implies that if $\mathbf{u} \in \mathbf{Z}$, then

$$\inf_{\mathbf{z}^h \in \mathbf{Z}^h} |\mathbf{u} - \mathbf{z}^h|_1 \leq \left(1 + \frac{\kappa_b}{\gamma_b}\right) \inf_{\mathbf{v}^h \in \mathbf{V}_0^h} |\mathbf{u} - \mathbf{v}^h|_1$$

so that the LBB condition implies that the error in the finite-element approximation depends only on the ability to approximate in the finite-element spaces.

4 Verifying the LBB condition

We now discuss four methods for verifying that a pair of finite-element spaces satisfy the LBB condition. First, we have *Fortin's method* (Fortin, 1977; Girault and Raviart, 1986). Fortin noted that the LBB condition is equivalent to the existence of an operator $\Pi^h : \mathbf{H}_0^1(\Omega) \to \mathbf{V}^h$ such that, given any $\mathbf{v} \in \mathbf{H}_0^1(\Omega)$,

$$b(\Pi^h \mathbf{v}, q^h) = b(\mathbf{v}, q^h) \quad \forall q^h \in S^h$$

and

$$|\Pi^h \mathbf{v}|_1 \leq C |\mathbf{v}|_1$$

Unfortunately, showing the existence of the operator Π^h is usually a difficult task.

Next, we have *Verfurth's method* (Verfurth, 1984) which applies to continuous pressure spaces, so that $S_0^h \subset H^1(\Omega) \cap L_0^2(\Omega)$. One combines the inverse inequality (Ciarlet, 1978)

$$|\mathbf{v}^h|_1 \leq (K_1/h) \|\mathbf{v}^h\|_0$$

and the 'easy' result

$$\sup_{0 \neq \mathbf{v}^h \in \mathbf{V}_0^h} (b(\mathbf{v}^h, q^h)/\|\mathbf{v}^h\|_0) \geq K_2 |q^h|_1$$

to yield

$$\sup_{0 \neq \mathbf{v}^h \in \mathbf{V}_0^h} (b(\mathbf{v}^h, q^h)/|\mathbf{v}^h|_1) \geq \frac{K_2}{K_1} h |q^h|_1 \quad \forall q^h \in S_0^h \tag{21}$$

Then, one combines the known result: for any $q^h \in S_0^h$, there exists $\mathbf{w} \in \mathbf{H}_0^1(\Omega)$ such that

$$\nabla \cdot \mathbf{w} = q^h \quad \text{and} \quad |\mathbf{w}|_1 \leq K_3 \|q^h\|_0$$

and the approximation theoretic result: for $k = 0, 1$ and $\mathbf{w} \in \mathbf{H}_0^1(\Omega)$, there exists $\mathbf{w}^h \in \mathbf{V}_0^h$ such that

$$|\mathbf{w} - \mathbf{w}^h|_k \leq K_4 h^{1-k} |\mathbf{w}|_1$$

to yield

$$\sup_{0 \neq \mathbf{v}^h \in \mathbf{V}_0^h} (b(\mathbf{v}^h, q^h)/|\mathbf{v}^h|_1) \geq K_5 - K_6 h |q^h|_1 \quad \forall q^h \in S_0^h \text{ with } \|q^h\|_0 = 1 \tag{22}$$

It can then be shown that (21) and (22) imply that the LBB condition holds.

The third method is the *Boland-Nicolaides method* (Boland and Nicolaides, 1983, 1984, 1985; Girault and Raviart, 1986). The LBB condition is difficult to verify because it is global in nature. Boland and Nicolaides showed how it may be reduced to a local test. First, subdivide Ω into disjoint macro-elements Ω_r, $r = 1, \ldots, R$. Each macro-element contains one or a few elements in the finite-element triangulation associated with $\mathbf{V}_0^h$ and S_0^h. The number of elements within a macro-element is independent of the grid size h. Now, suppose the LBB condition holds locally over macro-elements, i.e. let

$$\mathbf{V}_r^h = \{\, \mathbf{v} \in \mathbf{V}_0^h|_{\Omega_r} \mid \mathbf{v} = \mathbf{0} \text{ on } \partial\Omega_r \,\}$$

and

$$S_r^h = \{\, q \in L_0^2(\Omega_r) \cap S_0^h|_{\Omega_r} \,\}$$

where $\partial\Omega_r$ denotes the boundary of the macro-element Ω_r, and suppose that

$$\sup_{\mathbf{0} \neq \mathbf{v}^h \in \mathbf{V}_r^h} \frac{b(\mathbf{v}^h, q^h)}{|\mathbf{v}^h|_1} \geq \hat{\gamma}\|q^h\|_{0,\Omega_r} \quad \forall q^h \in S_r^h \tag{23}$$

where $\hat{\gamma}$ is independent of r and h. Since $\mathbf{V}_r^h$ and S_r^h have small dimension that is independent of h, (23) may often be verified by a direct computation. Next, suppose the LBB condition holds globally for comparison spaces $\tilde{\mathbf{V}}_0^h$ and $\tilde{S}_0^h$ such that

$$\tilde{\mathbf{V}}_0^h \subset \mathbf{V}_0^h$$

and

$$\tilde{S}_0^h \subset \{\, L_0^2(\Omega) \text{ piecewise constants with respect to the macro-elements } \Omega_r \,\},$$

i.e. suppose that

$$\sup_{\mathbf{0} \neq \mathbf{v}^h \in \tilde{\mathbf{V}}_0^h} \frac{b(\mathbf{v}^h, q^h)}{|\mathbf{v}^h|_1} \geq \tilde{\gamma}\|q^h\|_0 \quad \forall q^h \in \tilde{S}_0^h \tag{24}$$

where $\tilde{\gamma}$ is independent of h. Then, (23) and (24) imply that the LBB condition holds for the spaces $\mathbf{V}_0^h$ and S_0^h.

Finally, we have *Stenberg's method* (Stenberg, 1984, 1987). Again, one uses macro-elements and the local spaces $\mathbf{V}_r^h$ and S_r^h. Define the local null spaces (of the gradient operator)

$$\mathcal{N}_r^h = \{\, q^h \in S_0^h|_{\Omega_r} \mid b(\mathbf{v}^h, q^h) = 0 \quad \forall \mathbf{v}^h \in \mathbf{V}_r^h \,\}$$

Suppose that for every macro-element Ω_r the space $\mathcal{N}_r^h$ is one-dimensional, consisting of functions that are constant over Ω_r. Then, the LBB condition holds for $\mathbf{V}_0^h$ and S_0^h.

5 Ways in which the LBB condition fails

We now explore the different ways in which the LBB condition can fail to hold. First, it is possible for the only discretely solenoidal field in $\mathbf{V}_0^h$ to be the zero vector, i.e.

$$b(\mathbf{u}^h, q^h) = 0 \quad \forall q^h \in S_0^h \quad \Longrightarrow \quad \mathbf{u}^h = \mathbf{0}$$

Such an occurrence can usually be detected by a counting argument. For example, consider Ω = square that is divided into $N \times N$ boxes that are further divided into two triangles each as in Figure 1. Let Δ denote a triangle in the subdivision of Ω and let

$$\mathbf{V}_0^h = \{\, \mathbf{v} \in \mathbf{H}_0^1(\Omega) \,\, |\mathbf{v}|_\Delta \in [\mathcal{P}_1(\Delta)]^2 \,\}$$

and

$$S_0^h = \{\, q \in L_0^2(\Omega) \,\, |q|_\Delta \in \mathcal{P}_0(\Delta) \,\}$$

where $\mathcal{P}_k(\Delta)$, $k = 0, 1$, denotes the space of polynomials of degree less than or equal to k. Thus, here we use piecewise linear polynomials for the components of the velocity approximation and piecewise constant polynomials *with respect to the same grid* for the pressure approximation. Then, $\dim (S_0^h) = 2N^2 - 1$ and $\dim (\mathbf{V}_0^h) = 2(N-1)^2$. Then, $\dim (S_0^h) > \dim (\mathbf{V}_0^h)$ so that the divergence matrix B in (12) has more rows than columns and $B\vec{U} = 0$ implies that $\vec{U} = 0$.

The next way for the LBB condition to fail is in case that for one, or a few, $q^h \in S_0^h$, we have that

$$b(\mathbf{u}^h, q^h) = 0 \quad \forall \mathbf{u}^h \in \mathbf{V}_0^h \tag{25}$$

so that, in the LBB condition (17), $\gamma_b = 0$. Again, this is easy to detect since the divergence matrix is rank deficient. One can 'filter' the solution to remove

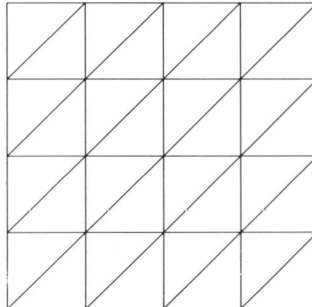

Figure 1 Grid for unstable linear/constant element.

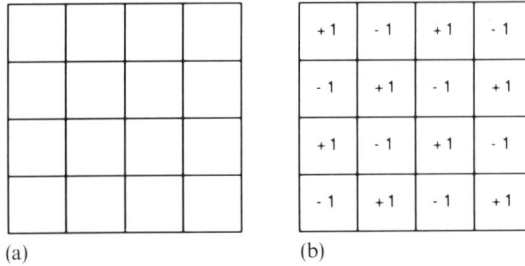

Figure 2 Grid for bilinear/constant element (on left) and checkerboard mode (on right).

offending modes. For example, consider the bilinear velocity-constant pressure element pair

$$\mathbf{V}_0^h = \{ \mathbf{v} \in \mathbf{H}_0^1(\Omega) \, |\mathbf{v}|_\square \in [\mathcal{Q}_1(\square)]^2 \}$$

and

$$S_0^h = \{ q \in L_0^2(\Omega) \, |q|_\square \in \mathcal{Q}_0(\square) \}$$

that is based on the subdivision of Ω given in Figure 2a. Here, $\square$ denote a rectangle in the subdivision of Ω and $\mathcal{Q}_k(\square)$, $k = 0, 1$, denotes the space of polynomials of degree less than or equal to k in each coordinate. Now, $\dim(S_0^h) = N^2 - 1$ and $\dim(\mathbf{V}_0^h) = 2(N^2 - 1)$ so that a simple counting argument yields no information. However, for the chequerboard mode depicted in Figure 2b, we have that (25) holds. This mode can be easily filtered out from the discrete pressure (see Boland and Nicolaides, 1984; Gresho, 1988).

A third way for the LBB condition to fail is in the case that for at least some $q^h \in S_0^h$, we have that

$$C_1 h \|q^h\|_0 \le \sup_{0 \ne \mathbf{v}^h \in \mathbf{V}_0^h} \frac{|b(\mathbf{v}^h, q^h)|}{|\mathbf{v}^h|_1} \le C_2 h \|q^h\|_0 \tag{26}$$

so that $\gamma_b = O(h)$. This case is hard to detect because, for finite h, no catastrophe, e.g. a singular matrix, occurs. However, in this case, there results a loss of accuracy. The bilinear–constant element pair suffers from this problem as well (Boland and Nicolaides, 1984). Indeed, examples have been given (Boland and Nicolaides, 1984) for which the pressure approximation does not converge and there is a loss of accuracy in the velocity approximation. These results are interesting since the bilinear–constant element pair is in widespread use in the engineering community! Perhaps one explanation for this is that the examples for which (26) holds for the bilinear–constant element pair involve very 'rough' data.

D Finite-element spaces that satisfy the LBB condition

Let

$$\mathcal{P}_k(\mathcal{D}) = \text{polynomials of degree} \leq k \text{ with respect to } \mathcal{D} \subset \mathbb{R}^n$$

$$\mathbf{P}_k(\mathcal{D}) = [\mathcal{P}_k(\mathcal{D})]^d = d\text{-vectors whose components belong to } \mathcal{P}_k(\mathcal{D})$$

$$\mathcal{Q}_k(\mathcal{D}) = \text{polynomials of degree} \leq k \text{ in each coordinate}$$

and

$$\mathbf{Q}_k(\mathcal{D}) = [\mathcal{Q}_k(\mathcal{D})]^d = d\text{-vectors whose components belong to } \mathcal{Q}_k(\mathcal{D})$$

We will assume that Ω is a polygonal domain in $\mathbb{R}^2$. The discussion that follows can be extended to curved boundaries by, e.g. using isoparametric elements (Ciarlet, 1978). For all the element pairs discussed, the LBB condition holds.

Although we only consider the two-dimensional case, many of the stable element pairs given below can be generalized to *three dimensions* (Thatcher, 1993).

1 Piecewise linear velocities

We subdivide Ω into triangles and let $\mathbf{V}_0^h \subset \mathbf{H}_0^1(\Omega)$ and $\mathbf{V}_0^h|_\Delta = \mathbf{P}_1(\Delta)$, where Δ denotes a triangle. For each of the pressure spaces discussed below, we have the error estimates:

if $\mathbf{u} \in \mathbf{H}^2(\Omega) \cap \mathbf{H}_0^1(\Omega)$ and $p \in H^1(\Omega) \cap L_0^2(\Omega)$, then

$$|\mathbf{u} - \mathbf{u}^h|_1 = O(h) \,, \quad \|\mathbf{u} - \mathbf{u}^h\|_0 = O(h^2) \,, \text{ and } \|p - p^h\|_0 = O(h)$$

The first pressure space consists of piecewise constants with respect to a triangulation twice the size of the velocity triangulation, as depicted below.

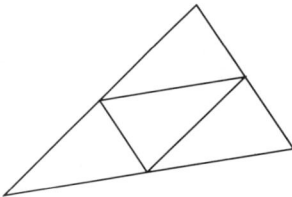

| Velocity triangles | Pressure triangle |

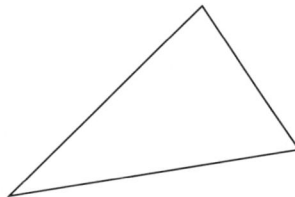

$$\mathbf{V}_0^h \subset \mathbf{H}_0^1(\Omega) \qquad S_0^h \subset L_0^2(\Omega)$$
$$\mathbf{V}_0^h|_{\Delta_{h/2}} = \mathbf{P}_1(\Delta_{h/2}) \qquad S_0^h|_{\Delta_h} = \mathcal{P}_0(\Delta_h)$$

The second pressure space associated with linear velocities also consists of piecewise constants, but now we have three out of the four possible pressures associated with the velocity triangulation, as depicted below.

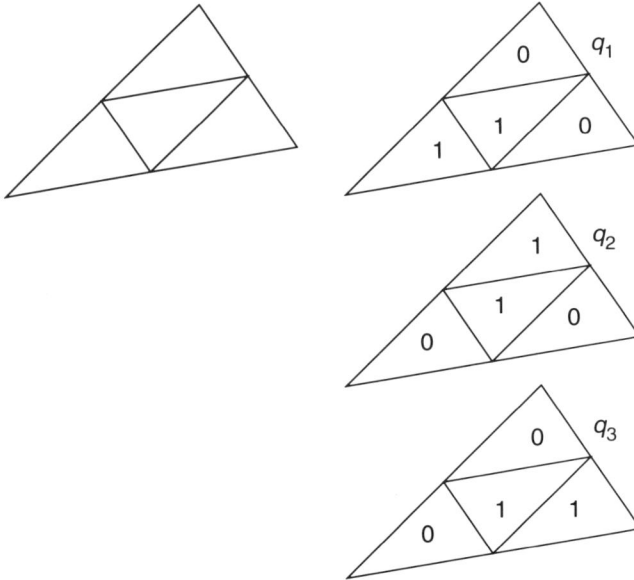

Velocity triangles   Pressure basis

$$\mathbf{V}_0^h \subset \mathbf{H}_0^1(\Omega)$$ $$S_0^h \subset L_0^2(\Omega)$$

$$\mathbf{V}_0^h|_{\Delta_{h/2}} = \mathbf{P}_1(\Delta_{h/2})$$ $$S_0^h|_{\Delta_h} = \text{span}\,\{q_1, q_2, q_3\}$$

A third pressure space that can be used with a linear velocity space is a continuous piecewise linear space as depicted below.

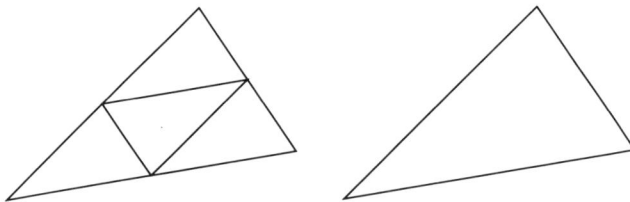

Velocity triangles   Pressure triangle

$$\mathbf{V}_0^h \subset \mathbf{H}_0^1(\Omega)$$ $$S_0^h \subset H^1(\Omega) \cap L_0^2(\Omega)$$

$$\mathbf{V}_0^h|_{\Delta_{h/2}} = \mathbf{P}_1(\Delta_{h/2})$$ $$S_0^h|_{\Delta_h} = \mathcal{P}_1(\Delta_h)$$

Because of the coupling of the pressure and velocity errors, one seemingly cannot take advantage of the better approximating ability of this pressure space.

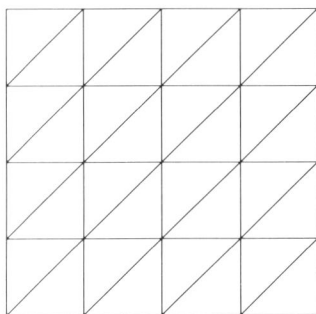

Figure 3 Pressure grid for stable linear velocity elements.

However, in practice, using piecewise linear pressures results in better approximations owing to smaller constants in the estimates. For example, we have that

$$|\mathbf{u} - \mathbf{u}^h|_1 \leq \underbrace{C_1 h}_{\text{velocity error}} + \underbrace{C_2 h}_{\text{pressure error}}$$

For the first two pressure spaces discussed above, we have that C_2 is 'large' so that the pressure error dominates. For the third pressure space, we have that $C_2 = O(h)$, so that pressure error is less dominant.

The linear–linear pair also results in fewer unknowns. For example, using $N \times N$ boxes for the pressure grid (as depicted in Figure 3) we have that, for all three element pairs discussed above, the number of velocity unknowns $= 2(2N - 1)^2$. However, the number of pressure unknowns is given by $(2N^2 - 1)$, $(6N^2 - 1)$, and $[(N + 1)^2 - 1]$, respectively. Thus, among stable element pairs having linear velocities, the linear pressure pair is preferable.

Note that a counting argument is generally not sufficient to show that an element pair satisfies the LBB condition; loosely speaking, one must get some control over $\mathbf{u} \cdot \mathbf{n}$ on the boundary of the element. For example, consider the element pair shown below:

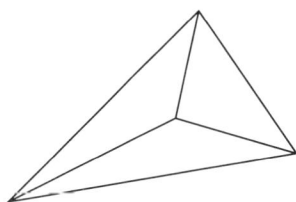

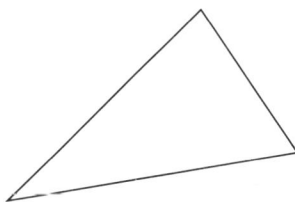

Velocity triangles   Pressure triangle

$$\mathbf{V}_0^h \subset \mathbf{H}_0^1(\Omega)$$ $$S_0^h \subset L_0^2(\Omega)$$

$$\mathbf{V}_0^h|_{\Delta_i} = \mathbf{P}_1(\Delta_i),\ i = 1, 2, 3$$ $$S_0^h|_{\Delta_h} = \mathcal{P}_0(\Delta_h)$$

5 Locally mass-conserving elements

If we can't have $\nabla \cdot \mathbf{u}^h = 0$ pointwise, we at least would like to have vanishing local averages of $\nabla \cdot \mathbf{u}^h$. Any element pair with discontinuous pressures is locally mass conserving in the sense that

$$\int_\Delta \nabla \cdot \mathbf{u}^h \, d\Omega = 0$$

where Δ denotes a pressure element.

One may modify the Taylor–Hood element so that it is locally mass conserving. One merely need enrich the Taylor–Hood pressure space $S^h_{(\text{th})}$ by one pressure per triangle:

$$S^h_0|_\Delta = \left(S^h_{(\text{th})}|_\Delta \oplus \mathcal{P}_0(\Delta)\right)$$

Now, $S^h_0 \not\subset H^1(\Omega)$, but still, $S^h_0 \subset L^2_0(\Omega)$. This locally mass-conserving element satisfies the same error estimates as those for Taylor–Hood, but involves more unknowns for the same triangulation (Silvester and Thatcher, 1986).

6 Some other elements of interest

There are many other element pairs that are known to be stable. For example, we have the *mini-element* (Arnold *et al,*. 1984)

$$\mathbf{V}^h_0 \subset \mathbf{H}^1_0(\Omega) \qquad \mathbf{V}^h_0|_\Delta = \mathbf{P}_1(\Delta) \oplus \text{span} \, (\lambda_1 \lambda_2 \lambda_3)$$

where $\lambda_i(\mathbf{x})$ is the linear polynomial that is unity at vertex i and vanishes at the other two vertices of Δ, and

$$S^h_0 \subset H^1(\Omega) \cap L^2_0(\Omega) \qquad \text{and} \qquad S^h_0|_\Delta = \mathcal{P}_1(\Delta)$$

We have continuous piecewise linear pressures and continuous piecewise linear velocities enriched with a cubic *bubble function* (a function with support over a single element) in each triangle. This element pair satisfies the error estimates

$$|\mathbf{u} - \mathbf{u}^h|_1 + \|p - p^h\|_0 = O(h) \qquad \text{and} \qquad \|\mathbf{u} - \mathbf{u}^h\|_0 = O(h^2)$$

Other element pairs using bubble functions are discussed in Mansfield (1982), Boland and Nicolaides (1983), Brezzi and Pitkaranta (1984) Girault and Raviart (1986), and Gunzburger (1989).

Another popular element pair is the $\mathbf{Q}_2 - \mathcal{P}_1$ *element* (Stenberg, 1984; Girault and Raviart, 1986) given by

$$\mathbf{V}^h_0 \subset \mathbf{H}^1_0(\Omega) \qquad \text{and} \qquad \mathbf{V}^h_0|_\square = \mathbf{Q}_2(\square)$$

and

$$S^h_0 \subset L^2_0(\Omega) \qquad \text{and} \qquad S^h_0|_\square = \mathcal{P}_1(\square)$$

We have continuous biquadratic velocities and discontinuous *linear* (not bilinear) pressures; we also have the error estimate

$$|\mathbf{u} - \mathbf{u}^h|_1 + \|p - p^h\|_0 = O(h^2) \qquad \text{and} \qquad \|\mathbf{u} - \mathbf{u}^h\|_0 = O(h^3)$$

E Inhomogeneous velocity boundary conditions

We now consider replacing the boundary condition (6) by the inhomogeneous condition (5), i.e.

$$\mathbf{u} = \mathbf{g} \quad \text{on } \Gamma \qquad \text{with} \qquad \int_\Gamma \mathbf{g} \cdot \mathbf{n} \, d\mathbf{x} = 0$$

Setting

$$\mathbf{V}_g = \{ \mathbf{u} \in \mathbf{H}^1(\Omega) \mid \mathbf{u} = \mathbf{g} \text{ on } \Gamma \}$$

we now seek $\mathbf{u} \in \mathbf{V}_g$ and $p \in L_0^2(\Omega)$ such that

$$a(\mathbf{u}, \mathbf{v}) + c(\mathbf{u}, \mathbf{u}, \mathbf{v}) + b(\mathbf{v}, p) = (\mathbf{f}, \mathbf{v}) \quad \forall \mathbf{v} \in \mathbf{H}_0^1(\Omega)$$

and

$$b(\mathbf{u}, q) = 0 \quad \forall q \in L_0^2(\Omega)$$

We now define finite-element spaces $\mathbf{V}^h \subset \mathbf{H}^1(\Omega)$ and $S_0^h \subset L_0^2(\Omega)$ and also choose an approximation $\mathbf{g}^h \in \mathbf{V}^h|_\Gamma$ to $\mathbf{g}$. The usual choice in engineering practice (because it is easy to implement) is the interpolant of $\mathbf{g}$ in $\mathbf{V}^h|_\Gamma$. Another possible choice for $\mathbf{g}^h$ is the $\mathbf{L}^2(\Gamma)$-projection of $\mathbf{g}$ onto $\mathbf{V}^h|_\Gamma$.

We consider *approximation by the boundary interpolant* in a little more detail. Let S_0^h and $\mathbf{V}^h$ be Lagrangian finite-element spaces and let

$$\{ q_j(\mathbf{x}) \}_{j=1}^J = \text{basis for } S_0^h$$

$$\{ \mathbf{v}_k(\mathbf{x}) \}_{k=1}^{\tilde{K}} = \text{basis for } \mathbf{V}^h$$

such that for $k = 1, \ldots, K < \tilde{K}$, $\mathbf{v}_k$ is associated with interior nodes so that $\mathbf{v}_k(\mathbf{x}) = \mathbf{0}$ for $\mathbf{x} \in \Gamma$ and $k \le K$. (This is not the best node numbering scheme in a practical implementation, but it simplifies our discussion.) For $k = K+1, \ldots, \tilde{K}$, $\mathbf{v}_k$ is associated with boundary nodes. Then, the $\mathbf{V}^h|_\Gamma$ interpolant $\mathbf{g}^h$ of $\mathbf{g}$ can be expressed in the form

$$\mathbf{g}^h(\mathbf{x}) = \sum_{k=K+1}^{\tilde{K}} \tilde{\beta}_k \mathbf{v}_k(\mathbf{x})$$

where $\tilde{\beta}_k$ is simply a component of $\mathbf{g}(\mathbf{x}_k)$, i.e. the $\tilde{\beta}_k$ are known and determined from the boundary data $\mathbf{g}$. Here, $\mathbf{x}_k$ denotes the position of the node on the boundary associated with the basis function $\mathbf{v}_k(\mathbf{x})$. The approximate velocity is then of the form

$$\mathbf{u}^h(\mathbf{x}) = \underbrace{\sum_{k=1}^{K} \beta_k \mathbf{v}_k(\mathbf{x})}_{\text{unknown coefficients}} + \underbrace{\sum_{k=K+1}^{\bar{K}} \tilde{\beta}_k \mathbf{v}_k(\mathbf{x})}_{\text{known coefficients}}$$

The non-linear discrete equations for $\vec{P}$ and $\vec{U}$, the vectors of unknowns having components α_j and β_k, respectively, are given by the system

$$\begin{pmatrix} A(\vec{U}) & B^{\mathrm{T}} \\ B & 0 \end{pmatrix} \begin{pmatrix} \vec{U} \\ \vec{P} \end{pmatrix} = \begin{pmatrix} \vec{F} \\ 0 \end{pmatrix}$$

where B is the same as in Eq. (12), A is the matrix having entries

$$A_{\ell k} = a(\mathbf{v}_k, \mathbf{v}_\ell) + \sum_{m=1}^{K} c(\mathbf{v}_m, \mathbf{v}_k, \mathbf{v}_\ell)\beta_m + \sum_{m=K+1}^{\bar{K}} c(\mathbf{v}_m, \mathbf{v}_k, \mathbf{v}_\ell)\tilde{\beta}_m$$

$$+ \sum_{m=K+1}^{\bar{K}} c(\mathbf{v}_k, \mathbf{v}_m, \mathbf{v}_\ell)\tilde{\beta}_m \quad \text{for } k, \ell = 1, \ldots, K$$

and $\vec{F}$ is the vector having components

$$F_\ell = (\mathbf{f}, \mathbf{v}_\ell) - a\left(\sum_{m=K+1}^{\bar{K}} \tilde{\beta}_m \mathbf{v}_m, \mathbf{v}_\ell \right)$$

$$- c\left(\sum_{m=K+1}^{\bar{K}} \tilde{\beta}_m \mathbf{v}_m, \sum_{m=K+1}^{\bar{K}} \tilde{\beta}_m \mathbf{v}_m, \mathbf{v}_\ell \right) \quad \text{for } \ell = 1, \ldots, K$$

Note that the entries of the matrix A are functions of the unknown coefficients $\beta_k, k = 1, \ldots, K$.

The error estimates for the inhomogeneous velocity boundary condition case are the same as (except for the constants) as the corresponding estimates for the analogous discretization in the homogeneous boundary condition case (Gunzburger and Hou, 1992).

F Other boundary conditions

1 Alternate weak formulations

Boundary conditions other than the velocity specification can be implemented as natural boundary conditions by a judicious choice for the viscous term. Recall that due to the incompressibility constraint $\nabla \cdot \mathbf{u} = 0$, we have that whenever $\nu = $ constant the four forms for the viscous term given in (3) are equivalent. (If $\nu \neq$ constant only the first form is correct.) Denote two segments of the boundary Γ by Γ_n and Γ_τ. These segments may be empty, are not necessarily connected, and are not necessarily disjoint; in fact they may be equal. Define the set

$$\mathbf{V}_g = \{\mathbf{v} \in \mathbf{H}^1(\Omega) \mid \mathbf{v} \cdot \mathbf{n} = g_n \text{ on } \Gamma_n \text{ and } \mathbf{n} \times \mathbf{u} \times \mathbf{n} = \mathbf{g}_\tau \text{ on } \Gamma_\tau \}$$

and the spaces

$$\mathbf{V}_z = \{\mathbf{v} \in \mathbf{H}^1(\Omega) \mid \mathbf{v} \cdot \mathbf{n} = 0 \text{ on } \Gamma_n \text{ and } \mathbf{n} \times \mathbf{u} \times \mathbf{n} = \mathbf{0} \text{ on } \Gamma_\tau \}$$

and

$$S = L_0^2(\Omega) \text{ if } \Gamma = \Gamma_n \quad \text{and} \quad S = L^2(\Omega) \quad \text{otherwise}$$

We then consider the weak formulation: for $i = 1$ or 2 or 3 or 4, given $\mathbf{f}$, g_n, $\mathbf{g}_\tau$, r and $\mathbf{s}$, seek $\mathbf{u} \in \mathbf{V}_g$ and $p \in S$ such that

$$a_i(\mathbf{u}, \mathbf{v}) + c(\mathbf{u}, \mathbf{u}, \mathbf{v}) + b(\mathbf{v}, p) = (\mathbf{f}, \mathbf{v}) + d(\mathbf{v}) \quad \forall \mathbf{v} \in \mathbf{V}_z$$

and

$$b(\mathbf{u}, q) = 0 \quad \forall q \in S$$

where

$$d(\mathbf{v}) = -\int_{\Gamma/\Gamma_n} r\mathbf{v} \cdot \mathbf{n} \, d\Gamma - \int_{\Gamma/\Gamma_\tau} \mathbf{s} \cdot \mathbf{v} \times \mathbf{n} \, d\Gamma$$

and where

$$a_1(\mathbf{u}, \mathbf{v}) = \frac{1}{2} \int_\Omega \nu \left[\nabla \mathbf{u} + (\nabla \mathbf{u})^\mathrm{T} \right] : \left[\nabla \mathbf{v} + (\nabla \mathbf{v})^\mathrm{T} \right] d\mathbf{x}$$

$$a_2(\mathbf{u}, \mathbf{v}) = \nu \int_\Omega \nabla \mathbf{u} : \nabla \mathbf{v} \, d\mathbf{x}$$

$$a_3(\mathbf{u}, \mathbf{v}) = \nu \int_\Omega (\nabla \times \mathbf{u}) \cdot (\nabla \times \mathbf{v}) \, d\mathbf{x}$$

$$a_4(\mathbf{u}, \mathbf{v}) = \nu \int_\Omega (\nabla \times \mathbf{u}) \cdot (\nabla \times \mathbf{v}) \, d\mathbf{x} + \nu \int_\Omega (\nabla \cdot \mathbf{u})(\nabla \cdot \mathbf{v}) \, d\mathbf{x}$$

For all four cases, the *essential boundary conditions* (conditions that are explicitly imposed on candidate solutions) are the same, namely the velocity components, i.e.

$$\mathbf{u} \cdot \mathbf{n} = g_n \quad \text{on } \Gamma_n$$

$$\mathbf{n} \times \mathbf{u} \times \mathbf{n} = \mathbf{g}_\tau \quad \text{on } \Gamma_\tau$$

2 Pressure, stress and vorticity boundary conditions

The *natural boundary conditions* (conditions that are not imposed on candidate solutions but which are automatically satsified by solutions of the weak formulation) differ according to the four cases of (3), i.e. we have for the four choices $a_i(\cdot,\cdot)$, $i = 1, \ldots, 4$:

$$-p + \nu \mathbf{n} \cdot [\nabla \mathbf{u} + (\nabla \mathbf{u})^{\mathrm{T}}] \cdot \mathbf{n} = r \quad \text{on } \Gamma/\Gamma_n$$
$$\nu \mathbf{n} \cdot [\nabla \mathbf{u} + (\nabla \mathbf{u})^{\mathrm{T}}] \times \mathbf{n} = \mathbf{s} \quad \text{on } \Gamma/\Gamma_\tau \tag{27}$$

$$-p + \nu \mathbf{n} \cdot \nabla \mathbf{u} \cdot \mathbf{n} = r \quad \text{on } \Gamma/\Gamma_n$$
$$\nu \mathbf{n} \cdot \nabla \mathbf{u} \times \mathbf{n} = \mathbf{s} \quad \text{on } \Gamma/\Gamma_\tau \tag{28}$$

$$-p = r \quad \text{on } \Gamma/\Gamma_n$$
$$-\mathbf{n} \times \omega \times \mathbf{n} = -\mathbf{n} \times (\nabla \times \mathbf{u}) \times \mathbf{n} = \mathbf{s} \quad \text{on } \Gamma/\Gamma_\tau \tag{29}$$

$$-p + \nu \nabla \cdot \mathbf{u} = r \quad \text{on } \Gamma/\Gamma_n$$
$$-\mathbf{n} \times \omega \times \mathbf{n} = -\mathbf{n} \times (\nabla \times \mathbf{u}) \times \mathbf{n} = \mathbf{s} \quad \text{on } \Gamma/\Gamma_\tau \tag{30}$$

For the first case of (27), the natural boundary conditions are the stresses. Specifically, we have that we are given the

velocity on $\Gamma_n \cap \Gamma_\tau$,
stress on $\Gamma/(\Gamma_n \cup \Gamma_\tau)$,
normal velocity, tangential stress on $\Gamma_n/(\Gamma_n \cap \Gamma_\tau)$, and
tangential velocity, normal stress on $\Gamma_\tau/(\Gamma_n \cap \Gamma_\tau)$.

For the second case of (28), the natural boundary conditions have no physical meaning, but are useful at quasi-unidirectional computational outflows. They can only be used in that case and whenever the velocity is specified on all other portions of the boundary.

For the third case of (29), the natural boundary conditions are the pressure and tangential vorticity. Unfortunately, the form $a_3(\cdot,\cdot)$ does not satisfy the fourth 'easy' condition unless $\mathbf{V}^h \subset \mathbf{Z}$, i.e. unless one has divergence-free approximate velocities. The boundary conditions in this case are that we are given the

velocity on $\Gamma_n \cap \Gamma_\tau$,
pressure, tangential vorticity on $\Gamma/(\Gamma_n \cup \Gamma_\tau)$,
normal velocity, tangential vorticity on $\Gamma_n/(\Gamma_n \cap \Gamma_\tau)$, and
tangential velocity, pressure on $\Gamma_\tau/(\Gamma_n \cap \Gamma_\tau)$.

For the fourth case of (30), the natural boundary conditions are again the pressure and tangential vorticity if $\nabla \cdot \mathbf{u} = 0$ holds up to the boundary on segments where the normal velocity is not specified, or if ν is small.

For these various combinations of boundary conditions, the error estimates that hold for velocity boundary conditions still hold for straight boundaries (Verfurth, 1985; Gunzburger, 1989). However, for some boundary conditions posed on curved boundaries, e.g. the combination of the normal velocity and the tangential stress, one gets a Babuska-type paradox, which may result in a loss of accuracy. This loss of accuracy can be remedied by using additional Lagrange multipliers on Γ. (The *Babuska paradox* states that the limit of solutions of problems posed on polygonal approximations to the boundary, as the polygonal approximations converge to the curved boundary, is not the solution of the problem posed on the curved domain.) See Verfurth (1985) and the references cited therein.

G Computation of the stress vector at the boundary

In many problems of practical interest one needs to determine the stress vector along portions of the boundary, especially where the velocity field is specified. The stress vector is given by

$$\tau = -p\mathbf{n} + \nu\mathbf{n} \cdot \left((\nabla\mathbf{u}) + (\nabla\mathbf{u})^{\mathrm{T}}\right) \tag{31}$$

so that one could determine an approximation of the stress vector by directly substituting into (31) approximations for the velocity and pressure. Two sources of additional errors arise when this technique is employed: the velocity field must be differentiated and the velocity derivatives and the pressure must be restricted to the boundary. At best, there will be at least a half-order of accuracy lost.

An alternative approach that leads to more accurate stress approximations is to solve for the approximate stress vector $\tau^h \in \mathbf{V}^h|_\Gamma$ (for simplicity we are assuming that the velocity is given on all of the boundary Γ) from the discrete equations

$$\int_\Gamma \tau^h \cdot \mathbf{v}^h \, \mathrm{d}\mathbf{x} = a(\mathbf{u}^h, \mathbf{v}^h) + c(\mathbf{u}^h, \mathbf{u}^h, \mathbf{v}^h) + b(\mathbf{v}^h, p^h) - (\mathbf{f}, \mathbf{v}^h) \quad \forall \mathbf{v}^h \in \mathbf{V}^h/\mathbf{V}_0^h \tag{32}$$

where $\mathbf{u}^h$ and p^h are approximate velocity and pressure fields, respectively, that have been determined from, e.g. (10) and (11). For example, if $\mathbf{V}^h$ is a Lagrangian finite-element space and if one uses a nodal basis, then the test functions in (32) are those basis functions associated with boundary nodes; note that the trial functions are the restriction of the functions in $\mathbf{V}^h$ to the boundary Γ. Approximations to the stress vector obtained from (32) can be shown to be optimally accurate in an appropriate sense. Certainly, force fields obtained from the approximate stresses determined from (32) are as accurate as one can expect, given the approximate velocity field $\mathbf{u}^h$ and pressure field p^h (see Gunzburger and Hou, 1992, for details).

H The effects of numerical integration

In practice, the integrals appearing in the discrete problem are not calculated exactly, but are evaluated using a quadrature rule. One wants to use a quadrature rule that is accurate enough so that the quality of the approximation is the same as when one uses exact integrations. On the other hand, the assembly cost is proportional to the number of quadrature points employed, so that one does not want to use a quadrature rule that is too accurate. For triangular elements, the rule of thumb for choosing an adequate quadrature rule is to choose a rule good enough to integrate the viscous term exactly; such a rule usually preserves the error estimates; an even less accurate rule is sufficient for quadrilateral elements. For a thorough discussion of the effects of numerical integration, the reader is referred to Crouzeix and Raviart (1973) and Jamet and Raviart (1973).

For example, for the Taylor–Hood pair, the viscous term has an integrand which is a quadratic polynomial so that the midside quadrature rule on triangles integrates it exactly. The convection term has a quintic integrand and is not integrated exactly. However, the use of the midside rule yields the same error estimates as does the use of exact integrations. Similarly, for linear velocity–constant pressure elements, the centroid rule integrates the constant integrand of the viscous term exactly, but does not exactly integrate the quadratic integrand of the convection term; however, this rule still results in optimal error estimates.

I Stabilized methods

All finite-element pairs satisfying the LBB condition involve different meshes and/or different degree polynomials for the velocity and pressure approximations. This creates considerable programming headaches. Methods have been developed and analysed for circumventing the LBB condition in the sense that equal-order polynomials with respect to the same meshes may be used for the velocity and pressure. Also, optimal error estimates have been obtained (see Brezzi and Pitkarnata 1984; Hughes *et al.*, 1986; Hughes and Franca, 1987; Brezzi and Douglas, 1988; Franca *et al.*, 1993).

These methods involve a modification of the discrete incompressibility constraint. For the Stokes problem (no convection terms) with homogeneous velocity boundary conditions, the simplest method is to replace the discrete incompressibility constraint

$$b(\mathbf{u}^h, q^h) = 0 \quad \forall q^h \in S_0^h$$

by

$$b(\mathbf{u}^h, q^h) + \alpha \sum_\Delta \int_\Delta \left(\nu \Delta \mathbf{u}^h - \nabla p^h + \mathbf{f} \right) \cdot \nabla q^h \, d\Omega = 0 \quad \forall q^h \in S_0^h$$

Refinements have been developed that yield symmetric discrete systems and other desirable properties.

J Penalty methods

It is not practical to uncouple the velocity calculation from the pressure calculation through the use of solenoidal elements. The coupling between the velocity and pressure may also be circumvented by other means, e.g. artificial compressibility (Chorin, 1967; Yanenko, 1971; Temam, 1979; Peyret and Taylor 1983 see also Chapter 2 of this book) and penalty methods (Hughes *et al.*, 1979; Glowinski, 1984; Fortin and Fortin, 1985; Cuvelier *et al.*, 1986; Segal, 1979). These involve a relaxation of the incompressibility constraint.

At the partial differential equation level, one may replace the Navier–Stokes equations by

$$-\nu\Delta\mathbf{u}_\epsilon + \mathbf{u}_\epsilon \cdot \nabla\mathbf{u}_\epsilon + \nabla p_\epsilon = \mathbf{f} \quad \text{in } \Omega$$

$$\nabla \cdot \mathbf{u}_\epsilon = -\epsilon p_\epsilon \quad \text{in } \Omega$$

and

$$\mathbf{u}_\epsilon = \mathbf{0} \quad \text{on } \Gamma$$

Eliminating the pressure, we get a system involving only the velocity:

$$-\nu\Delta\mathbf{u}_\epsilon + \mathbf{u} \cdot \nabla\mathbf{u}_\epsilon - \frac{1}{\epsilon}\nabla(\nabla \cdot \mathbf{u}_\epsilon) = \mathbf{f} \quad \text{in } \Omega$$

and

$$\mathbf{u}_\epsilon = \mathbf{0} \quad \text{on } \Gamma = \partial\Omega$$

After the velocity is calculated, one can determine the pressure from $p_\epsilon = -(1/\epsilon)\nabla \cdot \mathbf{u}_\epsilon$.

It can be shown that (Glowinski, 1984; Fortin and Fortin, 1985; Segal, 1979)

$$|\mathbf{u} - \mathbf{u}_\epsilon|_1 + \|p - p_\epsilon\|_0 = O(\epsilon)$$

Thus, if we apply a finite-element method to the penalty equations, we get that

$$|\mathbf{u}^h - \mathbf{u}_\epsilon|_1 = O(\epsilon) + O(h^s) \quad \text{for some } s$$

This estimate can be used to choose the values of ϵ and h so that the two terms on the right-hand-side are equilibrated. One can also apply the penalty method directly to discrete system, e.g.

$$\begin{pmatrix} A(\vec{U}) & B^{\mathrm{T}} \\ B & -\epsilon I \end{pmatrix} \begin{pmatrix} \vec{U} \\ \vec{P} \end{pmatrix} = \begin{pmatrix} \vec{F} \\ 0 \end{pmatrix}$$

One can also set up an iterated penalty method (Fortin and Fortin, 1985) such that the difference between the penalty solution and the exact solution can be made as small as one wishes for any $0 < \epsilon < 1$.

K Solution methods for the discrete equations

Equation (12), or equivalently, (10) and (11), is a finite-dimensional system of non-linear algebraic equations for the components of the vectors $\vec{U}$ and $\vec{P}$, i.e. for the coefficients in the expansions of $\mathbf{u}^h$ and p^h in terms of the basis functions of the corresponding finite-element spaces. In order to solve this system, some linearization procedure must be invoked. Here we discuss a few of these procedures. An alternative is to determine a steady-state solution as the large time limit of a transient problem by applying one of the methods discussed below in connection with time dependent problems.

1 Newton's method

Newton's method for the discrete system given by (10) and (11) is defined as follows. Given an initial guess $\mathbf{u}^{(0)} \in \mathbf{V}_0^h$, one generates the sequence $\{\mathbf{u}^{(m)}, p^{(m)}\}$ for $m = 1, 2, 3, \ldots$ by solving the sequence of *linear* problems

$$a(\mathbf{u}^{(m)}, \mathbf{v}^h) + c(\mathbf{u}^{(m-1)}, \mathbf{u}^{(m)}, \mathbf{v}^h) + c(\mathbf{u}^{(m)}, \mathbf{u}^{(m-1)}, \mathbf{v}^h) + b(\mathbf{v}^h, p^{(m)})$$
$$= (\mathbf{f}, \mathbf{v}^h) + c(\mathbf{u}^{(m-1)}, \mathbf{u}^{(m-1)}, \mathbf{v}^h) \qquad \forall \mathbf{v}^h \in \mathbf{V}_0^h$$

and

$$b(\mathbf{u}^{(m)}, q^h) = 0 \qquad \forall q^h \in S_0^h$$

Note that the initial guess is required to satisfy the boundary conditions but not the incompressibility constraint; also, no initial guess for the pressure is required.

For each m, these equations are equivalent to a linear system of algebraic equations. To see this, we again let $\{q_j(\mathbf{x})\}_{j=1}^J$ and $\{\mathbf{v}_k(\mathbf{x})\}_{k=1}^K$ denote bases for S_0^h and $\mathbf{V}_0^h$, respectively. Then, with $p^{(m)} = \sum_{j=1}^J \alpha_j^{(m)} q_j(\mathbf{x})$ and $\mathbf{u}^{(m)} = \sum_{k=1}^K \beta_k^{(m)} \mathbf{v}_k(\mathbf{x})$ we have for the vectors of unknowns $\vec{P}^{(m)}$ and $\vec{U}^{(m)}$

$$\begin{pmatrix} A^{(m)} & B^T \\ B & 0 \end{pmatrix} \begin{pmatrix} \vec{U}^{(m)} \\ \vec{P}^{(m)} \end{pmatrix} = \begin{pmatrix} \vec{F}^{(m)} \\ 0 \end{pmatrix}$$

where the matrix B is the same as in (12), the matrix $A^{(m)}$ has entries given by

$$A_{\ell k}^{(m)} = a(\mathbf{v}_k, \mathbf{v}_\ell) + c(\mathbf{u}^{(m-1)}, \mathbf{v}_k, \mathbf{v}_\ell) + c(\mathbf{v}_k, \mathbf{u}^{(m-1)}, \mathbf{v}_\ell) \qquad \text{for} \quad k, \ell = 1, \ldots, K$$

and the vector $\vec{F}$ has components given by

$$F_\ell = (\mathbf{f}, \mathbf{v}_\ell) + c(\mathbf{u}^{(m-1)}, \mathbf{u}^{(m-1)}, \mathbf{v}_\ell) \qquad \text{for} \quad \ell = 1, \ldots, K$$

Note that for each m, a different linear system must be solved. It can be shown that the Newton iterates $\{\mathbf{u}^{(m)}, p^{(m)}\}$ converge locally and quadratically to the solution $\{\mathbf{u}^h, p^h\}$ of (10) and (11) (Girault and Raviart, 1979; Karakashian, 1982; Gunzburger and Peterson, 1983)

Newton's method requires the evaluation and solution of a new Jacobian matrix at every iteration. At the price of a reduced convergence rate, one can lessen these costs by employing a quasi–Newton or update method such as Broyden's method or a sparse variant of this method (Matthies and Strang, 1977; Engelman *et al.*, 1981).

2 A simple iteration method

For sufficiently low values of the Reynolds number it can be shown that the following simple iterative method is *globally* and linearly convergent (Girault and Raviart, 1979; Karakashian, 1982; Gunzburger and Peterson, 1983). Given an initial guess $\mathbf{u}^{(0)} \in \mathbf{V}_0^h$, one generates the sequence $\{\mathbf{u}^{(m)}, p^{(m)}\}$ for $m = 1, 2, 3, \ldots$ by solving the sequence of *linear* problems

$$a(\mathbf{u}^{(m)}, \mathbf{v}^h) + c(\mathbf{u}^{(m-1)}, \mathbf{u}^{(m)}, \mathbf{v}^h) + b(\mathbf{v}^h, p^{(m)}) = (\mathbf{f}, \mathbf{v}^h) \qquad \forall \mathbf{v}^h \in \mathbf{V}_0^h$$

and

$$b(\mathbf{u}^{(m)}, q^h) = 0 \qquad \forall q^h \in S_0^h$$

Again, these equations are equivalent to a linear algebraic system of equations. Note that, as for Newton's method, the initial guess is required to satisfy the boundary conditions but not the incompressibility constraint; also, no initial guess for the pressure is required and a different linear system must be solved at each iteration.

A hybrid method that takes advantage of the global convergence properties of the simple iteration method and the local quadratic convergence property of Newton's method is easily defined. One merely starts the iteration with the simple iteration method, and once one is 'sufficiently' close to the solution, one switches over to Newton's method in order to speed up the convergence. This composite method may be implemented into an existing code using Newton's method since both methods have the form

$$a(\mathbf{u}^{(m)}, \mathbf{v}^h) + c(\mathbf{u}^{(m-1)}, \mathbf{u}^{(m)}, \mathbf{v}^h) + \sigma c(\mathbf{u}^{(m)}, \mathbf{u}^{(m-1)}, \mathbf{v}^h) + b(\mathbf{v}^h, p^{(m)})$$
$$= (\mathbf{f}, \mathbf{v}^h) + \sigma c(\mathbf{u}^{(m-1)}, \mathbf{u}^{(m-1)}, \mathbf{v}^h) \qquad \forall \mathbf{v}^h \in \mathbf{V}_0^h$$

and

$$b(\mathbf{u}^{(m)}, q^h) = 0 \qquad \forall q^h \in S_0^h$$

where $\sigma = 0$ for the simple iteration and $\sigma = 1$ for Newton's method.

3 Continuation methods

For moderate or high values of the Reynolds number, the convergence of all iterative methods for solving the non-linear system given by (10) and (11) hinges on the availability of a sufficiently accurate initial guess. One method for determining 'good enough' initial guesses is to use *continuation methods* (Ortega and Rheinboldt, 1970; Keller, 1978, 1987; Rheinboldt, 1980; den Heijer and Rheinboldt, 1981), i.e. to generate an initial guess by using information obtained at a lower value of the Reynolds number.

The simplest such method is to use the solution at a lower value of the Reynolds number as an initial guess for an iterative method at the desired Reynolds number. Thus, suppose one wishes to obtain a solution $\{\mathbf{u}^h(Re), p^h(Re)\}$ of (10) and (11) for $\nu = 1/Re$. One first obtains a solution $\{\mathbf{u}^h(Re_0), p^h(Re_0)\}$ of these equations for $\nu_0 = 1/Re_0$ where $Re_0 < Re$. One then uses $\mathbf{u}^h(Re_0)$ as the initial guess for the iteration that determines $\{\mathbf{u}^h(Re), p^h(Re)\}$, the solution at the desired value of the Reynolds number. If necessary, one solves a sequence of problems with increasing values of the Reynolds number $Re_0 < Re_1 < Re_2 < \ldots < Re_M = Re$; for each Re_m, $m \geq 1$, the iteration is started using the solution obtained for Re_{m-1}. By choosing Re_0 'sufficiently' small, one can always guarantee that the first solution $\{\mathbf{u}^h(Re_0), p^h(Re_0)\}$ is obtainable by the simple iteration method or the composite method. Furthermore, by choosing $Re_m - Re_{m-1}$, $m \geq 1$, 'sufficiently' small, the local convergence property of, say, Newton's method guarantees that the solution of the subsequent problems for $\{\mathbf{u}^h(Re_m), p^h(Re_m)\}$, $m \geq 1$, can also be obtained. Thus, in principle, the solution at 'any' value of the Reynolds number can be obtained by this continuation technique (see Gunzburger and Peterson, 1991).

A more sophisticated technique (Rheinboldt, 1980; den Heijer and Rheinboldt, 1981) that can also be adapted to work well in the presence of turning points and bifurcation points requires additional derivative (with respect to the Reynolds number) information. This scheme again starts with the solution of $\{\mathbf{u}^h(Re_0), p^h(Re_0)\}$ of (10) and (11) at the Reynolds number Re_0. One then solves the linear problem

$$a(\mathbf{u}', \mathbf{v}^h) + c(\mathbf{u}', \mathbf{u}^h(Re_0), \mathbf{v}^h) + c(\mathbf{u}^h(Re_0), \mathbf{u}', \mathbf{v}^h) + b(\mathbf{v}^h, p')$$
$$= \frac{1}{Re_0} a(\mathbf{u}^h(Re_0), \mathbf{v}^h) \quad \forall \mathbf{v}^h \in \mathbf{V}_0^h$$

and

$$b(\mathbf{u}', q^h) = 0 \quad \forall q^h \in S_0^h$$

for $\mathbf{u}' \in \mathbf{V}_0^h$ and $p' \in S_0^h$. (It is easily seen that $\mathbf{u}' = d\mathbf{u}/dRe$ evaluated at Re_0 and similarly for p'.) One then uses $\mathbf{u}^h|_{Re_0} + (Re_1 - Re_0)\mathbf{u}'$, i.e. a two-term Taylor polynomial, as the initial guess for the solution of (10) and (11) for $\nu = 1/Re_1$, where $Re_1 > Re_0$, by some iterative method. The process may be repeated for higher values of the Reynolds number. Again, this more sophisticated method can be guaranteed to yield solutions at 'any' value of the Reynolds number provided Re_0 and $Re_m - Re_{m-1}$ are both 'sufficiently' small.

4 Methods for solving linear systems

Ultimately, in obtaining solutions of (10) and (11), or even in the time-dependent case, very large, sparse, linear systems of algebraic equations must be solved. For two-dimensional simulations with low or moderate values of the Reynolds number, banded Gauss elimination solvers are often adequate. However, for three-dimensional problems and for high-Reynolds number two-dimensional problems (where fine grids are required), direct methods cannot be used due to the prohibitively large size of the linear systems. As a result, there have been numerous iterative methods proposed for solving the linear systems encountered in Navier–Stokes simulations. Included in this category are variants of conjugate gradient, multigrid, reduced basis, generalized minimum residual, and related algorithms. Although each of these algorithmic approaches has achieved some success, no general consensus has been reached with regard to which approach is 'best'.

L Time-dependent problems

1 Weak formulation

We consider the time-dependent problem given by (1), (2), (4) and (6) over the interval $[0, T]$. Loosely speaking, for the velocity we use the space $L^2(0, T; \mathbf{H}_0^1(\Omega))$ consisting of functions such that

$$\left(\int_0^T \|\mathbf{v}\|_1^2 \, dt \right)^{1/2} < \infty$$

i.e. having a bounded $L^2(0, T)$-norm of the $\mathbf{H}_0^1(\Omega)$-norm. Similarly, for the pressure we consider the space $L^2(0, T; L_0^2(\Omega))$ consisting of functions such that

$$\left(\int_0^T \|q\|_0^2 \, dt \right)^{1/2} < \infty$$

i.e. having a bounded $L^2(0, T)$-norm of the $L_0^2(\Omega)$-norm.

For the body force we require that $\mathbf{f} \in L^2(0, T; \mathbf{H}^{-1}(\Omega))$, with this space similarly defined; for the initial condition we require that

$$\mathbf{u}_0(\mathbf{x}) \in \mathbf{W} = \{ \mathbf{v} \in \mathbf{L}^2(\Omega) \; : \; \nabla \cdot \mathbf{u} = 0 \text{ a.e. in } \Omega \}$$

Then, a weak formulation of (1), (2), (4) and (6) is to seek $\mathbf{u} \in L^2(0, T; \mathbf{H}_0^1(\Omega))$ and $p \in L^2(0, T; L_0^2(\Omega))$ such that

$$(\mathbf{u}_t, \mathbf{v}) = \mathbf{F}(\mathbf{f}; \mathbf{u}, p; \mathbf{v})$$
$$= (\mathbf{f}, \mathbf{v}) - a(\mathbf{u}, \mathbf{v}) - c(\mathbf{u}, \mathbf{u}, \mathbf{v}) - b(\mathbf{v}, p) \quad \text{a.e. } t \in (0, T), \; \forall \, \mathbf{v} \in \mathbf{H}_0^1(\Omega)$$

$$b(\mathbf{u}, q) = 0 \quad \text{a.e. } t \in (0, T), \; \forall q \in L_0^2(\Omega)$$

and

$$\mathbf{u}(0, \mathbf{x}) = \mathbf{u}_0(\mathbf{x}) \quad \text{in } \mathbf{W}$$

2 Spatial semi-discretization

The spatial discretization is effected in exactly the same manner as that used in the steady-state case. Thus, one chooses finite dimensional subspaces $\mathbf{V}_0^h \in \mathbf{H}_0^1(\Omega)$ and $S_0^h \in L_0^2(\Omega)$ and then seeks functions $\mathbf{u}^h(t, \mathbf{x})$ and $p^h(t, \mathbf{x})$ such that for each $t \in [0, T]$, $\mathbf{u}^h \in \mathbf{V}_0^h$ and $p^h \in S_0^h$ and such that

$$(\mathbf{u}_t^h, \mathbf{v}^h) = \mathbf{F}(\mathbf{f}; \mathbf{u}^h, p^h; \mathbf{v}^h) \quad \text{a.e. } t \in (0, T), \ \forall \mathbf{v}^h \in \mathbf{V}_0^h \tag{33}$$

$$b(\mathbf{u}^h, q^h) = 0 \quad \text{a.e. } t \in (0, T), \ \forall q^h \in S_0^h \tag{34}$$

and

$$\mathbf{u}^h(0, \mathbf{x}) = \mathbf{u}_0^h(\mathbf{x}) \quad \text{for } \mathbf{x} \in \Omega \tag{35}$$

where $\mathbf{u}_0^h(\mathbf{x}) \in \mathbf{V}_0^h$ is an approximation to $\mathbf{u}_0(\mathbf{x})$, e.g. a projection onto that space. Any of the stable pair of spaces introduced for the steady-state case may be used for the spatial semi-discretization of a time dependent problem.

The system given by (33)–(35) is equivalent to an algebraic-(ordinary) differential system of equations. To see this, we again let $\{q_j(\mathbf{x})\}_{j=1}^J$ and $\{\mathbf{v}_k(\mathbf{x})\}_{k=1}^K$ denote bases for S_0^h and $\mathbf{V}_0^h$, respectively. Then, we have that

$$p^h(t, \mathbf{x}) = \sum_{j=1}^J \alpha_j(t) q_j(\mathbf{x}) \quad \text{and} \quad \mathbf{u}^h(t, \mathbf{x}) = \sum_{k=1}^K \beta_k(t) \mathbf{v}_k(\mathbf{x})$$

for some functions of time $\alpha_j(t)$ and $\beta_k(t)$. If $\vec{P}(t)$ and $\vec{U}(t)$ denote the vector of unknowns having components α_j and β_k, respectively, then the approximate problem is the algebraic-differential system

$$\begin{pmatrix} M\dot{\vec{U}} \\ \vec{0} \end{pmatrix} + \begin{pmatrix} A(\vec{U}) & B^{\mathrm{T}} \\ B & 0 \end{pmatrix} \begin{pmatrix} \vec{U} \\ \vec{P} \end{pmatrix} = \begin{pmatrix} \vec{F} \\ 0 \end{pmatrix} \tag{36}$$

where A and B are the matrices appearing in (12) and the mass matrix M has entries given by

$$M_{\ell k} = (\mathbf{v}_k, \mathbf{v}_\ell) = \int_\Omega \mathbf{v}_k \cdot \mathbf{v}_\ell \, \mathrm{d}\mathbf{x} \quad \text{for } k, \ell = 1, \ldots, K$$

It is usually not efficient to solve the system of (36), or equivalently, (33)–(35), by applying a library ODE routine. Instead, a finite-difference scheme is employed to effect the time discretization. In addition, various linearization schemes may be introduced in conjunction with the time discretization. In discussing these we will not explicitly choose a spatial discretization method; it is understood that any of the stable finite-element schemes disussed previously for the steady state case could be used in conjunction with the time discretization methods.

Throughout, δt denotes the time step and $\mathbf{u}^n$ and p^n respectively denote the approximations to the solutions of (33)–(35) at time $n\delta t$, i.e. to $\mathbf{u}^h(n\delta t, \mathbf{x})$ and $p^h(n\delta t, \mathbf{x})$. Also, $\mathbf{f}^n = \mathbf{f}(n\delta t, \mathbf{x})$.

3 Single-step fully implicit schemes

The simplest useful time discretization method is given by the *backward Euler method* or backward differentiation method which is defined as follows: given $\mathbf{u}^0$ (which may be chosen equal to $\mathbf{u}_0^h$), $\{\mathbf{u}^n, p^n\}$ for $n = 1, \ldots, N = T/\delta t$ is determined from

$$\frac{1}{\delta t}(\mathbf{u}^{n+1}, \mathbf{v}^h) - \mathbf{F}(\mathbf{f}^{n+1}; \mathbf{u}^{n+1}, p^{n+1}; \mathbf{v}^h) = \frac{1}{\delta t}(\mathbf{u}^n, \mathbf{v}^h) \quad \forall \mathbf{v}^h \in \mathbf{V}_0^h$$

and

$$b(\mathbf{u}^{n+1}, q^h) = 0 \quad \forall q^h \in S_0^h$$

This method is unconditionally stable and is first-order accurate in time, i.e. it has an error of $O(\delta t)$. Note that no initial condition for the pressure is needed to implement this scheme.

A nominally second-order accurate, unconditionally stable single-step fully implicit method is given by the *Crank–Nicolson*, or trapezoidal, method which is defined as follows: given a solenoidal $\mathbf{u}^0$ (which may be chosen equal to $\mathbf{u}_0$), $\{\mathbf{u}^{n+1}, p^{n+1}\}$ for $n = 1, \ldots, N = T/\delta t$ is determined from

$$\frac{1}{2\delta t}(\mathbf{u}^{n+1}, \mathbf{v}^h) - \mathbf{F}\left(\frac{\mathbf{f}^{n+1} + \mathbf{f}^n}{2}; \frac{\mathbf{u}^{n+1} + \mathbf{u}^n}{2}, \frac{p^{n+1} + p^n}{2}; \mathbf{v}^h\right)$$
$$= \frac{1}{2\delta t}(\mathbf{u}^n, \mathbf{v}^h) \quad \forall \mathbf{v}^h \in \mathbf{V}_0^h$$

and

$$b(\mathbf{u}^{n+1}, q^h) = 0 \quad \forall q^h \in S_0^h$$

Note that this scheme requires the definition of an initial pressure approximation $p^{(0)}$. Unfortunately, most common methods for defining such approximations result in a loss of accuracy. An alternative time discretization of the Navier–Stokes equations is given

$$\frac{1}{2\delta t}(\mathbf{u}^{n+1}, \mathbf{v}^h) - \mathbf{F}\left(\frac{\mathbf{f}^{n+1} + \mathbf{f}^n}{2}; \frac{\mathbf{u}^{n+1} + \mathbf{u}^n}{2}, p^{n+\frac{1}{2}}; \mathbf{v}^h\right)$$
$$= \frac{1}{2\delta t}(\mathbf{u}^n, \mathbf{v}^h) \quad \forall \mathbf{v}^h \in \mathbf{V}_0^h$$

6 A class of semi-implicit multistep methods

Multistep methods that are conditionally stable can be defined for which the coefficient matrices of the linear system that have to be solved at every time step does not change from time step to time step. For example, we have the backward-Euler-based method defined by

$$\frac{1}{\delta t}\left(\mathbf{u}^{n+1} - \mathbf{u}^n, \mathbf{v}^h\right) + a(\mathbf{u}^{n+1}, \mathbf{v}^h)$$
$$+ c(\mathbf{u}^n, \mathbf{u}^n, \mathbf{v}^h) + b(\mathbf{v}^h, p^{n+1}) = (\mathbf{f}^{n+1}, \mathbf{v}^h) \quad \forall \mathbf{v}^h \in \mathbf{V}_0^h$$

and

$$b(\mathbf{u}^{n+1}, q^h) = 0 \quad \forall q^h \in S_0^h$$

and the second-order method defined by

$$\frac{1}{2\delta t}\left(3\mathbf{u}^{n+1} - 4\mathbf{u}^n + \mathbf{u}^{n-1}, \mathbf{v}^h\right) + a(\mathbf{u}^{n+1}, \mathbf{v}^h)$$
$$+ c(2\mathbf{u}^n - \mathbf{u}^{n-1}, 2\mathbf{u}^n - \mathbf{u}^{(n-1)}, \mathbf{v}^h) + b(\mathbf{v}^h, p^{n+1})$$
$$= (\mathbf{f}^{n+1}, \mathbf{v}^h) \quad \forall \mathbf{v}^h \in \mathbf{V}_0^h$$

and

$$b(\mathbf{u}^{n+1}, q^h) = 0 \quad \forall q^h \in S_0^h$$

Higher-order methods may also be defined based on higher-order implicit multistep methods. Clearly, in these methods, the coefficient matrix of the linear system is independent of n, the time-step counter. Starting methods can also be devised that use the same coefficient matrix as well. These methods are optimally accurate but are only conditionally stable. The backward-Euler-based method is stable only if $\delta t \leq Ch$ and the second-order method is stable only if $\delta t \leq Ch^{4/5}$ for some constant C. Again, see Baker *et al.* (1982) for details.

M Pressure Poisson equation formulation

In the case $\nu = $ constant, taking the divergence of (1) and using (2) yields the *pressure Poisson equation*

$$-\Delta p = -\nabla \cdot \mathbf{f} + (\nabla \mathbf{u}) : (\nabla \mathbf{u})^{\mathrm{T}} \tag{37}$$

where the colon denotes the scalar product operation between two tensors. Thus, (1) and (37) provide an alternative set of governing equations for viscous, incompressible, laminar flows in terms of the velocity and pressure. Since (37) involves a differentiation of the Navier–Stokes equations, one needs to introduce an additional boundary condition. For example, on solid walls along which the velocity is specified, a boundary condition for the pressure is given by the normal component of the Navier–Stokes equation of (1). For the case of homogeneous velocity boundary conditions, i.e. (6), one would use (Gresho and Sani, 1988)

$$\frac{\partial p}{\partial n} = \nu \mathbf{n} \cdot \Delta \mathbf{u} - \mathbf{n} \cdot \mathbf{f} \tag{38}$$

as a boundary condition for the pressure when solving (37).

The pressure Poisson equation formulation is used in the following manner. Given an initial velocity field, one determines an initial pressure field from (37) and (38). This pressure field is then used to compute a preliminary velocity field at the next time step from a discretization of the Navier–Stokes equations of (1). In general, the preliminary velocity field so obtained may not satisfy the incompressibility constraint of (2). Thus, the velocity field at the new time step is obtained by projecting the preliminary velocity field onto the space of solenoidal (or at least discretely solenoidal) functions. The process is then repeated, starting with the determination of the pressure field at the new time step from (37) and (38). For a modern treatment of such projections schemes, see Löhner (1993) and the references cited therein.

III FORMULATIONS INVOLVING THE STREAMFUNCTION

A Mathematical models involving the streamfunction

1 The streamfunction-vorticity formulation for plane flows

A plane flow is one such that in a cartesian coordinate system (x, y, z), the velocity component in the z-direction vanishes throughout the flow and the remaining velocity components and the pressure are independent of z. Equation (2) then infers the existence of a streamfunction $\psi(t, x, y)$ such that

$$u(t, x, y) = \frac{\partial \psi}{\partial y} \quad \text{and} \quad v(t, x, y) = -\frac{\partial \psi}{\partial x} \tag{39}$$

where u and v denote the velocity components in the x and y directions, respectively. For plane flows, the vorticity has only one non-vanishing component which is given by $\omega = \partial v/\partial x - \partial u/\partial y$ so that, using (39),

$$\Delta \psi = \frac{\partial^2 \psi}{\partial x^2} + \frac{\partial^2 \psi}{\partial y^2} = -\omega \tag{40}$$

Assuming that $\nu = $ constant, taking the curl of (1) yields

$$\frac{\partial \omega}{\partial t} - \nu \left(\frac{\partial^2 \omega}{\partial x^2} + \frac{\partial^2 \omega}{\partial y^2} \right) + \frac{\partial \psi}{\partial y} \frac{\partial \omega}{\partial x} - \frac{\partial \psi}{\partial x} \frac{\partial \omega}{\partial y} = \frac{\partial f_2}{\partial x} - \frac{\partial f_1}{\partial y} \tag{41}$$

where f_1 and f_2 respectively denote the x and y components of $\mathbf{f}$.

Initial conditions are required for the vorticity; these may be obtained from initial conditions for the velocity using the definition of the vorticity. Thus, we have that

$$\omega(0, x, y) = \frac{\partial v_0}{\partial x} - \frac{\partial u_0}{\partial y} \qquad \text{for } (x, y) \in \Omega \tag{42}$$

where u_0 and v_0 denote the x and y components of the initial velocity field.

Boundary conditions are also obtained from the velocity whenever the latter is specified at the boundary. If g_n and g_τ denote the normal and tangential components of the velocity data at the boundary, we have, for *simply connected* domains, using (39), that

$$\psi = \int_{x_0}^{x} g_n \, dx \qquad \text{and} \qquad \frac{\partial \psi}{\partial n} = -g_\tau \tag{43}$$

where the integral is taken along a portion of the boundary on which the velocity is specified, starting at a fixed point x_0 on the boundary. Other boundary conditions, e.g. involving components of the stress vector, can also be formulated in terms of the streamfunction and vorticity.

Equations (40)–(43) are the *streamfunction-vorticity formulation for plane, incompressible, laminar flow*. A streamfunction, or vector potential, can also be defined in three-dimensions along with a vector-valued vorticity. However, formulations using these variables have not achieved great popularity, especially compared to the plane flow case for which the streamfunction-vorticity formulation is in very common use. However, in the special case of axisymmetric flows, i.e. flows such that in a cylindrical coordinate system (r, θ, z), the velocity component in the θ direction vanishes throughout the flow and the remaining velocity components and the pressure are independent of θ, again streamfunction-vorticity formulations are very popular.

2 The streamfunction formulation for plane flows

Eliminating ω from (40) and (41) yields a single fourth-order partial differential equation for the streamfunction:

$$-\frac{\partial}{\partial t}(\Delta \psi) + \nu \Delta^2 \psi - \frac{\partial \psi}{\partial y}\frac{\partial}{\partial x}(\Delta \psi) + \frac{\partial \psi}{\partial x}\frac{\partial}{\partial y}(\Delta \psi) = \frac{\partial f_2}{\partial x} - \frac{\partial f_1}{\partial y} \tag{44}$$

where the biharmonic operator is defined by

$$\Delta^2 \psi = \frac{\partial^4 \psi}{\partial x^4} + 2\frac{\partial^4 \psi}{\partial x^2 \partial y^2} + \frac{\partial^4 \psi}{\partial y^4}$$

Initial conditions are required for the Laplacian of the streamfunction. Since $\Delta \psi = -\omega$ for plane flows, the required initial conditions for $\Delta \psi$ can be obtained from (42). Boundary conditions for the streamfunction are given by, e.g. (43).

Equation (44), along with initial and boundary conditions, constitutes the *streamfunction formulation for plane, incompressible, laminar flow.*

The main advantage of the streamfunction formulation based on (44) is that one has to deal with only a single unknown field. However, the fact that (44) involves fourth derivatives presents difficulties. Since conforming finite-element methods require the use of continuously differentiable approximating functions, there has been considerable interest in developing non-conforming discretizations. Although numerous conforming and non-conforming finite element algorithms exist for determining approximate solutions of the streamfunction formulation, these have not become sufficiently popular to warrant any further discussion here. For more details on finite-element methods for the streamfunction formulation, one may consult Cayco and Nicolaides (1990) and the references cited therein.

B Finite-element methods for the streamfunction-vorticity formulation

The proper definition of a weak formulation of the streamfunction-vorticity equations requires the introduction of some additional function spaces. Here, we directly define the finite-element problem, thus avoiding these complications. We only consider the steady-state case, i.e. the term ω_t is omitted from (41), and also assume that homogeneous velocity boundary conditions hold on all of Γ.

1 Finite-element algorithms

We choose a finite-element space $V^h \subset C(\Omega)$ and then define $V_0^h = V^h \cap H_0^1(\Omega)$. We then seek $\psi^h \in V_0^h$ and $\omega^h \in V^h$ such that

$$\int_\Omega \omega^h \zeta^h \, d\mathbf{x} - \int_\Omega \nabla \psi^h \cdot \nabla \zeta^h \, d\mathbf{x} = 0 \qquad \forall \zeta^h \in V^h \qquad (45)$$

and

$$\nu \int_\Omega \nabla \omega^h \cdot \nabla \phi^h \, d\mathbf{x} + \int_\Omega \left(\frac{\partial \psi^h}{\partial y} \frac{\partial \omega^h}{\partial x} - \frac{\partial \psi^h}{\partial x} \frac{\partial \omega^h}{\partial y} \right) \phi^h \, d\mathbf{x}$$
$$= \int_\Omega \left(\frac{\partial f_2}{\partial x} - \frac{\partial f_1}{\partial y} \right) \phi^h \, d\mathbf{x} \qquad \forall \phi^h \in V_0^h \qquad (46)$$

For details, see Girault and Raviart (1979, 1986), Gunzburger (1989), Gunzburger and Peterson (1988) and Tezduyar *et al.* (1988).

2 Error estimates

Optimal error estimates can be obtained for continuously differentiable stream function approximating spaces. However, in this case, one could more easily solve the fourth-order problem for the streamfunction. If one uses merely continuous finite-element spaces for both the streamfunction and vorticity, then one can only obtain sub-optimal error estimates. For example, if one uses continuous piecewise polynomials of degree $\leq k$, then it can be shown that (Girault and Raviart, 1979, 1986)

$$|\psi - \psi^h|_1 + \|\omega - \omega^h\|_0 \leq Ch^{k-\frac{1}{2}} |\ln h|^\sigma \tag{47}$$

where $\sigma = 1$ for $k = 1$ and $\sigma = 0$ for $k > 1$, and where C is a constant independent of h. This estimate is not optimal, with regard to the power of h, for both the derivatives of the streamfunction and for the vorticity.

There is computational and some theoretical evidence that the estimate given in (47) is not sharp with respect to the streamfunction (Gunzburger and Peterson, 1988). In fact, for the linear Stokes problem, it can be shown (Fix *et al.*, 1984) that

$$|\psi - \psi^h|_1 \leq Ch^{k-\epsilon} \tag{48}$$

where $\epsilon = 0$ for $k > 1$ and $\epsilon > 0$ can be chosen arbitrarily small for $k = 1$. (It is likely that this estimate also holds for the non-linear case.) However, the estimate of (47) is probably sharp with respect to the vorticity error and thus vorticity approximations are, in general, very poor. For example, if piecewise linear elements are used for both the vorticity and streamfunction, the root mean square error in the vorticity is roughly of $O(h^{\frac{1}{2}})$ while, according to (48), the root mean square error in the derivatives of the streamfunction is roughly of $O(h)$.

It is possible, and, indeed, it may be desirable, to use different degree polynomials for the streamfunction and vorticity approximations (Habashi *et al.*, 1987; Hafez *et al.*, 1987; Peeters *et al.*, 1987).

C Recovery of the primitive variables

Once approximations to the vorticity and streamfunction have been obtained, one often wishes to obtain approximations to the velocity and pressure.

1 Recovery of the velocity

Velocity approximations can be obtained by directly applying (39) to the discrete streamfunction within each element. Thus, we obtain the approximate velocity components

$$u^h|_\Delta = \left.\frac{\partial \psi^h}{\partial y}\right|_\Delta \quad \text{and} \quad v^h|_\Delta = -\left.\frac{\partial \psi^h}{\partial x}\right|_\Delta$$

Equations (46) and (47) then imply that

$$\|\mathbf{u} - \mathbf{u}^h\|_0 \leq Ch^{k-\frac{1}{2}} |\ln h|^\sigma$$

for the non-linear case and

$$\|\mathbf{u} - \mathbf{u}^h\|_0 \leq Ch^{k-\epsilon}$$

for the linear case. The last estimate probably holds for the non-linear case as well.

2 Recovery of the pressure

The recovery of the pressure is a little more problematical. From (1) (for the steady-state case), one has that

$$\nabla \mathcal{H} = -\nu \begin{pmatrix} \dfrac{\partial \omega}{\partial y} \\ -\dfrac{\partial \omega}{\partial x} \end{pmatrix} + \begin{pmatrix} v \\ -u \end{pmatrix} \omega \tag{49}$$

where $\mathcal{H} = p + (\mathbf{u} \cdot \mathbf{u})/2$ is the total pressure. Clearly, once $\mathcal{H}$ and $\mathbf{u}$ are determined, so is the pressure p.

Given approximations for the velocity and vorticity, an approximation of the pressure p at any point $\mathbf{x} \in \bar{\Omega}$ may be recovered from (49) in various ways. For example, one could compute a line integral from a fixed point $\mathbf{x}_0$ on the boundary (at which we are free to set $\mathcal{H} = 0$) to the point $\mathbf{x}$, i.e.

$$\mathcal{H}^h(\mathbf{x}) = \int_{\mathbf{x}_0}^{\mathbf{x}} \left(-\nu \frac{\partial \omega^h}{\partial y} + v^h \omega^h \right) s_1 \, ds + \int_{\mathbf{x}_0}^{\mathbf{x}} \left(\nu \frac{\partial \omega^h}{\partial x} - u^h \omega^h \right) s_2 \, ds$$

where ds is the length element along the integration path and s_1 and s_2 are the x and y components of the tangent vector to that path. Then, of course, $p^h = \mathcal{H}^h - (\mathbf{u}^h \cdot \mathbf{u}^h)/2$. However, this approach is prone to errors since one must use derivatives of the approximate vorticity ω^h to evaluate the right-hand-side.

A second approach would start with taking the divergence of (49) in order to obtain a Poisson equation for the total pressure, i.e.

$$-\Delta \mathcal{H} = -\begin{pmatrix} v \\ -u \end{pmatrix} \cdot \nabla \omega - \omega^2$$

This approach, even after a weak formulation is introduced, not only requires derivatives of the approximate vorticity and velocity, but also requires a boundary condition for $\mathcal{H}$. The latter is usually chosen to be a Neumann condition which may be derived by taking the normal component of (49). The first of these problems leads to inaccuracies, and the second is difficult to justify.

A third approach is to embed (49) into a linear Stokes problem, i.e. to solve for $\mathbf{w}$ and $\mathcal{H}$ such that

$$-\Delta\mathbf{w} + \nabla\mathcal{H} = -\nu \begin{pmatrix} \dfrac{\partial\omega}{\partial y} \\ -\dfrac{\partial\omega}{\partial x} \end{pmatrix} + \begin{pmatrix} v \\ -u \end{pmatrix}\omega \quad \text{in } \Omega \tag{50}$$

$$\nabla\cdot\mathbf{w} = 0 \quad \text{in } \Omega \tag{51}$$

and

$$\mathbf{w} = \mathbf{0} \quad \text{on } \Gamma \tag{52}$$

It can be shown that $\mathbf{w}$ and $\mathcal{H}$ satisfying (50)–(52) also satisfy $\mathbf{w} = \mathbf{0}$ and (49). The advantages of using this approach is that no boundary condition need be specified for $\mathcal{H}$ and that a weak form of (50)–(52) may be devised that does not require derivatives of the vorticity. Indeed, an approximate solution $\mathbf{w}^h \in \mathbf{V}_0^h$ and $\mathcal{H}^h \in S_0^h$ may be found from

$$\int_\Omega \nabla\mathbf{w}^h : \nabla\tilde{\mathbf{v}}^h \, d\mathbf{x} - \int_\Omega \mathcal{H}^h \nabla\cdot\tilde{\mathbf{v}}^h \, d\mathbf{x}$$
$$= \nu\int_\Omega \omega^h\left(\frac{\partial\tilde{u}^h}{\partial y} - \frac{\partial\tilde{v}^h}{\partial x}\right) d\mathbf{x} + \int_\Omega \omega^h(v^h\tilde{u}^h - u^h\tilde{v}^h)\, d\mathbf{x} \quad \forall\tilde{\mathbf{v}}^h \in \mathbf{V}_0^h \tag{53}$$

and

$$\int_\Omega q^h\nabla\cdot\mathbf{w}^h \, d\mathbf{x} = 0 \quad \forall q^h \in S_0^h \tag{54}$$

where u^h and v^h are the components of the approximate velocity $\mathbf{u}^h$ and $\tilde{u}^h$ and $\tilde{v}^h$ are the components of the test function $\tilde{\mathbf{v}}^h$. For the finite-element spaces $\mathbf{V}_0^h \subset \mathbf{H}_0^1(\Omega)$ and $S_0^h \subset L_0^2(\Omega)$ one may choose any of the stable pairs of finite-element spaces discussed in connection with the primitive variable formulation. It can be shown that, as $h \to 0$, $\mathbf{w}^h \to \mathbf{0}$ and that the root mean square, i.e. the $L^2(\Omega)$, error in the approximate pressure $p^h = \mathcal{H}^h - \mathbf{u}^h\cdot\mathbf{u}^h/2$ cannot be better than the corresponding error in the vorticity. Thus, provided the spaces used in (53) and (54) can approximate the pressure to $O(h^k)$, one has that the approximate pressure obtained from that system satisfies

$$\|p - p^h\|_0 \leq Ch^{k-\frac{1}{2}}|\ln h|^\sigma \, ;$$

see (47). The loss of accuracy in the pressure is due to the fact that in the right-hand-side of (53) one uses the approximate vorticity. As an example, consider the use of piecewise linear elements for the vorticity and streamfunction, i.e. $k = 1$. Then, in (53) and (54), one should merely use finite element spaces that yield at best first-order accurate pressure approximations to the Stokes equations in the case of a known right-hand-side. See Cayco and Nicolaides (1986) and Gunzburger (1989) for details.

D Multiply connected domains

1 Boundary conditions

For the streamfunction-vorticity formulation, *multiply connected* domains pose additional difficulties. (The discussion so far for this formulation was for simply connected domains.) For simplicity, we only consider the steady-state case; we also assume that the boundary Γ of the flow domain Ω consists of the disjoint segments Γ_i, $i = 0, \dots, m$, and that the velocity is specified on all of these segments, i.e. we have that

$$\mathbf{u} = \mathbf{g}_i \quad \text{on } \Gamma_i \quad \text{for } i = 0, \dots, m \tag{55}$$

where $\mathbf{g}_i$ for $i = 0, \dots, m$ are given functions. A necessary condition for a streamfunction to exist is that

$$\int_{\Gamma_i} \mathbf{g}_i \cdot \mathbf{n} \, d\mathbf{x} = 0 \qquad \text{for } i = 0, \dots, m \tag{56}$$

These simply state that there is no net mass flow through any of the boundary pieces.

In terms of the streamfunction, (55) and (56) imply that

$$\psi = q_0 \quad \text{and} \quad \frac{\partial \psi}{\partial n} = -\mathbf{g}_0 \cdot \mathbf{s} \quad \text{on } \Gamma_0 \tag{57}$$

and

$$\psi = q_i + a_i \quad \text{and} \quad \frac{\partial \psi}{\partial n} = -\mathbf{g}_i \cdot \mathbf{s} \quad \text{on } \Gamma_i \text{ for } i = 1, \dots, m \tag{58}$$

where $\mathbf{s}$ denotes the unit counterclockwise tangent vector to Γ. For $i = 0, \dots, m$, q_i is a function such that

$$\frac{\partial q_i}{\partial s} = \mathbf{g}_i \cdot \mathbf{n} \qquad \text{on } \Gamma_i \text{ for } i = 1, \dots, m$$

where $\partial(\cdot)/\partial s$ denotes the tangential derivative in the direction of $\mathbf{s}$. These functions may be determined in terms of integrals of $\mathbf{g}_i \cdot \mathbf{n}$ which may often be explicitly evaluated or, in any case, can be evaluated to arbitrary accuracy through numerical integration procedures.

In (58), a_i for $i = 1, \dots, m$ are constants that are to be determined. Their appearance results from the well-known fact that the streamfunction is uniquely determined up to a *single* additive constant. We have fixed that constant by fixing the streamfunction at a point on Γ_0. Then, a_i denotes the unknown mass flow across any curve joining Γ_i and Γ_0.

Owing to the appearance of the unknown constants a_i in (58), additional data must be supplied. These data, which may be derived from the requirement that the pressure be a single-valued function, are given by

$$\int_{\Gamma_i} \left(\nu \frac{\partial \omega}{\partial n} - \omega \mathbf{g}_i \cdot \mathbf{n} + \mathbf{f} \cdot \mathbf{s} \right) dx = 0 \qquad \text{for } i = 1, \ldots, m \qquad (59)$$

Thus, the streamfunction-vorticity formulation for the steady-state, multiply connected case we are considering is given by (40) and (41) (with the time derivative term omitted from the latter) along with (57)–(59).

2 Finite-element discretizations

We choose a finite-element space $V^h \subset C(\Omega)$ and then define

$$V_a^h = \{ \phi^h \in V^h \mid \phi^h = q_0^h \quad \text{on } \Gamma_0$$

$$\text{and} \quad \phi^h = q_i^h + a_i \quad \text{on } \Gamma_i, i = 1, \ldots, m \}$$

and

$$V_0^h = \{ \phi^h \in V^h \mid \phi^h = 0 \quad \text{on } \Gamma_0 \quad \text{and} \quad \phi^h = a_i \quad \text{on } \Gamma_i, i = 1, \ldots, m \}$$

where, for $i = 1, \ldots, m$, $a_i \in \mathbb{R}$ is arbitrary and $q_i^h \in V^h|_{\Gamma_i}$ are suitably chosen approximations, e.g. boundary interpolants, to the corresponding q_i. We then seek $\psi^h \in V_a^h$, $\omega^h \in V^h$, and for $i = 1, \ldots, m$, $a_i \in \mathbb{R}$ such that

$$\int_\Omega \omega^h \zeta^h \, dx - \int_\Omega \nabla \psi^h \cdot \nabla \zeta^h \, dx = \sum_{i=0}^m \int_{\Gamma_i} \zeta^h \mathbf{g}_i \cdot \mathbf{s} \, dx \qquad \forall \zeta^h \in V^h \qquad (60)$$

and

$$\nu \int_\Omega \nabla \omega^h \cdot \nabla \phi^h \, dx + \int_\Omega \left(\frac{\partial \psi^h}{\partial y} \frac{\partial \phi^h}{\partial x} - \frac{\partial \psi^h}{\partial x} \frac{\partial \phi^h}{\partial y} \right) \omega^h \, dx$$

$$= -\int_\Omega \left(\frac{\partial f_2}{\partial x} - \frac{\partial f_1}{\partial y} \right) \phi^h \, dx + \sum_{i=1}^m \int_{\Gamma_i} \phi^h \mathbf{f} \cdot \mathbf{s} \, dx \qquad \forall \phi^h \in V_0^h \qquad (61)$$

For this formulation, the normal derivative boundary conditions of (57) and (58) and the auxiliary conditions of (59) are naturally satisfied. The other boundary conditions of (57) and (58), i.e. the ones that specify the streamfunction, are essential boundary conditions to the weak formulation of (60) and (61).

Although no theoretical results exist, computational evidence (Gunzburger and Peterson, 1988) indicates that the errors resulting from the use of the finite element problem of (60) and (61) are similar to those obtained in the simply connected case.

3 Solution algorithms

A direct application of Newton's method, or some other linearization method, for the solution of the nonlinear discrete system of (60) and (61) will require the use of *semi-local* basis functions, i.e. finite-element functions that are non-zero on all of a boundary segment Γ_i, $i = 1, \ldots, m$. This results from the fact that on each of these boundary segments the streamfunction is *a priori* known only up to an additive constant a_i which must be determined as part of the solution process. Some care must therefore be exercised in the node numbering scheme so that the semi-local property of some of the basis functions does not increase the bandwidth of the coefficient matrices in the linear systems one must solve.

An algorithm using only *local* basis functions can also be devised. The success of this algorithm rests on the facts that the conditions of (57)–(59) are all *linear* in ψ and ω and also on the fact that, after one linearizes (60) and (61), *linear* systems of algebraic equations are encountered. We describe the algorithm in the context of Newton's method. Suppose that one has in hand the Newton iterate $\{\psi^{(\ell-1)}, \omega^{(\ell-1)}, [a_i^{(\ell-1)}, i = 1, \ldots, m]\}$. Then, the next Newton iterate $\{\psi^{(\ell)}, \omega^{(\ell)}, [a_i^{(\ell)}, i = 1, \ldots, m]\}$ is determined as follows.

(i) For $i = 1, \ldots, m$ and $j = 0, \ldots, m$, choose numbers α_{ij} that are arbitrary except for the requirement that the $(m+1) \times (m+1)$ matrix whose j-th row is given by $(1, \alpha_{1j}, \ldots, \alpha_{mj})$ is non-singular.
(ii) Choose a finite-element space $V^h \subset C(\Omega)$ and then, for each $j, j = 0, \ldots, m$, define the sets

$$V_j^h = \{\, \phi^h \in V^h \mid \phi^h = q_0^h \text{ on } \Gamma_0$$

$$\text{and } \phi^h = q_i^h + \alpha_{ij} \text{ on } \Gamma_i, i = 1, \ldots, m \,\}$$

and the space

$$V_0^h = \{\, \phi^h \in V^h \mid \text{ and } \phi^h = 0 \text{ on } \Gamma_i, i = 0, \ldots, m \,\}$$

We then seek $\psi_j \in V_j^h$ and $\omega_j \in V^h$ such that

$$\int_\Omega \omega_j \zeta^h \, d\mathbf{x} - \int_\Omega \nabla \psi_j \cdot \nabla \zeta^h \, d\mathbf{x} = \sum_{i=0}^m \int_{\Gamma_i} \zeta^h \mathbf{g}_i \cdot \mathbf{s} \, d\mathbf{x} \qquad \forall \zeta^h \in V^h \qquad (62)$$

and

$$\nu \int_\Omega \nabla \omega_j \cdot \nabla \phi^h \, d\mathbf{x} + \int_\Omega \left(\frac{\partial \psi^{(\ell-1)}}{\partial y} \frac{\partial \phi^h}{\partial x} - \frac{\partial \psi^{(\ell-1)}}{\partial x} \frac{\partial \phi^h}{\partial y} \right) \omega_j \, d\mathbf{x}$$

$$+ \int_\Omega \left(\frac{\partial \psi_j}{\partial y} \frac{\partial \phi^h}{\partial x} - \frac{\partial \psi_j}{\partial x} \frac{\partial \phi^h}{\partial y} \right) \omega^{(\ell-1)} \, d\mathbf{x}$$

$$= \int_\Omega \left(\frac{\partial \psi^{(\ell-1)}}{\partial y} \frac{\partial \phi^h}{\partial x} - \frac{\partial \psi^{(\ell-1)}}{\partial x} \frac{\partial \phi^h}{\partial y} \right) \omega^{(\ell-1)} \, d\mathbf{x} \qquad (63)$$

$$- \int_\Omega \left(\frac{\partial f_2}{\partial x} - \frac{\partial f_1}{\partial y} \right) \phi^h \, d\mathbf{x} \qquad \forall \phi^h \in V_0^h$$

Note that, for each j, the boundary conditions for ψ_j^h for these problems are completely prescribed so that one may use the usual local finite-element basis functions everywhere. Also note that the linear systems used to solve (62) and (63) for (ψ_j, ω_j) for each $j = 0, \ldots, m$ have different right-hand-sides, but all have the same coefficient matrix. One should also note that, since the numbers α_{ij} were chosen arbitrarily, the solutions (ψ_j, ω_j), $j = 0, \ldots, m$, of (62) and (63) will not satisfy (59), even in an approximate manner.

(iii) Determine the numbers β_j, $j = 0, \ldots, m$, by solving the linear system of $(m+1)$ equations

$$\sum_{j=0}^{m} \beta_j = 1$$

and

$$\sum_{j=0}^{m} \int_{\Gamma_i} \beta_j \left(\omega_j \mathbf{g}_i \cdot \mathbf{n} - \nu \frac{\partial \omega_j}{\partial n} \right) d\mathbf{x} = \int_{\Gamma_i} \mathbf{f} \cdot \mathbf{s} \ d\mathbf{x} = 0 \qquad \text{for } i = 1, \ldots, m$$

(iv) Set

$$\psi^{(\ell)} = \sum_{j=0}^{m} \beta_j \psi_j, \quad \omega^{(\ell)} = \sum_{j=0}^{m} \beta_j \omega_j, \quad \text{and} \quad a_i^{(\ell)} = \sum_{j=0}^{m} \beta_j \alpha_{ij} \text{ for } i = 1, \ldots, m$$

The sequence of Newton iterates $\{\psi^{(\ell)}, \omega^{(\ell)}, [a_i^{(\ell)}, i = 1, \ldots, m]\}$, $\ell > 1$, will converge to the solution $\{\psi^h, \omega^h, [a_i, i = 1, \ldots, m]\}$ of (60) and (61), provided, of course, that a sufficiently accurate initial iterate is available. Further details may be found in Gunzburger and Peterson (1988).

IV LEAST-SQUARES FINITE-ELEMENT METHODS

Least-squares finite-element methods can be devised for any formulation of the governing equations. However, for practical reasons, such methods are best developed in the context of first-order formulations of those equations. The use of such formulations allows for the use of merely continuous finite-element functions and results in discrete systems whose condition numbers are comparable to those obtained from Galerkin methods for the primitive variable formulation. Even among first-order formulations, one has several choices, e.g. velocity–pressure–stress tensor, velocity–pressure–acceleration, velocity–vorticity–pressure, etc. Here we will only consider the velocity–vorticity–pressure formulation which has proven to be the most popular and useful. For simplicity, we only consider the steady-state case with homogeneous velocity boundary conditions.

A Least-squares principles for the Navier-Stokes equations

1 The velocity–vorticity–pressure formulation

Equations (1), (2), and (6) may be written, in the steady-state case, as a first-order system in the form

$$
\begin{aligned}
\nabla \times \mathbf{u} - \boldsymbol{\omega} &= 0 \quad \text{in } \Omega \\
\nabla \cdot \mathbf{u} &= 0 \quad \text{in } \Omega \\
\nu \nabla \times \boldsymbol{\omega} + \nabla p + \boldsymbol{\omega} \times \mathbf{u} &= \mathbf{f} \quad \text{in } \Omega \\
\nabla \cdot \boldsymbol{\omega} &= 0 \quad \text{in } \Omega \\
\mathbf{u} &= \mathbf{0} \quad \text{on } \Gamma
\end{aligned}
\tag{64}
$$

This formulation involves seven scalar fields in three dimensions, i.e. the velocity $\mathbf{u}$, vorticity $\boldsymbol{\omega}$, and the total pressure p. (Thus, here p is not the pressure, but the pressure summed with $\mathbf{u} \cdot \mathbf{u}/2$. However, for simplicity, we will henceforth refer to p as the pressure.) For plane flows, we have four scalar fields since in that case the vorticity may be viewed as a scalar; also, for plane flows, the fourth equation in (64) is omitted. In three dimensions, the equation $\nabla \cdot \boldsymbol{\omega} = 0$ is redundant in view of the first equation in (64); however, its inclusion in the least-squares functional given below is necessary to the success of least-squares finite-element methods (Bochev and Gunzburger, 1994).

2 Least-squares principles

Let $Y = \mathbf{H}_0^1(\Omega) \times \mathbf{H}^1(\Omega) \times [H^1(\Omega) \cap L_0^2(\Omega)]$. For $(\mathbf{u}, \boldsymbol{\omega}, p) \in Y$ one can define the least-squares functional

$$
\begin{aligned}
\mathcal{J}(\mathbf{u}, p, \boldsymbol{\omega}) &= \int_\Omega \left[(\nabla \times \mathbf{u} - \boldsymbol{\omega})^2 + (\nabla \cdot \mathbf{u})^2 \right] d\mathbf{x} \\
&+ \int_\Omega \left[(\nu \nabla \times \boldsymbol{\omega} + \nabla p + \boldsymbol{\omega} \times \mathbf{u} - \mathbf{f})^2 + (\nabla \cdot \boldsymbol{\omega})^2 \right] d\mathbf{x}
\end{aligned}
\tag{65}
$$

and then pose the problem:

$$
\min \left\{ \mathcal{J}(\mathbf{u}, p, \boldsymbol{\omega}) \mid (\mathbf{u}, p, \boldsymbol{\omega}) \in Y \right\}
\tag{66}
$$

Finite-element methods based on (65) do not always yield optimal accurate approximations (Bochev and Gunzburger, 1994). For this reason, we also define the *weighted* least-squares functional

$$
\begin{aligned}
\mathcal{J}_h(\mathbf{u}, p, \boldsymbol{\omega}) &= h^{-2} \int_\Omega \left[(\nabla \times \mathbf{u} - \boldsymbol{\omega})^2 + (\nabla \cdot \mathbf{u})^2 \right] d\mathbf{x} \\
&+ \nu^{-2} \int_\Omega \left[(\nu \nabla \times \boldsymbol{\omega} + \nabla p + \boldsymbol{\omega} \times \mathbf{u} - \mathbf{f})^2 + (\nabla \cdot \boldsymbol{\omega})^2 \right] d\mathbf{x}
\end{aligned}
\tag{67}
$$

and pose the problem

$$\min \{ \mathcal{J}_h(\mathbf{u}, p, \boldsymbol{\omega}) \mid (\mathbf{u}, p, \boldsymbol{\omega}) \in Y \} \tag{68}$$

In Eq. (67), h is, as always, a measure of the grid size.

B Finite-element approximations

1 Discrete least-squares principles

We choose a finite-element subspace $Y^h = \mathbf{V}_0^h \times \mathbf{W}^h \times S_0^h \subset Y$, where $\mathbf{V}_0^h \subset \mathbf{H}_0^1(\Omega)$, $\mathbf{W}^h \subset \mathbf{H}^1(\Omega)$, and $S_0^h \subset H^1(\Omega) \cap L_0^2(\Omega)$. All of these finite-element spaces can be based on the same mesh and the same degree polynomials. Then, we can pose the discrete problems

$$\min \{ \mathcal{J}(\mathbf{u}^h, p^h, \boldsymbol{\omega}^h) \mid (\mathbf{u}^h, p^h, \boldsymbol{\omega}^h) \in Y^h \} \tag{69}$$

and

$$\min \{ \mathcal{J}_h(\mathbf{u}^h, p^h, \boldsymbol{\omega}^h) \mid (\mathbf{u}^h, p^h, \boldsymbol{\omega}^h) \in Y^h \} \tag{70}$$

corresponding to the problems of (66) and (68), respectively.

Solutions of (69) are suboptimally accurate (Bochev and Gunzburger, 1994) in the case we are considering here, i.e. velocity boundary conditions. For some other cases, e.g. the pressure and normal velocity specified on the boundary, (69) does produce optimally accurate approximations. However, since the velocity boundary condition case is of such great importance, we will henceforth only consider approximations determined from (70).

2 The discrete equations

The necessary conditions for $(\mathbf{u}^h, p^h, \boldsymbol{\omega}^h) \in Y^h$ to be a solution of (70) are easily determined to be

$$h^{-2} \int_\Omega \left((\nabla \times \mathbf{u}^h - \boldsymbol{\omega}^h) \cdot (\nabla \times \mathbf{v}^h) + (\nabla \cdot \mathbf{u}^h)(\nabla \cdot \mathbf{v}^h) \right) d\mathbf{x}$$
$$+ \nu^{-2} \int_\Omega (\nu \nabla \times \boldsymbol{\omega}^h + \nabla p^h + \boldsymbol{\omega}^h \times \mathbf{u}^h - \mathbf{f}) \cdot (\boldsymbol{\omega}^h \times \mathbf{v}^h) d\mathbf{x} = 0 \quad \forall \mathbf{v}^h \in \mathbf{V}_0^h \tag{71}$$

$$- h^{-2} \int_\Omega (\nabla \times \mathbf{u}^h - \boldsymbol{\omega}^h) \cdot \boldsymbol{\zeta}^h \, d\mathbf{x}$$
$$+ \nu^{-2} \int_\Omega (\nu \nabla \times \boldsymbol{\omega}^h + \nabla p^h + \boldsymbol{\omega}^h \times \mathbf{u}^h - \mathbf{f}) \cdot (\nu \nabla \times \boldsymbol{\zeta}^h + \boldsymbol{\zeta}^h \times \mathbf{u}^h) \, d\mathbf{x} \tag{72}$$
$$+ \nu^{-2} \int_\Omega (\nabla \cdot \boldsymbol{\omega}^h)(\nabla \cdot \boldsymbol{\zeta}^h) \, d\mathbf{x} = 0 \qquad \forall \boldsymbol{\zeta}^h \in \mathbf{W}^h$$

and

$$\nu^{-2} \int_{\Omega} (\nu \nabla \times \boldsymbol{\omega}^h + \nabla p^h + \boldsymbol{\omega}^h \times \mathbf{u}^h - \mathbf{f}) \cdot \nabla q^h \, d\mathbf{x} \qquad \forall q^h \in S_0^h \qquad (73)$$

For plane flows, the terms involving $\nabla \cdot \boldsymbol{\omega}^h$ are omitted. Equations (71)–(73) are equivalent to a non-linear system of algebraic equations. These may be linearized by Newton's method or by some other method for solving non-linear equations.

The discrete system of (71)–(73) has many advantageous properties. Suppose, for example, that Newton's method is used to linearize these equations. (Our discussion holds equally well for any symmetry-preserving linearization procedure.) Since the Newton jacobian matrix is the hessian matrix for the functional $\mathcal{J}_h$, this matrix is always *symmetric*. Moreover, since solutions of (71)–(73) are minimizers of the functional $\mathcal{J}_h$, the Newton jacobian matrix is, at least in a neighbourhood of the solution, *positive definite*. These results are independent of the Reynolds number, except that the size of the neighbourhood for which the Newton matrix is positive definite shrinks as the value of the Reynolds number increases. Thus, if a 'good enough' initial guess for the Newton iteration is available, e.g. through the use of a continuation procedure, one can obtain a solution of the Navier–Stokes system for any value of the Reynolds number through the solution of a sequence of positive definite linear systems. This opens the possibility of using *robust iterative methods* such as the conjugate gradient method for the solution of linear systems. Moreover, such methods can be implemented so that *no matrix assembly,* even at the element level, is needed.

3 Error estimates

Error estimates are available only for quadratic or higher-order velocity spaces (Bochev and Gunzburger, 1994). The spaces for the vorticity and pressure may be chosen to be based on the same triangulation and to consist of polynomials of the same degree or one degree less than that used for the velocity space. If, for example, continuous piecewise polynomials of degree k, $k \geq 2$, are used for the velocity approximation and continuous piecewise polynomials of degree $(k - 1)$ are used for the vorticity and pressure approximations, we have, for smooth enough solutions, that the solution $(\mathbf{u}^h, \boldsymbol{\omega}^h, p^h)$ of (71)–(73) satisfies the error estimate

$$\|\mathbf{u} - \mathbf{u}^h\|_1 + \|\boldsymbol{\omega}^h - \boldsymbol{\omega}\|_0 + \|p - p^h\|_0 = O(h^k) \qquad (74)$$

where $(\mathbf{u}, \boldsymbol{\omega}, p)$ denotes the exact solution of (64). This estimate is optimal.

One can choose the underlying finite-element space for all of the variables to be based on the same grid and same degree polynomials. For example, one could use continuous piecewise polynomials of degree k, $k \geq 2$, for the velocity, vorticity, and pressure approximations, all based on the same grid. In this case, one still obtains the error estimate of (74). This estimate is suboptimal for the vorticity and pressure approximations. However, there is computational evidence that (74) is not sharp in this case, i.e. in the case of equal-order interpolation for all variables. In fact, pressure and vorticity approximations obtained from (71)–(73) seemingly are better than is indicated by (74). The case of $k = 1$, i.e. linear velocities, is not covered by the theory that results in (74). Computational studies (Bochev and Gunzburger, 1993) indicate that this case may be used in practice and that if any loss of accuracy occurs, it is very slight. Thus, unlike other methods involving the vorticity, e.g. the streamfunction–vorticity or velocity–vorticity formulations, the least-squares method discussed here yields accurate vorticity and pressure approximations. Moreover, no boundary conditions for the vorticity need be devised at solid walls. One may then conclude that these advantages, along the other advantages discussed above, make least-squares finite-element methods very attractive for large-scale, three-dimensional computations.

V THE VELOCITY–VORTICITY FORMULATION

A The mathematical model

1 Equations derivable from the primitive variable formulation

A popular model for incompressible fluid mechanics is the velocity–vorticity formulation. For steady-state problems with velocity boundary conditions, we have the incompressibility constraint

$$\nabla \cdot \mathbf{u} = 0 \qquad \text{in } \Omega \tag{75}$$

the definition of the vorticity

$$\nabla \times \mathbf{u} - \boldsymbol{\omega} = \mathbf{0} \qquad \text{in } \Omega \tag{76}$$

and, by taking the curl of (1) (with the time derivative term deleted), the vorticity transport equation

$$-\nu \Delta \boldsymbol{\omega} + \nabla \times (\boldsymbol{\omega} \times \mathbf{u}) = \nabla \times \mathbf{f} \qquad \text{in } \Omega \tag{77}$$

Note that the pressure is eliminated in this step. Equations (75) and (76) may be combined to yield

$$-\Delta \mathbf{u} = \nabla \times \nabla \times \mathbf{u} - \nabla(\nabla \cdot \mathbf{u}) = \nabla \times \boldsymbol{\omega} \qquad \text{in } \Omega \tag{78}$$

We also have the boundary condition

$$\mathbf{u} = \mathbf{g} \qquad \text{on } \Gamma \tag{79}$$

This boundary condition immediately yields a boundary condition for the normal component of $\boldsymbol{\omega}$, i.e.

$$\boldsymbol{\omega} \cdot \mathbf{n} = (\nabla \times \mathbf{g}) \cdot \mathbf{n} \qquad \text{on } \Gamma \tag{80}$$

The right-hand-side of (80) is well-defined since $(\nabla \times \mathbf{g}) \cdot \mathbf{n}$ only involves tangential (to Γ) derivatives of the components of $\mathbf{g}$.

For plane flow problems, the vorticity may be viewed as a scalar function ω or as the vector $\omega\mathbf{k}$, where $\mathbf{k}$ denotes the unit vector in the z direction. In any case, we have that $\boldsymbol{\omega}$ and $\mathbf{n}$ are always orthogonal, so that the boundary condition of (80) is vacuous in this case.

One is immediately faced with the choice of using (75), (76), (77), (79), and (80) or (77)–(80) as the mathematical model. Although both choices have found favour in the literature, we will only consider the second choice since it is in more common use. One of the attractions of the second choice is that (77) and (78) are a system of non-linear Poisson equations to which one can presumably bring to bear some well-known solution methods. However, there are some difficulties to be discussed.

2 Additional boundary conditions

The boundary condition of (79) is sufficient for the differential equation of (78) in the sense that, should $\boldsymbol{\omega}$ be known, then (78) and (79) are uniquely solvable for $\mathbf{u}$. Note that (79) by itself is sufficient for solving the primitive variable formulation, but since (77) and (78) are derived by differentiating the primitive variable equations, one needs additional boundary conditions. One of these is provided by (80); however (80) is not by itself 'sufficient' for solving (77) and, in fact, (79) and (80) are not sufficient for solving the coupled system of (77) and (78).

Exact additional conditions can be derived from the primitive variable equations. Unfortunately, these are *non-local* in character; see Quartapelle (1993) and the references cited therein. They relate boundary values of the vorticity to interior values of the velocity field. Although there is considerable interest in using these non-local conditions, their implementation into computational algorithms is cumbersome. One would prefer to use a purely *local* boundary condition, i.e. one relating the vorticity or derivatives of the vorticity at a point on the boundary to the velocity or its derivatives at the same point. No exact boundary condition of this type is known to exist. However, this has not stopped researchers from applying a variety of such conditions. Here, we discuss the one that is, by far, the most popular (Gunzburger *et al.*, 1990), namely, using the definition of the vorticity, given in (76), on the boundary. Thus, since we have already decided to impose (80), we have the additional local condition

$$\mathbf{n} \times \boldsymbol{\omega} \times \mathbf{n} = \mathbf{n} \times (\nabla \times \mathbf{u}) \times \mathbf{n} \qquad \text{on } \Gamma \tag{81}$$

Equations (80) and (81) may be combined into the simpler relation $\boldsymbol{\omega} = \nabla \times \mathbf{u}$ on Γ, but we choose to keep the normal and tangential components separate since the first, i.e. (80), is a relation that is derivable from the data $\mathbf{g}$, while the second, i.e. (81), remains a relation between the unknown vorticity and velocity fields.

B Finite-element discretizations

1 The discrete equations

We now want to formulate finite-element algorithms for determining approximate solutions of (77)–(81). Equations (77) and (78), both being non-linear second-order elliptic equations, suggest that one can look for velocity and vorticity approximations in subspaces of $\mathbf{H}^1(\Omega)$. In fact, this is desirable since then one can use merely continuous piecewise polynomial functions based on a single triangulation and of the same degree and one might hope to achieve the same accuracy for both variables. Unfortunately, the latter hope is not easily realized.

We choose a finite-element subspace $\mathbf{V}^h \subset \mathbf{H}^1(\Omega)$ and let $\mathbf{V}_0^h = \mathbf{V}^h \cap \mathbf{H}_0^1(\Omega)$. We the seek $\mathbf{u}^h \in \mathbf{V}^h$ and $\boldsymbol{\omega}^h \in \mathbf{V}^h$ such that

$$
\nu \int_\Omega \left((\nabla \times \boldsymbol{\omega}^h) \cdot (\nabla \times \boldsymbol{\zeta}^h) + (\nabla \cdot \boldsymbol{\omega}^h)(\nabla \cdot \boldsymbol{\zeta}^h) \right) \, \mathrm{d}\mathbf{x}
$$
$$
+ \int_\Omega (\boldsymbol{\omega}^h \times \mathbf{u}^h) \cdot (\nabla \times \boldsymbol{\zeta}^h) \, \mathrm{d}\mathbf{x} = \int_\Omega \mathbf{f} \cdot (\nabla \times \boldsymbol{\zeta}^h) \qquad \forall \boldsymbol{\zeta}^h \in \mathbf{V}_0^h \tag{82}
$$

and

$$
\nu \int_\Omega \left((\nabla \times \mathbf{u}^h) \cdot (\nabla \times \mathbf{v}^h) + (\nabla \cdot \mathbf{u}^h)(\nabla \cdot \mathbf{v}^h) \right) \, \mathrm{d}\mathbf{x}
$$
$$
- \int_\Omega \boldsymbol{\omega}^h \cdot (\nabla \times \mathbf{v}^h) \, \mathrm{d}\mathbf{x} = 0 \qquad \forall \mathbf{v}^h \in \mathbf{V}_0^h \tag{83}
$$

In addition, we need to impose, at least in an approximate sense, the boundary conditions of (79)–(81). Corresponding to the first two of these, we impose the boundary conditions

$$
\mathbf{u}^h = \mathbf{g}^h \qquad \text{on } \Gamma \tag{84}
$$

and

$$
\boldsymbol{\omega}^h \cdot \mathbf{n} = (\nabla \times \mathbf{g}^h) \cdot \mathbf{n} \qquad \text{on } \Gamma \tag{85}
$$

where $\mathbf{g}^h \in \mathbf{V}^h|_\Gamma$ is an approximation, e.g. the boundary interpolant, of $\mathbf{g}$. If we use merely continuous finite-element spaces, then, in (85), $(\nabla \times \mathbf{g}^h)$ should be interpreted as a smoothed approximation to $\nabla \times \mathbf{g}^h$.

There remains the implementation of (81) which relates the boundary value of the tangential vorticity to some known and some unknown derivatives of the velocity components on the boundary. The most obvious way of implementing (81) is to use a difference quotient to approximate the derivatives of the velocity. For example, suppose our grid is a cartesian grid and suppose we have a boundary segment that coincides with the (x_1, x_2) plane and such that the flow domain lies above that plane. Suppose $\mathbf{x}_0$ denotes a grid point on that boundary segment. We then denote the adjacent grid points to $\mathbf{x}_0$ lying on the boundary by $\mathbf{x}_E$, $\mathbf{x}_W$, $\mathbf{x}_N$, $\mathbf{x}_S$ (with east–west in the x_1 direction and north–south in the x_2 direction) and the *interior* point directly above $\mathbf{x}_0$ by $\mathbf{x}_I$. Then, we could implement (81) through the relations

$$\omega_1^h(\mathbf{x}_0) = \frac{g_3^h(\mathbf{x}_N) - g_3^h(\mathbf{x}_S)}{2\delta x_2} - \frac{u_2^h(\mathbf{x}_I) - g_2^h(\mathbf{x}_0)}{\delta x_3} \tag{86}$$

and

$$\omega_2^h(\mathbf{x}_0) = \frac{u_1^h(\mathbf{x}_I) - g_1^h(\mathbf{x}_0)}{\delta x_3} - \frac{g_3^h(\mathbf{x}_W) - g_3^h(\mathbf{x}_E)}{2\delta x_1} \tag{87}$$

where ω_i^h, for $i = 1, 2, 3$, is the x_i component of $\boldsymbol{\omega}^h$ and similarly for g_i^h and u_i^h; also, δx_i denotes the grid length in the x_i direction. Note that these equations relate the value of the vorticity at a point on the boundary to known components of the velocity at nearby points on the boundary and to the unknown values of the velocity components at a nearby interior point. Equations (86) and (87), or variants of these, are often used with bilinear elements for the velocity and vorticity.

Equations (86) and (87) can be derived from a projection-type implementation of the boundary condition of (81), along with a suitably chosen quadrature rule. Specifically, one imposes (81) weakly (since a straightforward pointwise implementation does not make sense) by requiring that

$$\int_\Gamma \boldsymbol{\omega}^h \cdot \boldsymbol{\phi}^h \, d\mathbf{x} = \sum_{\Gamma_i \in \mathcal{F}} \int_\Gamma \boldsymbol{\phi}^h \cdot (\nabla \times \mathbf{u}^h) \, d\mathbf{x} \qquad \forall \boldsymbol{\phi}^h \in \mathbf{V}^h|_\Gamma \tag{88}$$

where $\mathcal{F} = \{\Gamma_j, j = 1, \ldots, J\}$ denotes the set of all element faces that coincide with the boundary Γ. Then, one approximates the integrals in (88) by using accurate enough integration rules. For a specific choice for the integration rule, one obtains (86) and (87) for a boundary segment that is perpendicular to the x_3-axis and for a cartesian grid. For more general boundaries and more general grids, the relation between the boundary values of the vorticity and the interior values of the velocity resulting from the application of (88) along with a quadrature rule is more complicated than that given in (86) and (87) (Gunzburger *et al.*, 1990).

2 Accuracy considerations

No theory exists for estimating the errors in the approximate solution obtained from (82)–(85) and generalizations of (86) and (87). However, from ample

using quadrature rules. One must make sure that the quadrature rule is accurate enough so that approximations are not polluted by the quadrature error. On the other hand, using a too accurate rule is wasteful since needless work is being performed. Composite Gauss–Legendre rules for which the quadrature points are interior to an element Ω_k are used to evaluate the terms involving pressure test and trial functions, i.e. the terms in Eqs. (89) and (90) involving p^r and q^r. The remaining terms are evaluated by composite Gauss–Lobatto rules for which the set of quadrature points include boundary points of the subdomains Ω_k.

4 Error estimates

For smooth, e.g. locally analytic, solutions of the Navier–Stokes system the spectral-element method achieves spectral accuracy. In fact, it can be shown that

$$|\mathbf{u} - \mathbf{u}^r|_1 + \|p - p^r\|_0 = O(r^{-\sigma})$$

where $\sigma > 0$ is related to the smoothness of the exact solution $\mathbf{u}$ and p and data $\mathbf{f}$. (Recall that r and $(r - 2)$ are the degrees of the polynomials used for the velocity and pressure approximations.)

5 Bases for spectral-element subspaces

The choice of bases for the spectral-element spaces $\mathbf{V}^r$ and S^r does not affect the error estimates, since the latter assume exact precision arithmetic is used. However, this choice greatly affects the conditioning and sparsity of the discrete systems. For the rectangular element example discussed above, one can use Lagrange interpolating polynomials through the tensor product Gauss–Lobatto points for $\mathbf{V}^r$ and through the tensor product Gauss–Legendre points for S^r.

ACKNOWLEDGEMENTS

The preparation of this chapter was partially supported by the U.S. Office of Naval Research under grant number N00014-91-J-1493 and by U.S. Air Force Office of Scientific Research under grant number AFOSR-93-1-0280.

REFERENCES

Adams, R. (1975). *Sobolev Spaces.* Academic Press, New York.
Arnold, D., Brezzi, F. and Fortin, M. (1984). *Calcolo* **21**, 337–344.
Babuska, I. (1973). *Numer. Math.* **16**, 322–333.

Babuska, I. and Aziz, K. (1972) In *The Mathematical Foundations of the Finite Element Method with Application to Partial Differential Equations* (A. Aziz, ed.), pp. 1–359. Academic Press, New York.

Baker, A. (1985). *Finite Element Computational Fluid Mechanics.* McGraw-Hill, New York.

Baker, G., Dougalis, V. and Karakshian, O. (1982). *Math. Comput.* **39**, 339–375.

Bochev, P. and Gunzburger, M. (1993). *Computers and Fluids* **22**, 549–563.

Bochev, P. and Gunzburger, M. (1994). *Math. Comput.* **63**, 479–506.

Boland, J. and Nicolaides, R. (1983). *SIAM J. Numer. Anal.* **20**, 722–731.

Boland, J. and Nicolaides, R. (1984). *Numer. Math.* **44**, 219–222.

Boland, J. and Nicolaides, R. (1985). *SIAM J. Numer. Anal.* **22**, 474–492.

Brezzi, F. (1974). *Rairo Anal. Numér.* **8**, 129–151.

Brezzi, F. and Douglas, J. (1988). *Numer. Math.* **53**, 225–235.

Brezzi, F. and Fortin, M. (1991). *Mixed and Hybrid Finite Element Methods.* Springer-Verlag, Berlin.

Brezzi, F. and Pitkaranta, J. (1984). In *Efficient Solutions of Elliptic Systems* (W. Hackbush, ed.), pp. 11–19. Vieweg-Verlag, Braunschweig.

Brezzi, F. Rappaz, J., and Raviart, P.-A. (1980). *Numer. Math.* **36**, 1–25.

Canuto, C. Hussaini, M., Quarteroni, A., and Zang, T. (1988) *Spectral Methods in Fluid Dynamics.* Springer-Verlag, Berlin.

Cayco, M. and Nicolaides, R. (1986). *Math. Comput.* **46**, 371–377.

Cayco, M. and Nicolaides, R. (1990). *Comput. Math. Appl.* **41**, 213–234.

Chorin, A. (1967). *J. Comput. Phys.* **2**, 12–26.

Ciarlet, P. (1978). *The Finite Element Methods for Elliptic Problems.* North-Holland Publishing, Amsterdam.

Crouzeix, M., and Raviart, P.-A. (1973). *RAIRO Anal. Numér.* **7**, 33–76.

Cuvelier, C., Segal, A., and van Steenhoven, A. (1986). *Finite Element Methods and Navier–Stokes Equations.* Reidel Publishing, Dordrecht.

den Heijer, C., and Rheinboldt, W. (1981). *SIAM J. Numer. Anal.* **18**, 925–948.

Engelman, M., Strang, G., and Bathe, K.-J. (1981). *Int. J. Numer. Methods. Engrg.* **17**, 707–718.

Fix, G., Gunzburger, M., Nicolaides, R., and Peterson, J. (1984). In *Proc. 5th International Symposium on Finite Element and Flow Problems* (G. Carey and J. Oden, eds.), pp. 281–286. Univ. of Texas Press, Austin.

Fletcher, C. (1988). *Computational Techniques for Fluid Dynamics.* Springer-Verlag, Berlin.

Franca, L., Hughes, T., and Stenberg, R. (1993). In '*Incompressible Computational Fluid Dynamics: Trends and Advances* (M. Gunzburger and R. Nicolaides, eds), pp. 87–107. Cambridge Univ. Press, Cambridge.

Fortin, M. (1977). *Rairo Anal. Numér.* **11**, 341–354.

Fortin, M., and Fortin, A. (1985). *Comm. Appl. Numer. Meth.* **1**, 205–208.

Girault, V., and Raviart, P.-A. (1979). *Finite Element Approximation of the Navier–Stokes Equations.* Springer-Verlag, Berlin.

Girault, V., and Raviart, P.-A. (1986). *Finite Element Methods for Navier–Stokes Equations.* Springer-Verlag, Berlin.

Glowinski, R. (1984). *Numerical Methods for Nonlinear Variational Problems.* Springer-Verlag, Berlin.

Gresho, P. (1988). *Report Number UCR-99221.* Lawrence Livermore National Laboratory, Livermore.

Gresho, P., and Sani, R. (1988). *Int. J. Numer. Methods Fluids* **7**, 1111–1145.

Gunzburger, M. (1989). *Finite Element Methods for Viscous Incompressible Flows: A Guide to Theory, Practice, and Algorithms.* Academic Press, Boston.

Gunzburger, M., and Hou, L. (1992). *SIAM J. Numer. Anal.* **29**, 390–424.

Gunzburger, M. and Nicolaides, R. (eds) (1993). *Incompressible Computational Fluid Dynamics: Trends and Advances.* Cambridge Univ. Press, Cambridge.

Gunzburger, M., and Peterson, J. (1983). *Numer. Math.* **22**, 173–181.

Gunzburger, M., and Peterson, J. (1988). *SIAM J. Sci. Stat. Comput.* **9**, 650–668.

Gunzburger, M., and Peterson, J. (1991). *Comput. Math. Appl.* **42**, 173–194.

Gunzburger, M., Mundt, M., and Peterson, J. (1990). In *Computational Methods in Viscous Aerodynamics* (T. Murthy and C. Brebbia, eds), pp. 229–271. Elsevier Science Publishers, Amsterdam.

Habashi, W., Peeters, M., Guevremont, G., and Hafez, M. (1987). *AIAA J.* **25**, 944–948.

Hafez, M., Habashi, W., Przybytkowski, S., and Peeters, M. (1987). *AIAA Paper 87-0644.* AIAA, New York.

Heywood, J., and Rannacher, R. (1982). *SIAM J. Numer. Anal.* **19**, 275–311.

Hughes, T., and Franca, L. (1987). *Comput. Methods Appl. Mech. Engrg.* **65**, 85–96.

Hughes, T., Franca, L., and Ballestra, M. (1986). *Comput. Methods Appl. Mech. Engrg.* **59**, 85–99.

Hughes, T., Liu, W., and Brooks, A. (1979). *J. Comput. Phys.* **30**, 1–60.

Jamet, P., and Raviart, P.-A. (1973). In *Computing Methods in Science and Engineering* (R. Glowinski and J.-L. Lions, eds), pp.193–223. Springer-Verlag, Berlin.

Karakashian, O. (1982). *SIAM J. Numer. Anal.* **19**, 909–923.

Karniadakis, G., Orszag, S., Ronquist, E., and Patera, A. (1993). In *Incompressible Computational Fluid Dynamics: Trends and Advances* (M. Gunzburger and R. Nicolaides, eds), pp. 203–266. Cambridge Univ. Press, Cambridge.

Keller, H. (1978). In *Recent Advances in Numerical Analysis* (C. de Boor and G. Golub, eds), pp. 73–94. Academic Press, New York.

Keller, H. (1987). *Numerical Methods in Bifurcation Problems.* Springer-Verlag, Berlin.

Ladyzhenskaya, O. (1969). *Mathematical Theory of Viscous Incompressible Flows.* Gordon and Breach Science Publishers, New York.

Löhner, R. (1993). In *Incompressible Computational Fluid Dynamics: Trends and Advances* (M. Gunzburger and R. Nicolaides, eds), pp. 267–293. Cambridge Univ. Press, Cambridge.

Mansfield, L. (1982). *RAIRO Anal. Numer.* **16**, 49–66.

Matthies, H., and Strang, G. (1977). *Int. J. Numer. Methods Engrg.* **14**, 1613–1626.

Ortega, J., and Rheinboldt, W. (1970). *Iterative Solution of Nonlinear Equations in Several Variables.* Academic Press, New York.

Peeters, M., Habashi, W., and Dueck, E. (1987). *Int. J. Numer. Methods. Fluids* **7**, 17–27.

Peyret, R., and Taylor, T. (1983). *Computational Methods for Fluid Flow.* Springer Verlag, New York.

Pironneau, O. (1989). *Finite Element Methods for Fluids.* J. Wiley & Sons, New York.

Quartapelle, L. (1993). *Numerical Solution of the Incompressible Navier–Stokes Equations.* Birkhauser-Verlag, Basel.

Rheinboldt, W. (1980). *SIAM J. Numer Anal.* **17**, 221–237.

Scott, R., and Vogelius, M. (1985a). *Math. Model. Numer. Anal.* **19**, 111–143.

Scott, R., and Vogelius, M. (1985b). In *Large Scale Computations in Fluid Dynamics* (B. Enquist, S. Osher, and R. Somerville, eds), vol. 2, pp. 221–244. American Mathematical Society Publications, Providence.

Segal, A. (1979). *Comput. Methods Appl. Mech. Engrg.* **19**, 165–185.

Serrin, J. (1959). In *Encyclopedia of Physics* (S. Flugge and C. Truesdell, eds), vol. 2, VIII/1, pp. 125–263. Springer-Verlag, Berlin.

Silvester, D., and Thatcher, R. (1986). *Int. J. Numer. Methods Fluids* **6**, 841–853

Stenberg, R. (1984). *Math. Comput.* **42**, pp. 9–23.

Stenberg, R. (1987). *Comput. Methods Appl. Mech. Engrg.* **63**, 261–269.

Taylor, C. and Hood, P. (1973). *Computers and Fluids* **1**, 73–100.

Temam, R. (1979). *Navier–Stokes Equations: Theory and Numerical Analysis.* North-Holland Publishing, Amsterdam.

Temam, R. (1983). *Navier Stokes Equations and Nonlinear Functional Analysis.* SIAM Publications, Philadelphia.

Tezduyar, T., Glowinski, R., and Liou, J. (1988). *Int. J. Numer. Methods. Fluids* **8**, 1269–1290.

Thatcher, R. (1993). In *Incompressible Computational Fluid Dynamics: Trends and Advances* (M. Gunzburger and R. Nicolaides, eds), pp. 427–445. Cambridge Univ. Press, Cambridge.

Thomasset, F. (1981). *Implementation of Finite Element Methods for the Navier–Stokes Equations.* Springer-Verlag, New York.

Verfurth, R. (1984). *Math. Model. Numer. Anal.* **18**, 175–182.

Verfurth, R. (1985). *Math. Model. Numer. Anal.* **19**, 461–475.

Vogelius, M. (1983). *Numer. Math.* **41**, 19–37.

Yanenko, N. (1971). *The Method of Fractional Steps.* Springer-Verlag, Berlin.

4 Euler and Navier–Stokes equations for compressible flows: finite-volume methods

Francesco Grasso and Carlo Meola

HANDBOOK OF COMPUTATIONAL FLUID MECHANICS
ISBN 0-12-532200-3

INTRODUCTION

The objective of this chapter is to guide the reader toward an understanding of the numerical techniques best suited to simulate high Reynolds number compressible flows and to lay the fundamentals for developing novel approaches. The main emphasis is on the critical issues that arise when solving the Euler and Navier–Stokes equations following a finite volume approach on structured meshes.

It is important to note that compressible flows differ substantially from incompressible ones in many aspects, the main differences stemming from the density variations. In the incompressible models the pressure loses its thermodynamic meaning and it contributes to the enforcement of the divergence-free constraint, while in the compressible case the pressure maintains its thermodynamic meaning and is related to density and internal energy variations. The governing equations for the simulation of compressible laminar flows can be formally closed. However, the mathematical closure as well as the numerical solution is a formidable task, particularly when simulating transonic and high Mach number flows. Indeed, the nature of the flow may change throughout the domain from an elliptic to hyperbolic-type behaviour, and for a given flow problem the numerical scheme must be able to resolve phenomena of different characters.

In the presence of high Reynolds numbers, convective effects dominate, and the techniques employed for the solution of the Navier–Stokes equations are generally derived in a straightforward manner from those used for the solution of the inviscid equations together with a central discretization of the stress tensor contribution. Most of this chapter is devoted to analysing numerical schemes for the inviscid terms.

In general, most of the techniques that will be discussed in this chapter are also consistent with non-steady flow solutions. However, special emphasis is on steady flow solutions that play an important role for assessing the validity of any computational technique, especially in multidimensions. As a consequence, we will focus on time marching approaches that allow any (steady) fluid dynamic flow problem to be formulated as a pseudo-unsteady one, where the time is to be interpreted as an evolution coordinate such that the use of a time marching strategy is equivalent to an iterative process, whose convergence is indeed equivalent to the existence of a steady state (Jameson, 1983; Peyret and Taylor, 1983; Hirsch, 1990).

Notwithstanding the complexity of the Euler and/or Navier–Stokes equations, the analysis and the design of most of the successful numerical approaches are intimately connected with the progress in the numerical treatment of simple linear and non-linear model equations, such as the (one-dimensional) advection-diffusion equation. The milestones for the modern development of schemes for solving the compressible flow equations are the Courant–Isaacson–Rees (Courant *et al.*, 1952) and Lax–Wendroff (1960, 1964) schemes, together with the theory of Lax on the weak solutions of systems of conservation laws (Lax and Wendroff, 1960, 1964; Lax, 1972).

In the past 15–20 years a large effort has been put in extending the results of those studies to multidimensional problems. A classical example is the theory based on operator splitting for solving partial differential equations mainly developed by the Russian school (Yanenko, 1971). A direct formulation to solve multidimensional problems is the finite-volume approach; this has the merit of having a clear physical interpretation, yielding at the same time a discretized weak solution of the equations. Furthermore, the approach is in principle well suited to handling complex geometries, and requires a careful analysis of the generation of the control volumes; this has been discussed elsewhere in this book.

Further developments have been necessary in order to extend the theoretical results obtained by Lax and others for one-dimensional non-linear problems to multidimensional situations. A critical survey of such effort with the emphasis on the Euler and Navier–Stokes equations is the subject of the following sections.

Many schemes have been derived from the Lax–Wendroff scheme, the most popular being MacCormack's (1969) (and its variants), which date back to the 1960s. Several methods have been proposed, assuming the separability of space and time discretizations, giving rise to the following (rather general) classification: explicit or implicit schemes; single- or multistep methods as far as time integration is concerned; centred or uncentred (so called upwind methods) schemes when referring to space discretization. In the last 5–10 years different categories of schemes preventing the generation of numerical oscillations have been proposed (and successfully employed) following the Flux Corrected Transport (FCT) concept developed by Boris and Book (1973): the Total Variation Diminishing (TVD) concept first developed by Harten (1983), with the recent interpretation of Jameson as a Local Extremum Diminishing (LED) principle (Jameson 1993), the monotone upstream discretization MUSCL concept of Van Leer (1977, 1979), etc.

In this chapter most of these schemes and their applications will be discussed. Moreover, special topics regarding the treatment of the boundary conditions, the related stability analysis, and efficiency and accuracy improvements will also be introduced.

II GOVERNING EQUATIONS

For a given thermodynamic system having two intensive degrees of freedom, its fluid dynamic behaviour can generally be described by means of the system of conservation laws corresponding to the conservation of total mass, momentum and energy.

Let **W** be the vector unknown defined as

$$\mathbf{W} = [\rho, \rho u, \rho v, \rho w, \rho E]^{\mathrm{T}} = [w_1, w_2, w_3, w_4, w_5]^{\mathrm{T}} \tag{1}$$

where E is the total energy ($E = e + u^2/2$). Let V be any volume with bounding surface ∂V and outward unit normal **n**, assuming that the volume does not vary with time, **W** satisfies the following integral conservation laws

$$\frac{\mathrm{d}}{\mathrm{d}t} \int_V \mathbf{W} \, \mathrm{d}V = \int_V \frac{\partial}{\partial t} \mathbf{W} \, \mathrm{d}V = -\oint_{\partial V} \mathbf{F} \cdot \mathbf{n} \, \mathrm{d}S \tag{2}$$

that (for an inertial reference system) in the equivalent differential form is

$$\frac{\partial}{\partial t} \mathbf{W} = -\nabla \cdot \mathbf{F} \tag{3}$$

where **F** is the flux of **W** across the bounding surface ∂V, which accounts for the inviscid ($\mathbf{F}_E$) and viscous ($\mathbf{F}_V$) contributions, i.e.

$$\mathbf{F} = \mathbf{F}_E - \mathbf{F}_V, \tag{4}$$

where

$$\mathbf{F}_E = [\rho \mathbf{u}, \rho \mathbf{u}\mathbf{u} + p\mathbf{I}, \rho \mathbf{u}H]^{\mathrm{T}}$$
$$\mathbf{F}_V = [0, \boldsymbol{\sigma}, -(\mathbf{q} - \mathbf{u} \cdot \boldsymbol{\sigma})]^{\mathrm{T}}$$

and $p, H, \boldsymbol{\sigma}, \mathbf{q}$ are, respectively, the thermodynamic pressure, the total enthalpy, the stress tensor and the heat flux, and **I** is the unit tensor. For an ideal gas the pressure is related to the density and temperature according to the equation of state

$$p = \rho RT \tag{5}$$

In the present chapter we are only concerned with simulations of laminar flows, hence the stress tensor and the heat flux are determined according to Newton's and Fourier's laws

$$\boldsymbol{\sigma} = \mu(\nabla \mathbf{u} + \nabla \mathbf{u}^{\mathrm{T}}) - \tfrac{2}{3}\mu \nabla \cdot \mathbf{u}\mathbf{I}$$
$$\mathbf{q} = -\lambda \nabla T .$$

From kinetic theory it can be shown that, for a thermodynamic system having only two intensive degrees of freedom, the viscosity (μ) and the heat conduction (λ) coefficients depend only upon the gas temperature. In particular, μ can be evaluated according to Sutherland's law

$$\mu = \mu_r \frac{T_r + 110}{T + 110} (T/T_r)^{\frac{3}{2}} \tag{6}$$

where T is in Kelvin, and $\lambda = c_p \mu / Pr$, where Pr is the the Prandtl number and c_p the specific heat at constant pressure (which can be assumed constant for temperatures below 1000 K), and μ_r and T_r are reference values ($\mu_r = 1.789 \times 10^{-5}$ kg m^{-1} s^{-1}, $T_r = 288$ K).

The conservation equations are formally closed once the rate of all fluxes on the boundary ∂V and the initial state of $\mathbf{W}$ are known. However, the determination of the boundary fluxes and the mathematical closure is a difficult task. Indeed, the problem of closure is not yet fully resolved; moreover, the numerical treatment of the boundary conditions is a very critical issue that deserves a detailed analysis; this will be discussed in a later section.

It is important to point out that for steady flows the right-hand-side of the governing equations must vanish: in computational fluid dynamics this is equivalent to saying that the residual of the steady equations goes to zero.

The character of the Navier–Stokes is parabolic/hyperbolic, or more precisely incompletely parabolic in the unsteady case (Strikwerda, 1977; Gustafsson and Sundström, 1978), while it is parabolic/elliptic in the steady state. The Euler equations can be interpreted as the limit of vanishing μ and λ, and their character is hyperbolic in the unsteady case, while the associated steady potential problem is either elliptic, hyperbolic or mixed depending upon the Mach number (i.e. depending upon whether the flow is either subsonic, supersonic or transonic).

A Quasilinear conservative/non-conservative formulations of Euler equations

Most of the flow situations of practical interest correspond to high Reynolds numbers, i.e. flows which are convection dominated. Hence, in the present section we first discuss different forms of the Euler equations for a better understanding of the schemes that will be analysed in later sections. In particular, the matrices that diagonalize the inviscid flux jacobian and the characteristic variables associated with either the conservative or non-conservative formulations are derived.

From a physical–mathematical point of view the Euler equations can be viewed as the limit of the Navier–Stokes equations for vanishing diffusion effects (i.e. $\mu \rightarrow 0$ and $\lambda \rightarrow 0$), and can be simply obtained by setting $\mathbf{F}_V = 0$ in (2) or (3). In cartesian coordinates the non-conservative differential form of the Euler equations is

$$\frac{\partial \mathbf{W}}{\partial t} + \mathbf{A}\frac{\partial \mathbf{W}}{\partial x} + \mathbf{B}\frac{\partial \mathbf{W}}{\partial y} + \mathbf{C}\frac{\partial \mathbf{W}}{\partial z} = 0 \qquad (7)$$

where $\mathbf{A}, \mathbf{B}$ and $\mathbf{C}$ are, respectively, the jacobian matrices of the inviscid flux vector (where the subscript E has been dropped) with respect to the conservative variable vector

$$(\mathbf{A}, \mathbf{B}, \mathbf{C}) = \left(\frac{\partial \mathbf{F}_x}{\partial \mathbf{W}}, \frac{\partial \mathbf{F}_y}{\partial \mathbf{W}}, \frac{\partial \mathbf{F}_z}{\partial \mathbf{W}} \right)$$

and $\mathbf{F}_x$, $\mathbf{F}_y$ and $\mathbf{F}_z$ represent, respectively, the x, y and z components of $\mathbf{F}$.
For a thermodynamic system whose equation of state is of the form

$$p = \rho f(e) \tag{8}$$

where e is the internal energy per unit mass, the inviscid flux is a homogeneous function of degree 1 of the conservative variable vector $\mathbf{W}$, i.e.

$$\mathbf{F} = \frac{\partial \mathbf{F}}{\partial \mathbf{W}} \mathbf{W} \tag{9}$$

and the Euler equations can also be cast in the following form

$$\frac{\partial}{\partial t} \mathbf{W} + \frac{\partial}{\partial x} (\mathbf{A}\mathbf{W}) + \frac{\partial}{\partial y} (\mathbf{B}\mathbf{W}) + \frac{\partial}{\partial z} (\mathbf{C}\mathbf{W}) = 0$$

For example, from (9) and assuming for equation of state $p = (\gamma - 1)\rho e$, one obtains

$$\mathbf{F}_x = \begin{pmatrix} w_2 \\ \dfrac{3-\gamma}{2} \dfrac{w_2^2}{w_1} - \dfrac{\gamma-1}{2} \dfrac{w_3^2 + w_4^2}{w_1} + (\gamma-1)w_5 \\ \dfrac{w_2 w_3}{w_1} \\ \dfrac{w_2 w_4}{w_1} \\ w_2 H \end{pmatrix} \tag{10}$$

while the jacobian matrix $\mathbf{A}$ is

$$\mathbf{A} = \begin{pmatrix} 0 & 1 & 0 & 0 & 0 \\ \dfrac{\gamma-1}{2} q^2 - \dfrac{w_2^2}{w_1^2} & (3-\gamma)\dfrac{w_2}{w_1} & (1-\gamma)\dfrac{w_3}{w_1} & (1-\gamma)\dfrac{w_4}{w_1} & \gamma-1 \\ -\dfrac{w_2 w_3}{w_1^2} & \dfrac{w_3}{w_1} & \dfrac{w_2}{w_1} & 0 & 0 \\ -\dfrac{w_2 w_4}{w_1^2} & \dfrac{w_4}{w_1} & 0 & \dfrac{w_2}{w_1} & 0 \\ \left(\dfrac{\gamma-1}{2} q^2 - H \right)\dfrac{w_2}{w_1} & H + (1-\gamma)\dfrac{w_2^2}{w_1^2} & (1-\gamma)\dfrac{w_2 w_3}{w_1^2} & (1-\gamma)\dfrac{w_2 w_4}{w_1^2} & \gamma\dfrac{w_2}{w_1} \end{pmatrix}$$

where γ is the specific heat ratio, and H and q^2 are, respectively, the total enthalpy and the square of the velocity modulus

$$H = \gamma \frac{w_5}{w_1} - \frac{\gamma - 1}{2} q^2; \quad q^2 = \frac{(w_2^2 + w_3^2 + w_4^2)}{w_1^2}$$

Likewise, $\mathbf{F}_y$, $\mathbf{F}_z$, $\mathbf{B}$ and $\mathbf{C}$ are given by the following expressions

$$\mathbf{F}_y = \begin{pmatrix} w_3 \\ \dfrac{w_2 w_3}{w_1} \\ \dfrac{3-\gamma}{2}\dfrac{w_3^2}{w_1} - \dfrac{\gamma-1}{2}\dfrac{w_2^2 + w_4^2}{w_1} + (\gamma-1)w_5 \\ \dfrac{w_3 w_4}{w_1} \\ w_3 H \end{pmatrix} \tag{11}$$

$$\mathbf{F}_z = \begin{pmatrix} w_4 \\ \dfrac{w_2 w_4}{w_1} \\ \dfrac{w_3 w_4}{w_1} \\ \dfrac{3-\gamma}{2}\dfrac{w_4^2}{w_1} - \dfrac{\gamma-1}{2}\dfrac{w_2^2 + w_3^2}{w_1} + (\gamma-1)w_5 \\ w_4 H \end{pmatrix} \tag{12}$$

$$\mathbf{B} = \begin{pmatrix} 0 & 0 & 1 & 0 & 0 \\ -\dfrac{w_2 w_3}{w_1^2} & \dfrac{w_3}{w_1} & \dfrac{w_2}{w_1} & 0 & 0 \\ \dfrac{\gamma-1}{2}q^2 - \dfrac{w_3^2}{w_1^2} & (1-\gamma)\dfrac{w_2}{w_1} & (3-\gamma)\dfrac{w_3}{w_1} & (1-\gamma)\dfrac{w_4}{w_1} & \gamma-1 \\ -\dfrac{w_3 w_4}{w_1^2} & 0 & \dfrac{w_4}{w_1} & \dfrac{w_3}{w_1} & 0 \\ \left(\dfrac{\gamma-1}{2}q^2 - H\right)\dfrac{w_3}{w_1} & (1-\gamma)\dfrac{w_2 w_3}{w_1^2} & H+(1-\gamma)\dfrac{w_3^2}{w_1^2} & (1-\gamma)\dfrac{w_3 w_4}{w_1^2} & \gamma\dfrac{w_3}{w_1} \end{pmatrix}$$

$$\mathbf{C} = \begin{pmatrix} 0 & 0 & 0 & 1 & 0 \\[2mm] -\dfrac{w_2 w_4}{w_1^2} & \dfrac{w_4}{w_1} & 0 & \dfrac{w_2}{w_1} & 0 \\[2mm] -\dfrac{w_3 w_4}{w_1^2} & 0 & \dfrac{w_4}{w_1} & \dfrac{w_3}{w_1} & 0 \\[2mm] \dfrac{\gamma-1}{2}q^2 - \dfrac{w_4^2}{w_1^2} & (1-\gamma)\dfrac{w_2}{w_1} & (1-\gamma)\dfrac{w_3}{w_1} & (3-\gamma)\dfrac{w_4}{w_1} & \gamma-1 \\[2mm] \left(\dfrac{\gamma-1}{2}q^2 - H\right)\dfrac{w_4}{w_1} & (1-\gamma)\dfrac{w_2 w_4}{w_1^2} & (1-\gamma)\dfrac{w_4 w_3}{w_1^2} & H + (1-\gamma)\dfrac{w_4^2}{w_1^2} & \gamma\dfrac{w_4}{w_1} \end{pmatrix}$$

The design of most of the numerical schemes for the inviscid flux requires the evaluation of the eigenvalues of the jacobian matrices as well as their left and right eigenvector matrices. For that purpose, introducing the primitive variable vector $\mathbf{U}$ defined as $\mathbf{U} = [\rho, u, v, w, p]^{\mathbf{T}} = [u_1, u_2, u_3, u_4, u_5]^{\mathbf{T}}$, it is easy to show that the conservative variables are related to $\mathbf{U}$ according to the following transformation $\mathbf{W}(\mathbf{U})$

$$w_1 = u_1$$

$$w_2 = u_1 u_2$$

$$w_3 = u_1 u_3$$

$$w_4 = u_1 u_4$$

$$w_5 = u_1 \frac{(u_2^2 + u_3^2 + u_4^2)}{2} + \frac{u_5}{(\gamma - 1)}$$

Then, from (7) one obtains

$$\mathbf{M}\frac{\partial \mathbf{U}}{\partial t} + \mathbf{AM}\frac{\partial \mathbf{U}}{\partial x} + \mathbf{BM}\frac{\partial \mathbf{U}}{\partial y} + \mathbf{CM}\frac{\partial \mathbf{U}}{\partial z} = 0$$

or, equivalently

$$\frac{\partial \mathbf{U}}{\partial t} + \tilde{\mathbf{A}}\frac{\partial \mathbf{U}}{\partial x} + \tilde{\mathbf{B}}\frac{\partial \mathbf{U}}{\partial y} + \tilde{\mathbf{C}}\frac{\partial \mathbf{U}}{\partial z} = 0$$

where

$$
\mathbf{M} \equiv \frac{\partial \mathbf{W}}{\partial \mathbf{U}} = \begin{pmatrix} 1 & 0 & 0 & 0 & 0 \\ u & \rho & 0 & 0 & 0 \\ v & 0 & \rho & 0 & 0 \\ w & 0 & 0 & \rho & 0 \\ \dfrac{q^2}{2} & \rho u & \rho v & \rho w & \dfrac{1}{\gamma - 1} \end{pmatrix}
$$

$$
\mathbf{M}^{-1} = \left(\frac{\partial \mathbf{W}}{\partial \mathbf{U}}\right)^{-1} = \begin{pmatrix} 1 & 0 & 0 & 0 & 0 \\ -\dfrac{u}{\rho} & \dfrac{1}{\rho} & 0 & 0 & 0 \\ -\dfrac{v}{\rho} & 0 & \dfrac{1}{\rho} & 0 & 0 \\ -\dfrac{w}{\rho} & 0 & 0 & \dfrac{1}{\rho} & 0 \\ (\gamma - 1)\dfrac{q^2}{2} & (1 - \gamma)u & (1 - \gamma)v & (1 - \gamma)w & \gamma - 1 \end{pmatrix}
$$

and the matrices $(\widetilde{\mathbf{A}}, \widetilde{\mathbf{B}}, \widetilde{\mathbf{C}})$ are related to $(\mathbf{A}, \mathbf{B}, \mathbf{C})$ by the similarity transformation

$$
\widetilde{\mathbf{A}} = \mathbf{M}^{-1}\mathbf{A}\mathbf{M}
$$
$$
\widetilde{\mathbf{B}} = \mathbf{M}^{-1}\mathbf{B}\mathbf{M}
$$
$$
\widetilde{\mathbf{C}} = \mathbf{M}^{-1}\mathbf{C}\mathbf{M}
$$

Let $\mathbf{k}$ be any arbitrary direction, the hyperbolicity of the transformed Euler equations requires that the $(\mathbf{k})$ projected matrix $(\widetilde{\mathbf{A}}, \widetilde{\mathbf{B}}, \widetilde{\mathbf{C}}) \cdot \mathbf{k}$ has real eigenvalues and a set of linearly independent eigenvectors. Let $\widetilde{\mathbf{J}}$ be such a matrix

$$\widetilde{\mathbf{J}} = \widetilde{\mathbf{A}}k_x + \widetilde{\mathbf{B}}k_y + \widetilde{\mathbf{C}}k_z = \begin{pmatrix} V_k & \rho k_x & \rho k_y & \rho k_z & 0 \\ 0 & V_k & 0 & 0 & \dfrac{k_x}{\rho} \\ 0 & 0 & V_k & 0 & \dfrac{k_y}{\rho} \\ 0 & 0 & 0 & V_k & \dfrac{k_z}{\rho} \\ 0 & \rho c^2 k_x & \rho c^2 k_y & \rho c^2 k_z & V_k \end{pmatrix} \tag{13}$$

where $c^2 = (\gamma - 1)\left(H - (q^2/2)\right)$, and $V_k = \mathbf{u} \cdot \mathbf{k}$. It is easy to verify that $\widetilde{\mathbf{J}}$ has the following eigenvalues

$$\begin{aligned} \lambda_1 &= V_k - c \\ \lambda_2 &= V_k \\ \lambda_3 &= V_k + c \\ \lambda_4 &= V_k \\ \lambda_5 &= V_k \end{aligned} \tag{14}$$

which are also the eigenvalues of the matrix $\mathbf{J} = \mathbf{A}k_x + \mathbf{B}k_y + \mathbf{C}k_z$. As a consequence, the eigenvectors of the conservative jacobian are immediately determined once those of the transformed (non-conservative) jacobian are known. Indeed, let $\mathbf{L}$ and $\mathbf{R}$ be the left and right eigenvector matrices of $\mathbf{J}$ (satisfying $\mathbf{LR} = \mathbf{I}$), and let $\widetilde{\mathbf{L}}$ and $\widetilde{\mathbf{R}}$ indicate those of $\widetilde{\mathbf{J}}$, then

$$\widetilde{\mathbf{J}} = \mathbf{M}^{-1}\mathbf{JM} \tag{15}$$

and

$$\Lambda = \widetilde{\mathbf{L}}\widetilde{\mathbf{J}}\widetilde{\mathbf{R}} \tag{16}$$

Substituting (15) into (16) yields

$$\Lambda = \widetilde{\mathbf{L}}\mathbf{M}^{-1}\mathbf{JM}\widetilde{\mathbf{R}}$$

which shows that the left and right eigenvector matrices of $\mathbf{J}$ are related to those of $\widetilde{\mathbf{J}}$ according to

$$\mathbf{L} = \widetilde{\mathbf{L}}\mathbf{M}^{-1}$$

$$\mathbf{R} = \mathbf{M}\widetilde{\mathbf{R}} \tag{17}$$

Observe that due to the multiplicity of the eigenvalue V_k one has two additional degrees of freedom for determining the set of linearly independent left eigenvectors (the rows of $\mathbf{L}$). Following Hirsch (1990), a particular choice yields

$$
\widetilde{\mathbf{L}} = \begin{pmatrix}
0 & -k_x & -k_y & -k_z & \dfrac{1}{\rho c} \\[2mm]
k_x & 0 & k_z & -k_y & \dfrac{-k_x}{c^2} \\[2mm]
0 & k_x & k_y & k_z & \dfrac{1}{c} \\[2mm]
k_y & -k_z & 0 & k_x & \dfrac{-k_y}{c^2} \\[2mm]
k_z & k_y & -k_x & 0 & \dfrac{-k_z}{c^2}
\end{pmatrix}
\tag{18}
$$

$$
\widetilde{\mathbf{R}} = \begin{pmatrix}
\dfrac{\rho}{2c} & k_x & \dfrac{\rho}{2c} & k_y & k_z \\[2mm]
\dfrac{-k_x}{2} & 0 & \dfrac{k_x}{2} & -k_z & k_y \\[2mm]
\dfrac{-k_y}{2} & k_z & \dfrac{k_y}{2} & 0 & -k_x \\[2mm]
\dfrac{-k_z}{2} & -k_y & \dfrac{k_z}{2} & k_x & 0 \\[2mm]
\dfrac{\rho c}{2} & 0 & \dfrac{\rho c}{2} & 0 & 0
\end{pmatrix}
\tag{19}
$$

Other expressions can be obtained. For example, introducing a local orthonormal basis $(\mathbf{l}, \mathbf{m}, \mathbf{n})$ and choosing $\mathbf{k} = \mathbf{n}$, Gnoffo (1989), Yee (1989), Manna (1992) etc. exploit the multiplicity of the eigenvalue $\mathbf{V} \cdot \mathbf{n}$ in order to obtain different $\widetilde{\mathbf{L}}$ and $\widetilde{\mathbf{R}}$, which corresponds to a simple wave decomposition in the directions $\mathbf{l}, \mathbf{m}$ and $\mathbf{n}$ (as will be shown in a later section, this choice allows the reduction of a multidimensional formulation to a locally one-dimensional one). In particular, in a finite-volume formulation $\mathbf{n}$ is identified with the outward unit normal to the control surface, and the matrices $\widetilde{\mathbf{L}}$ and $\widetilde{\mathbf{R}}$ become

$$
\widetilde{\mathbf{L}} = \begin{pmatrix}
0 & -\dfrac{n_x \rho}{2c} & -\dfrac{n_y \rho}{2c} & -\dfrac{n_z \rho}{2c} & \dfrac{1}{2c^2} \\[2mm]
1 & 0 & 0 & 0 & -\dfrac{1}{c^2} \\[2mm]
0 & \dfrac{n_x \rho}{2 c} & \dfrac{n_y \rho}{2 c} & \dfrac{n_z \rho}{2 c} & \dfrac{1}{2c^2} \\[2mm]
0 & \rho \ell_x & \rho \ell_y & \rho \ell_z & 0 \\[2mm]
0 & \rho m_x & \rho m_y & \rho m_z & 0
\end{pmatrix}
\tag{20}
$$

$$\widetilde{\mathbf{R}} = \begin{pmatrix} 1 & 1 & 1 & 0 & 0 \\ -\dfrac{n_x c}{\rho} & 0 & \dfrac{n_x c}{\rho} & \dfrac{\ell_x}{\rho} & \dfrac{m_x}{\rho} \\ -\dfrac{n_y c}{\rho} & 0 & \dfrac{n_y c}{\rho} & \dfrac{\ell_y}{\rho} & \dfrac{m_y}{\rho} \\ -\dfrac{n_z c}{\rho} & 0 & \dfrac{n_z c}{\rho} & \dfrac{\ell_z}{\rho} & \dfrac{m_z}{\rho} \\ c^2 & 0 & c^2 & 0 & 0 \end{pmatrix} \tag{21}$$

The conservative left and right eigenvector matrices can easily be determined using the transformation relation of (17) thus obtaining

$$\mathbf{L} = \begin{pmatrix} \dfrac{1}{2}\left(b_1 + \dfrac{q_n}{c}\right) & -\dfrac{1}{2}\left(b_2 u + \dfrac{n_x}{c}\right) & -\dfrac{1}{2}\left(b_2 v + \dfrac{n_y}{c}\right) & -\dfrac{1}{2}\left(b_2 w + \dfrac{n_z}{c}\right) & \dfrac{b_2}{2} \\ 1 - b_1 & b_2 u & b_2 v & b_2 w & -b_2 \\ \dfrac{1}{2}\left(b_1 - \dfrac{q_n}{c}\right) & -\dfrac{1}{2}\left(b_2 u - \dfrac{n_x}{c}\right) & -\dfrac{1}{2}\left(b_2 v - \dfrac{n_y}{c}\right) & -\dfrac{1}{2}\left(b_2 w - \dfrac{n_z}{c}\right) & \dfrac{b_2}{2} \\ -q_\ell & \ell_x & \ell_y & \ell_z & 0 \\ -q_m & m_x & m_y & m_z & 0 \end{pmatrix} \tag{22}$$

$$\mathbf{R} = \begin{pmatrix} 1 & 1 & 1 & 0 & 0 \\ u - cn_x & u & u + cn_x & \ell_x & m_x \\ v - cn_y & v & v + cn_y & \ell_y & m_y \\ w - cn_z & w & w + cn_z & \ell_z & m_z \\ H - cq_n & \dfrac{q^2}{2} & H + cq_n & q_\ell & q_m \end{pmatrix} \tag{23}$$

where

$$q_m = um_x + vm_y + wm_z$$
$$q_\ell = u\ell_x + v\ell_y + w\ell_z$$
$$q_n = un_x + vn_y + wn_z$$

and

$$b_1 = b_2 \frac{q^2}{2}; \ b_2 = \frac{\gamma - 1}{c^2}$$

It is important to note that in general the (x, y, z) components $(\mathbf{A}, \mathbf{B}, \mathbf{C})$ of the flux jacobian do not commute. As a consequence, they cannot be simultaneously diagonalized by $\mathbf{L}$ and $\mathbf{R}$, and the system of equations in characteristic form cannot be decoupled. However, one can define a pfaffian characteristic form δW associated with a direction of propagation $\mathbf{n}$ as

$$\delta W = \mathbf{L}\delta \mathbf{W} \tag{24}$$

where $\delta \mathbf{W}$ is the time variation of $\mathbf{W}$. From the equation of state we have

$$\delta \rho E = \frac{1}{\gamma - 1}\delta p + \mathbf{u} \cdot \delta \rho \mathbf{u} - \frac{q^2}{2}\delta \rho \tag{25}$$

and (24) becomes

$$\delta W = \begin{bmatrix} \dfrac{1}{2}\left(-\dfrac{\rho}{c}\delta q_n + \dfrac{\delta p}{c^2}\right) \\[2ex] \delta \rho - \dfrac{\delta p}{c^2} \\[2ex] \dfrac{1}{2}\left(\dfrac{\rho}{c}\delta q_n + \dfrac{\delta p}{c^2}\right) \\[2ex] \rho \delta q_\ell \\[2ex] \rho \delta q_m \end{bmatrix} \tag{26}$$

It must be pointed out that a pfaffian characteristic form δW (sometimes called the difference of characteristic variables; Yee, 1989) can always be introduced even though the characteristic variables W may not exist (for example, in the case where $\mathbf{L}$ is not a constant coefficient matrix or in general in the presence of multidimensional flows). Keeping in mind that whenever we talk of characteristic variables we actually refer to the difference of characteristic variables, simple wave-like solutions correspond to waves propagating with speed λ_i and intensity δW_i, where λ_i is the i-th eigenvalue and δW_i is the i-th component of δW. In particular:

- the wave propagating in the direction $\mathbf{n}$ with speed $q_n - c$ (resp. $q_n + c$) is an acoustic wave of intensity δW_1 (resp. δW_3) that produces a normal velocity variation;
- the wave propagating with speed q_n and intensity δW_2 is defined as an entropy wave associated with entropy variations;
- the waves propagating with speed q_n and intensities δW_4 and δW_5 are shear waves (also defined as vorticity waves) that produce velocity variations.

The importance of a simple wave decomposition will become clearer in later sections where the treatment of the boundary conditions is addressed and where upwind methods are discussed.

III FINITE VOLUME FORMULATION

In the present section the problem associated with the approximate solutions of
the viscous (NS) or non-viscous (E) conservation equations in their integral
formulation is addressed. From a mathematical point of view these solutions
can be identified with those of the so-called weak formulation supplemented (for
the Euler equations) with an entropy condition, to discard the weak solutions
that cannot be viewed as the limit of NS solutions for vanishing diffusion
coefficients. Obviously, the functional space containing the weak solutions is
generally larger than the one where the strong solutions are sought, owing to the
change in the character and the order of the mathematical operator involved in
the formulation.

In general finding an approximate solution is equivalent to restricting the
functional space where the solution is sought. The approximation space is
defined without ambiguities in finite-element and/or spectral methods.
Conversely, such a space is not immediately and/or not univocally recognizable
(at least in a traditional functional meaning) for finite difference (FD) and/or
finite-volume (FV) methods. It is typical of FV approximations to limit the
verification of the balance equations only on a finite set (S) of all possible
subsets of $\mathbb{R}^3 \times [0, t]$, and refer to average quantities. Of course, S can be
thought of as being made out of all possible unions of elementary subsets (the
control volumes). If the chosen mathematical structure of the elementary
balance satisfies an additive property, then it suffices to verify the conservation
only on the elementary control volumes. Observe that it is also possible to
interpret the FV as a projection and/or weighted residual approach by the use of
suitable test functions associated with each elementary control volume.

The weak fully discretized formulation of the system of conservation laws
corresponding to (2) is then equivalent to requiring that

$$\int_{C\Omega} \mathbf{W}(\mathbf{x}, t^{n+1}) \, dC\,\Omega - \int_{C\Omega} \mathbf{W}(\mathbf{x}, t^n) \, dC\,\Omega + \int_{t^n}^{t^{n+1}} \oint_{BC\Omega} \mathbf{F}(\mathbf{W}, \mathbf{x}, t) \cdot \mathbf{n} \, dS \, dt = 0$$

$$(27)$$

where $C\Omega$, and $BC\Omega$ are, respectively, the control volume and its bounding
surface. The above equation can be recast as

$$\widetilde{\mathbf{W}}^{n+1} - \widetilde{\mathbf{W}}^n + \lambda \oint_{\partial V} \overline{\mathbf{F}}(\mathbf{W}) \cdot \mathbf{n} \, dS = 0 \qquad (28)$$

where

$$\widetilde{\mathbf{W}} m(C\Omega) = \int_{C\Omega} \mathbf{W} \, dC\Omega, \quad m(C\Omega) = \int_{C\Omega} dC\Omega = V$$

and

$$\overline{\mathbf{F}}(\mathbf{W}) = \frac{1}{t^{n+1} - t^n} \int_{t^n}^{t^{n+1}} \mathbf{F}(\mathbf{W}(\mathbf{x}_B, t)) \, dt \tag{29}$$

with $\mathbf{x}_B$ the boundary surface coordinates, and $\lambda = (t^{n+1} - t^n)/V$.

The semidiscretized formulation is easily obtained by taking the limit for $t^{n+1} \to t^n$ of (27), yielding

$$\frac{d}{dt} \int_{C\Omega} \mathbf{W}(\mathbf{x}, t) \, dC \, \Omega + \oint_{BC\Omega} \mathbf{F}(\mathbf{W}, \mathbf{x}, t) \cdot \mathbf{n} \, dS = 0 \tag{30}$$

that, introducing the volume averaged quantities, reduces to

$$\frac{d}{dt} \widetilde{\mathbf{W}} + \frac{1}{V} \oint_{\partial V} \mathbf{F}(\mathbf{W}(\mathbf{x}, t)) \cdot \mathbf{n} \, dS = 0 \tag{31}$$

The flux $\overline{\mathbf{F}}(\mathbf{W})$ depends continuously on $\mathbf{x}_B$ even in the case that the weak solution $\mathbf{W}(\mathbf{x}, t)$ presents first-order discontinuities. Indeed, a discontinuity (on F) may appear only at one isolated instant within a sufficiently small time interval $(t^{n+1} - t^n)$. In fact, the permanence of the discontinuity over the entire interval would correspond to a zero shock velocity; however, from the Rankine–Hugoniot jump conditions, this would imply either that the jump in F is zero or that the jump conditions are violated. This observation also sheds light on the smoothing mechanism by which the space–time averaging, peculiar of the finite-volume fully discretized approach, allows the capturing of a shock.

Then, for the fully discretized control volume approach corresponding to (28) a central role is played by the evaluation of a time averaged flux $\overline{\mathbf{F}}$. The latter can be determined only if the time evolution of the solution is 'known'! Hence, one must somehow predict such evolution in terms of the (time discretized) state vectors. This is generally the key point that characterizes finite-volume methods. Such an observation does not apply directly to the semidiscretized formulation. However, the issue arises in any case when one introduces the time integration algorithms for the solution of the system of ordinary differential equations (ODE) associated with (31).

From the above considerations it is clear that the implementation of a finite-volume method amounts to solving different classes of problems: (i) control volume decomposition; (ii) evaluation of the fluxes through internal control surfaces; (iii) evaluation of the fluxes at the (physical) domain boundaries (i.e. treatment of boundary conditions); and (iv) numerical solution of control volume schemes dealing with the design of algorithms for the solution of the discretized equations. It is important to note that the selection of the solution methods and their chronological implementation are intimately correlated, even though conceptually the four classes of problems are distinct. For example, the decomposition in control volumes may depend adaptively on the solution; likewise, for multidimensional problems when using a so-called splitting approach, points (ii)–(iv), are not solved separately, etc. In this chapter we are mainly concerned with the topics related to points (ii)–(iv), which will be discussed in separate sections, and we only briefly discuss few of the issues related to point (i).

A Control volume decomposition and surface vector definition

In order to produce an approximate solution one must obviously limit the number of variables (and equations) to finite values. This amounts to adopting a suitable set of state vectors of finite dimensions and a set of control volumes. Then, a grid needs to be generated in order to partition the physical volume in a finite set of elementary regions (or cells) characterized by vertices, interfaces, surface vectors, volumes, etc.

Once the unknown state vectors are chosen, the algebraic equations corresponding to (28), (31) can be constructed. Following the early practice of FD methods used for solving partial differential equations, one can assume as vector unknowns $\{\mathbf{W}_i^n\}$ (the set of local cell-centred values of $\mathbf{W}(\mathbf{x}, t)$), and then express and/or approximate the quantities $(\cdot)$, $\overline{(\cdot)}$, as well as the other integrals involved in the equations, all in terms of $\{\mathbf{W}_i^n\}$. In the same fashion, one can assume as unknowns $\{\widetilde{\mathbf{W}}_i^n\}$; however, the evolution of the integrals appearing in (28) and (31) requires a reconstruction of the unknown function $\mathbf{W}(\mathbf{x}, t)$.

The (approximate) reconstruction can, in principle, be either local or global. From a functional point of view, it remains in finite dimension spaces and it involves algebraic relations between the average values and the reconstruction parameters. As yet there is not a well founded theory for multidimensions that directly correlates the reconstruction parameters with the average values rather than the local ones. Therefore, to fill such a gap one (more or less) formally introduces the nodal values so as to exploit the results of numerical interpolation and theory of integration, although the distinction (until the advent of more modern schemes such as MUSCL, ENO, etc.) between the two approaches (based on the nodal and central values) was in some sense neglected. Indeed, for smooth solutions

$$\left| \widetilde{\mathbf{W}}_i - \mathbf{W}_i \right| = O(h^p)$$

where h is a representative diameter of the averaging control volume, and p is in general related to the order of the adopted reconstruction formula and the mesh geometry. Nevertheless, such a distinction is fundamental in characterizing most algorithms. An alternative discretization (the so-called vertex formulation) uses the set of vertex defined values, where the correspondence between vertices and cells may not be obvious.

In this section we briefly outline the procedure for surface and control volume definitions for a structured mesh (Chapter 7 addresses the problem of mesh generation). Assuming, for simplicity, that the mesh is made out of hexahedra, an example of surface and volume evaluation is now given. Referring to Figure 1 the algebraic equation corresponding to the fully discretized equation (28) is

$$\widetilde{\mathbf{W}}_{ijk}^{n+1} - \widetilde{\mathbf{W}}_{ijk}^{n} + \lambda \sum_{i_s=1,6} \left(\overline{\mathbf{F}} \cdot \mathbf{n}S \right)_{i_s} = 0 \tag{32}$$

where i_s is the index of the six hexahedron faces, and

$$\left(\overline{\mathbf{F}} \cdot \mathbf{n}S \right)_{i_s} = \int_{S_{i_s}} \overline{\mathbf{F}}(\mathbf{W}) \cdot \mathbf{n} \, dS$$

Likewise, referring to the semidiscretized formulation (Eq. (31)) one obtains

$$\Delta t \frac{d}{dt} \widetilde{\mathbf{W}}_{ijk} + \lambda \sum_{i_s=1,6} (\widetilde{\mathbf{F}} \cdot \mathbf{n}S)_{i_s} = 0 \tag{33}$$

where now $\widetilde{\mathbf{F}}$ stands for

$$\left(\widetilde{\mathbf{F}} \cdot \mathbf{n}S \right)_{i_s} = \int_{S_{i_s}} \mathbf{F}(\mathbf{W}) \cdot \mathbf{n} \, dS$$

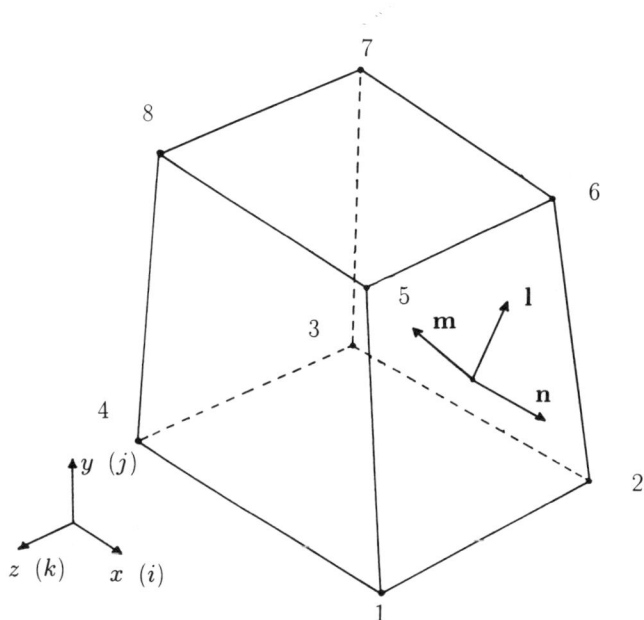

Figure 1 Three-dimensional computational volume.

The surface vector $(\mathbf{n}S)_{i_S}$ of outward unit normal $\mathbf{n}$ is defined as

$$(\mathbf{n}S)_{i_S} = \int_{S_{i_S}} \mathbf{n} \, dS$$

Let us introduce the notation $\mathbf{S}_{i_1,i_2,...,i_k} = (\mathbf{n}S)_{i_S}$, where $(i_1, i_2 \ldots i_k)$ are the node indices of the cell face of index i_S. Referring to Figure 1, the surface vector defined by the four corner points (5, 6, 7, 8) is estimated as follows.

In general, the four corners may not belong to the same planar surface. A unique definition is obtained by evaluating $\mathbf{S}_{5678}$ as the vector sum of the surfaces of the two triangles 567 and 578, where the positive sign of $\mathbf{S}_{5678}$ coincides with the positive outward direction. Let $\mathbf{r}_i$ be the vector position of the generic node i

$$\mathbf{r}_i = x_i \mathbf{i} + y_i \mathbf{j} + z_i \mathbf{k} \tag{34}$$

where $\mathbf{i}$, $\mathbf{j}$ and $\mathbf{k}$ are the unit vectors in the x, y, z coordinate directions

$$\mathbf{S}_{567} = \tfrac{1}{2}(\mathbf{r}_5 - \mathbf{r}_7) \times (\mathbf{r}_6 - \mathbf{r}_7)$$

$$\tag{35}$$

$$\mathbf{S}_{578} = \tfrac{1}{2}(\mathbf{r}_8 - \mathbf{r}_7) \times (\mathbf{r}_6 - \mathbf{r}_8)$$

The vector sum of the above equations yields an expression for the vector surface $\mathbf{S}_{5678}$

$$\mathbf{S}_{5678} = \tfrac{1}{2}(\mathbf{r}_5 - \mathbf{r}_7) \times (\mathbf{r}_6 - \mathbf{r}_8), \tag{36}$$

which shows that the vector surface is the vector product of the vectors connecting opposite vertices. A simple inspection of (36) shows that the formula is invariant to a cyclic permutation of the indices, i.e. to the partitioning in triangles. It is also easy to verify that the formula is still valid if one introduces the partitioning in four triangles having a common arbitrary vertex.

The evaluation of the volume is more critical (Kordulla and Vinokur, 1983). In general, the volume of a hexahedron is evaluated as the sum of the volumes of the tetrahedra corresponding to its partitioning. Such a partitioning is not unique and may give rise to different algebraic expressions that differ in the cost (i.e. number of operations) and accuracy. Kordulla and Vinokur (1983) eliminated this inconvenience by using a symmetric partitioning of the faces decomposing the hexahedron into three tetrahedra, thus obtaining

$$V_{12345678} = \tfrac{1}{3}(\mathbf{r}_5 - \mathbf{r}_1) \cdot (\mathbf{S}_{1234} + \mathbf{S}_{3762} + \mathbf{S}_{3784}) \tag{37}$$

In two dimensions, referring to Figure 2, the 'line' normal vector and the cell area become

$$\mathbf{l}_{12} = (y_2 - y_1)\mathbf{i} - (x_2 - x_1)\mathbf{j}$$

$$S_{1234} = \tfrac{1}{2}[(x_1 - x_3)(y_2 - y_4) - (y_1 - y_3)(x_2 - x_4)] \tag{38}$$

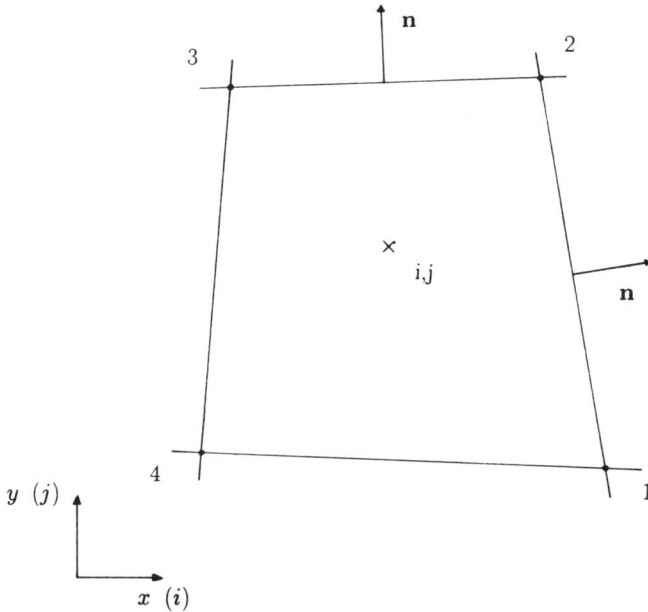

Figure 2 Two-dimensional computational volume.

IV DISCRETIZATION OF THE INVISCID FLUX

In the present section the problem of the discretization of the inviscid flux is discussed in detail, with particular emphasis on multidimensional flows. However, since most of the analysis can only be done thoroughly for one-dimensional problems, we first discuss the concept of upwind discretization formulas, higher-order methods, central differencing, etc. assuming one-dimensional behaviour. A generalization of the results to two- and three-dimensional flows is then presented.

A Basic concepts

Let us consider a one-dimensional scalar conservation equation

$$\frac{\partial w}{\partial t} + \frac{\partial}{\partial x} f(w) = 0 \tag{39}$$

and let us assume as control volume a suitable cell whose measure is (see Figure 3)

$$m(C\Omega_i) = x_{i+\frac{1}{2}} - x_{i-\frac{1}{2}}$$

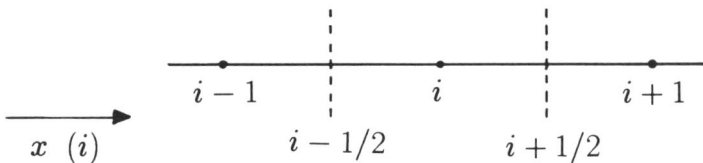

Figure 3 One-dimensional computational volume.

where $x_{i\pm\frac{1}{2}}$ identify the boundary of the *i-th* cell. Then, assuming uniform mesh spacing ($\Delta x = x_{i+\frac{1}{2}} - x_{i-\frac{1}{2}}$), the volume averaged value $\widetilde{w}_i$ becomes

$$\widetilde{w}_i = \frac{1}{\Delta x} \int_{x_i - \Delta x/2}^{x_i + \Delta x/2} w(x, t)\ \mathrm{d}x$$

In general, one can correlate the $\widetilde{w}_i$ to a set of nodal values w_i by means of a numerical integration formula

$$\widetilde{w}_i = \sum C_{ij} w_j \tag{40}$$

that, when $C_{ij} = \delta_{ij}$ (the Kronecker delta), yields the simplest (but often hidden and somewhat confusing) relation $\widetilde{w}_i = w_i$. Obviously (40) can also be inverted to obtain the w_i from the $\widetilde{w}_i$.

It is important to note that the real goal of any finite volume formulation is the design of the numerical flux. If we assume as state vectors the $\widetilde{w}_i$ (the unknowns in computational jargon), then all other quantities involved must be expressed as a function of $\widetilde{w}_i$ itself. Such quantities are mainly the fluxes at the boundaries. Let us define as numerical flux function ($h_{i\pm\frac{1}{2}}$) the relation that furnishes the approximation of the fluxes at the bounding control surfaces. Then, the semidiscretized equation corresponding to (31) is rewritten as

$$\frac{\mathrm{d}}{\mathrm{d}t} \widetilde{w}_i = -\frac{1}{\Delta x}(h_{i+\frac{1}{2}} - h_{i-\frac{1}{2}}) \tag{41}$$

where in principle $h_{i\pm\frac{1}{2}}$ may be a function of the entire set of $\widetilde{w}_i$. A suitable representation uses a symmetric form around $i \pm \frac{1}{2}$. For example, the numerical flux function at $i + \frac{1}{2}$ is generally written as

$$h_{i+\frac{1}{2}} = h(w_{i-m+1}, \ldots, w_{i+m}) \tag{42}$$

where the $\widetilde{\ }$ has been dropped.

Of course, such a symmetric notation does not exclude the vanishing of some of the partial derivatives $\partial h/\partial w_j$, i.e. h may not depend upon the entire set of the $2m$ values of w_j and therefore both centred and non-centred formulas are contemplated by (42). If the $2m$ components are all equal to w_c, then the numerical flux function must satisfy the consistency requirement

$$h(w_c, w_c, \ldots, w_c) = f(w_c) \qquad \forall w_c \tag{43}$$

If h is a differentiable function of the w's, consistency also implies

$$\sum_{j=-m+1}^{m} \frac{\partial h}{\partial w_j}\bigg|_{w_c} = \frac{df}{dw}\bigg|_{w_c} = a(w_c) \qquad \forall w_c \qquad (44)$$

The question now arises as to how to define a numerical flux function. Several approaches can be envisaged. For example, one could use an average and/or interpolation of $f(w_{i+m}) \ldots f(w_{i-m+1})$. The flux function could also be defined by evaluating the flux at an averaged and/or interpolated quantity w^* (of course a hybrid approach might also be envisaged). Setting $m = 1$ in (42), the flux function $h_{i+\frac{1}{2}}$ could then be defined as

(i) $h_{i+\frac{1}{2}}^{(f)} = \dfrac{f(w_i) + f(w_{i+1})}{2}$

(ii) $h_{i+\frac{1}{2}}^{(w)} = f(w^*)$

where, for example, $w^* = \mathcal{I}(w_i, w_{i+1})$ is an intermediate state between w_i and w_{i+1}, $\mathcal{I}$ being an interpolation operator.

(iii) $h_{i+\frac{1}{2}} = H(h_{i+\frac{1}{2}}^{(f)}, h_{i+\frac{1}{2}}^{(w)})$

where the superscripts (f) and (w) stand respectively for flux and variable interpolation approaches, and H is any continuous function such that $H(p,p) = p$; a natural choice might be $H(p,q) = (p + q)/2$.

The fully discretized formulation (32) becomes

$$\widetilde{w}_i^{n+1} - \widetilde{w}_i^n = -\lambda(\overline{h}_{i+\frac{1}{2}} - \overline{h}_{i-\frac{1}{2}}) \qquad (45)$$

where $\lambda = \Delta t / \Delta x$, and

$$\overline{h}_{i+\frac{1}{2}} = \frac{1}{\Delta t} \int_0^{\Delta t} f[w(x_{i+\frac{1}{2}}, t^n + \tau] \, d\tau \qquad (46)$$

The evaluation of the numerical flux function therefore requires two types of approximations associated with space averaging and time evolution. Observe that the evaluation of the time integral of (46) can be reduced to an 'upwind' space integration at fixed time t^n. Indeed, a solution of the linearized form of (39) in the neighbourhood of $x_{i+1/2}$ is a wave-like solution of the form $w(x_{i+\frac{1}{2}} + \zeta, t^n + \tau) = w(x_{i+\frac{1}{2}} + \zeta - a\tau, t^n)$, with $a = df/dw|_{i+\frac{1}{2}}$. Then, introducing the variable $s = x_{i+\frac{1}{2}} - a\tau$, (46) becomes

$$\overline{h} = \frac{1}{a\Delta t} \int_{x_{i+1/2}-a\Delta t}^{x_{i+1/2}} f(w(s, t^n)) \, ds \qquad (47)$$

The use of (47) for determining the numerical flux leads naturally to the so-called upwind schemes.

Alternatively, one may use a two-point numerical integration in time yielding

$$\bar{h}_{i+1/2} = h(\tilde{w}^n_{i+m}, \ldots, \tilde{w}^n_{i-m+1}; \tilde{w}^{n+1}_{i+m}, \ldots, \tilde{w}^{n+1}_{i-m+1}) \tag{48}$$

The numerical flux function must satisfy consistency and differentiability properties. Dropping again for (formal) simplicity $^-$ and $^\sim$, the (simplest) choice

$$\frac{\partial h}{\partial w^{n+1}_j} = 0 \, \forall j \tag{49}$$

yields the explicit form

$$w^{n+1}_i - w^n_i = -\lambda(h^n_{i+\frac{1}{2}} - h^n_{i-\frac{1}{2}}) \tag{50}$$

where $h^n_{i+\frac{1}{2}}$ is only a function of w^n.

In order to construct the numerical flux function it is useful to recall some of the properties of the exact solution of an initial value problem of a scalar hyperbolic conservation equation in $\mathbb{R} \times [0, t]$. Any weak solution verifies (Lax, 1972, LeVeque, 1992):

(i)   jump conditions across lines of discontinuities (Rankine–Hugoniot)

$$[f] = s[w]$$

where [ ] is the jump and s is the propagation velocity of the discontinuity;

(ii) $\|w(\cdot, t_2)\|_1 \le \|w(\cdot, t_1)\|_1$ for $t_2 \ge t_1$,
where

$$\|(\cdot)\|_1 = \int_{-\infty}^{\infty} |(\cdot)| \, dx$$

(iii) $TV(w(\cdot, t_2)) \le TV(w(\cdot, t_1))$ for $t_2 \ge t_1$,
where

$$TV(\cdot) = \int_{-\infty}^{\infty} \left| \frac{\partial \cdot}{\partial x} \right| \, dx$$

and $\partial/\partial x$ must be interpreted as a distribution derivative, which includes delta functions at points where w is discontinuous, or equivalently

$$TV(w(\cdot, t)) = \sup \sum_j |w(x_j, t) - w(x_{j-1}, t)|$$

for any arbitrary set $\{x_j\}$ of $\mathbb{R}$.

The Burgers equation $u_t + uu_x = 0$ with initial conditions

$$u(x, 0) = \begin{cases} 0 & x \le 0 \\ 1 & x > 0 \end{cases}$$

easily shows the nonuniqueness of weak solutions. Indeed, one may have either

$$u(x, t) = \begin{cases} 0 & x < 0 \\ x/t & 0 \leq x \leq 1 \\ 1 & x > 1 \end{cases} \quad \text{(fan of straight characteristic lines)}$$

or (for instance) the discontinuous solution

$$u(x, t) = \begin{cases} 0 & x < t/2 \\ 1 & x > t/2 \end{cases}$$

i.e. a discontinuity propagating with the speed $s = 1/2$. It is easy to verify that a small perturbation of the initial condition

$$u(x, 0) = \begin{cases} 0 & x < -\delta \\ f(x) & -\delta \leq x \leq \delta \\ 1 & x > \delta \end{cases} \quad \text{with } f(x) \text{ monotonic and } f(-\delta) = 0, f(\delta) = 1$$

would produce a small perturbation on the first solution, and a dramatic change of the discontinuous solution even for an infinitesimal δ, thus having an unphysical character.

The initial value problem admits a unique physically relevant weak solution w_S (also called entropy solution) defined as

$$w_S = \lim_{\epsilon \to 0} w_\epsilon$$

where w_ϵ satisfies the modified conservation equation

$$\frac{\partial w}{\partial t} + \frac{\partial f}{\partial x} = \epsilon \frac{\partial^2 w}{\partial x^2} \qquad \text{(with } \epsilon > 0)$$

For more details on the concept and definitions of entropy solutions refer for example to LeVeque (1992).

Entropy solutions verify the following properties:

(i) the monotonicity, i.e. given two different initial data $w_S(x, 0) \geq v_S(x, 0)$ then

$$w_S(x, t) \geq v_S(x, t) \qquad \forall x, t$$

(ii) the ℓ_1-contraction, i.e.

$$\|w_S(\cdot, t_2) - v_S(\cdot, t_2)\|_1 \leq \|w_S(\cdot, t_1) - v_S(\cdot, t_1)\|_1 \qquad \text{for } t_2 \geq t_1$$

(iii) the total variation diminishing property, i.e.

$$TV[w_S(\cdot, t_2)] \leq TV([w_S(\cdot, t_1)] \quad \text{for } t_2 \geq t_1$$

(iv) the monotonicity of given (monotone) initial conditions is preserved.

Observe that for any class of functions satisfying the above properties the ordering is strictly hierarchical, i.e.

$$\text{(i)} \Rightarrow \text{(ii)} \Rightarrow \text{(iii)} \Rightarrow \text{(iv)}$$

and the monotonicity property is the strongest.

Let us now see what is generally required to a numerical solution. If w is the true solution and $\widetilde{w}$ its 'local' average, then

$$\widetilde{w}(x,t) = \frac{1}{\Delta x} \int_{-\Delta x/2}^{\Delta x/2} w(x_i + s, t) \, ds \qquad \text{for} \quad x_i - \Delta x/2 \leq x \leq x_i + \Delta x/2$$

and the monotonicity property is also verified by $\widetilde{w}$, i.e.

$$w(x,0) \geq v(x,0) \Rightarrow \widetilde{w}(x,0) \geq \widetilde{v}(x,0) \quad \text{and} \quad \widetilde{w}(x,t) \geq \widetilde{v}(x,t) \quad \forall x,t$$

Hence, cell averages of the true solution verify the same properties (i)–(iv). However, even though obvious, it is still important to point out (for the understanding of the requirements of numerical approximations) that, given two local averaged exact solutions that satisfy $\widetilde{w}(x,0) \geq \widetilde{v}(x,0)$, then, $\widetilde{w}(x,t) \geq \widetilde{v}(x,t)$ is not implied $\forall(x,t)$.

Similar considerations can also be repeated for the set of discrete values $\{w(x_i,t)\}$ and $\{\widetilde{w}(x_i,t)\}$ for an arbitrary set $\{x_i\}$ of $\mathbb{R}$.

With some arbitrariness it is reasonable to require that cell-averaged numerical solutions satisfy the monotonicity property and hence the ℓ_1-contraction, the TVD and the monotonicity-preserving properties. The arbitrariness stems from the fact that the correspondence between an initial discretized data and the continuous initial conditions is not one-to-one. The question then arises as to which of the above properties should be satisfied by an approximate grid solution of a non-linear hyperbolic conservation equation to guarantee convergence, accuracy, stability, etc.

The theorem of Lax–Wendroff (1960) extends to the non-linear scalar hyperbolic conservation equations the milestone Lax equivalence theorem, which states that a consistent stable finite-difference method converges to the exact solution (Richtmyer and Morton, 1967). Lax–Wendroff theorem ensures that a consistent conservative stable method converges to a weak solution. Two issues must then be addressed: (i) the stability; and (ii) the convergence to a physically relevant weak solution.

A useful form of stability is the total variation (TV-) stability: given a consistent conservative flux function, TV-stability is ensured if

$$TV(w^n) \leq R \qquad \forall n, \Delta t \text{ with} \qquad \Delta t < \Delta t_0, \ t_n \leq T$$

with R a non-negative constant that depends only on the initial data (see Theorem 15.1 of LeVeque, 1992). If a numerical method is TV-stable, then the convergence toward a weak solution is guaranteed.

The convergence toward the physically relevant solution can be ensured if the approximate weak solution satisfies the monotonicity property, i.e. the numerical scheme is monotone. However, the monotonicity property is not a necessary condition (as can be argued from the previous considerations). Moreover, monotone schemes are at most first order accurate as demonstrated by Godunov (1959) for the linear case, and by Harten *et al.* (1976) for the non-linear one. Recall that a numerical scheme is said to be r-th order accurate if its local truncation error satisfies pointwise

$$w(\cdot, t + \Delta t) = E_h(\Delta t)w(\cdot, t) + O(\Delta x^{r+1})$$

for all smooth exact solutions $w(\cdot, t)$, with E_h the time shift operator associated to the numerical scheme and $\Delta t = O(\Delta x)$.

The construction of methods that yield physically relevant solutions (of first and possibly higher-order accuracy) is a critical point that will be discussed in the sections that follow.

1 Low-order schemes for scalars

Before entering into the analysis of the more advanced approaches for the Euler and/or Navier–Stokes equations, we briefly summarize (for the reader's convenience) some formal properties of numerical schemes, and review the most popular ones (cast in conservation form) and the concepts of low-order flux function definition. In particular, we recall the scheme of Courant–Isaacson–Rees (Courant *et al.*, 1952), which was developed mainly for linear hyperbolic conservation equations, and the scheme of Godunov (1959), which represents the first genuine attempt to solve non-linear hyperbolic conservation laws.

Courant–Isaacson–Rees (CIR)
The scheme was developed in 1952 and is probably the first upwind scheme developed for solving a linear hyperbolic scalar equation. The use of CIR for the numerical solution of the linear form of (39) ($f = aw, a = \mathrm{d}f/\mathrm{d}w = \mathrm{constant}$) corresponding to (50), and assuming $\widetilde{w}_i = w_i$ (i.e. $C_{ij} = \delta_{ij}$ in (40)), yields

$$w_i^{n+1} = w_i^n - a\frac{\Delta t}{\Delta x}\begin{cases} w_{i+1}^n - w_i^n & \text{if } a < 0 \\ w_i^n - w_{i-1}^n & \text{otherwise} \end{cases} \tag{51}$$

The scheme has formally a three-point stencil, it is first-order accurate and satisfies the consistency requirement, and in conservative form it becomes

$$w_i^{n+1} = w_i^n - \lambda\left[\left(a\frac{w_{i+1}^n + w_i^n}{2} - \frac{1}{2}\operatorname{sign}(a)a\delta^+ w_i^n\right) \right.$$
$$\left. - \left(a\frac{w_i^n + w_{i-1}^n}{2} - \frac{1}{2}\operatorname{sign}(a)a\delta^+ w_{i-1}^n\right)\right] \tag{52}$$

where $\delta^+(\cdot) = (\cdot)_{i+1} - (\cdot)_i$, and $\lambda = \Delta t/\Delta x$.

Hence, the numerical flux function can be defined as

$$h_{i+\frac{1}{2}} = a\frac{w_{i+1}^n + w_i^n}{2} - \frac{1}{2} \text{ sign } (a)a\delta^+ w_i^n = a^- w_{i+1}^n + a^+ w_i^n \tag{53}$$

where

$$a^+ = a[1 + \text{ sign } (a)]/2 = \max (a,0) = (a + |a|)/2$$

$$\tag{54}$$

$$a^- = a[1 - \text{ sign } (a)]/2 = \min (a,0) = (a - |a|)/2$$

Note that the use of multiple (equivalent) expressions of a^+, a^- is useful for the understanding of the (not always equivalent) extensions of the scheme to the non-linear case and to systems of conservation laws.

For example, a generalization to the non-linear case can be obtained by defining the numerical flux function as follows

$$h_{i+\frac{1}{2}} = \tfrac{1}{2}(f_{i+1} + f_i) - \tfrac{1}{2}d_{i+\frac{1}{2}} \tag{55}$$

where the second term of (55) is the so-called dissipation flux

$$d_{i+\frac{1}{2}} = \text{ sign } (a_{i+\frac{1}{2}})a_{i+\frac{1}{2}}\delta^+ w_i \tag{56}$$

which satisfies the consistency conditions, i.e. $d(w_i, w_i) = 0$, and $a_{i+\frac{1}{2}} = (df/dw)_{i+\frac{1}{2}}$ can be evaluated numerically by employing either one of the following formulas

$$a_{i+\frac{1}{2}} = \begin{cases} \delta^+ f_i/\delta^+ w_i & \text{if } \delta^+ w_i \neq 0 \\ \\ \text{or} \\ \\ (df/dw)_{w^*} & \text{with } w^* \text{ an interpolated value of } w_i \text{ and } w_{i+1} \\ \\ \text{or} \\ \\ \text{hybrid formulas} \end{cases}$$

Godunov (G)

The famous scheme proposed by Godunov (1959) is the result of research activity conducted by Godunov and other Russian researchers that dates back to 1953 (Godunov et al., 1979). The scheme is based on a fully discretized numerical flux which is exact under two main assumptions: (i) piecewise (locally) constant reconstruction equal to volume averaged values; (ii) solution of a Riemann problem at each cell interface with time integration over a time interval small enough (for the validity of self-similar solutions of the Riemann initial value problem) to avoid the interaction of the waves originating at cell interfaces. As a consequence of the self-similarity the time integration becomes trivial. Indeed, the solution at cell interface $i + \frac{1}{2}$ is constant within the time interval and depends only on (w_i, w_{i+1}). Then, as proposed by Osher (1984), the numerical flux function of Godunov's scheme in compact form is

$$
h_{i+\frac{1}{2}} = \begin{cases} \min_{(w_i,w_{i+1})} f(w) & \text{if } w_i < w_{i+1} \\ \max_{(w_{i+1},w_i)} f(w) & \text{if } w_i > w_{i+1} \end{cases}
$$

or equivalently

$$
h_{i+\frac{1}{2}} = \frac{1}{2}\left[\min_{(w_i,w_{i+1})} f(w) + \max_{(w_{i+1},w_i)} f(w) \right] - \frac{s_i}{2}\left[\max_{(w_{i+1},w_i)} f(w) - \min_{(w_i,w_{i+1})} f(w) \right]
$$

where $s_i = $ sign $(\delta^+ w_i)$, and min/max refer to the interval of boundaries w_i, w_{i+1}. When f is a monotone function of w in the interval, the min/max values of f coincide with the flux evaluated at the end values; at a sonic point (defined by the vanishing of the first derivative of f) then either the sonic or an end value is taken.

2 Schemes for one-dimensional systems of conservation laws

In this section we present a formal generalization of the CIR scheme to one-dimensional systems of conservation laws, and we introduce some more recent approaches.

Let us consider the conservative and non-conservative formulations of the one-dimensional Euler system of conservation laws corresponding to (3) and (7)

$$
\frac{\partial \mathbf{W}}{\partial t} + \frac{\partial \mathbf{F}}{\partial x} = 0
$$

$$
\frac{\partial \mathbf{W}}{\partial t} + \mathbf{A}\frac{\partial \mathbf{W}}{\partial x} = 0
$$

(57)

where

$$
\mathbf{W} = [\rho, \rho u, \rho E]^{\mathrm{T}} = [w_1, w_2, w_3]^{\mathrm{T}}
$$

$$
\mathbf{F} = \begin{bmatrix} w_2 \\ \dfrac{3 - \gamma}{2}\dfrac{w_2^2}{w_1} + (\gamma - 1)w_3 \\ w_2 H \end{bmatrix}
$$

(58)

$$
\mathbf{A} = \begin{pmatrix} 0 & 1 & 0 \\ \dfrac{\gamma - 3}{2}u^2 & (3 - \gamma)u & \gamma - 1 \\ \left(\dfrac{\gamma - 1}{2}u^2 - H\right)u & H + (1 - \gamma)u^2 & \gamma u \end{pmatrix}
$$

and the total enthalpy is

$$H = \gamma \frac{w_3}{w_1} - \frac{\gamma - 1}{2} \frac{w_2^2}{w_1^2}$$

The flux jacobian $\mathbf{A}$ has a set of real and distinct eigenvalues and a complete set of eigenvectors, and its diagonalization matrices $\mathbf{L}$ and $\mathbf{R}$ are easily obtained from (22) and (23).

$$\mathbf{L} = \begin{pmatrix} \frac{1}{2}\left(b_1 + \frac{u}{c}\right) & -\frac{1}{2}\left(b_2 u + \frac{1}{c}\right) & \frac{b_2}{2} \\ 1 - b_1 & b_2 u & -b_2 \\ \frac{1}{2}\left(b_1 - \frac{u}{c}\right) & -\frac{1}{2}\left(b_2 u - \frac{1}{c}\right) & \frac{b_2}{2} \end{pmatrix}$$

$$\mathbf{R} = \begin{pmatrix} 1 & 1 & 1 \\ u - c & u & u + c \\ H - cu & \frac{u^2}{2} & H + cu \end{pmatrix}$$

(59)

where $b_1 = b_2 u^2/2; b_2 = (\gamma - 1)/c^2$ and $c^2 = (\gamma - 1)(H - u^2/2)$, and the eigenvalues of $\mathbf{A}$ are

$$\lambda_1 = u - c; \quad \lambda_2 = u; \quad \lambda_3 = u + c$$

It is useful to note that, for a linear system with constant coefficients, Riemann invariants exist, and (52) can be applied to each component of the vector of characteristic variables, i.e. to the fully diagonalized system of equations. In the present case we can apply (52) to each component of the pfaffian of the characteristic variables. Then, by means of the compatibility relation $\delta \mathbf{W} = \mathbf{L}^{-1} \delta \boldsymbol{W}$, one easily obtains

$$\mathbf{W}_i^{n+1} = \mathbf{W}_i^n - \lambda \left[\left(\mathbf{A} \frac{\mathbf{W}_{i+1}^n + \mathbf{W}_i^n}{2} - \tfrac{1}{2} \mathbf{L}^{-1} \boldsymbol{\Phi}_{i+\frac{1}{2}} \right) \right.$$
$$\left. - \left(\mathbf{A} \frac{\mathbf{W}_i^n + \mathbf{W}_{i-1}^n}{2} - \tfrac{1}{2} \mathbf{L}^{-1} \boldsymbol{\Phi}_{i-\frac{1}{2}} \right) \right]$$

(60)

where $\boldsymbol{\Phi}_{i+\frac{1}{2}} = |\boldsymbol{\Lambda}|\delta^+ \boldsymbol{W}_i^n$, and $\lambda = \Delta t/\Delta x$. Equation (60) can also be rewritten in the equivalent form

$$\mathbf{W}_i^{n+1} = \mathbf{W}_i^n - \lambda \left[\left(\mathbf{A} \frac{\mathbf{W}_{i+1}^n + \mathbf{W}_i^n}{2} - \tfrac{1}{2} |\mathbf{A}|\delta^+ \mathbf{W}_i^n \right) \right.$$
$$\left. - \left(\mathbf{A} \frac{\mathbf{W}_i^n + \mathbf{W}_{i-1}^n}{2} - \tfrac{1}{2} |\mathbf{A}|\delta^+ \mathbf{W}_{i-1}^n \right) \right]$$

(61)

In general, the extension to systems of non-linear conservation laws is a weak point in the design of numerical flux functions. Indeed, the natural approach of transforming the equations in an equivalent scalar system of decoupled equations is frustrating for two reasons. First of all the diagonalization leads to the problem of the existence of the Riemann invariants, which generally do not exist for the set of Euler equations that include the energy conservation law. Moreover, the equivalence between the weak formulation of the original system and that of the transformed system does not exist.

In order to introduce the commonly accepted definitions of invariants, it is useful to recall some elements of the theory of systems of hyperbolic equations. The Riemann invariant $\mathbf{G}(\mathbf{W})$ is a primitive of the left eigenvector matrix $\mathbf{L}$ such that the jacobian $\partial \mathbf{G} / \partial \mathbf{W}$ satisfies

$$\frac{\partial \mathbf{G}}{\partial \mathbf{W}} \mathbf{A} \left(\frac{\partial \mathbf{G}}{\partial \mathbf{W}} \right)^{-1} = \Lambda$$

For the linear case (when $\mathbf{A}$ is a constant coefficient matrix) and for the linearized case (corresponding to the evolution of a perturbation) $\partial \mathbf{G} / \partial \mathbf{W}$ admits a primitive. The Riemann invariant satisfies

$$\frac{\partial \mathbf{G}}{\partial t} + \Lambda \frac{\partial \mathbf{G}}{\partial x} = 0$$

thus showing that $\mathbf{G}$ is invariant along the characteristic lines (as long as strong solutions exist). Another way of obtaining a diagonal form of (57) consists in finding a self-similar solution $\mathbf{W} = \mathbf{W}_k(\sigma(x, t))$, where the functional relationship $\mathbf{W}_k(\cdot)$ is determined by solving

$$\frac{d \mathbf{W}_k}{d \sigma} = \mathbf{r}_k (\mathbf{A}(\mathbf{W}_k))$$

with proper initial conditions in a suitable interval of σ; $\mathbf{r}_k$ is the $k - th$ right eigenvector. In turn the scalar function σ satisfies $\partial \sigma / \partial t + \lambda_k \partial \sigma / \partial x = 0$. Note that the relation $\lambda_k = \lambda_k(\mathbf{W}_k(\sigma))$ is in principle known, even though complicated; $\mathbf{W}_k$ is the so-called k-invariant, which is invariant along the characteristic line $dx/dt = \lambda_k$ where σ is constant. The above considerations then show that perturbations (i.e. signals) and/or k-invariants propagate along the characteristics.

The problem of uniqueness and/or physical relevance of weak solutions of systems of conservation laws is generally not yet resolved. However, physical and mathematical reasons impose that physically admissible weak solutions should also satisfy an additional conservation law for the entropy (or a generalized entropy) for which the production terms should have a (physically) correct sign. The common practice in designing the numerical flux functions for systems of conservation laws is then to consider the diagonal form of the matrix $\mathbf{A}$, so as to exploit the analogy between its eigenvalues and the characteristic velocity df/dw of a non-linear scalar equation. This amounts to associating with any function $g(df/dw)$, which appears in the schemes for a scalar equation, a matrix $g(\mathbf{A})$ obtained as $\mathbf{L}^{-1}\mathbf{D}\mathbf{L}$ where $\mathbf{D}$ is a diagonal matrix of non-zero elements $g(\lambda_i)$.

For example, a generalization of the formulas corresponding to (53), (55) yields

$$\mathbf{H}_{i+\frac{1}{2}} = \mathbf{A}^-_{i+\frac{1}{2}}\mathbf{W}_{i+1} + \mathbf{A}^+_{i+\frac{1}{2}}\mathbf{W}_i \tag{62}$$

$$\mathbf{H}_{i+\frac{1}{2}} = \tfrac{1}{2}(\mathbf{F}_{i+1} + \mathbf{F}_i) - \tfrac{1}{2}|\mathbf{A}_{i+\frac{1}{2}}|\delta^+\mathbf{W}_i \tag{63}$$

where $\mathbf{A}^\pm_{i+\frac{1}{2}} = (A_{i+\frac{1}{2}} \pm |\mathbf{A}|_{i+\frac{1}{2}})/2$.

The resulting schemes are upstream schemes in the sense of Harten, Lax and Van Leer (Harten *et al.*, 1983). Indeed, if $\mathbf{A}^\pm_{i+\frac{1}{2}}$ are evaluated in terms of an intermediate state $\mathbf{W}^*$ between $\mathbf{W}_i$ and $\mathbf{W}_{i+1}$, then

$$\mathbf{H}_{i+\frac{1}{2}} = \mathbf{F}(\mathbf{W}^*) + \mathbf{A}^+(\mathbf{W}^*)(\mathbf{W}_i - \mathbf{W}^*)$$

$$+ \mathbf{A}^-(\mathbf{W}^*)(\mathbf{W}_{i+1} - \mathbf{W}^*) \tag{64}$$

$$+ O(|\mathbf{W}_i - \mathbf{W}^*| + |\mathbf{W}_{i+1} - \mathbf{W}^*|)$$

If one chooses $\mathbf{W}^* = (\mathbf{W}_i + \mathbf{W}_{i+1})/2$ and observing that

$$\mathbf{F}(\mathbf{W}^*) = \tfrac{1}{2}[\mathbf{F}(\mathbf{W}_i) + \mathbf{F}(\mathbf{W}_{i+1})] + O(|\mathbf{W}_{i+1} - \mathbf{W}_i|) \tag{65}$$

then, (64) reduces to (62) and (63), which can be cast as

$$\mathbf{H}_{i+\frac{1}{2}} = \tfrac{1}{2}[\mathbf{F}(\mathbf{W}_i) + \mathbf{F}(\mathbf{W}_{i+1})] - \tfrac{1}{2}\mathbf{d}(\mathbf{W}_i, \mathbf{W}_{i+1}) \tag{66}$$

where the dissipation flux $\mathbf{d}$ satisfies the consistency conditions (i.e. $\mathbf{d}(\mathbf{W}, \mathbf{W}) = 0$) and is given by

$$\mathbf{d}(\mathbf{W}_i, \mathbf{W}_{i+1}) = |\mathbf{A}(\mathbf{W}_i, \mathbf{W}_{i+1})|(\mathbf{W}_{i+1} - \mathbf{W}_i) \tag{67}$$

Flux vector splitting

These schemes are based on a splitting of the governing equations. In principle, such a splitting can be implemented on either the conservative or the non-conservative forms of the equations.

- *Conservative splitting*

   Conservative splitting consists of decomposing the flux in two parts $\mathbf{F}^+$ and $\mathbf{F}^-$ which will be specified later

$$\mathbf{F} = \mathbf{F}^+ + \mathbf{F}^- \tag{68}$$

and the conservative equations become

$$\frac{\partial \mathbf{W}}{\partial t} + \frac{\partial \mathbf{F}^+}{\partial x} + \frac{\partial \mathbf{F}^-}{\partial x} = 0 \tag{69}$$

- *Non-conservative splitting*
  Analogously, non-conservative splitting amounts to splitting the jacobian of the flux in two contributions $\mathbf{A}^+$ and $\mathbf{A}^-$

$$\mathbf{A} = \frac{\partial \mathbf{F}}{\partial \mathbf{W}} = \mathbf{A}^+ + \mathbf{A}^- \tag{70}$$

and the quasilinearized equation is cast in the following non-conservative form

$$\frac{\partial \mathbf{W}}{\partial t} + \mathbf{A}^+ \frac{\partial \mathbf{W}}{\partial x} + \mathbf{A}^- \frac{\partial \mathbf{W}}{\partial x} = 0 \tag{71}$$

A trivial but nevertheless subtle point is the following: any conservative splitting can be cast in an equivalent non-conservative splitting by assuming

$$\mathcal{A}^\pm = \frac{\partial \mathbf{F}^\pm}{\partial \mathbf{W}}$$

However, a non-conservative splitting cannot be cast as a conservative flux splitting. Indeed, let

$$\mathbf{A}^\pm = \frac{\mathbf{A} \pm \mathcal{G}}{2}$$

in order for $\mathbf{A}^\pm$ to verify $\mathbf{A}^\pm = \partial \mathbf{F}^\pm / \partial \mathbf{W}$, the matrix function $\mathcal{G}$ must be chosen as an exact pfaffian form, i.e.

$$\exists \mathcal{F} : \mathcal{G} = \frac{\partial \mathcal{F}}{\partial \mathbf{W}} \tag{72}$$

yielding

$$\mathbf{F}^\pm = \frac{\mathbf{F} \pm \mathcal{F}}{2}$$

On the other hand, for a system of conservation laws $\mathbf{F}$ satisfies the homogeneity property (9), i.e. $\mathbf{F} = \mathbf{A}\mathbf{W}$. If the following conservative flux splitting is adopted

$$\mathbf{F}^\pm = \mathbf{A}^\pm \mathbf{W}$$

where

$$\mathbf{A}^\pm = \frac{\mathbf{A} \pm \mathcal{G}}{2}$$

and $\mathcal{G} = \mathcal{G}(\mathbf{A})$ is any matrix function of $\mathbf{A}$, then, the matrices of the non-conservative form of the splitting are

$$\mathcal{A}^{\pm} = \frac{\partial \mathbf{F}^{\pm}}{\partial \mathbf{W}} = \frac{\partial}{\partial \mathbf{W}} \left(\mathbf{A}^{\pm} \mathbf{W} \right)$$

Obviously, the difference

$$\mathcal{A}^{\pm} - \mathbf{A}^{\pm} = \frac{\partial \mathbf{A}^{\pm}}{\partial \mathbf{W}} \mathbf{W} = \pm \frac{\partial \mathcal{G}(\mathbf{A})}{\partial \mathbf{W}} \mathbf{W}$$

does not vanish in general. Such a difference can be made to vanish if one selects $\mathcal{G}$ as a polynomial matrix of $\mathbf{A}$ with coefficients that are homogeneous functions of $\mathbf{W}$ of degree zero.

Numerical flux splitting

We now discuss how to exploit the flux vector splitting in a control volume approach. Following the current definition, the numerical flux function associated with a flux vector splitting is expressed as

$$\mathbf{H}_{i+\frac{1}{2}} = \mathbf{F}^{+}(\mathbf{W}_i) + \mathbf{F}^{-}(\mathbf{W}_{i+1}) \tag{73}$$

The above equation can also be cast in the form

$$\mathbf{H}_{i+\frac{1}{2}} = \tfrac{1}{2}(\mathbf{F}_i + \mathbf{F}_{i+1}) - \tfrac{1}{2}\mathbf{d}_{i+\frac{1}{2}} \tag{74}$$

where, by means of (68), the antidiffusion flux becomes

$$\mathbf{d}_{i+\frac{1}{2}} = (\mathbf{F}_{i+1}^{+} - \mathbf{F}_{i+1}^{-}) - (\mathbf{F}_i^{+} - \mathbf{F}_i^{-}) \tag{75}$$

It is obvious that a blind splitting (for example $\mathbf{F}^{\pm} = \mathbf{F}/2$) would inaccurately estimate the time integral of the flux through the control volume boundary points.

Conversely, the construction of a splitting where $\mathbf{F}^{+}$ and $\mathbf{F}^{-}$ coincide with contributions that depend, respectively, upon $\mathbf{W}_i$ and $\mathbf{W}_{i+1}$ in an adaptive way is the consequence of a better evaluation of the time integral of the flux. This requires that the eigenvalues of the jacobians of the split flux $\mathbf{F}^{+}$(resp. $\mathbf{F}^{-}$) are real and positive (resp. negative). However, a flux splitting that automatically verifies the above requirement is not straightforward.

If the flux satisfies homogeneity conditions, one may introduce a flux splitting equivalent to an upwind approach ((62) and (63)), thus obtaining (Steger and Warming, 1981)

$$\mathbf{F}^{\pm}(\mathbf{W}) = \mathbf{A}^{\pm}\mathbf{W}$$

A larger class of flux decomposition can be obtained by defining $\mathbf{A}^{\pm}$ as

$$\mathbf{A}^{\pm} = \frac{\mathbf{A} \pm g(\mathbf{A})}{2} \tag{76}$$

where $g(\mathbf{A})$ is any matrix having the right and left eigenvectors coinciding with those of $\mathbf{A}$, i.e.

$$g(\mathbf{A}) = \mathbf{L}^{-1}\Lambda_g\mathbf{L}$$

and Λ_g is a diagonal matrix whose non-zero coefficients are g_i. Let us represent these coefficients with the improper functional notation $g(\lambda_i)$ to indicate the same ordering of λ_i. Then, the eigenvalues of $\mathbf{A}^\pm$ are

$$\lambda_i^\pm = \frac{\lambda_i \pm g(\lambda_i)}{2}$$

In order to ensure the correct adaptive character of (73) (i.e. $\lambda_i^+ \geq 0$ and $\lambda_i^- \leq 0$), the eigenvalues of $g(\mathbf{A})$ must verify

$$g(\lambda_i) \geq |\lambda_i| \tag{77}$$

A possible representation of the matrix $g(\mathbf{A})$ is a polynomial matrix of $\mathbf{A}$. If k is the number of eigenvalues of $\mathbf{A}$ and if they are all distinct, then a polynomial matrix of degree $k - 1$ may represent any $g(\mathbf{A})$. For the one-dimensional system of conservation laws we have

$$g(\mathbf{A}) = \alpha\mathbf{A}^2 + \beta\mathbf{A} + \gamma\mathbf{I} \tag{78}$$

where α, β, and γ are the solution of the system

$$\alpha\lambda_i^2 + \beta\lambda_i + \gamma = g_i \qquad i = 1, 3 \tag{79}$$

It is easy to recognize that the split fluxes can equally be expressed in terms of a polynomial matrix of $\mathbf{A}$

$$\mathbf{F}^\pm = \pm\tfrac{1}{2}\left[\alpha\mathbf{A}^2 + (\beta \pm 1)\mathbf{A} + \gamma\mathbf{I}\right]\mathbf{W} \tag{80}$$

Steger–Warming splitting
The most popular splitting of Steger and Warming (1981) amounts to choosing for $g(\mathbf{A})$ the following

$$g(\lambda_1) = |\lambda_1| = |u - c| = c|M - 1|$$

$$g(\lambda_2) = |\lambda_2| = |u| = c|M|$$

$$g(\lambda_3) = |\lambda_3| = |u + c| = c|M + 1|$$

where the Mach number $M = u/c$.

The solution of (79) furnishes the following expressions for the coefficients

α, β, γ

$$\alpha = \frac{1}{2c}[|M-1| - 2|M| + |M+1|]$$

$$\beta = \frac{1}{2}[-(2M+1)|M-1| + 4M|M| - (2M-1)|M+1|]$$

$$\gamma = \frac{c}{2}[M(M+1)|M-1| - 2(M^2-1)|M| + M(M-1)|M+1|]$$

Observe that

$$\alpha = 0; \quad \gamma = 0; \quad \beta = +1 \quad \text{for } M \geq 1$$

$$\alpha = 0; \quad \gamma = 0; \quad \beta = -1 \quad \text{for } M \leq -1$$

so that when $M \geq 1$ the positive split flux coincides with the total flux, while for $M \leq -1$ the negative split flux is equal to the total flux.

In the range $M = [-1, 1]$ one obtains

$$\alpha = (1 - |M|)/c; \quad \beta = M(2|M| - 1); \quad \gamma = -c(M^2 - 1)|M|$$

The above expressions show that α and γ are even functions of M, while β is an odd function of M. Moreover, they are not continuous functions of the Mach number. In particular, α and γ are not continuous at sonic ($M = \pm 1$) and stagnation ($M = 0$) points, while β is discontinuous at $M = \pm 1$ (see Figure 4).

In the range $-1 \leq M \leq 1$ the split fluxes are then expressed as follows

$$\mathbf{F}^{\pm} = \begin{pmatrix} \left(\dfrac{M \pm |M|}{2} \pm \dfrac{1 - |M|}{2}\dfrac{1}{\gamma}\right)\rho c \\[2ex] \left[M^2 + \dfrac{1}{\gamma} \pm M\left(\dfrac{2}{\gamma} + |M|\dfrac{\gamma-1}{\gamma}\right)\right]\dfrac{\rho c^2}{2} \\[2ex] \left[\dfrac{M^2}{2}\left(M \pm |M|\dfrac{\gamma-1}{\gamma}\right) \pm \left(\dfrac{3}{2\gamma}M^2 \pm \dfrac{M}{\gamma-1} + \dfrac{1}{\gamma(\gamma-1)}\right)\right]\dfrac{\rho c^3}{2} \end{pmatrix}$$

The eigenvalues of the split jacobians ($\partial \mathbf{F}^+/\partial \mathbf{W}$) and ($-\partial \mathbf{F}^-/\partial \mathbf{W}$) are all positive except at sonic and stagnation points where the split fluxes are discontinuous (see Figure 5).

A less popular Steger–Warming type splitting assumes

$$g(\lambda_1) = |u| + c; \quad g(\lambda_2) = |u|; \quad g(\lambda_3) = |u| + c$$

which still verifies the condition of adaptivity corresponding to (73).

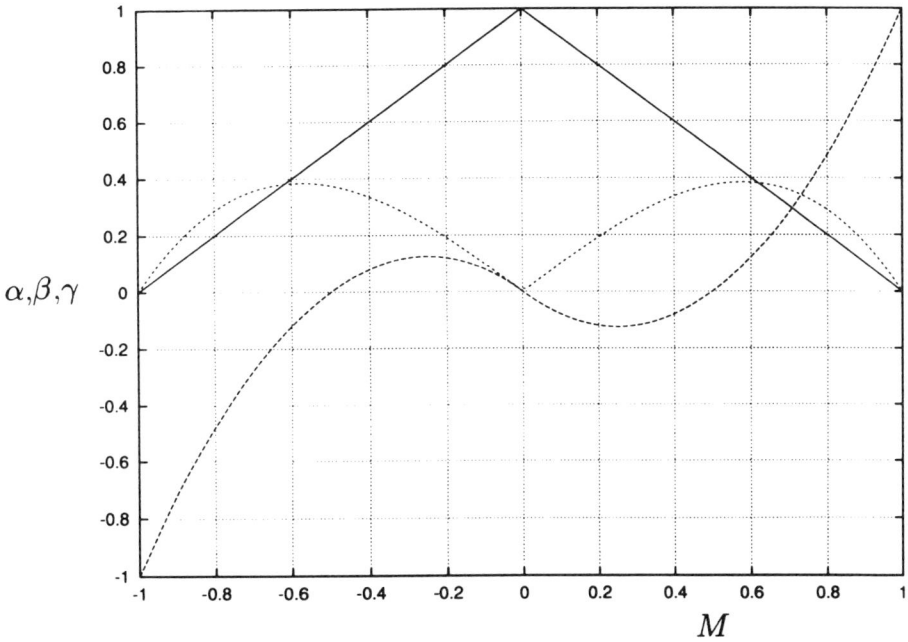

Figure 4 Mach number dependence of the coefficients of the polynomial approximation of $|A|$. Key: —, α; - - -, β; $\cdots$, γ.

Van Leer splitting

Van Leer (1982) has proposed a splitting that removes the discontinuous behaviour of the split fluxes by modifying their functional dependence upon M. In particular, the split fluxes are represented by a polynomial in M that gives the same function value and slope of the unsplit fluxes at $M = \pm 1$ (Eberle *et al.*, 1992). He further requires that the symmetry properties of each split flux component should be the same as those of the unsplit one, i.e.

$$F_j^+(M) = \pm F_j^-(M) \qquad \text{if } F_j(M) = \pm F_j(-M)$$

where the mass, momentum and energy flux components expressed as functions of (ρ, c, M) are

$$F_1 = \rho c M; \quad F_2 = \rho c^2 \left(M^2 + \frac{1}{\gamma}\right); \quad F_3 = \rho c^3 \left(\frac{1}{\gamma - 1} + \frac{M^2}{2}\right)$$

(a)

(b)

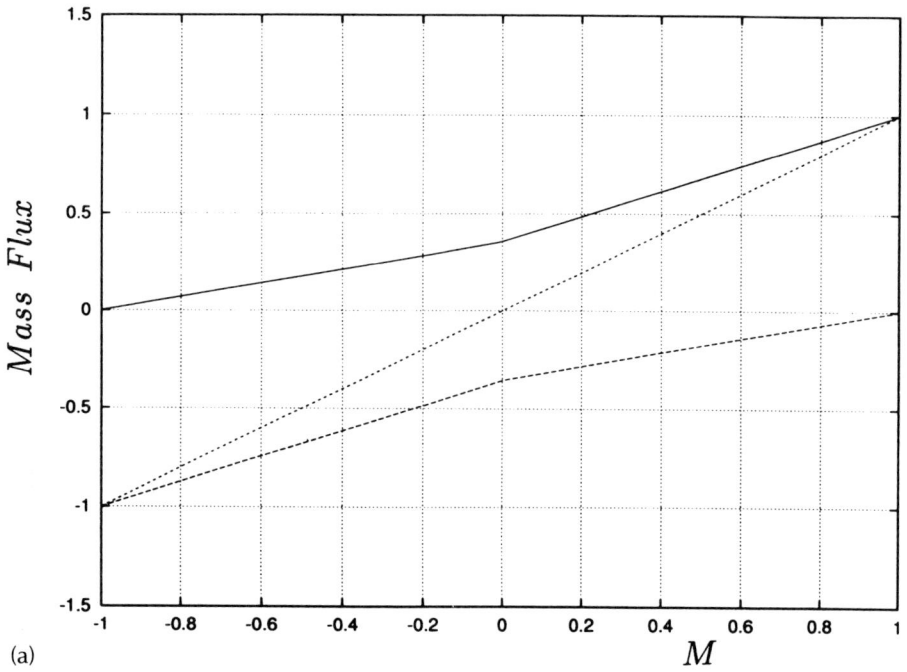

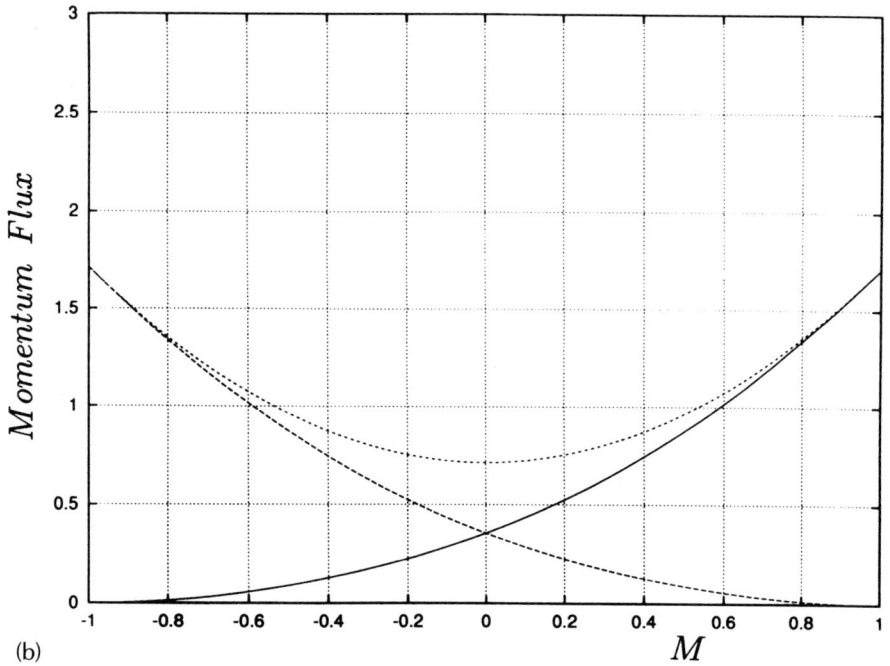

Figure 5 Mach number dependence of the split fluxes corresponding to polynomial approximation of $|A|$. Key: —, F^+; - - -, F^-; $\cdots$, $F^+ + F^-$.

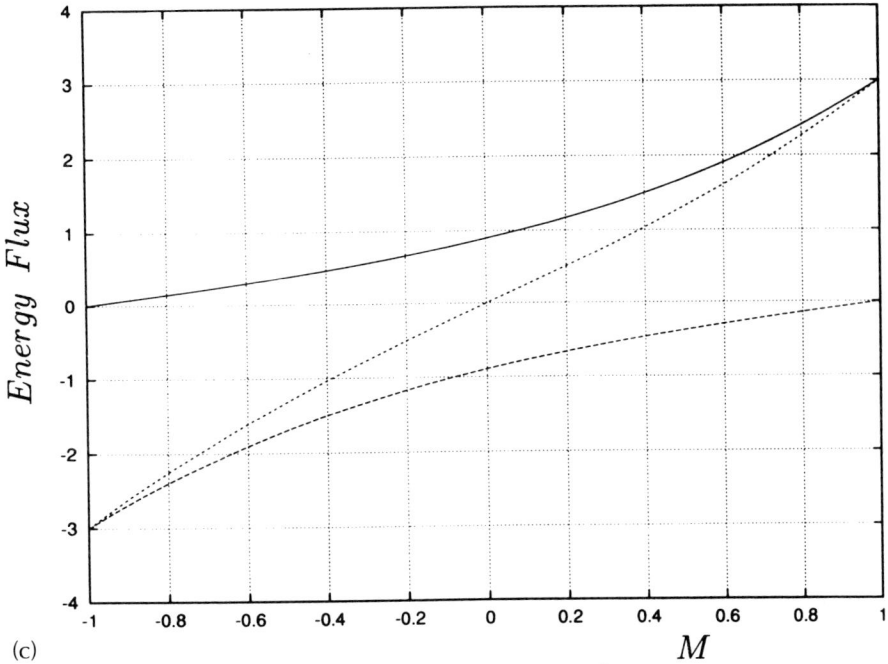

(c)

Figure 5 (Continued)

The splitting proposed by Van Leer is then the following (see Figure 6)

$$
\mathbf{F}^{\pm} = \pm\rho c\,\frac{(1 \pm M)^2}{4}
\begin{bmatrix}
1 \\
u \pm (2 \mp M)\dfrac{c}{\gamma} \\
\dfrac{\gamma^2}{2(\gamma^2 - 1)}\left[u \pm (2 \mp M)\dfrac{c}{\gamma}\right]^2
\end{bmatrix}
$$

It must be pointed out that, for steady flows, the use of Van Leer splitting may result in the violation of the constant total enthalpy constraint. Hänel has proposed a variation to the Van Leer splitting (Hänel *et al.*, 1987) requiring that the split mass and energy flux components are scaled by the total enthalpy, arriving at the following formulas

$$
\mathbf{F}^{\pm} = \pm\rho c\,\frac{(1 \pm M)^2}{4}
\begin{bmatrix}
1 \\
u \pm (2 \mp M)\dfrac{c}{\gamma} \\
\dfrac{c^2}{\gamma - 1}\left(1 + \dfrac{\gamma - 1}{2}M^2\right)
\end{bmatrix}
$$

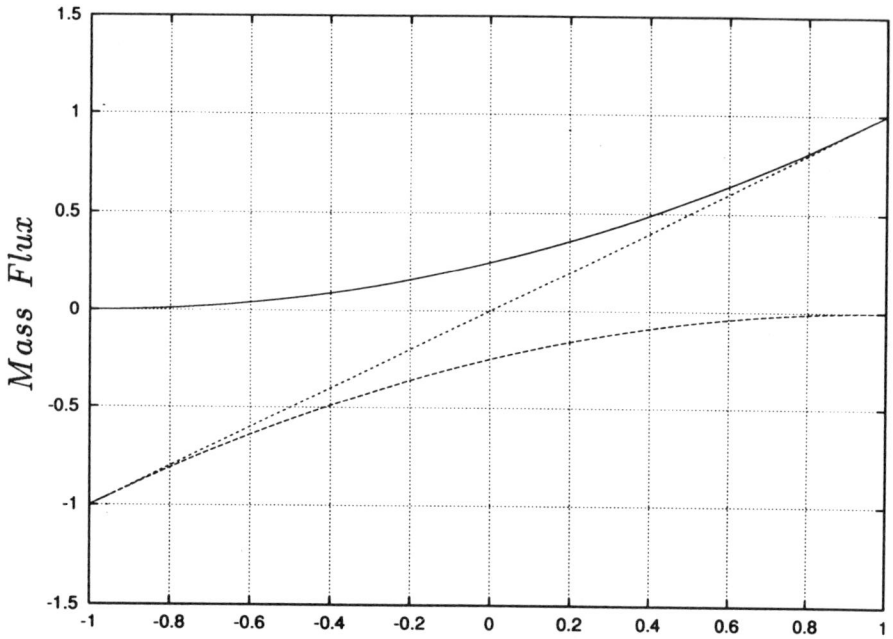

(a)

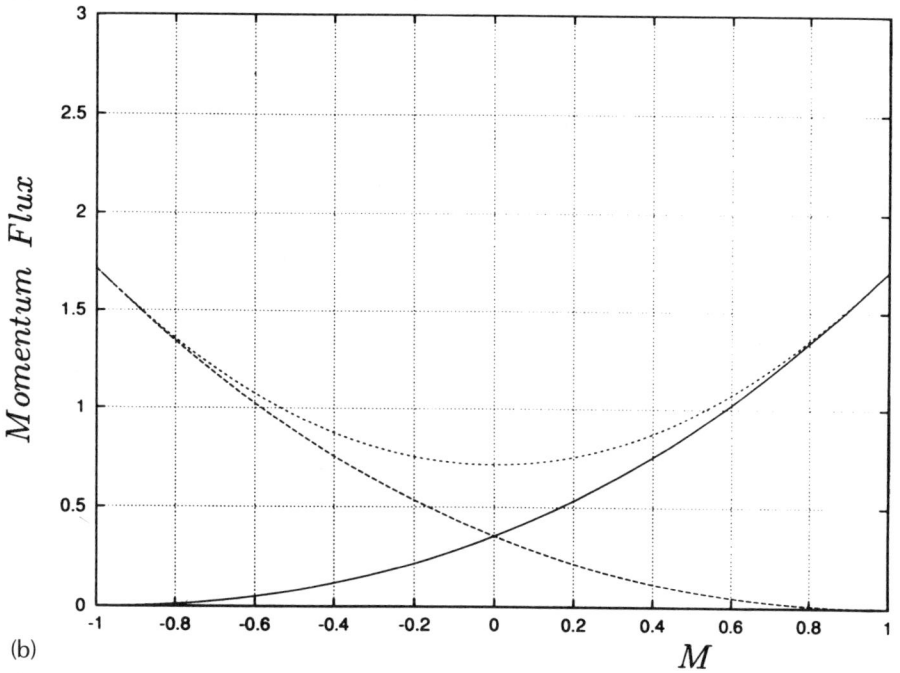

(b)

Figure 6 Mach number dependence of Van Leer's split fluxes. Key: —, F^+; - - -, F^-; $\cdots$, $F^+ + F^-$.

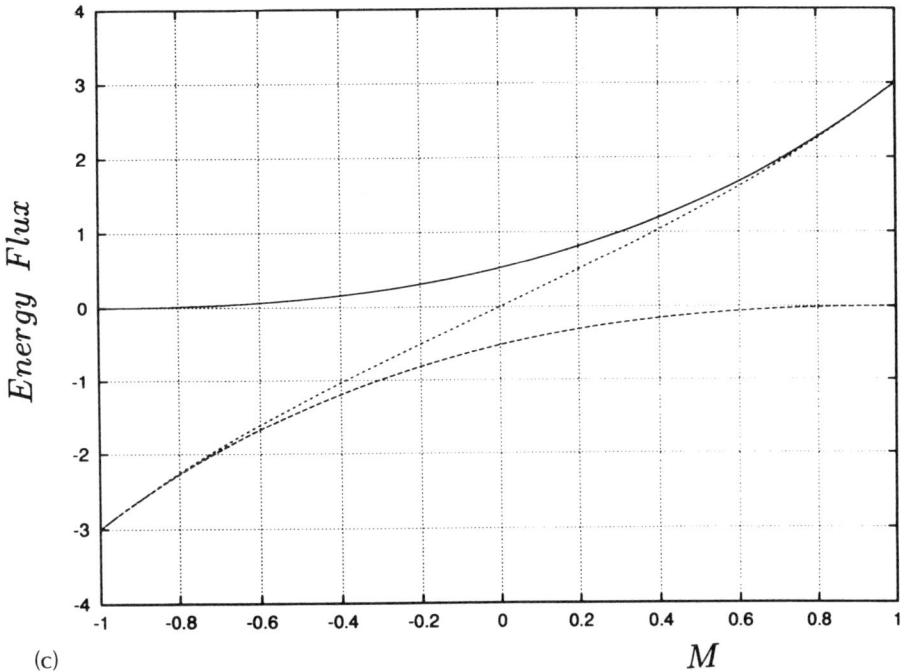

(c)

Figure 6 (Continued)

Advection Upwind Splitting Method (AUSM)

Liou (Liou, 1993; Liou and Steffen, 1993; Wada and Liou, 1994) has proposed splitting by introducing the cell-face advection velocity (or the Mach number) to determine the upwind extrapolation of the 'convected' variables ϕ. Separating the pressure from the 'convective' flux, the inviscid flux is written as

$$\mathbf{F} = \mathcal{U}\phi + \mathcal{P}\mathbf{b} \qquad (81)$$

where $\mathcal{U}\phi$ represents the 'improperly' defined convective flux (being ϕ different from the conservative variable $\mathbf{W}$), and

$$\mathcal{U} = \mathbf{\Lambda} = u\mathbf{I}$$
$$\phi = [\rho, \rho u, \rho H]^{\mathrm{T}}$$
$$\mathcal{P} = p\mathbf{I}$$
$$\mathbf{b} = [0, 1, 0]^{\mathrm{T}}$$

At a generic cell interface the numerical flux function is obtained by applying a numerical flux splitting to (81) yielding

$$\mathbf{H}_{i+\frac{1}{2}} = \mathbf{F}^{+}(\mathbf{W}_i) + \mathbf{F}^{-}(\mathbf{W}_{i+1})$$

where

$$\mathbf{F}^{\pm} = \mathcal{U}^{\pm}_{i+\frac{1}{2}}\boldsymbol{\phi}^{\mp} + \mathcal{P}^{\pm}\mathbf{b}$$

and

$$\mathcal{U}^{\pm}_{i+\frac{1}{2}} = \tfrac{1}{2}(\mathcal{U}(\mathbf{W}_i, \mathbf{W}_{i+1}) \pm |\mathcal{U}(\mathbf{W}_i, \mathbf{W}_{i+1})|)$$

$$\mathcal{U}(\mathbf{W}_i, \mathbf{W}_{i+1}) = \Lambda^+(\mathbf{W}_i) + \Lambda^-(\mathbf{W}_{i+1})$$

$$= \{s_i u_i^+ + (1 - s_i)\hat{u}_i^+ + s_i u_{i+1}^- + (1 - s_i)\hat{u}_{i+1}^-\}\mathbf{I}$$

$$\mathcal{P}^{\pm} = \left\{ s\frac{1 \pm \text{ sign } (u)}{2} + (1 - s)\frac{(M \pm 1)^2(2 \mp M)}{4} \right\}p\mathbf{I}$$

with

$$u^{\pm} = \frac{1 \pm \text{sign } (u)}{2}u; \quad \hat{u}^{\pm} = \pm\frac{1}{4}\left(\sqrt{|M|} \pm \frac{1}{\sqrt{|M|}}\right)^2 u; \quad s = \text{sign } (|M| - 1)$$

The AUSM splitting (like Van Leer splitting) does not require any diagonalization owing to the diagonal character of the matrices $\mathcal{U}$ and $\mathcal{P}$. Moreover, it captures a stationary contact discontinuity over two cells and ensures conservation of enthalpy for steady flows. However, it may cause post-shock oscillations (Wada and Liou, 1994) and even/odd point decoupling (Quirk, 1994).

To cure the overshoot problem Wada and Liou (1994) have recently proposed a variation of the method by recasting the flux in the form

$$\mathbf{F}(\mathbf{W}) = \mathcal{R}\boldsymbol{\Psi} + \mathcal{P}\mathbf{b}$$

where

$$\mathcal{R} = \Lambda = \rho u\mathbf{I}; \qquad \boldsymbol{\Psi} = [1, u, H]^{\mathrm{T}}$$

and the numerical flux function is again defined as

$$\mathbf{H}_{i+\frac{1}{2}} = \mathbf{F}^+(\mathbf{W}_i) + \mathbf{F}^-(\mathbf{W}_{i+1})$$

where

$$\mathbf{F}^{\pm} = \mathcal{R}^{\pm}_{i+\frac{1}{2}}\boldsymbol{\Psi}^{\mp} + \mathcal{P}^{\pm}\mathbf{b}$$

and

$$\mathcal{R}^{\pm}_{i+\frac{1}{2}} = \tfrac{1}{2}(\mathcal{R}(\mathbf{W}_i, \mathbf{W}_{i+1}) \pm |\mathcal{R}(\mathbf{W}_i, \mathbf{W}_{i+1})|)$$

$$\mathcal{R}(\mathbf{W}_i, \mathbf{W}_{i+1}) = \{\rho_i[\tilde{s}_i u_i^+ + (1 - \tilde{s}_i)\hat{u}_i^+] + \rho_{i+1}[\tilde{s}_i u_{i+1}^- + (1 - \tilde{s}_i)\hat{u}_{i+1}^-]\}\mathbf{I}$$

where $\hat{u}^{\pm}$ is defined so as to correctly capture stationary/moving contact discontinuities with the lowest level of numerical dissipation, i.e.

$$\hat{u}^{\pm} = \pm\alpha\left[\frac{1}{4}\left(\sqrt{|\tilde{M}|} \pm \frac{1}{\sqrt{|\tilde{M}|}}\right) - \frac{1 \pm \text{sign}(u)}{2}\right]u + \frac{1 \pm \text{sign}(u)}{2}u$$

with

$$\tilde{M} = \frac{u}{\max(c_i, c_{i+1})}; \quad \alpha = \frac{2c^2}{c_i^2 + c_{i+1}^2}; \quad \tilde{s} = \text{sign}(|\tilde{M}| - 1)$$

and

$$\mathcal{P}^{\pm} = \left\{\tilde{s}\frac{1 \pm \text{sign}(u)}{2} + (1 - \tilde{s})\frac{(\tilde{M} \pm 1)^2(2 \mp \tilde{M})}{4}\right\}p\mathbf{I}$$

A more general variant to the advection upstream splitting method may be obtained by redefining the flux as

$$\mathbf{F} = \mathcal{R}\tilde{\mathbf{\Psi}} + \mathcal{U}\tilde{\mathbf{\phi}} + \mathcal{P}\mathbf{b}$$

where

$$\tilde{\mathbf{\Psi}} = [1, 0, H]^{\mathrm{T}}$$
$$\tilde{\mathbf{\phi}} = [0, \rho u, 0]^{\mathrm{T}}$$

and the numerical split fluxes are

$$\mathbf{F}^{\pm} = \mathcal{R}^{\pm}_{i+\frac{1}{2}}\tilde{\mathbf{\Psi}}^{\mp} + (\alpha\mathcal{R}^{\pm}_{i+\frac{1}{2}} + (1 - \alpha)\mathcal{U}^{\pm})\tilde{\mathbf{\phi}}^{\mp} + \mathcal{P}^{\pm}\mathbf{b}$$

where

$$\mathcal{U}^{+} = \Lambda^{+}(\mathbf{W}_i) = (s_i u_i^{+} + (1 - s_i)\hat{u}_i^{+})\mathbf{I}$$
$$\mathcal{U}^{-} = \Lambda^{-}(\mathbf{W}_{i+1}) = (s_{i+1}u_{i+1}^{-} + (1 - s_{i+1})\hat{u}_{i+1}^{-})\mathbf{I}$$

As a final remark, observe that the methods based on the advection upstream splittings (in all their variants) are easy to implement.

Approximate Riemann solver

The method developed by Roe (1981a) is based on a characteristic decomposition of the flux difference and represents a successful attempt to extend the exact linear wave decomposition to non-linear hyperbolic equations. The idea consists of determining the solution by solving a modified equation, where the flux $\mathbf{F}$ is quasilinearized by introducing a matrix $\widehat{\mathbf{A}}$ and assuming $\mathbf{F} = \widehat{\mathbf{A}}\mathbf{W}$ (where $\widehat{\mathbf{A}}$ is only formally a jacobian).

Assuming that the discretized solution of the system of conservation laws is piecewise constant within each computational volume, the solution of (57) is then equivalent to solving approximately a Riemann-type problem at cell interfaces. At the interface $i + \frac{1}{2}$ the matrix $\widehat{\mathbf{A}}$ then satisfies the following properties

(i) $\delta\mathbf{F} = \widehat{\mathbf{A}}(\mathbf{W}_L, \mathbf{W}_R)\delta\mathbf{W}$ where $\delta = (\ )_R - (\ )_L$;

(ii) $\widehat{\mathbf{A}}(\mathbf{W}, \mathbf{W}) = \mathbf{A}(\mathbf{W}) = \dfrac{\partial\mathbf{F}}{\partial\mathbf{W}}$;

(iii) $\widehat{\mathbf{A}}$ has a set of real eigenvalues and linearly independent eigenvectors.

From property (iii) the difference of characteristic variables $\delta\mathcal{W}$ can be defined

$$\delta\mathcal{W} = \mathbf{L}\delta\mathbf{W} \tag{82}$$

where $\mathbf{L}$ is the left eigenvector matrix, and from property (i) one obtains

$$\delta\mathbf{F} = \mathbf{L}^{-1}\Lambda\delta\mathcal{W} \tag{83}$$

The property (i) corresponds to the Rankine–Hugoniot jump condition if $i + \frac{1}{2}$ separates the left and right states of a discontinuity (to a first-order approximation those states coincide with the values at cell centers i and $i + 1$).
The numerical flux function can alternatively be expressed as

$$\mathbf{H}_{i+\frac{1}{2}} = \mathbf{F}(\mathbf{W}_i) + \mathbf{L}_{i+\frac{1}{2}}^{-1}\Lambda^-_{i+\frac{1}{2}}\delta^+\mathcal{W}_i$$
$$\mathbf{H}_{i+\frac{1}{2}} = \mathbf{F}(\mathbf{W}_{i+1}) - \mathbf{L}_{i+\frac{1}{2}}^{-1}\Lambda^+_{i+\frac{1}{2}}\delta^+\mathcal{W}_i \tag{84}$$

and averaging the two expressions yields

$$\mathbf{H}_{i+\frac{1}{2}} = \tfrac{1}{2}(\mathbf{F}(\mathbf{W}_i) + \mathbf{F}(\mathbf{W}_{i+1})) - \mathbf{L}_{i+\frac{1}{2}}^{-1}[\Lambda^+_{i+\frac{1}{2}} - \Lambda^-_{i+\frac{1}{2}}]\delta^+\mathcal{W}_i$$
$$= \tfrac{1}{2}(\mathbf{F}(\mathbf{W}_i) + \mathbf{F}(\mathbf{W}_{i+1})) - \frac{|\widehat{\mathbf{A}}_{i+\frac{1}{2}}|}{2}(\mathbf{W}_{i+1} - \mathbf{W}_i) \tag{85}$$

which shows that an approximate Riemann solver is, indeed, an upwind method based on simple wave decomposition at cell interfaces, and it requires an appropriate definition of an intermediate state for evaluating $\widehat{\mathbf{A}}_{i+\frac{1}{2}}$.
As shown by Roe (1981a) the vector unknown $\mathbf{W}$ and the flux vector $\mathbf{F}$ are quadratic functions of a vector variable $\mathbf{Z}$ defined as

$$\mathbf{Z} = [\sqrt{\rho}, \sqrt{\rho}u, \sqrt{\rho}H]^{\mathrm{T}} \tag{86}$$

Then, it can be easily shown that $\widehat{\mathbf{A}}_{i+1/2}$ coincides with the jacobian matrix $\mathbf{A}$ evaluated at a state defined in terms of the so-called Roe average

$$\overline{(\cdot)}_{i+\frac{1}{2}} = \frac{\sqrt{\rho_i}(\cdot) + \sqrt{\rho_{i+1}}(\cdot)}{\sqrt{\rho_i} + \sqrt{\rho_{i+1}}} \tag{87}$$

and

$$\rho_{i+1/2} = \sqrt{\rho_i \rho_{i+1}} \tag{88}$$

The approximate Riemann solver of Roe yields an exact solution if the discontinuous solution is a shock. The major pitfall of the method is its inability to distinguish between physical and non-physical solutions, with the consequence that in some circumstances it may produce expansion shocks. As proposed by Harten and Hyman (1983), the only means for preventing expansion shocks from occurring is to modify the numerical flux function by introducing the so-called entropy condition; this point will be discussed in a later section.

Other approximate Riemann solvers have been proposed in the literature, for more details see for example Engquist and Osher (1980), Osher (1984), Osher and Solomon (1982), Van Leer (1984a,b) and Pandolfi (1989).

B High-order methods

In this section we discuss the issues related to the design of higher-order schemes such as the Total Variation Diminishing (TVD), the Monotone Upstream Schemes for Conservation Laws (MUSCL), the Local Extremum Diminishing (LED) and the Essentially Non-Oscillatory (ENO) methods.

1 Schemes for scalar equations

Lax–Wendroff (LW)

This scheme, developed in 1960, is probably the most famous second-order scheme for solving a hyperbolic scalar equation. For the linear case the Lax–Wendroff is second-order also in time, and the numerical flux function is

$$h_{i+\frac{1}{2}} = \tfrac{1}{2} a (w_{i+1} + w_i) - \tfrac{1}{2} \lambda a^2 \delta^+ w_i \tag{89}$$

where $\lambda = \Delta t / \Delta x$ and $\delta^+(\cdot) = (\cdot)_{i+1} - (\cdot)_i$. A generalization to the non-linear case gives

$$h_{i+\frac{1}{2}} = \tfrac{1}{2} (f_{i+1} + f_i) - \tfrac{1}{2} d_{i+\frac{1}{2}} \tag{90}$$

where the dissipation flux is

$$d_{i+\frac{1}{2}} = \lambda a_{i+\frac{1}{2}} \left(\frac{\mathrm{d}f}{\mathrm{d}w} \right)_{i+\frac{1}{2}} \delta^+ w_i \tag{91}$$

To understand the design of high-order numerical flux functions of more advanced approaches one can easily show that a high-order scheme can always be constructed by defining the flux function as a linear combination of low-(h^ℓ)

and high-(h^h) order flux functions. Indeed,

$$h_{i+\frac{1}{2}} = h^\ell_{i+\frac{1}{2}} + \left(h^h_{i+\frac{1}{2}} - h^\ell_{i+\frac{1}{2}}\right) = h^\ell_{i+\frac{1}{2}} + h^a_{i+\frac{1}{2}} \tag{92}$$

where h^a is the so called antidiffusive flux.

For example, using the CIR scheme for the low-order flux function and the (LW) method for the high-order flux function the antidiffusive flux becomes

$$h^a_{i+\frac{1}{2}} = \frac{1}{2}\left(\text{sign }(a_{i+\frac{1}{2}}) - \lambda a_{i+\frac{1}{2}}\right)\left(\frac{df}{dw}\right)_{i+\frac{1}{2}} \delta^+ w_i \tag{93}$$

which shows that the artificial diffusion ($\sim \text{sign }(a) df/dw\, \delta^+ w$) is limited by means of h^a.

Total Variation Diminishing (TVD) schemes

In a previous section we recalled the important result that a monotone scheme yields a solution that converges toward the physically admissible one. If $\tilde{u}_i$ is the state vector, a numerical method is said to be monotone if, given two approximate weak solutions $\tilde{u}$ and $\tilde{v}$ (of initial value problem) that satisfy

$$\tilde{u}^n_i \geq \tilde{v}^n_i \qquad \forall i$$

then

$$\tilde{u}^{n+1}_i \geq \tilde{v}^{n+1}_i \qquad \forall i$$

The monotonicity also implies that a non-negative perturbation (δ) at time t_n produces a non-negative perturbation at a successive time t_{n+1}, e.g.

$$\delta v^n_i = \epsilon_i \geq 0 \Rightarrow \delta_\epsilon v^{n+1}_i \geq 0$$

where δ_ϵ stands for the induced perturbation. Useful algebraic relations can be obtained by assuming $\epsilon_i = \epsilon \delta_{ij}$ (with δ_{ij} the Kronecker delta and j fixed) and taking the limit for $\epsilon \to 0$. In fact, if one recasts a numerical scheme in the form

$$u^{n+1}_i = H(u_{i-m}, \ldots, u_i, \ldots, u_{i+m})$$

where the ~ has been dropped, the monotonicity amounts to requiring that

$$\frac{\partial H}{\partial u_{i+j}} \geq 0 \qquad \forall i; j = -m, m \tag{94}$$

Referring to the fully discretized formulation of (45), the above condition becomes

$$\delta_{j0} - \lambda\left(\frac{\partial h_{i+\frac{1}{2}}}{\partial u_{i+j}} - \frac{\partial h_{i-\frac{1}{2}}}{\partial u_{i+j}}\right) \geq 0 \qquad \forall i; j = -m, m \tag{95}$$

where ~ has been dropped.

The monotonicity constraint can in principle be extended to the semidiscretized formulation of (41) as well, yielding

$$-\left(\frac{\partial h_{i+\frac{1}{2}}}{\partial u_{i+j}} - \frac{\partial h_{i-\frac{1}{2}}}{\partial u_{i+j}}\right) \geq 0 \qquad \forall i; j = -m, m \text{ with } j \neq 0$$

When the above condition is verified one can expect that for a sufficiently small τ the solution $u_i(t + \tau; u_{i+m}(t), \ldots, u_i(t), \ldots, u_{i-m}(t))$ satisfies a constraint equivalent to (94)

$$\frac{\partial u_i(t + \tau)}{\partial u_{i+j}(t)} \geq 0$$

Godunov (1959) and Harten $et\ al.$ (1976) have shown that monotone schemes are at most first-order accurate. To achieve higher accuracy, the monotonicity must be relaxed, while the monotonicity preserving property must still be verified. This is achieved by requiring the schemes to satisfy the Total Variation Diminishing (TVD) property.

TVD schemes are defined (Harten, 1983, 1984; LeVeque, 1992) as those schemes which satisfy the TVD property in a discrete sense. Let the total variation be

$$TV(u) = \sum_j |u_{j+1} - u_j| = \sum_j |V_j|$$

where V_j stands for the j-th cell variation. For a fully discretized numerical method the TVD condition amounts to requiring that

$$\sum_j |u_{j+1}^{n+1} - u_j^{n+1}| \leq \sum_j |u_{j+1}^n - u_j^n|$$

i.e.

$$\phi_1(\mathbf{V}^{n+1}) \leq \phi_1(\mathbf{V}^n)$$

where ϕ_1 is the L_1-norm, and $\mathbf{V}$ is the cell variation state vector. From (50) it is easy to show that the cell variation satisfies the following equation

$$V_i^{n+1} = V_i^n - f_i(V_j^n) \tag{96}$$

where

$$f_i(V_j^n) = \lambda(\delta h_{i+1} - \delta h_i)$$

and $\delta(\cdot)_i = (\cdot)_{i+\frac{1}{2}} - (\cdot)_{i-\frac{1}{2}}$. If one can set

$$\mathbf{f}(\mathbf{V}) = \mathbf{C}(\mathbf{V})\mathbf{V} \tag{97}$$

where $\mathbf{C}$ is a finite dimension matrix, then the TVD condition requires

$$\|\mathbf{I} - \mathbf{C}(\mathbf{V})\|_1 \leq 1 \tag{98}$$

where $\|A\|_1 \equiv \max_j \left(\sum_i |a_{ij}| \right)$ is the geometric norm induced by ϕ_1.

The above inequality can be exploited to obtain design constraints for the numerical flux function. Referring to (48) and assuming only a dependence on the n-th values, the numerical fluxes involved in the conservation of the i-th cell can be recast as follows

$$h_{i-\frac{1}{2}} = h(\mathbf{u}) = h\left(E^{-(m-1)}u_{i-1}, \ldots, u_{i-1}, E^1 u_{i-1}, \ldots, E^{m-1}u_{i-1}, E^m u_{i-1} \right)$$

$$h_{i+\frac{1}{2}} = h(\mathbf{u} + \mathbf{V}) = h\left(E^{-(m-1)}(u_{i-1}+V_{i-1}), \ldots, (u_{i-1}+V_{i-1}), E^1(u_{i-1}+V_{i-1}), \ldots, \right.$$
$$\left. E^{m-1}(u_{i-1}+V_{i-1}), E^m(u_{i-1}+V_{i-1}) \right)$$

where E is the shift operator, i.e. $E^m u_i = u_{i+m}$. Hence, one obtains

$$\delta h_i = h_{i+\frac{1}{2}} - h_{i-\frac{1}{2}} = \sum_{j=1,m} \left(\frac{\partial h}{\partial u_{i-j}} \right)_{\mathbf{u}+\theta\mathbf{V}} V_{i-j} + \sum_{j=1,m} \left(\frac{\partial h}{\partial u_{i+j-1}} \right)_{\mathbf{u}+\theta\mathbf{V}} V_{i+j-1}$$

where $\theta \in (0,1)$. It is then easy to show that the equation for the i-th cell variation is

$$V_i^{n+1} = \lambda C_{i-m+\frac{1}{2}}^{+(m)} V_{i-m} - \lambda \sum_{j=1,m-1} (C_{i-j+\frac{1}{2}}^{+(j+1)} - C_{i-j+\frac{1}{2}}^{+(j)}) V_{i-j}$$
$$+ \left[1 - \lambda(C_{i+\frac{1}{2}}^{+(1)} - C_{i+\frac{1}{2}}^{-(1)}) \right] V_i$$
$$- \lambda \sum_{j=1,m-1} (C_{i+j+\frac{1}{2}}^{-(j)} - C_{i+j+\frac{1}{2}}^{-(j+1)}) V_{i+j}$$
$$- \lambda C_{i+m+\frac{1}{2}}^{-(m)} V_{i+m} \tag{99}$$

where

$$C_{i-j+\frac{1}{2}}^{+(j)} = \left(\frac{\partial h}{\partial u_{i-j}} \right)_{\mathbf{u}+\theta\mathbf{V}}; \quad C_{i+j-\frac{1}{2}}^{-(j)} = \left(\frac{\partial h}{\partial u_{i+j-1}} \right)_{\mathbf{u}+\theta\mathbf{V}}$$

From the right-hand-side of (99) it can be seen that, for a periodic initial value problem, the i-th row of the matrix $\mathbf{I} - \mathbf{C}(\mathbf{V})$ is also the i-th column of the matrix. Consequently, the TVD condition corresponding to (98) is ensured if

$$\lambda \left(C_{i+\frac{1}{2}}^{+(1)} - C_{i+\frac{1}{2}}^{-(1)} \right) \leq 1$$

$$C_{i+\frac{1}{2}}^{+(1)} \geq C_{i+\frac{1}{2}}^{+(2)} \geq \cdots \geq 0$$

$$-C_{i+\frac{1}{2}}^{-(1)} \geq -C_{i+\frac{1}{2}}^{-(2)} \geq \cdots \geq 0$$

For the semidiscretized formulation of (41), the cell variation satisfies the following equation

$$\frac{\mathrm{d}}{\mathrm{d}t} V_i = -g_i(V_j) \tag{100}$$

where

$$g_i(V_j) = \lambda \frac{1}{\Delta t} (\delta h_{i+1} - \delta h_i)$$

and the TVD property amounts to

$$\frac{\mathrm{d}}{\mathrm{d}t} \phi_1(\mathbf{V}) \leq 0 \tag{101}$$

If one sets $\mathbf{g}(\mathbf{V}) = \mathbf{G}(\mathbf{V})\mathbf{V}$, and recalling that

$$\frac{\mathrm{d}}{\mathrm{d}t} \phi_1(\mathbf{V}) = \lim_{\Delta t \to 0} \left(\frac{\phi_1\{[\mathbf{I} - \mathbf{G}(\mathbf{V})\Delta t]\mathbf{V}\} - \phi_1(\mathbf{V})}{\Delta t} + O(\Delta t) \right)$$

then, for sufficiently small Δt, (101) is verified if,

$$\phi_1[(\mathbf{I} - \mathbf{G}(\mathbf{V})\Delta t)\mathbf{V}] - \phi_1(\mathbf{V}) \leq 0 \tag{102}$$

Thus, in analogy with the condition that ensures (98), it suffices that

$$G_{ii} - \sum_{j \neq i} |G_{ji}| \geq 0$$

and, as outlined for the fully discretized formulation, one can obtain the same type of design requirements for the numerical flux function (TVD requirements for semidiscretized schemes are described in more detail in Jameson and Lax, 1984).

A total variation diminishing scheme is also monotonicity preserving (i.e. if $u_i^n \geq u_{i-1}^n \Rightarrow u_i^{n+1} \geq u_{i-1}^{n+1} \ \forall i$). However, unlike monotone schemes, TVD methods do not guarantee the convergence of the numerical solution to a weak physically admissible one (Harten, 1983, 1984; Hirsch, 1990; Osher, 1984; LeVeque, 1992). Following Harten (Harten, 1983; Harten and Hyman, 1983), one can construct, for example, a three-point TVD scheme consistent with an entropy inequality. Let the numerical flux function be defined as

$$h_{i+\frac{1}{2}} = \tfrac{1}{2}(f_i + f_{i+1}) - \tfrac{1}{2\lambda} Q_{i+\frac{1}{2}} V_i \tag{103}$$

The equation for the cell variation is

$$V_i^{n+1} = \lambda C_{i-\frac{1}{2}}^{+(1)} V_{i-1}^n + \left[1 - \lambda \left(C_{i+\frac{1}{2}}^{+(1)} - C_{i+\frac{1}{2}}^{-(1)} \right) \right] V_i^n - \lambda C_{i+\frac{3}{2}}^{-(1)} V_{i+1}^n$$

and the TVD property is ensured by

$$\lambda\left(C_{i+\frac{1}{2}}^{+(1)} - C_{i+\frac{1}{2}}^{-(1)}\right) = Q_{i+\frac{1}{2}} \leq 1$$

$$-\lambda C_{i+\frac{1}{2}}^{-(1)} = \tfrac{1}{2}\left[-\lambda a_{i+\frac{1}{2}} + Q_{i+\frac{1}{2}}\right] \geq 0$$

$$\lambda C_{i+\frac{1}{2}}^{+(1)} = \tfrac{1}{2}\left[\lambda a_{i+\frac{1}{2}} + Q_{i+\frac{1}{2}}\right] \geq 0$$

i.e.

$$|\lambda a_{i+\frac{1}{2}}| \leq Q_{i+\frac{1}{2}} \leq 1$$

Non-admissible discontinuities are inhibited if

$$Q(|x|; \epsilon) = \frac{1 + \operatorname{sign}\,(|x| - \epsilon)}{2}|x| + \frac{1 - \operatorname{sign}\,(|x| - \epsilon)}{4}\left(\frac{x^2}{\epsilon} + \epsilon\right) \quad (104)$$

where ϵ is a small positive number (Harten, 1983, 1984; Harten and Hyman, 1983; Hirsch, 1990; Osher, 1984; LeVeque, 1992).

Osher (1984) has defined a more general class of TVD schemes that ensure the convergence to unique entropy solutions: the so-called E-schemes. By definition the latter are consistent schemes whose numerical flux function satisfies

$$\operatorname{sign}\,(V_i)[h_{i+\frac{1}{2}} - f(u)] \leq 0 \qquad \forall u \in [u_i, u_{i+1}]$$

However, E-schemes are at most first-order accurate (Osher, 1984; Hirsch, 1990), and thus face the same limitations encountered in the construction of monotone schemes.

A possible remedy consists of no longer accounting for second- and higher-order flux contribution everywhere. There are many ways for estimating the higher-order corrections and/or the regions where these corrections should be accounted for. The flux – and slope – limiter methods are the most commonly used approaches.

Flux limiter method

Flux limiter methods are based on the idea of including the higher-order contributions mainly in smooth regions, where they are evaluated by means of an algebraic expansion. These methods originate from the modified flux approach of Harten (1984), which closely resembles the Flux Corrected Transport method of Boris and Book (1973). A higher-order numerical flux function is constructed by simply applying a low-order formula to a modified flux function. The latter is such that the low-order dissipation terms of the 'equivalent equation' (that corresponds to the selected low-order formula) are 'cancelled' out (Lerat and Peyret, 1974).

As an example, let us consider the CIR method applied to a scalar linear hyperbolic equation with constant positive advection velocity for which $f = au$

and $h_{i+\frac{1}{2}} = au_i$. Then, the equivalent equation corresponding to the chosen numerical flux function is

$$u_t + au_x = \left[a\frac{\Delta x}{2}(1 - \sigma)u_x \right]_x + O(\Delta x^2) \tag{105}$$

where $\sigma = a\Delta t/\Delta x$ is the Courant (or CFL) number. Equation (105) shows that the solution of the original equation obtained with CIR corresponds to an equation for which the flux is $f - g$ where

$$g = \frac{a\Delta x}{2}(1 - \sigma)u_x$$

Then, if one introduces a modified flux $\tilde{f} = f + g$, a higher-order numerical flux function can be defined, i.e.

$$h_{i+\frac{1}{2}}^h = au_i + \frac{a}{2}(1 - \sigma)\delta^+ u_i$$

which is nothing other than the Lax–Wendroff scheme. Obviously the procedure can also be extended to the non-linear case.

A straightforward use of a high-order numerical flux function can violate TV-stability and entropy conditions. As a remedy the higher-order numerical flux function can be defined as

$$h_{i+\frac{1}{2}} = h_{i+\frac{1}{2}}^\ell + \psi_{i+\frac{1}{2}}\left(h_{i+\frac{1}{2}}^h - h_{i+\frac{1}{2}}^\ell \right) \tag{106}$$

where the superscripts h and ℓ stand, respectively, for high and low order, and $\psi_{i+\frac{1}{2}}$ is the limiter function that makes $h_{i+\frac{1}{2}}$ degenerate into low order accuracy near a discontinuity. For consistency, the function ψ must also be a bounded function of the ratio of consecutive gradients (a measure of the smoothness of the solution), i.e. $\psi_{i+\frac{1}{2}} = \psi(r_i)$, where

$$r_i = \frac{\delta^- u_i}{\delta^+ u_i} \ , \quad \delta^-(\cdot) = (\cdot)_i - (\cdot)_{i-1} \quad \text{and} \quad \delta^+(\cdot) = (\cdot)_{i+1} - (\cdot)_i$$

In order to clarify the role of r_i observe that (for constant mesh spacing)

- $r_i < 0$
  local extremum
- $0 < r_i < 1$
  local slope increase
- $r_i \sim 1$
  local 'constant slope'
- $r_i > 1$
  local slope decrease

Moreover, a succession of values r_i of the same class indicates:

- $r_i < 0$
  oscillatory behaviour (wiggling if $|r_i| << 1$)
- $0 < r_i < k < 1$
  monotone exponential-like behaviour typical of a
  discontinuity capturing region
- $r_i \sim 1$
  monotone behaviour typical of a smooth region
- $r_i > k > 1$
  monotone behaviour typical of a constant state reaching region

Therefore, the limiting functions are generally constructed by requiring

$$\psi(r) = O(r) \quad r < 1$$
$$= 0 \quad r < 0$$
$$\sim O(1) \quad r \geq 1$$

The enforcement of the TVD property yields further constraints on ψ. For example, for the linear case illustrated here we have

$$h_{i+\frac{1}{2}}^h - h_{i+\frac{1}{2}}^\ell = \frac{a}{2}(1 - \sigma)\delta^+ u_i \tag{107}$$

and in order to guarantee the TVD property (Davis, 1984; Roe, 1984; Sweby, 1984) the limiter must lie in the region where

$$0 < \psi \leq 2 \min (r, 1)$$

provided the CFL condition $\sigma < 1$ holds. Sweby (1984) has also shown that the scheme must be a convex combination of the Lax–Wendroff and the Warming and Beam scheme (Warming and Beam, 1976) to maintain second order accuracy, i.e.

$$\psi(r) = 1 + \theta(r)(r - 1) \tag{108}$$

where $\theta = 0$ yields the Lax–Wendroff scheme, while $\theta = 1$ gives the Warming and Beam scheme. Several limiters have been proposed in the literature (Roe, 1981b, 1985; Van Leer, 1977, 1979; Van Albada et al., 1982; Chakravarthy and Osher, 1983; Sweby, 1984; Yee, 1989; Hirsch, 1990; LeVeque, 1992); the most popular are (see Figure 7)

$$\psi = \max [0, \min (r, 1)] \qquad \text{(Roe's minmod)}$$
$$\psi = \max [0, \min (2r, 1), \min (r, 2)] \quad \text{(Roe's superbee)}$$
$$\psi = (r + |r|)/(1 + r) \qquad \text{(Van Leer's)}$$
$$\psi = (r^2 + r)/(1 + r^2) \qquad \text{(Van Albada's)}$$

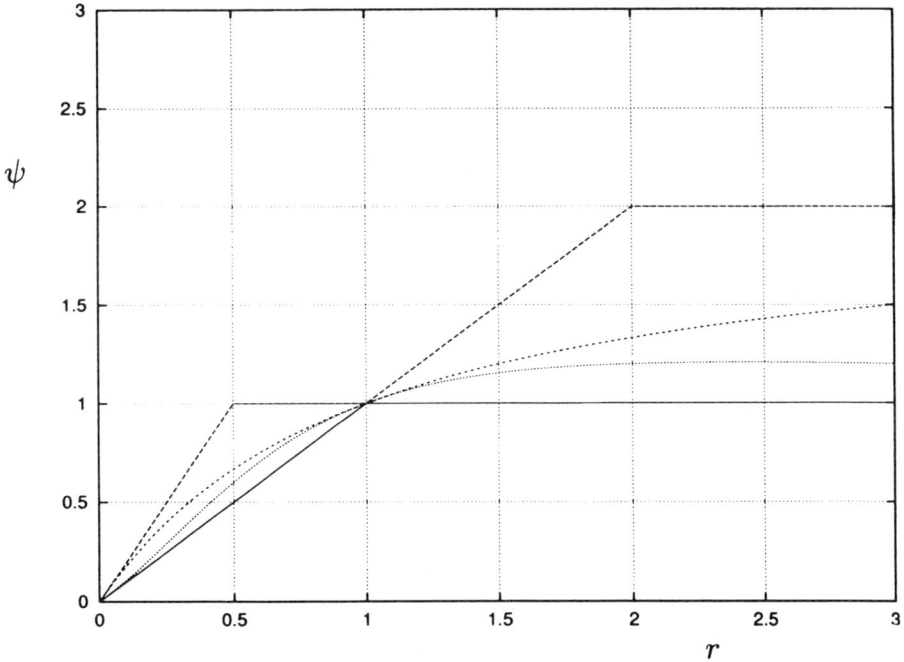

Figure 7 Popular limiter function. Key: —, minmod; - - -, Roe Superbee; ----, Van Leer; · · ·, Van Albada.

A comparison of the different limiters (applied to the inviscid Burgers equation and/or Sod one-dimensional test problem; Sod 1978) shows that Roe's minmod is too dissipative while Roe's superbee is overcompressive. Van Leer's and Van Albada's limiters are defined in terms of differentiable functions and are smooth; this last property has been shown by Venkatakrishnan (1993) to be very important for robustness and convergence toward steady-state solutions. Moreover, they behave very similarly and generate a low level (smaller than minmod) of dissipation. More general limiters have been proposed by Lacor *et al.*, (1993); these will be discussed in the next section.

A question now arises regarding the construction of high-order schemes and the convergence toward physically admissible solutions. To our knowledge there seems to be no rigorous theory for designing a full entropy-satisfying higher-order explicit scheme (other than E-schemes). A weak condition has been stated by Osher and Chakravarthy (1984) who have proved that for a single entropy function (*de facto* an energy) a second order TVD scheme can be constructed provided the numerical flux function satisfies the inequality

$$\int_{u_i}^{u_{i+1}} s_{uu}[h_{i+\frac{1}{2}} - f(u)]\ \mathrm{d}u \le 0$$

where s and s_{uu} are the entropy function and its second derivative with respect to u.

In a heuristic construction of higher-order schemes, the use of a first-order entropy satisfying scheme with limited antidiffusion is expected to yield a higher order (nearly) entropy-satisfying scheme (Sweby, 1984). For example, if one uses as a low order numerical flux function the scheme corresponding to (103), then a (heuristically) higher-order entropy satisfying scheme is

$$h_{i+\frac{1}{2}} = \tfrac{1}{2}(f_i + f_{i+1}) + \tfrac{1}{2}Q(a_{i+\frac{1}{2}})(g_i + g_{i+1}) - \frac{1}{2}\frac{Q}{\lambda}(\tilde{a}_{i+\frac{1}{2}})\delta^+ u_i$$

where

$$\tilde{a}_{i+\frac{1}{2}} = a_{i+\frac{1}{2}} + \gamma_{i+\frac{1}{2}}$$

$$a_{i+\frac{1}{2}} = \begin{cases} \delta^+ f_i / \delta^+ u_i & \text{if } \delta^+ u_i \neq 0 \\ df/du & \text{otherwise} \end{cases}$$

$$\gamma_{i+\frac{1}{2}} = \begin{cases} \delta^+ g_i / \delta^+ u_i & \text{if } \delta^+ u_i \neq 0 \\ 0 & \text{otherwise} \end{cases}$$

and popular expressions for the limited antidiffusion flux are

$$g_i = \max\,[0, \min\,(r_i, 1)]\delta^+ u_i$$
$$g_i = \max\,[0, \min\,(2r_i, 1), \min\,(r_i, 2)]\delta^+ u_i$$
$$g_i = (r_i + |r_i|)/(1 + r_i)\delta^+ u_i$$
$$g_i = (r_i^2 + r_i)/(1 + r_i^2)\delta^+ u_i$$

A more detailed overview of high-order schemes can be found in Yee (1989).

Slope limiter methods

Keeping in mind that in a finite-volume formulation one generally assumes as state variables the cell averages $(\tilde{u})$, higher-order methods can be constructed by higher-order reconstruction of the local values.

According to Van Leer's point of view, upwind methods can be interpreted as a projection phase followed by an evolution phase (Godunov, 1959; Van Leer, 1977, 1985). In the projection (or reconstruction) phase the piecewise continuous initial values are interpolated to yield a continous distribution within each computational cell, while the evolution phase corresponds to the updating of $\tilde{u}$ exploiting the reconstructed solution (Van Leer, 1985).

The simplest idea underlying the design principle for increasing the order of accuracy consists in considering a linear variation within each cell. Then, for a one-dimensional problem, the projection amounts to defining local interpolation laws of the type

$$u^n(x) = \tilde{u}_i^n + p_i(x - x_i) \qquad x_i - \Delta x/2 \le x \le x_i + \Delta x/2$$

where p_i is the 'slope' of u at cell i.

Observing that the cell averaging is invariant with respect to the reconstruction parameter p_i, one can exploit the freedom in the choice of p_i to ensure the total variation diminishing property and high accuracy in the updating of $\tilde{u}$. Then, in the projection phase one may require

$$TV(u^n) \le TV(\tilde{u}^n) \tag{109}$$

and, provided the evolution phase is total variation diminishing, the method is globally TVD (Van Leer, 1985; LeVeque, 1992). However, the total variation of the projected solution may even violate (109), but still obtain globally a TVD scheme as long as the evolution phase is TVD.

The satisfaction of (109) is ensured by introducing limitations on the slope, thus obtaining

$$u^n(x) = \tilde{u}_i + (x - x_i)\psi_i \hat{p}_i \tag{110}$$

where ψ is the limiter, subject to the constraint that the scheme is TVD, and $\hat{p}_i$ is for example a first-order approximation of the slope expressed in terms of $\tilde{u}$. All the limiter functions previously given for the flux limiter method are good candidates for defining ψ, while, assuming constant mesh spacing, for $\hat{p}_i$ one may assume

$$\hat{p}_i = \frac{1}{\Delta x} \delta^+ \tilde{u}_i$$

We now show how to construct a slope limiter TVD method for a scalar hyperbolic equation. Let the initial data be the cell average solution $\tilde{u}_i^n$, the updated solution is

$$\tilde{u}_i^{n+1} = \tilde{u}_i^n - \lambda \left[\frac{1}{\Delta t} \int_{t_n}^{t_{n+1}} f[u^n(x_{i+\frac{1}{2}}, t)] \, dt - \frac{1}{\Delta t} \int_{t_n}^{t_{n+1}} f[u^n(x_{i-\frac{1}{2}}, t)] \, dt \right] \tag{111}$$

where $u^n(x, t)$ indicates the reconstructed solution evaluated at time t by projection of the initial data $\tilde{u}^n$, and $\lambda = \Delta t/\Delta x$.

At time t_n an upwind limited reconstruction yields

$$u^n(x_{i+\frac{1}{2}}, t_n) = \frac{1 + s_i}{2} \left[\tilde{u}_i^n + \frac{\psi_i^+}{2} \delta^+ \tilde{u}_i \right] + \frac{1 - s_i}{2} \left[\tilde{u}_{i+1}^n - \frac{\psi_i^-}{2} \delta^+ \tilde{u}_i \right] \tag{112}$$

where

$$s_i = \text{sign}\left(\frac{df}{du}\right)_i, \quad \psi_i^+ = \psi\left(\frac{\delta^+ \tilde{u}_{i-1}^n}{\delta^+ \tilde{u}_i^n}\right), \quad \psi_i^- = \psi\left(\frac{\delta^+ \tilde{u}_{i+1}^n}{\delta^+ \tilde{u}_i^n}\right)$$

The evolution of the reconstructed solution amounts to solving locally and 'exactly' (or as accurately as possible) the governing equation. In order to produce an 'exact' solution, LeVeque (1992) proposed to approximate the flux by a linear spline with nodal values $\{\tilde{u}_i^n\}$, thus obtaining

$$u^n(x_{i+\frac{1}{2}}, t) = u^n[x_{i+\frac{1}{2}} - \tilde{a}_{i+\frac{1}{2}}(t - t_n), t_n]$$

where

$$\tilde{a}_{i+\frac{1}{2}} = \begin{cases} \dfrac{f(\tilde{u}_{i+1}^n) - f(\tilde{u}_i^n)}{\tilde{u}_{i+1} - \tilde{u}_i^n} & \text{if } \delta^+ \tilde{u}_i^n \neq 0 \\[2ex] \left(\dfrac{df}{du}\right)_i & \text{otherwise} \end{cases}$$

Hence

$$u^n(x_{i+\frac{1}{2}}, t) = \frac{1 + s_i}{2}\left[\tilde{u}_i^n + \psi_i^+ \delta^+ \tilde{u}_i\left(1 - 2|\sigma_i|\frac{t - t_n}{\Delta t}\right)\right]$$

$$+ \frac{1 - s_i}{2}\left[\tilde{u}_{i+1}^n - \psi_i^- \delta^+ \tilde{u}_i\left(1 - 2|\sigma_i|\frac{t - t_n}{\Delta t}\right)\right] \tag{113}$$

where $\sigma_i = \lambda \tilde{a}_{i+\frac{1}{2}}$ is the local Courant number.

The above solution is 'exact' as long as the shocks that form at cell centres (due to discontinuities in the derivatives of the spline approximation of $f(u)$) do not reach the cell interfaces, i.e. provided $\max(|\sigma_i|) \leq \frac{1}{2}$. Consistent with the approximation used for the evolution of the reconstructed solution one can easily evaluate the time integral of the flux function maintaining the linear spline approximation, i.e.

$$f[u^n(x_{i+\frac{1}{2}}, t)] = \frac{1 + s_i}{2}f(\tilde{u}_i^n) + \frac{1 - s_i}{2}f(\tilde{u}_{i+1}^n)$$

$$+ |\tilde{a}_i|\left(1 - 2|\sigma_i|\frac{t - t_n}{\Delta t}\right)\left(\frac{1 + s_i}{2}\frac{\psi_i^+}{2} + \frac{1 - s_i}{2}\frac{\psi_i^-}{2}\right)\delta^+ \tilde{u}_i^n$$

and

$$\tilde{u}_i^{n+1} = \tilde{u}_i^n - \lambda\left[\frac{1 + s_i}{2}f(\tilde{u}_i^n) + \frac{1 - s_i}{2}f(\tilde{u}_{i+1}^n)\right]$$

$$- |\sigma_i|\frac{1 - |\sigma_i|}{2}\left(\frac{1 + s_i}{2}\psi_i^+ + \frac{1 - s_i}{2}\psi_i^-\right)\delta^+ \tilde{u}_i^n$$

$$+ \lambda\left[\frac{1 + s_{i-1}}{2}f(\tilde{u}_{i-1}^n) + \frac{1 - s_{i-1}}{2}f(\tilde{u}_i^n)\right] \tag{114}$$

$$+ |\sigma_{i-1}|\frac{1 - |\sigma_{i-1}|}{2}\left(\frac{1 + s_{i-1}}{2}\psi_{i-1}^+ + \frac{1 - s_{i-1}}{2}\psi_{i-1}^-\right)\delta^+ \tilde{u}_{i-1}^n$$

Then, the numerical flux function is

$$h_{i+\frac{1}{2}} = \frac{1 + s_i}{2} f(\tilde{u}_i^n) + \frac{1 - s_i}{2} f(\tilde{u}_{i+1}^n)$$

$$+ \frac{|\tilde{a}_i|}{2}(1 - |\sigma_i|)\left(\frac{1 + s_i}{2}\psi_i^+ + \frac{1 - s_i}{2}\psi_i^-\right)\delta^+\tilde{u}_i^n \tag{115}$$

It is easy to verify that: (i) the TVD property is satisfied provided ψ is a limiter function of the type previously discussed; and (ii) in the linear case (114) reduces to the expression given by Davis (1984) for the TVD Lax–Wendroff scheme, i.e.

$$\tilde{u}_i^{n+1} = \tilde{u}_i^n - \sigma\left(\frac{1 + \sigma}{2}\right)\delta^+\tilde{u}_{i-1}^n - \sigma\left(\frac{1 - \sigma}{2}\right)\delta^+\tilde{u}_i^n$$

$$+ |\sigma|\frac{1 - |\sigma|}{2}\left(1 - \frac{1 + s}{2}\psi_i^+ - \frac{1 - s}{2}\psi_i^-\right)\delta^+\tilde{u}_i^n \tag{116}$$

$$- |\sigma|\frac{1 - |\sigma|}{2}\left(1 - \frac{1 + s}{2}\psi_{i-1}^+ - \frac{1 - s}{2}\psi_{i-1}^-\right)\delta^+\tilde{u}_{i-1}^n$$

Lacor–Hirsch limiters
Lacor *et al.* (1993) have introduced another class of limiter functions. Assuming constant mesh spacing, from (110) the left and right states are

$$u_L = \tilde{u}_i + \hat{\delta}^+\tilde{u}_i/2; \quad u_R = \tilde{u}_{i+1} - \overline{\delta}^+\tilde{u}_i/2$$

If one defines the limited slopes as harmonic averages of successive slopes, i.e.

$$\frac{1}{\hat{\delta}^+\tilde{u}_i} = \frac{1}{2}\left(\frac{1}{\delta^-\tilde{u}_i} + \frac{1}{\delta^+\tilde{u}_i}\right); \quad \frac{1}{\overline{\delta}^+\tilde{u}_i} = \frac{1}{2}\left(\frac{1}{\delta^-\tilde{u}_{i+1}} + \frac{1}{\delta^+\tilde{u}_{i+1}}\right)$$

then

$$\hat{\delta}^+\tilde{u}_i = \psi(r_i^+)\delta^+\tilde{u}_i; \quad \overline{\delta}^+\tilde{u}_i = \psi(r_i^-)\delta^+\tilde{u}_i$$

where $r_i^+ = \delta^-\tilde{u}_i/\delta^+\tilde{u}_i; r_i^- = \delta^+\tilde{u}_{i+1}/\delta^-\tilde{u}_{i+1}$ and $\psi(r)$ coincides with Van Leer's limiter.

Lacor *et al.* show that more general formulas can be obtained by defining the limiter as a non-separable function of both ratios r_i^- and r_i^+. Introducing an extended harmonic mean, i.e.

$$\frac{1}{\hat{\delta}^+\tilde{u}_i} = \frac{1}{\overline{\delta}^+\tilde{u}_i} = \frac{1}{2}\left[\frac{1}{2}\left(\frac{1}{\delta_i^-\tilde{u}_i} + \frac{1}{\delta_i^+\tilde{u}_i}\right) + \frac{1}{2}\left(\frac{1}{\delta_i^-\tilde{u}_{i+1}} + \frac{1}{\delta_i^+\tilde{u}_{i+1}}\right)\right]$$

one obtains

$$\hat{\delta}^+ \tilde{u}_i = \bar{\delta}^+ \tilde{u}_i = \psi(r_i^+, r_i^-)\delta^+ \tilde{u}_i$$

where

$$\psi(r_i^+, r_i^-)_i = \frac{4r_i^+ r_i^-}{r_i^+ + r_i^- + 2r_i^+ r_i^-}$$

is the harmonic mean of Van Leer's limiter applied to r_i^+ and r_i^-. Then, a class of (more general) smooth limiters is defined in terms of the harmonic mean of both ratios r_i^+ and r_i^-, i.e.

$$\psi(r_i^+, r_i^-) = \frac{(r_i^*)^\alpha + r_i^*}{1 + (r_i^*)^\alpha} \tag{117}$$

where

$$\frac{1}{r_i^*} = \frac{1}{2}\left(\frac{1}{r_i^+} + \frac{1}{r_i^-}\right)$$

2 Schemes for systems of conservation laws

Total variation diminishing schemes (TVD)
In the case of a hyperbolic system of conservation laws the total variation of the solution is not necessarily a monotonic decreasing function of time (Lax, 1972, Harten, 1983; Yee, 1989; Hirsch, 1990; LeVeque, 1992). However, even if the total variation increases when waves interact, the boundedness of the total variation is still enforced to design schemes for systems of conservation laws. In particular, to construct the numerical flux function a local characteristic decomposition is introduced and a TVD scheme is applied 'scalarly' to each of the fields. Then, the numerical flux function is generally written as

$$\mathbf{H}_{i+\frac{1}{2}} = \tfrac{1}{2}(\mathbf{F}_i + \mathbf{F}_{i+1}) - \tfrac{1}{2}\mathbf{L}_{i+\frac{1}{2}}^{-1}\mathbf{\Phi}_{i+\frac{1}{2}} \tag{118}$$

where $\mathbf{\Phi}$ is the antidiffusive flux and $\mathbf{L}$ is the left eigenvector matrix.

Many schemes have been proposed in the literature, several of which are described by Yee (1989). The so-called upwind total variation diminishing schemes identify for example a class of very robust not too dissipative schemes, widely used to solve Euler and Navier–Stokes equations, where the ℓ-th component of $\mathbf{\Phi}$ is

$$\phi_{i+\frac{1}{2}}^\ell = Q(a_{i+\frac{1}{2}}^\ell + \gamma_{i+\frac{1}{2}}^\ell)\alpha_{i+\frac{1}{2}}^\ell - \tfrac{1}{2}Q(a_{i+\frac{1}{2}}^\ell)(g_i^\ell + g_{i+1}^\ell) \tag{119}$$

where Q is Harten's entropy function (Harten, 1983, 1984)

$$Q(z) = \begin{cases} \dfrac{z^2 + \epsilon^2}{2\epsilon} & |z| < \epsilon \\ |z| & |z| > \epsilon \end{cases}$$

and

$$\alpha_{i+\frac{1}{2}} = \mathbf{L}_{i+\frac{1}{2}} \delta^+ \mathbf{W}_i$$
$$g_i = \psi(\alpha_i^\ell, \alpha_{i+1}^\ell)$$

with ψ a limiter function, and $a_{i+1/2}^\ell$ is the ℓ-th eigenvalue of the inviscid flux jacobian and $\gamma_{i+\frac{1}{2}}^\ell$ is defined according to

$$\gamma_{i+\frac{1}{2}}^\ell = Q(a_{i+\frac{1}{2}}^\ell) \begin{cases} \dfrac{g_{i+1}^\ell - g_i^\ell}{\alpha_{i+\frac{1}{2}}^\ell} & \alpha_{i+\frac{1}{2}}^\ell \neq 0 \\ 0 & \text{otherwise} \end{cases}$$

The left and right eigenvector matrices are evaluated either in terms of the geometric or Roe's average. However, the latter is preferable if the solution to the system of conservation laws admits a shock. As shown by Sweby (1984), good results are generally obtained if different limiters are employed for each of the components of the difference of characteristic vector. In particular, the use of Roe's superbee limiter gives good resolution of contact discontinuities when applied to the linear field in conjunction with Van Leer's for the non-linear one.

Monotone upstream schemes for conservation laws (MUSCL)

The monotonic upstream schemes for systems of conservation laws (the so-called MUSCL schemes) are a generalization of slope limiter methods that in the projection phase add to the slope parameter another degree of freedom (related to the curvature of the reconstructed solution). In order to maintain the invariance with respect to the cell averages, independently of the chosen reconstruction parameters, Van Leer (1979) proposes to cast the polynomial reconstruction in terms of the Legendre polynomial expansion. For a constant mesh spacing, introducing a local non-dimensional coordinate

$$\zeta = 2(x - x_i)/\Delta x \qquad \text{for} \qquad x_i - \Delta x/2 \leq x \leq x_i + \Delta x/2$$

the reconstructed solution is

$$u^n(\zeta) = \sum_{\ell=0,m} c_\ell P_\ell(\zeta) \tag{120}$$

where $P_\ell(\zeta)$ are the Legendre polynomials of degree ℓ that are mutually orthogonal in L_2 and verify

$$\int_{-1}^{1} P_\ell(\zeta)P_0(\zeta) \, d\zeta = \int_{-1}^{1} P_\ell(\zeta) \, d\zeta = 0 \qquad \ell \neq 0$$

where $P_0(\zeta) = 1$. As a consequence, (120) can be suitably rewritten as

$$u^n(\zeta) = \tilde{u}_i^n + \sum_{\ell=1,m} c_\ell P_\ell(\zeta) \tag{121}$$

The freedom in the choice of the coefficients c_ℓ must be exploited to ensure the accuracy and the control of the total variation of the projected solution. Then, it is convenient to set $c_\ell = \hat{c}_\ell \psi$, where the $\hat{c}$'s are determined by accuracy requirements (in the correlation between reconstructed and cell average values) and ψ is a limiter function. Restricting (121) to second-order expansion and assuming for the moment $\psi = 1$, one has

$$u^n(\zeta) = \tilde{u}_i^n + \hat{c}_1\zeta + \hat{c}_2(3\zeta^2 - 1)/2 \tag{122}$$

It is clear that $\hat{c}_1 = (\partial u/\partial\zeta)_{\zeta=0}, \hat{c}_2 = \frac{1}{3}\left(\partial^2 u/\partial\zeta^2\right)_{\zeta=0}$; however, in order to determine the $\hat{c}$'s in terms of cell averages one can extend the correlation between the local reconstruction and the average values to the adjacent cells, and require conditions such as

$$\frac{1}{2}\int_{1}^{3} u^n(\zeta) \, d\zeta = \tilde{u}_{i+1}^n \tag{123}$$

$$\frac{1}{2}\int_{-3}^{-1} u^n(\zeta) \, d\zeta = \tilde{u}_{i-1}^n \tag{124}$$

Equations (123) and (124) also imply the consistency of the forward and backward derivatives in terms of cell averages determined by reconstruction extrapolation. Then, substituting (122) in (123) and (124), one obtains

$$2\hat{c}_1 + 6\hat{c}_2 = \delta^+\tilde{u}_i^n \tag{125}$$

$$2\hat{c}_1 - 6\hat{c}_2 = \delta^+\tilde{u}_{i-1}^n \tag{126}$$

We now show how one can exploit the freedom in the choice of the coefficients $\hat{c}_\ell$ to recover some well-known reconstruction formulas (Van Leer, 1985).

(i) *Third-order upwind biased scheme*

$$\hat{c}_1 = \tfrac{1}{4}(\delta^+\tilde{u}_i^n + \delta^+\tilde{u}_{i-1}^n), \ \hat{c}_2 = \tfrac{1}{12}(\delta^+\tilde{u}_i^n - \delta^+\tilde{u}_{i-1}^n)$$

and

$$u^n(\zeta) = \tilde{u}_i + \frac{1}{4}\left[\left(\zeta + \left(\frac{\zeta^2}{2} - \frac{1}{6}\right)\right)\delta^+\tilde{u}_i^n + \left(\zeta - \left(\frac{\zeta^2}{2} - \frac{1}{6}\right)\right)\delta^+\tilde{u}_{i-1}^n\right] \quad (127)$$

(ii) *Three-point central difference scheme*

Setting $\hat{c}_2 = 0$, and the satisfaction of (125) gives

$$u^n(\zeta) = \tilde{u}_i^n + \frac{\zeta}{2}\delta^+\tilde{u}_i^n$$

that can be formally recast as

$$u^n(\zeta) = \tilde{u}_i^n + \frac{1}{4}\left\{[\zeta + (\zeta)]\delta^+\tilde{u}_i^n + [\zeta - (\zeta)]\delta^+\tilde{u}_{i-1}^n\right\} \quad (128)$$

(iii) *Fully upwind scheme*

Setting $\hat{c}_2 = 0$ and only satisfying (126) yields

$$u^n(\zeta) = \tilde{u}_i^n + \frac{1}{4}\left\{[\zeta + (-\zeta)]\delta^+\tilde{u}_i^n + [\zeta - (-\zeta)]\delta^+\tilde{u}_{i-1}^n\right\} \quad (129)$$

(iv) *Fromm scheme* (Fromm, 1968)

Setting $\hat{c}_2 = 0$ and satisfying the sum of (125) and (126) gives

$$u^n(\zeta) = \tilde{u}_i^n + \frac{1}{4}\left(\zeta\delta^+\tilde{u}_i^n + \zeta\delta^+\tilde{u}_{i-1}^n\right). \quad (130)$$

Only the third-order upwind biased reconstruction formula introduces the curvature in the cell by cell projection.

For the evolution phase one is mainly interested in the left and right states at the cell interface $(i + \frac{1}{2})$, which correspond to the reconstructed solutions of the two adjacent cells i and $i + 1$ (see Figure 8). These states are easily obtained by setting $\zeta = 1$ (i-th cell reconstruction) for the left state and $\zeta = -1$ ($i + 1$-st cell reconstruction) for the right state in (127)–(130). If one also enforces the boundedness of the total variation of the reconstructed solution (by means of limiter functions), then in a compact form one obtains

$$u_L^n = \tilde{u}_i^n + \frac{1}{4}[(1 - \eta)\overline{\delta}^+\tilde{u}_{i-1}^n + (1 + \eta)\hat{\delta}^+\tilde{u}_i^n]$$

$$(131)$$

$$u_R^n = \tilde{u}_{i+1}^n - \frac{1}{4}[(1 - \eta)\overline{\delta}^+\tilde{u}_{i+1}^n + (1 + \eta)\hat{\delta}^+\tilde{u}_i^n]$$

where $\eta = \frac{1}{3}$ yields (127)
       $\eta = 1$ yields (128)
       $\eta = -1$ yields (129)
       $\eta = 0$ yields (130)

and $\hat{\delta}^+, \overline{\delta}^+$ are the limited slopes

$$\hat{\delta}^+(\cdot)_i = \mathcal{L}(\delta^+(\cdot)_i, \beta\delta^+(\cdot)_{i-1}); \quad \overline{\delta}^+(\cdot)_i = \mathcal{L}(\delta^+(\cdot)_i, \beta\delta^+(\cdot)_{i+1})$$

where $1 \leq \beta \leq (3 - \eta)/(1 - \eta)$ (with $\eta \neq -1$), and $\mathcal{L}$ is a limiter function that

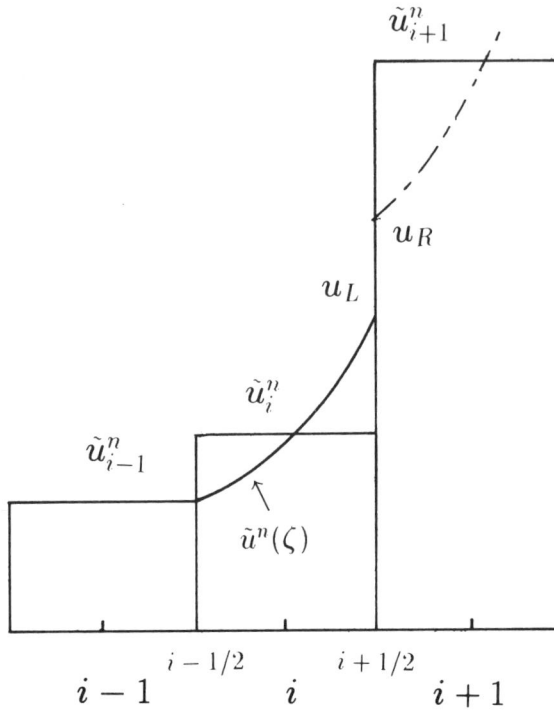

Figure 8 One-dimensional slope limited reconstruction.

satisfies

$$\mathcal{L}(a,b) = \mathcal{L}(b,a)$$

$$\mathcal{L}(a,a) = a$$

$$\mathcal{L}(a,b) = 0 \qquad \text{if } ab < 0$$

Any of the limiters previously discussed can be taken for $\mathcal{L}$.

For variable mesh spacing one may introduce the following local non-dimensional coordinate

$$\zeta = 4 \frac{x - x_i}{\delta^+ x_i + \delta^+ x_{i-1}} - \frac{\delta^+ x_i - \delta^+ x_{i-1}}{\delta^+ x_i + \delta^+ x_{i-1}}$$

$$\text{for} \quad x_i - \frac{\delta^+ x_i + \delta^+ x_{i-1}}{4} \le x \le x_i + \frac{\delta^+ x_i + \delta^+ x_{i-1}}{4}$$

and slope limited reconstruction yields

$$u_L = \tilde{u}_i + \frac{1}{4}\left[(d_i^- - (2d_i^- - 1)\eta)\overline{\delta}^+\tilde{u}_{i-1} + (d_{i-1}^+ + (2 - d_{i-1}^+)\eta)\hat{\delta}^+\tilde{u}_i\right]$$

$$u_R = \tilde{u}_{i+1} - \frac{1}{4}\left[(d_{i+1}^- + (2 - d_{i+1}^-)\eta)\hat{\delta}^+\tilde{u}_i + (d_i^+ - (2d_i^+ - 1)\eta)\overline{\delta}^+\tilde{u}_{i+1}\right]$$

where

$$d_{i-1}^+ = \frac{\delta^+ x_{i-1}}{\delta^+ x_i}, \quad d_i^- = \frac{1}{d_{i-1}^+}$$

$$\hat{\delta}^+(\cdot)_i = \mathcal{L}\left(\delta^+(\cdot)_i, \beta_i^-\,\delta^+(\cdot)_{i-1}\right), \quad \overline{\delta}^+(\cdot)_i = \mathcal{L}\left(\delta^+(\cdot)_i, \beta_i^+\,\delta^+(\cdot)_{i+1}\right)$$

with $1 \leq \beta_i \leq \dfrac{4 - d_i - (2 - d_i)\eta}{1 - (2 - d_i)\eta}$ and $\eta \neq -1$.

For a scalar conservation law the order of accuracy can then be increased by defining the numerical flux function in terms of the left and right states. For example, referring to (55) and (56) one has

$$h_{i+1/2} = \frac{1}{2}[f(u_L) + f(u_R)] - \frac{1}{2}|a(u_L, u_R)|(u_R - u_L) \qquad (132)$$

By generalizing the above equation to a system of conservation laws, one may then increase the order of accuracy of any Godunov-type first-order scheme by defining the numerical flux function in terms of symmetric averages of the left and right states (Van Leer, 1979; Yee, 1989; Hirsch, 1990; LeVeque, 1992). For example, the first-order approximate Riemann solver (85) can be made second-order by defining the flux function as

$$\mathbf{H}_{i+1/2} = \frac{1}{2}(\mathbf{F}(\mathbf{W}_L) + \mathbf{F}(\mathbf{W}_R)) - \frac{1}{2}|\hat{\mathbf{A}}|(\mathbf{W}_R - \mathbf{W}_L) \qquad (133)$$

where $\hat{\mathbf{A}}$ is the jacobian matrix evaluated at a Roe-type average state

$$\overline{(\cdot)}_{i+\frac{1}{2}} = \frac{\sqrt{\rho_L}(\cdot)_L + \sqrt{\rho_R}(\cdot)_R}{\sqrt{\rho_L} + \sqrt{\rho_R}}$$

and the left and right states are obtained by means of MUSCL-type reconstruction. Extending, for example, the limiters of Lacor et al. (1993) to a system of conservation laws, one may recast (133) in the form

$$\mathbf{H}_{i+\frac{1}{2}} = \frac{1}{2}(\mathbf{F}(\mathbf{W}_L) + \mathbf{F}(\mathbf{W}_R)) - \frac{1}{2}|\mathbf{A}(\mathbf{W}_L, \mathbf{W}_R)|\delta^+\mathbf{W}_i - \frac{1}{2}\mathbf{L}_{i+\frac{1}{2}}^{-1}\mathbf{\Phi}_{i+\frac{1}{2}} \qquad (134)$$

where the elements of $\mathbf{\Phi}$ are

$$\phi_{i+\frac{1}{2}}^\ell = \psi(r_\ell^*)|a_{i+\frac{1}{2}}^\ell|\alpha_{i+\frac{1}{2}}^\ell$$

and $\psi(r_\ell^*)$ is obtained from (117), and

$$\frac{1}{r_\ell^*} = \frac{1}{2}\left(\frac{1}{r_\ell^+} + \frac{1}{r_\ell^-}\right); \quad r_\ell^+ = \frac{\alpha_{i-\frac{1}{2}}^\ell}{\alpha_{i+\frac{1}{2}}^\ell}; \quad r_\ell^- = \frac{\alpha_{i+\frac{3}{2}}^\ell}{\alpha_{i+\frac{1}{2}}^\ell}$$

Jameson flux function

A well-known numerical flux function is the high-resolution Jameson–Schmidt–Turkel (JST) scheme (Jameson et al., 1981), which introduces an anti-diffusion flux in a controlled manner. At cell interface $(i + \frac{1}{2})$ the numerical flux function for the one-dimensional system of conservation laws is

$$\mathbf{H}_{i+\frac{1}{2}} = \frac{1}{2}(\mathbf{F}(\mathbf{W}_i) + \mathbf{F}(\mathbf{W}_{i+1}))$$

$$- \frac{1}{2}|r_{i+\frac{1}{2}}|\left[\epsilon_{i+\frac{1}{2}}^{(2)}\delta^+\mathbf{W}_i - \epsilon_{i+\frac{1}{2}}^{(4)}(\delta^+\mathbf{W}_{i+1} - 2\delta^+\mathbf{W}_i + \delta^+\mathbf{W}_{i-1})\right] \tag{135}$$

where $r_{i+\frac{1}{2}}$ is the spectral radius of the jacobian flux, and $\epsilon^{(2)}$ and $\epsilon^{(4)}$ are the so-called first- and third-order adaptive dissipation coefficients defined as

$$\epsilon_{i+\frac{1}{2}}^{(2)} = \min(1, \alpha S); \quad \epsilon_{i+\frac{1}{2}}^{(4)} = \max(0, \tfrac{1}{2} - \beta\epsilon_{i+\frac{1}{2}}^{(2)}) \tag{136}$$

The parameter S makes the scheme switch to first-order near shocks depending on the values of α and β. In the original formulation of Jameson (Jameson et al., 1981) S is taken as the maximum of a 'shock sensor' in a neighbourhood of $(i + \frac{1}{2})$, i.e.

$$S = \max(\nu_{i-1}, \nu_i, \nu_{i+1}, \nu_{i+2}) \tag{137}$$

where the 'shock sensor' is defined in terms of the pressure p

$$\nu_i = \frac{|\delta^+ p_i - \delta^- p_i|}{p_{i+1} + 2p_i + p_{i-1} + \epsilon_0} \tag{138}$$

An improved switch that gives better shock resolution has recently been proposed by Swanson and Turkel (1992)

$$\nu_i = \frac{|\delta^+ p_i - \delta^- p_i|}{\epsilon(p_{i+1} + 2p_i + p_{i-1}) + (1 - \epsilon)(|\delta^+ p_i| + |\delta^- p_i|) + \epsilon_0} \tag{139}$$

where ϵ_0 is a threshold value that ensures that the denominator is not zero, and ϵ is typically $\frac{1}{2}$.

Swanson and Turkel (1993) have shown the importance of the form of the numerical dissipation model for obtaining a high-resolution scheme for the Navier–Stokes equations. In particular, they propose a matrix dissipation model (Swanson and Turkel, 1992) which is a variant of the JST scheme that is simply obtained by replacing the spectral radius with the jacobian matrix. From the results they report of flows over flat plates, around transonic airfoils and over hypersonic compression ramps, one may conclude that matrix dissipation is essential in achieving high accuracy.

Symmetric/upstream limited positive schemes

Jameson (Jameson, 1993; Tatsumi et al., 1994) has conceived a new class of schemes derived by enforcing a design principle based on a local extremum diminishing (LED) property. According to the definition of Jameson (1993), LED appears to be equivalent to the TVD property, at least for multi-dimensional problems with structured grids. Indeed, the construction of the numerical flux function closely follows the one described for the TVD schemes, yielding

$$\mathbf{H}_{i+\frac{1}{2}} = \tfrac{1}{2}(\mathbf{F}(\mathbf{W}_i) + \mathbf{F}(\mathbf{W}_{i+1})) - \tfrac{1}{2}\mathbf{L}_{i+\frac{1}{2}}^{-1}\mathbf{\Phi}_{i+\frac{1}{2}} \tag{140}$$

For the symmetric limited positive scheme (SLIP), the elements of $\mathbf{\Phi}$ corresponding to the ℓ-th difference of characteristic variable α^ℓ are

$$\phi_{i+\frac{1}{2}}^\ell = \psi(a_{i+\frac{1}{2}}^\ell)\left[\alpha_{i+\frac{1}{2}}^\ell - \mathcal{L}(\alpha_{i+\frac{3}{2}}^\ell, \alpha_{i-\frac{1}{2}}^\ell)\right] \tag{141}$$

where $\mathcal{L}$ is a limiter function of the same type as previously discussed.

For the upstream limited schemes (USLIP) the elements of the antidiffusive flux are

$$\phi_{i+\frac{1}{2}}^\ell = \psi(a_{i+\frac{1}{2}}^\ell)\left[\alpha_{i+\frac{1}{2}}^\ell - \frac{1+s^\ell}{2}\mathcal{L}(\alpha_{i+\frac{1}{2}}^\ell, \alpha_{i-\frac{1}{2}}^\ell) - \frac{1-s^\ell}{2}\mathcal{L}(\alpha_{i+\frac{1}{2}}^\ell, \alpha_{i+\frac{3}{2}}^\ell)\right] \tag{142}$$

where $s^\ell = \text{sign } (a_{i+\frac{1}{2}}^\ell)$.

Higher order SLIP or USLIP schemes can also be constructed by simply adding to the lower order ones the limited differences between a high and low order flux function expression. For a more detailed discussion the reader is referred to Jameson (1993), and Tatsumi et al. (1994), where applications to transonic wing flows and supersonic flat plate flows are reported that show that these schemes are good candidates for constructing efficient high order schemes.

Essentially Non–Oscillatory schemes (ENO)

The solution of (complex) flows characterized by shear layers, shock–shock interaction and shock-wave boundary-layer interaction, require efficient and high-order methods for obtaining high-order accurate non-oscillatory solutions. Total variation diminishing methods do meet such a requirement. However, as previously pointed out TVD schemes are not uniformly high-order accurate: they are generally first-order accurate at local solution extrema and second-order in smooth regions.

ENO schemes have been introduced by Harten (1987), Harten et al. (1987), Shu and Osher (1988) and Chakravarthy (1990) for achieving uniformly high accuracy. The schemes which are a generalization of Godunov's and Van Leer's MUSCL approaches, use a piecewise polynomial reconstruction based on an adaptive stencil to avoid interpolation across discontinuities so as to inhibit Gibbs-like phenomena. We now briefly describe the ENO reconstruction as originally developed by Harten *et al* (1987) and applied by Casper (1992), Atkins (1991), Zhang (1994) and Godfrey *et al.* (1993).

For the sake of simplicity we first consider the one-dimensional case. Given cell averages $\{\tilde{w}_j\}$, a piecewise-smooth function $w(x)$ can be determined by interpolating its primitive function $W(x)$ defined as (Harten *et al.*, 1987; Harten, 1991)

$$W(x) = \int_{x_{j_0-\frac{1}{2}}}^{x} w \, dx \qquad (143)$$

Then, assuming for simplicity constant mesh spacing, the point value W $(x_{j+\frac{1}{2}})$ can be determined according to

$$W(x_{j+\frac{1}{2}}) = \Delta x \sum_{k=j_0}^{j} \tilde{w}_k$$

Once the point values of the primitive function are computed, an interpolation technique is used to obtain an r-th order polynomial interpolation $H_r(x; W)$ of W and a piecewise-polynomial reconstruction of degree $r - 1$ that satisfies

$$R(x; \tilde{w}) = w(x) + O(h^r) = \frac{d}{dx} H_r(x; W) \quad \text{for } x_{j-1/2} \le x \le x_{j+\frac{1}{2}}$$

The polynomial $H_r(x; W)$ interpolates the values of W over an adaptive stencil $S_{i(j;r)} = \{i(j; r) < \cdots < i(j; r) + r\}$ that is chosen so that H_r is the 'smoothest'. The reconstruction polynomial satisfies the conservation property

$$\frac{1}{\Delta x} \int_{x_j-\Delta x/2}^{x_j+\Delta x/2} R(x; \tilde{w}) \, dx = \tilde{w}_j$$

and in addition it also satisfies the relation

$$TV(R(\cdot; \tilde{w})) \le TV(\tilde{w}) + O(h^{1+p}) \qquad (144)$$

where TV is the total variation, and $0 < p \le r - 1$ (if w is sufficiently smooth).

Such a reconstruction is said to be essentially non-oscillatory in the sense that it may inhibit Gibbs-like phenomena near discontinuities (oscillations that are of $O(h)$); however, it may still produce $O(h^r)$ oscillations in the smooth part of $w(x)$. Referring to a fully discretized formulation, a scheme is essentially non-oscillatory if

$$\tilde{w}_j^{n+1} = \frac{1}{\Delta x} \int_{x_j-\Delta x/2}^{x_j+\Delta x/2} E \cdot R(x; \tilde{w}) \, dx$$

and

$$TV(\tilde{w}^{n+1}) \leq TV(\tilde{w}^n) + O(h^{1+p})$$

where E is the evolution operator. The latter inequality follows from (144), and from the fact that, for a scalar case, the cell averaging and the evolution operators are order preserving (and also total variation diminishing), i.e. $TV(\tilde{w}^{n+1}) \leq TV(R(\cdot; \tilde{w}^n))$.

The condition on the total variation is ensured by defining a stencil S_i that satisfies the following properties: (i) the conservation imposes that j belongs to $S_{i(j;r)}$; (ii) the stencil must be adapted to the local structure of the function to be reconstructed, in particular any singular point should be outside the interval $[i(j;r)\Delta x, (i(j;r) + r)\Delta x]$ as far as possible. It is known that one can detect a singular behaviour of a piecewise smooth function f by examining its table of divided differences. Let $\{x_{j+\frac{1}{2}} < x_{j+\frac{1}{2}+1} < \ldots < x_{j+\frac{1}{2}+r}\}$ be a stencil and let $f[x_{j+\frac{1}{2}}, \ldots, x_{j+\frac{1}{2}+r}]$ the $(r + 1)$-st divided difference of f. If f is a continuously differentiable function up to its r-th derivative $f^{(r)}$, then

$$f[x_{j+\frac{1}{2}}, \ldots, x_{j+\frac{1}{2}+r}] = \frac{1}{r!} f^{(r)}(\zeta) \quad \text{for } \zeta \in [x_{j+\frac{1}{2}}, x_{j+\frac{1}{2}+r}]$$

where

$$f[x_{j+\frac{1}{2}}, x_{j+\frac{3}{2}}, \ldots, x_{j+\frac{1}{2}+r}] = \sum_{\ell=1}^{r+1} \frac{f(x_{\ell+\frac{1}{2}})}{\prod_{m \neq \ell}(x_\ell - x_m)}$$

If $f^{(p)}$, with $p < r$, is discontinuous in $[x_{j+\frac{1}{2}}, x_{j+\frac{1}{2}+r}]$ then

$$f[x_{j+\frac{1}{2}}, \ldots, x_{j+\frac{1}{2}+r}] = O(h^{p-r})$$

These two results show that $f[x_{j+\frac{1}{2}}, \ldots, x_{j+\frac{1}{2}+r}]$ remains bounded when the mesh spacing tends to zero when f is a smooth function, while it tends to infinity in the other case. However, these two results are true provided the mesh is regular enough; an extension to general meshes has been presented by Abgrall (1992).

Then, following Harten, $S_{i(j;r)}$ is heuristically determined so that the absolute value of the divided difference $W[S_{i(j;r)}]$ is the smallest over all possible $r + 1$ point stencils containing j. The determination of the stencil interval $(x_{i(j;r)+\frac{1}{2}}, x_{i(j;r)+\frac{1}{2}+r})$ is obtained recursively in r steps.

First set $i(j; 1) = j - 1$; hence for $\ell = 2, r$, let

$$i\ell = i(j; \ell - 1) - 1 \ , \ ir = i(j; \ell - 1)$$

then

$$i(j; \ell) = i\ell \quad \text{if } |W[S_{i(i\ell;\ell)}]| < |W[S_{i(ir;\ell)}]|$$

$$i(j; \ell) = ir \quad \text{otherwise}$$

We point out that the use of the divided differences allows a fast evaluation of the successive stencils $S_{i(j;\ell)}$.

An essentially non-oscillatory scheme can then be constructed by evaluating the inviscid numerical flux function by using any Godunov-type scheme (for example Roe's approximate Riemann solver with an entropy correction) based on left and right states obtained by an ENO reconstruction at the left and right sides of the cell interfaces, respectively.

It must be observed that high-order ENO schemes are very satisfactory for truly unsteady problems. However, for steady flows the convergence may not be easily attained since the changing of the stencil at each time step may produce a reinforcement rather than a damping of high-frequency errors produced by different sources (e.g. boundary conditions or truncation errors) (Godfrey *et al.*, 1993).

C Multidimensional extension

The formulas previously described hold for one-dimensional systems of conservation laws. Their extension to multidimensional problems is a critical point in the theory of finite difference approximations. The early (and most commonly used) approach is based on the splitting method, which is a general methodology for reducing the solution of a complicated problem to successive solutions of simpler problems (Yanenko, 1971). The character of splitting may be geometrical and/or physical. The former reduces a multidimensional problem to a temporal sequence of multidimensional ones, while the latter reduces the original physical problem to a sequence of simpler physical processes. Of course, the geometrical splitting is directly related to the multidimensional extension. However, the main purpose of the early development of splitting techniques was to reduce the computational effort of implicit formulations, this will be discussed in a later section.

The formal extension to multidimensions is rather straightforward in the framework of a finite-volume approach. Indeed, the numerical flux function at a cell face represents the numerical approximation of the flux through the bounding control surface of outward unit normal **n**. The problem can then be reduced to a locally one-dimensional problem in the direction normal to the cell face, and most of the formulas discussed are unchanged if one replaces the flux with the normal flux and the jacobian with the normal flux jacobian.

Despite the apparent simplicity of constructing multidimensional control volume schemes, the treatment of multidimensional convective (or highly convective) transport phenomena is still a difficult task. In fact, even the incompressible convective transport of a scalar may not be accurately described on a fixed grid: anisotropic changes in the wavelengths of the scalar distributions produced by a stretching of material volumes may not be well resolved. In the absence of stretching the description of a passive convective transport may still be critical (when not impossible), contrary to a one-dimensional situation. A confirmation of the difficulties is demonstrated by the large volume of published results for the simulation of convective transport for scalars, the so-called Crowley test (Crowley, 1968; Orszag, 1971; De Felice *et al.*, 1993a, b). Indeed, most of the methods obtained by a straightforward extension of one-dimensional formulas yield poor resolution, unless an adaptive mesh is used (Cavaliere *et al.*, 1993; De Felice *et al.*, 1993b). The compressibility, the non-linearity and the appearance of non-smooth solutions further complicate the requirements for the design of the numerical flux function in multidimensions. Furthermore, in order to obtain a unique weak solution an entropy inequality must be enforced.

Crandall and Majda (1980) have shown that the numerical solution of a scalar multidimensional conservation law converges to a unique weak solution if the numerical method is constructed by means of Strang (1968) splitting with monotone (one-dimensional) fractional steps. For multidimensions, the low accuracy of monotone schemes is even more critical than in the one-dimensional case; a single order reduction in mesh size may correspond an $O(10^3-10^4)$ increase in computational effort. However, one can substitute the monotonicity property of the numerical scheme with the TVD requirement in order to guarantee at least the convergence toward a weak solution. Unfortunately, Goodman and LeVeque (1985) have shown that in the cases of real interest multidimensional TVD schemes are at most first-order accurate.

Despite the lack of theoretical proofs, it is reasonable to construct multidimensional schemes that maintain most of the one-dimensional requirements for each fractional step and/or numerical flux component.

Van Leer (1992) has pointed out that a reduction to locally one-dimensional problems in the directions normal to cell faces is correct in the presence of grid-aligned or nearly aligned discontinuities. A loss of accuracy may occur if the waves are far from being aligned with the grid, as is the case, for example, of an oblique shock. In this case, selecting the normal to the cell face as the direction for wave propagation amounts to a grid-aligned wave decomposition with a misrepresentation of the waves and a consequent loss of resolution (yielding, for example, a smearing of the waves).

An increase in the resolution can be obtained by introducing a rotated coordinate system aligned with the shock front. In two dimensions one may then solve two approximate Riemann problems in the direction normal ($\mathbf{n}_\perp$) and parallel ($\mathbf{n}_\parallel$) to the wave (Levy *et al.*, 1989; Dadone and Grossman, 1992). For example, referring to Figure 9, the flux at the interface between cells i and $i+1$ is

$$\mathbf{H}_n = \mathbf{F}_\perp \cdot \mathbf{n} + \mathbf{F}_\parallel \cdot \mathbf{n} \tag{145}$$

where $\mathbf{F}_\perp$ and $\mathbf{F}_\parallel$ are the flux components in the two directions $\mathbf{n}_\perp$ and $\mathbf{n}_\parallel$ that are evaluated in terms of the left $(L_\perp, L_\parallel)$ and the right states $(R_\perp, R_\parallel)$

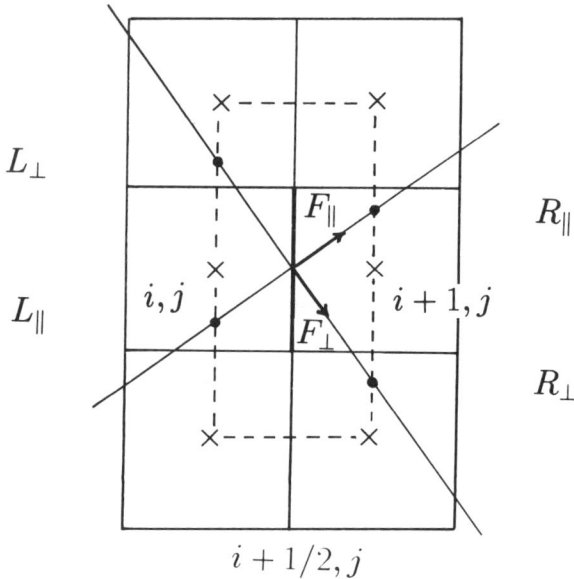

Figure 9 Multidimensional construction of numerical flux function.

interpolated along the two directions. Note that the increase in the resolution may be accompanied by a loss of monotonicity and convergence to a steady state (Van Leer, 1992).

Another promising method, at least for the Euler equations, is the fluctuation splitting method of Roe (1986a), Catalano *et al.* (1992), Deconinck (Deconinck *et al.*, 1991; Strujis *et al.*, 1991) and Hirsch (Hirsch *et al.*, 1987; Hirsch, 1990; Hirsch and Lacor, 1989), which is based on a truly multidimensional wave model.

In the present section we address the aspect of multidimensional numerical flux function definition in terms of locally one-dimensional problems in the direction normal to cell faces. In particular we limit the discussion to the multidimensional counterpart of some of the formulas analyzed above. The reader should be warned that the formulas listed below are only the skeleton of possible schemes. Indeed, they simply refer to a generic left and right state on the control surface: lower- and higher-order schemes can be recovered (or at least one may attempt to recover them) by exploiting reconstruction algorithms along the lines previously indicated.

1 Van Leer splitting

The numerical flux function approximating the flux at a cell face of outward unit normal $\mathbf{n}$ is

$$\mathbf{H}_n = \mathbf{F}_n^+(\mathbf{W}_L) + \mathbf{F}_n^-(\mathbf{W}_R)$$

where $\mathbf{W}_L$ $(\mathbf{W}_R)$ is the value of the variable upstream (downstream) of the bounding control surface (of outward unit normal $\mathbf{n}$) that represents the left (right) state transported by $\mathbf{F}_n^+$ $(\mathbf{F}_n^-)$.

An extension of the Van Leer-type splitting to multidimensional flows is not quite trivial. As proposed by Eberle *et al.* (1992), the split flux component can be constructed by retaining the one-dimensional structure. For example, the Hänel (Hänel *et al.*, 1987) split (normal) flux in three dimensions becomes

$$
\mathbf{F}_n^{\pm} = \pm\rho c\frac{(1\pm M_n)^2}{4}
\begin{bmatrix}
1 \\
u \pm (2 - M_n)\dfrac{c}{\gamma}n_x \\
v \pm (2 - M_n)\dfrac{c}{\gamma}n_y \\
w \pm (2 - M_n)\dfrac{c}{\gamma}n_z \\
\dfrac{c^2}{\gamma - 1}\left(1 + \dfrac{\gamma - 1}{2}M_n^2\right)
\end{bmatrix}
$$

where $M_n = \mathbf{u} \cdot \mathbf{n}/c$.

2 Advection upstream splitting

The advection upstream-type splitting can be extended to multidimensions by formulating the normal flux function in terms of the 'convective' velocity $\mathbf{u} \cdot \mathbf{n}$ (Liou, 1993; Wada and Liou, 1994), i.e.

$$
\mathbf{F}_n(\mathbf{W}) = \mathcal{U}_n\Phi + \mathcal{P}\mathbf{b}_n
$$

$$
\mathcal{U}_n = \mathbf{\Lambda}_n = \mathbf{u} \cdot \mathbf{n}\mathbf{I}
$$

$$
\Phi = [\rho, \rho u, \rho v, \rho w, \rho H]^{\mathrm{T}}
$$

$$
\mathcal{P} = p\mathbf{I}
$$

$$
\mathbf{b}_n = [0, n_x, n_y, n_z, 0]^{\mathrm{T}}
$$

The split flux components corresponding to the multidimensional extension of the AUSM are

$$
\mathbf{F}_n^{\pm} = \mathcal{U}_n^{\pm}(\mathbf{W}_L, \mathbf{W}_R)\Phi + \mathcal{P}^{\pm}\mathbf{b}_n
$$

and

$$\mathcal{U}_n^\pm(\mathbf{W}_L, \mathbf{W}_R) = \frac{\mathcal{U}_n(\mathbf{W}_L, \mathbf{W}_R) \pm |\mathcal{U}_n(\mathbf{W}_L, \mathbf{W}_R)|}{2}$$

$$\mathcal{U}_n(\mathbf{W}_L, \mathbf{W}_R) = \Lambda_n^+(\mathbf{W}_L) + \Lambda_n^-(\mathbf{W}_R)$$

$$= \{s_L u_L^+ + (1 - s_L)\hat{u}_L^+ + s_R u_R^- + (1 - s_R)\hat{u}_R^-\}\mathbf{I}$$

$$\mathcal{P}^\pm = \left\{ s\frac{1 \pm \ \text{sign}\ (|M_n|)}{2} + (1 - s)\frac{(M_n \pm 1)^2(2 \mp M_n)}{4} \right\}p\mathbf{I}$$

where

$$u^\pm = \frac{1 \pm \ \text{sign}\ (M_n)}{2}\ \mathbf{u} \cdot \mathbf{n}$$

$$\hat{u}^\pm = \pm\frac{1}{4}\left(\sqrt{|M_n|} \pm \frac{1}{\sqrt{|M_n|}} \right)^2 \mathbf{u} \cdot \mathbf{n}$$

$$s = \ \text{sign}\ (|M_n| - 1)$$

All previously discussed variants can be obtained in a similar manner.

3 Approximate Riemann solver

The approximate Riemann solver of Roe can be extended to multidimensions by replacing the jacobian matrix with the normal flux jacobian (Hirsch, 1990). Then, the numerical flux function simply reads

$$\mathbf{H}_n = \frac{\mathbf{F}_n(\mathbf{W}_L) + \mathbf{F}_n(\mathbf{W}_R)}{2} - \frac{|\tilde{\mathbf{A}}(\mathbf{W}_L, \mathbf{W}_R)|}{2}(\mathbf{W}_R - \mathbf{W}_L)$$

where $\mathbf{W}_L$ ($\mathbf{W}_R$) represents the value of the variable upstream (downstream) of the bounding control surface, and $\tilde{\mathbf{A}}$ is the normal flux jacobian $(\mathbf{A}n_x + \mathbf{B}n_y + \mathbf{C}n_z)$ whose expression is obtained from (13) if one substitutes the arbitrary direction $\mathbf{k}$ with the normal one.

4 Total variation diminishing methods

A multidimensional extension of the flux function 'corresponding' to a total variation diminishing formulation can be obtained by introducing a 'characteristic' decomposition in the normal direction (Yee, 1989), thus obtaining

$$\mathbf{H}_n = \frac{\mathbf{F}_n(\mathbf{W}_L) + \mathbf{F}_n(\mathbf{W}_R)}{2} - \frac{1}{2}\mathbf{L}^{-1}\mathbf{\Phi}$$

where **L** is the left eigenvector matrix of the normal flux jacobian (see for example (22)) evaluated by means of Roe's average, and Φ is defined in terms of the difference of characteristic variable α by an extension of (119) to multidimensions, where α is defined as

$$
\alpha = \begin{bmatrix} \dfrac{1}{2}\left(-\dfrac{\bar{\rho}}{\bar{c}}\delta^+ q_n + \dfrac{\delta^+ p}{\bar{c}^2}\right) \\[2ex] \delta^+\rho - \dfrac{\delta^+ p}{\bar{c}^2} \\[2ex] \dfrac{1}{2}\left(\dfrac{\bar{\rho}}{\bar{c}}\delta^+ q_n + \dfrac{\delta^+ p}{\bar{c}^2}\right) \\[2ex] \bar{\rho}\delta^+ q_\ell \\[1ex] \bar{\rho}\delta^+ q_m \end{bmatrix}
$$

where $\overline{(\cdot)}$ represents the values at cell interface evaluated by means of Roe's average and

$$
q_\ell = \mathbf{u} \cdot \mathbf{l} \quad ; \quad q_m = \mathbf{u} \cdot \mathbf{m} \quad ; \quad q_n = \mathbf{u} \cdot \mathbf{n}
$$

with $(\mathbf{l}, \mathbf{m}, \mathbf{n})$ a local orthonormal basis.

The pressure difference $\delta^+ p$ is computed through the equation of state, i.e.

$$
\delta^+ p = (\gamma - 1)(\delta^+ \rho E + \tfrac{1}{2}q^2 \delta^+\rho - \bar{q}_n \delta^+ \rho q_n)
$$

with $q^2 = \bar{q}_n^2 + \bar{q}_m^2 + \bar{q}_\ell^2$.

5 Essentially Non-Oscillatory schemes (ENO)

A multidimensional ENO reconstruction can in principle be achieved by using d (with d the number of space dimensions) one-dimensional interpolation stencils based on cell averages. However, such a dimension-by-dimension procedure yields ENO schemes that are at most second-order accurate (Casper, 1992). A fully multidimensional ENO reconstruction has been proposed by Casper (1992), who has extended the one-dimensional interpolation by introducing the primitive functions of the line averages in the d directions.

For example in two-dimensions one may define the following x and y averages, i.e.

$$
\bar{w}_j(x) = \frac{1}{\Delta y} \int_{y_j - \Delta y/2}^{y_j + \Delta y/2} w(x, y)\, \mathrm{d}y \tag{146}
$$

$$
\tilde{w}_{ij} = \frac{1}{\Delta x} \int_{x_i - \Delta x/2}^{x_i + \Delta x/2} \bar{w}_j(x)\, \mathrm{d}x \tag{147}
$$

Then, from cell averages $\tilde{w}_{ij}$ the one-dimensional ENO reconstruction along the x direction yields the pointwise value of the y-average $\overline{w}_j(x)$, and the pointwise values $w(x, y)$ are finally determined from $\overline{w}_j(x)$ by using the one-dimensional ENO reconstruction along the y direction (Atkins, 1991; Casper, 1992; Casper and Atkins, 1993; Zhong, 1994).

Different ENO multidimensional extensions can be obtained, for example based on the k-exact reconstruction or on the dimensionally split-ENO reconstruction (for more detail see Barth and Frederikson, 1990; Godfrey *et al.*, 1993). A more general approach for unstructured meshes has been presented by Abgrall (1994).

V DISCRETIZATION OF THE VISCOUS FLUX

As pointed out in the introductory sections, this chapter is mainly concerned with the numerical solution of the Navier–Stokes equations at high Reynolds numbers. As a consequence, the numerical flux function can be constructed as the sum of two contributions: one associated with the inviscid flux ($\mathbf{H}^E$) and the other coming from the approximation of the viscous terms ($\mathbf{H}^V$). Hence, referring to cell face $(i + \frac{1}{2}, j, k)$ of Figure 1 and recalling (4), by dropping the indices (j, k) one can formally write

$$\mathbf{H}_{i+\frac{1}{2}} = \mathbf{H}^E_{i+\frac{1}{2}} - \mathbf{H}^V_{i+\frac{1}{2}}$$

where, referring for example to a fully discretized formulation (see (28) and (29)) in analogy with the definition of $\mathbf{H}^E$, one has

$$\mathbf{H}^V_{i+\frac{1}{2}} = \frac{1}{\Delta t} \int_{t_n}^{t_{n+1}} \mathbf{F}^V [\mathbf{W}(x_{i+\frac{1}{2}}, t)] \, \mathrm{d}t$$

If one assumes a weak dependence of the transport coefficients on the independent variables, the viscous fluxes are linear functions of the independent variables. For purely viscous problems the convergence toward the strong solution of a consistent scheme is ensured by its stability. Consequently, the construction of the numerical viscous flux function is in principle a less complicated task than for the inviscid contribution.

In general, viscous fluxes depend on the gradients of the unknowns and their evaluation requires the numerical approximation of such gradients on the control surfaces in terms of the adopted state variables. If one renounces a local solution reconstruction and assumes that the nodal values coincide with the volume-averaged ones (i.e. C_{ij} in (40) is the Kronecker delta δ_{ij}), one is faced with the problem of accuracy of numerical differentiation.

Two-point central space discretization is straightforward for evaluating gradients on a uniform rectangular grid. For practical applications the grid is generally irregular and highly stretched in the proximity of the wall (with mesh aspect ratios $O(10^2–10^3))$, thus making the problem of space discretization on non-uniform grids rather critical.

In the framework of a finite-volume method the use of integral theorems leads to a straightforward evaluation of the gradients. In particular, the application of Gauss theorem to a volume V_g (where the subscript g identifies the volume used for gradient evaluation) gives for scalar or vector quantities the following

$$\int_{V_g} \nabla q \, dV = \oint_{\partial V_g} q\mathbf{n} \, dS, \quad \int_{V_g} \nabla \cdot \mathbf{q} \, dV = \oint_{\partial V_g} \mathbf{q} \cdot \mathbf{n} \, dS$$

An algebraic expression for partial derivatives can then be simply obtained by numerical approximations of the above integrals. For example, for a two-dimensional case

$$\frac{\hat{\partial} q}{\partial x} = \frac{1}{V_g} \sum_{\alpha_g} (q n_x \, dS)_{\alpha_g} \tag{148}$$

$$\frac{\hat{\partial} q}{\partial y} = \frac{1}{V_g} \sum_{\alpha_g} (q n_y S)_{\alpha_g} \tag{149}$$

where $\hat{\ }$ represents the numerical approximation of the derivative in the centre of V_g, and α_g identifies the generic face of the volume V_g.

The selected volume V_g and the quadrature formulas employed to evaluate the surface integrals on the right-hand-side of (148) and (149) affect the order of accuracy of the discretization formulas. The accuracy can be increased at the expense of computational efficiency. For example, referring to Figure 1 to evaluate the derivatives at the cell face $(i + \frac{1}{2}, j, k)$, whose normal is in the i direction, additional geometric quantities are needed such as the volume V_g, its bounding surface vector and the values of the variables at the points selected for numerical integration along the bounding surfaces. For cartesian meshes, simple expressions can be obtained for example by defining the derivatives at $(i + \frac{1}{2}, j, k)$ as an average of the derivatives evaluated at the centres of the adjacent cells (i, j, k) and $(i + 1, j, k)$. Such an approach obviously does not require any additional geometric quantity evaluation. For instance, the derivatives of $\tilde{u}$ in the ℓ-th direction at cell centre i, j, k can be simply evaluated as

$$\frac{\hat{\partial} \tilde{u}}{\partial x_\ell}\bigg|_{ijk} = [\hat{u}(x_i)\mathbf{i}_\ell \cdot \mathbf{S}_{1584} + \hat{u}(x_{i-1})\mathbf{i}_\ell \cdot \mathbf{S}_{2673}$$

$$+ \, \bar{u}(y_j)\mathbf{i}_\ell \cdot \mathbf{S}_{1562} + \bar{u}(y_{j-1})\mathbf{i}_\ell \cdot \mathbf{S}_{4873}$$

$$+ \, \tilde{u}(z_k)\mathbf{i}_\ell \cdot \mathbf{S}_{5678} + \tilde{u}(z_{k-1})\mathbf{i}_\ell \cdot \mathbf{S}_{1234}]/\mathbf{V}_{12345678} \tag{150}$$

where $\hat{u}(x_i)$, $\bar{u}(y_j)$ and $\tilde{u}(z_k)$ represent, respectively, interpolated values at the centres of the surfaces having (positive) outward unit normals in the i, j and k

directions, while $\mathbf{i}_\ell$ is the unit vector in the $\mathbf{x}_\ell$ direction and the $\mathbf{S}$ values are the surface area vectors. A low-order efficient definition of the above quantities is

$$\hat{u}(x_i) = \tfrac{1}{4}(u_1 + u_5 + u_4 + u_8)$$

$$\bar{u}(y_j) = \tfrac{1}{4}(u_1 + u_5 + u_6 + u_2) \tag{151}$$

$$\tilde{\tilde{u}}(z_k) = \tfrac{1}{4}(u_5 + u_6 + u_7 + u_8)$$

where u_ℓ indicates the values at the cell vertices. Again, having assumed as unknown state vectors the cell averages, an interpolation is required. Other discretization formulas have been proposed and interested readers are referred, for example, to the work of Kordulla (1987), Hirsch (1990) and Manna (1992).

It is important to note that the order of accuracy of the above formulas is strongly dependent on the smoothness properties of the mesh rather than the staggering. This has been shown by Swanson and Radespiel (1991) who have investigated the influence of the so-called cell-centre or cell-vertex representations.

Another approach that can be followed consists of using a local polynomial reconstruction so as to evaluate 'analytically' the space derivatives at any control surface. In multidimensional space any interpolation requires the determination of n_p coefficients of a polynomial of degree p

$$n_p = \frac{(p+d)(p+d-1)\ldots p}{d!}$$

where d is the number of dimensions.

As is well known from finite-element methods, in general it may not be possible to interpolate n_p values by means of a polynomial of degree p in a d-dimensional space. This is the case, for example, in two dimensions when the number of aligned nodal points is greater than $p+1$. Likewise, in three dimensions the number of coplanar nodal points must be less than $(p+1)(p+2)/2$. In general, it may be difficult to generate a lagrangian simplex distribution of points; as a consequence, it is common practice to interpolate a number of points less than n_p by means of an incomplete polynomial. For instance, in two dimensions the full second-order polynomial is replaced by a bilinear approximation. The polynomial approximation may even be defined in terms of a best fit polynomial, rather than an interpolatory one, by devising a best fit based on the fitting of the averages.

Let $(i + \tfrac{1}{2}, j, k)$ be the index of the control surface where the viscous flux needs to be determined and let $P_{i+\frac{1}{2},j,k}$ be the polynomial. The numerical approximation of the x-derivatives, for example, is then

$$\left(\frac{\partial u}{\partial x}\right)_{i+\frac{1}{2},j,k} = \left(\frac{\partial P}{\partial x}\right)_{i+\frac{1}{2},j,k}$$

More detailed discussion on multidimensional interpolatory reconstruction is provided by Oden and Reddy (1976).

It is important to stress that no matter how one devises high-order formulas for space discretization of gradients, from a practical point of view the primary concern remains the robustness and the computational cost.

VI TREATMENT OF BOUNDARY CONDITIONS

In a control volume approach the number of state vector unknowns is equal to the number of cells. However, to close the system of algebraic equations (either in the semi- or fully discretized formulation) the numerical fluxes must be evaluated at all cell surfaces, and in particular also at cell faces lying on the domain boundaries. In principle, these fluxes cannot be defined with the formulas used at interior cells owing to (possible) lacking information and/or definition of either 'left' or 'right' state variables that would correspond to states associated with cells external to the computational domain. With the exception of a periodic initial value problem the lack of information could be remedied by an extrapolation procedure from the interior. Intuitively such an approach could even be acceptable at an outflow boundary, i.e. a boundary where $\mathbf{u} \cdot \mathbf{n} > 0$, with $\mathbf{n}$ the positive outward unit normal. However, as shown by many authors (e.g. see Cambier et al., 1984; Hirsch, 1990), a blind extrapolation procedure would not account for the nature of the governing equations and for the physical phenomena occurring on the boundaries.

Following recent works on the subject (Strikwerda, 1977; Hedström, 1979; Rudy and Strikwerda, 1981; Thompson, 1987, 1990; Hirsch, 1990; Poinsot and Lele, 1992), we will focus on the aspect of time-dependent boundary conditions in order to properly account for the interaction between inner and outer phenomena. The first issue that arises is the assessment of how the initial inner (region) conditions contribute to the determination of the time derivatives of the state variables on the boundaries. The second and more critical issue is related to providing the missing information required to determine correctly and univocally $\partial \mathbf{W}/\partial t$: once $\partial \mathbf{W}/\partial t$ is determined, the $\mathbf{W}$ and the fluxes on the boundaries can easily be evaluated.

The determination of $\partial \mathbf{W}/\partial t$ is strictly related to the theoretical problem of the closure of Euler and/or Navier–Stokes equations, i.e. defining the set of boundary conditions that, together with the initial conditions, can ensure, at least locally, a stable (well-posed) solution. To explain the concept of well posedness, let us consider an initial boundary value problem in the form (Gustafsson, 1991; Majda, 1984)

$$\frac{\partial \mathbf{U}}{\partial t} = \mathcal{L}\mathbf{U} \qquad x \in \Omega \qquad t \geq 0$$

$$B\mathbf{U} = 0 \qquad x \in \partial \Omega$$

$$\mathbf{U}(x, 0) = f(x) \qquad t = 0$$

First of all the differential operator $\mathcal{L}$ is said to be semibounded if

$$(\mathbf{U}, \mathcal{L}\mathbf{U}) \leq \beta(\mathbf{U}, \mathbf{U})$$

where $(\cdot, \cdot)$ is a suitable scalar product, for instance

$$(\mathbf{U}, \mathbf{V}) = \int_\Omega \mathbf{U}\mathbf{V}\, \mathrm{d}x$$

Then, the problem is well posed if $\mathcal{L}$ is semibounded and

$$\frac{\mathrm{d}}{\mathrm{d}t}(\mathbf{U}, \mathbf{U}) \leq 2\beta(\mathbf{U}, \mathbf{U})$$

The above inequality shows that: (i) the energy $(\mathbf{U}, \mathbf{U})$ evolves in time at a rate that is bounded by an exponential growth, and (ii) a unique solution is obtained that satisfies the condition

$$\|\mathbf{U}\| = (\mathbf{U}, \mathbf{U})^{1/2} \leq K\, \mathrm{e}^{\beta t}\|f\|$$

It is important to observe that for one-dimensional Euler problems the assessment of well posedness is quite straightforward, and leads to the conclusion that the number of required boundary conditions depends on the possibility of expressing the ingoing characteristic variables in terms of the outgoing ones.

Navier–Stokes equations are neither hyperbolic nor parabolic and the assessment of well posedness is far from being satisfactory. The presence of the viscous terms (in all equations but the continuity one) makes the distinction between ingoing and outgoing quantities not rigorous with respect to the determination of the number of boundary conditions. In general extra (with respect to the inviscid case) boundary conditions are needed that may include space derivatives.

As yet only a limited number of results are available for the general closure of the Navier–Stokes equations (Strikwerda, 1977; Dutt, 1988; Kreiss, 1989; Poinsot and Lele, 1992); while for the Euler equations the state of theoretical knowledge is more complete (Engquist and Majda, 1977, 1979; Oliger and Sundström, 1978; Gustafsson, 1982; Kreiss, 1989). A generally accepted indication on the number of required boundary conditions is reported in Table 1.

From hereon, following Thompson (1990), for boundary condition we mean a single mathematical relation that: (i) can be exploited to determine $\partial\mathbf{W}/\partial t$ along the boundary, and (ii) is not obtainable solely from the conservation equations of $\mathbf{W}$ in the interior of the region of integration.

From a computational point of view it is then clear that one is faced with the following problems: (i) identify the possible (or compatible) sets of necessary and sufficient boundary conditions for determining $\partial\mathbf{W}/\partial t$; (ii) select the most suitable set of boundary conditions by exploiting all possible information (theoretical and/or experimental); (iii) construct the numerical flux function at

Table 1
Number of boundary conditions required for Euler (E) and Navier–Stokes (NS) equations in a d-dimensional space

Type of boundary condition	Flow conditions	E	NS
Wall	No slip	1	$1 + d$
Wall	Slip	1	$1 + d$
Inflow	$M < 1$	$1 + d$	$2 + d$
$(\mathbf{u} \cdot \mathbf{n} < 0)$	$M > 1$	$2 + d$	$2 + d$
Outflow	$M < 1$	1	$1 + d$
$(\mathbf{u} \cdot \mathbf{n} > 0)$	$M > 1$	0	$1 + d$

the boundaries with a given accuracy; and (iv) verify the stability of the global algorithm.

A Identification of compatible boundary conditions

We first consider the Euler equations and then we will attempt an extension of the analysis to the Navier–Stokes equations

1 Euler equations

Referring to (7) we assume, without loss of generality, that at the boundary the x direction coincides with the positive outward unit normal. Then, the equation for any vector (either conservative or primitive) variable $\mathbf{U}$ can be cast in the following form

$$\frac{\partial \mathbf{U}}{\partial t} + \mathbf{A}\frac{\partial \mathbf{U}}{\partial x} + \mathbf{Q} = 0 \tag{152}$$

where for a three-dimensional case $\mathbf{Q} = \mathbf{B}\partial \mathbf{U}/\partial y + \mathbf{C}\partial \mathbf{U}/\partial z$, and $(\mathbf{A}, \mathbf{B}, \mathbf{C})$ are either the jacobian matrices of the Euler flux or those obtained by a similarity transformation as shown in Section II.A.

Equation (152) can be cast in the so-called compatibility form by premultiplying the equation by a left eigenvector matrix $\mathbf{L}$ of $\mathbf{A}$

$$\mathbf{L}\frac{\partial \mathbf{U}}{\partial t} + \mathbf{\Lambda}\,\mathbf{L}\frac{\partial \mathbf{U}}{\partial x} + \mathbf{L}\mathbf{Q} = 0 \tag{153}$$

where Λ is the diagonal matrix

$$\Lambda = \begin{pmatrix} \mathbf{v} \cdot \mathbf{n} - c & 0 & 0 & 0 & 0 \\ 0 & \mathbf{v} \cdot \mathbf{n} & 0 & 0 & 0 \\ 0 & 0 & \mathbf{v} \cdot \mathbf{n} + c & 0 & 0 \\ 0 & 0 & 0 & \mathbf{v} \cdot \mathbf{n} & 0 \\ 0 & 0 & 0 & 0 & \mathbf{v} \cdot \mathbf{n} \end{pmatrix}$$

Let us then consider the reduced set of equations obtained by removing from (153) those equations associated with the incoming waves which correspond to the negative eigenvalues, thus obtaining

$$\mathbf{L}_r \frac{\partial \mathbf{U}}{\partial t} + \Lambda_o \mathbf{L}_r \frac{\partial \mathbf{U}}{\partial x} + \mathbf{L}_r \mathbf{Q} = 0 \tag{154}$$

where $\mathbf{L}_r$ is a rectangular matrix of dimensions $(n_r \times n)$ and Λ_o is the square diagonal matrix of the non-negative eigenvalues, with n_r the number of non-negative eigenvalues and $n = 2 + d$.

Equation (154) represents the immovable relations that must be used in conjunction with the set of boundary conditions in order to determine $\partial \mathbf{U}/\partial t$ on the boundary (being reasonable to evaluate the second term either by means of one-sided discretization formulas or by extrapolation from the interior), and it is the key for identifying possible sets of boundary conditions.

The number of boundary conditions needed (given in Table 1) is $n_{bc} = n - n_r$. Then, given a set of boundary conditions of the form

$$\mathbf{L}_c \frac{\partial \mathbf{U}}{\partial t} = \mathbf{B} \tag{155}$$

where $\mathbf{L}_c$ is rectangular matrix of dimensions $(n_{bc} \times n)$, the boundary conditions are compatible if the system given by (155), together with (154), allows the determination of $\partial \mathbf{U}/\partial t$.

In order to clarify the above concepts let us consider some typical boundary problems. Assuming for simplicity a one-dimensional flow, choosing for $\mathbf{U}$ the primitive variables $[\rho, u, p]$, and selecting

$$\mathbf{L} = \begin{pmatrix} 0 & 1 & -\dfrac{1}{\rho c} \\ 1 & 0 & -\dfrac{1}{c^2} \\ 0 & 1 & \dfrac{1}{\rho c} \end{pmatrix} \tag{156}$$

the matrix $\mathbf{L}_r$ will depend on the possible flow conditions at the boundary that will be discussed in the following.

- *Subsonic inflow* $(\mathbf{v} \cdot \mathbf{n} < 0,\ |\mathbf{v} \cdot \mathbf{n}| < c)$
  In this case $n_r = 1$ and

$$\mathbf{L}_r = \begin{bmatrix} 0 & 1 & \dfrac{1}{\rho c} \end{bmatrix} \tag{157}$$

The matrix $\mathbf{L}_c$ is a(2×3) matrix, i.e. one needs two boundary conditions, and the selection of $\mathbf{L}_c$ is constrained by

$$\det \begin{bmatrix} \mathbf{L}_r \\ \mathbf{L}_c \end{bmatrix} \neq 0$$

Then, the first column of $\mathbf{L}_c$ cannot be identically zero because there is a zero in the first column of $\mathbf{L}_r$. In other words, the boundary condition must contain information on the density (either explicitly or implicitly).

- *Supersonic inflow* $(\mathbf{v} \cdot \mathbf{n} < 0,\ |\mathbf{v} \cdot \mathbf{n}| > c)$
  In this case $n_r = 0$ and three boundary conditions are needed satisfying the constraint

$$\det |\mathbf{L}_c| \neq 0$$

- *Subsonic outflow* $(\mathbf{v} \cdot \mathbf{n} > 0,\ |\mathbf{v} \cdot \mathbf{n}| < c)$
  The number of non-negative eigenvalues is $n_r = 2$ and

$$\mathbf{L}_r = \begin{bmatrix} 1 & 0 & -\dfrac{1}{c^2} \\ 0 & 1 & \dfrac{1}{\rho c} \end{bmatrix} \tag{158}$$

The matrix $\mathbf{L}_c$ is a (1×3) matrix, and only one boundary condition can be imposed such that

$$\det \begin{bmatrix} \mathbf{L}_r \\ \mathbf{L}_c \end{bmatrix} \neq 0$$

It is easy to see that in principle one can assign any one of the primitive variables (i.e. only one coefficient of the row matrix $\mathbf{L}_c$ can be non-zero).

- *Supersonic outflow* $(\mathbf{v} \cdot \mathbf{n} > 0,\ |\mathbf{v} \cdot \mathbf{n}| > c)$
  In this case $n_r = 3$ and $\mathbf{L}_c$ coincides with $\mathbf{L}$, hence boundary conditions are not required.

- *Wall* $(\mathbf{v} \cdot \mathbf{n} = 0)$
  Along a wall the reduced matrix $\mathbf{L}_r$ coincides with that corresponding to the subsonic outflow case and $\mathbf{L}_c = [0, 1, 0]$.

2 Navier–Stokes equations

A first look at Table 1 shows that the number of boundary conditions that are expected to be necessary for the closure of the Navier–Stokes equations is generally greater than in the inviscid situation. It is clear that the meaning of wave propagation for the Navier–Stokes equations must be rediscussed; indeed, the truly outgoing wave is that associated with the continuity equation. Following a mathematical classification the equations are no longer hyperbolic, but are incompletely parabolic (Strikwerda, 1977; Gustafsson and Sundström, 1978). However, waves also appear in viscous flows (if this were not the case they would not be observed in laboratory experiments!).

For high Reynolds number flows one can assume that the behaviour of these waves does not differ from that of waves in the inviscid case. Then, in order to extend the concept of compatibility relations (at boundaries) to the Navier–Stokes equations, following Poinsot and Lele (1992), one can recast the latter in the so-called singular perturbation form

$$\frac{\partial \mathbf{U}}{\partial t} + \mathbf{A}\frac{\partial \mathbf{U}}{\partial x} + \mathbf{Q} - \epsilon \mathbf{Q}_V = 0$$

where $\mathbf{Q}_V$ accounts for the viscous flux contribution and ϵ is a small parameter (assuming that the diffusion coefficients are proportional to ϵ). In analogy to the procedure discussed for the Euler equations, we define a Navier–Stokes 'compatible' set of boundary conditions when $\partial \mathbf{U}/\partial t$ on the boundary verifies the following system of equations (suitably reordered)

$$\mathbf{L}\frac{\partial \mathbf{U}}{\partial t} + \widetilde{\mathbf{\Lambda}}_O \mathbf{L}\frac{\partial \mathbf{U}}{\partial x} + \mathbf{LQ} - \epsilon \mathbf{LQ}_V + \widetilde{\mathbf{B}} = 0 \tag{159}$$

where the diagonal square matrix $\widetilde{\mathbf{\Lambda}}_O$ is

$$\widetilde{\mathbf{\Lambda}}_O = \begin{pmatrix} \mathbf{\Lambda}_O & \mathbf{0} \\ \mathbf{0} & \mathbf{0} \end{pmatrix}$$

and $\widetilde{\mathbf{B}} = [\mathbf{0}, \mathbf{B}]^T$ with $\mathbf{B}$ a column vector of dimensions $n - n_r$, while $\mathbf{\Lambda}_O$ is the diagonal matrix of non-negative eigenvalues of (154).

The quantities $\mathbf{L}, \partial \mathbf{U}/\partial x$ and $\mathbf{Q}$ can be evaluated from the interior, while the $n - n_r$ components of $\mathbf{B}$ and the $n - 1$ components of $\mathbf{Q}_V$ account, respectively, for the 'inviscid' and 'viscous' interactions between inner and outer phenomena.

For example, at an open boundary $\mathbf{B}$ represents the contribution due to the inviscid incoming waves. If the exterior state were known at the boundary, one could evaluate $\widetilde{\mathbf{B}}$ as

$$\widetilde{\mathbf{B}} = \widetilde{\mathbf{\Lambda}}_I \mathbf{L}\left(\frac{\partial \mathbf{U}}{\partial x}\right)_{outer}$$

where $\widetilde{\mathbf{\Lambda}}_I$ is the square matrix associated with the incoming waves (i.e., those that have negative eigenvalues)

$$\widetilde{\Lambda}_I = \begin{pmatrix} 0 & 0 \\ 0 & \Lambda_I \end{pmatrix} .$$

However, in general the exterior state vector is not available for all components; on the other hand one only needs $n - n_r$ linear combination of exterior data. Conversely, to define $\mathbf{Q}_V$ one always needs (in principle) $n - 1$ conditions. The issue related to the selection of suitable boundary conditions will be addressed and better clarified in the next section.

It is easy to verify that (159) relaxes to (154) and (155) for $\epsilon \to 0$, so that the number of conditions required for determining $\partial \mathbf{U}/\partial x$ at the boundary is equal to the number indicated in Table 1.

B Choice of suitable boundary conditions

It should be clear at this point that a compatible set of boundary conditions does not necessarily yield a well-posed problem, i.e. a stable unique solution of the (Euler and/or Navier–Stokes) Initial Boundary Value Problem (IBVP) is not guaranteed. In Tables 2 and 3 we report a synopsis (Poinsot and Lele, 1992) of theoretical results about boundary conditions that may ensure well-posedness (when not otherwise specified in the Tables).

In the case of a subsonic inviscid inflow, for example, one can choose from a

Table 2
Physical boundary conditions for Euler and Navier–Stokes equations (subsonic inflow)

Well posedness	Euler	Navier–Stokes	
		Inviscid conditions	Viscous conditions
—	$\mathbf{u}, T$	$\mathbf{u}, T$	—
No proof for NS	$\mathbf{u}, \rho$	$\mathbf{u}, \rho$	Constant normal stress
Numerically unstable for NS	$u - 2c/(\gamma - 1)$ v, w, s (entropy)	$u - 2c/(\gamma - 1)$ v, w, s (entropy)	Constant normal stress
No proof for E and NS	$\Lambda_1 L_c \dfrac{\partial \mathbf{U}}{\partial x} = 0$	$\Lambda_1 L_c \dfrac{\partial \mathbf{U}}{\partial x} = 0$	Constant normal stress

E, Euler; NS, Navier–Stokes.

Table 3
Physical boundary conditions for Euler and Navier–Stokes equations (subsonic outflow)

| | Euler | Navier–Stokes | |
		Inviscid conditions	Viscous conditions
Outflow non-reflecting	p_∞	p_∞	Constant tangential stresses and normal heat flux
Outflow reflecting	p	p	Constant tangential stresses and normal heat flux

four-fold set of boundary conditions (see Table 2), while at the outflow one may specify the pressure (see Table 3). At the wall one generally imposes either no-slip (resp. slip) conditions on the velocity for the Navier–Stokes (resp. Euler) equations, and either isothermal ($T = T_w$) or adiabatic ($\partial T/\partial n = 0$, with n the wall normal direction) conditions.

It is useful to note that, if one is interested in the steady solution, obtained through a transient approach both for the interior and exterior schemes, then the chosen set of compatible boundary conditions (at their steady state) must ensure a unique state for the stationary problem. In practice, this implies that the boundary conditions cannot be fixed arbitrarily (Hirsch, 1990). Hence, in selecting the set of compatible conditions one must account for the available experimental (or even heuristic) information. Unfortunately, in most cases the experimental information of a variable is not known exactly at the boundary, but at some distance away from it (and/or in correspondence of the steady state). In this case, as originally proposed by Hedström (1979), it may be preferable to use the so-called non-reflecting boundary conditions. This corresponds to annihilating the contribution associated with the incoming waves, i.e. assuming that $(\partial U/\partial x)_{outer} \sim 0$. In a one-dimensional inviscid case this is also equivalent to assuming that the amplitudes of the incoming waves are constant in time. However, as indicated in Table 2, neglecting the gradients (normal to the boundary) in the outer region may lead to ill-posed problems.

Non-reflecting boundary conditions for the Euler and/or Navier–Stokes equations may require some modifications in the case that one knows *a priori* the steady-state condition of a boundary variable (typically the pressure p_∞ in the outer outflow region not far from the boundary). For such cases Rudy and Strikwerda (1981) model the term $\widetilde{\mathbf{B}}$ of (159) in such a way that at steady state $\widetilde{\mathbf{B}} = 0$ implies $p = p_\infty$. More precisely, for a one-dimensional problem, referring to the set of primitive variables, the modified non-reflecting condition amounts to

$$\frac{\partial u}{\partial t} - \frac{1}{\rho c}\frac{\partial p}{\partial t} - \frac{\alpha}{\rho c}(p - p_\infty) = 0$$

Non-reflecting outflow boundary conditions yield faster convergence to the steady state in comparison to a constant pressure outflow boundary condition. Optimal values (for convergence) of α are found to depend on the flow conditions (typically $\alpha = O(0.2 - 0.4)$); while $\alpha = 0$ is appropriate for transient calculations.

In principle, the procedure can be generalized to any number of steady-state conditions by introducing a suitable norm of all deviations from the steady-state values. One must always verify the well posedness of the problem and the convergence toward a steady state.

The viscous conditions should be assigned in terms of exterior data. In most circumstances, owing to the lack of exact information, extrapolated (or approximate) data may be used. However, the additional (with respect to the inviscid case) conditions may have a large effect on the solution, for example an artificially thick viscous layer at the boundary. As a remedy Dutt (1988) shows that one should use a set of boundary conditions that involve the viscous fluxes, so as to ensure L_2-boundedness of the solution and at the same time guarantee a well posed boundary value problem for the hyperbolic part of the Navier–Stokes equations. However, if the open boundaries are located in regions where viscous effects are negligible, then one may neglect the viscous conditions altogether. This approach is now the most commonly used in practical computations (Poinsot and Lele, 1992).

C Construction of numerical fluxes at boundaries

The state vector at the boundaries can be updated once the set of boundary conditions for determining $\partial U/\partial t$ are specified. Then, formulas for numerical integration, derivation and extrapolation are needed.

A question as to the required accuracy of the formulas arises; of course, the numerical accuracy at the boundary must be correlated with the accuracy of the interior scheme. For mixed initial boundary value problems Gustafsson (1975, 1981) has shown that the accuracy at the boundary can be one order lower than the inner accuracy without reducing the global order of accuracy. An equivalent statement is generally accepted also for non-linear problems; however, there seems to be no rigorous theoretical indication on this point.

As a clarifying example let us determine the conditions at a subsonic outflow for a one-dimensional inviscid case. Assuming a primitive variable formulation and closing the outflow compatibility equations with a non-reflecting boundary condition, from (156) and (158) one obtains

$$\frac{\partial \rho}{\partial t} - \frac{1}{c^2}\frac{\partial p}{\partial t} = -u\left(\frac{\partial \rho}{\partial x} - \frac{1}{c^2}\frac{\partial p}{\partial x}\right)_{\text{int}}$$

$$\frac{\partial u}{\partial t} + \frac{1}{\rho c}\frac{\partial p}{\partial t} = -(u+c)\left(\frac{\partial u}{\partial x} + \frac{1}{\rho c}\frac{\partial p}{\partial x}\right)_{\text{int}} \qquad (160)$$

$$\frac{\partial u}{\partial t} + \frac{1}{\rho c}\frac{\partial p}{\partial t} - \frac{\alpha}{\rho c}p = -\frac{\alpha}{\rho c}p_\infty$$

where $(\cdot)_{\text{int}}$ indicates that the $(\cdot)$ can be determined from the interior. Then, to update the boundary values of (ρ, u, p) one needs to:

(i) Evaluate the right-hand-side by means of non-symmetric space derivatives that can be expressed either explicitly, referring to the n-th values, or implicitly, referring to the $(n+1)$-th values, or to an average of the n-th and $(n+1)$-th value (in a Crank–Nicolson approach). This step might also require reconstruction from cell-averaged (internal) quantities.
(ii) Express (in a fully discretized formulation) the left-hand-side of (160) in terms of the boundary state values. Then, the solution of the algebraic system of equations (so obtained) yields the updated boundary values

or

Cast (in a semidiscretized formulation) (160) as a system of ordinary differential equations

$$\frac{\mathrm{d}}{\mathrm{d}t}\mathbf{U} = f(\mathbf{U})$$

that needs to be solved together with the system of ordinary differential equations for the interior.

If one is interested in the steady state (assuming it exists), simplified relations for steady boundary conditions can easily be obtained from (160). Indeed, setting $\partial(\cdot)/\partial t = 0$ one obtains

$$p = p_\infty$$

$$\left(\frac{\partial \rho}{\partial x}\right)_{\text{int}} = \frac{1}{c^2}\left(\frac{\partial p}{\partial x}\right)_{\text{int}} \qquad (161)$$

$$\left(\frac{\partial u}{\partial x}\right)_{\text{int}} = -\frac{1}{\rho c}\left(\frac{\partial p}{\partial x}\right)_{\text{int}}$$

Again, by means of non-symmetric space derivatives one can evaluate the boundary values in terms of the inner ones. This observation is the basis for the so-called extrapolation boundary conditions.

D Stability

The theoretical analysis of the correlation between the treatment of the boundary conditions and the stability behaviour of time marching algorithms is far from being satisfactory for many reasons:

(i) Very few theoretical results are available even for a general multidimensional nonlinear initial value problem (IVP)

(ii) Despite a large collection of results on the stability of the analytical solutions of mixed linear hyperbolic problems (Majda, 1984; Kreiss, 1989), the analysis of the numerical treatment of (linear hyperbolic) initial boundary value problem (IBVP) may be difficult.

(iii) The information on the well posedness of non-linear incompletely parabolic (Navier–Stokes) IBVP is not complete (as highlighted in Tables 2 and 3); hence, well-founded numerical stability requirements are lacking.

(iv) As yet the existence of optimal (where optimal refers to convergence and accuracy) finite-volume interior schemes for a general multidimensional IVP is somewhat dubious. Consequently, even the systematic collection of results of numerical experiments may not furnish indications as to the different treatment of boundary conditions and the stability. In fact, most of the analyses have been developed for the linear case with central, upwind or Lax–Wendroff schemes (Hirsch, 1990; Warming and Beam, 1990; Gustafsson, 1991).

(v) The analysis of the coupled effects produced by the boundary conditions and the non-linearities requires an interdisciplinary knowledge of both numerical analysis and non-linear dynamics. For example, fluid dynamicists, who are aware of the intricate relationship between the boundaries and the occurrence of turbulence, would not be surprised to find a dependence of the numerical solutions on the boundary conditions for non-linear models.

In this section we limit the discussion to linearized problems and report a synthesis of classical results. In particular, we refer to the stability of a numerical procedure with respect to the perturbation of the initial conditions.

1 The linear evolution scenario

Let $\mathbf{u}$ be the state vector or else the perturbation with respect to any arbitrary initial state; then its evolution can be cast as follows.

- *For fully discretized formulation:*

$$\mathbf{u}^{n+1} = \mathbf{C}\mathbf{u}^n = (\mathbf{I} + \Delta t \mathbf{D})\mathbf{u}^n \tag{162}$$

- *For semidiscretized formulation:*

$$\frac{d\mathbf{u}}{dt} = \mathbf{A}\mathbf{u} \tag{163}$$

where $\mathbf{C}$ and $\mathbf{A}$ account for the interior and boundary formulas.

If $\mathbf{A}$ and $\mathbf{C}$ do not vary with time, the state at $t = n\Delta t$ can be related to the initial state $\mathbf{u}^0$, i.e.

$$\mathbf{u}^n = \mathbf{C}^n\mathbf{u}^0; \quad \mathbf{u}(t) = e^{\mathbf{A}t}\mathbf{u}^0$$

where $e^{\mathbf{A}t} = \sum_{r=0}^{\infty} t^r\mathbf{A}^r/r! = \mathbf{I} + \sum_{r=1}^{\infty} t^r\mathbf{A}^r/r!$, and $\mathbf{u}^n$ and $\mathbf{u}(t)$ stand, respectively, for the fully discretized or semidiscretized solution.

In the following we briefly highlight the importance of some of the properties of the matrices. $\mathbf{C}$ and $\mathbf{A}$ on the state evolution. Let us first assume that $\mathbf{C}$ and $\mathbf{A}$ are diagonalizable. Then

$$\mathbf{u}^n = \mathbf{R}_C\mathbf{w}^n = \mathbf{R}_C\mathbf{\Lambda}_C^n\mathbf{w}^0 \tag{164}$$

$$\mathbf{u}(t) = \mathbf{R}_A\mathbf{w}(t) = \mathbf{R}_A e^{\mathbf{\Lambda}_A t}\mathbf{w}^0 \tag{165}$$

where $\mathbf{\Lambda}_C = \mathbf{R}_C^{-1}\mathbf{C}\mathbf{R}$ and $\mathbf{\Lambda}_A = \mathbf{R}_A^{-1}\mathbf{A}\mathbf{R}_A$, $\mathbf{R}_A$ and $\mathbf{R}_C$ being the right eigenvector matrices, and $\mathbf{w} = \mathbf{R}_C^{-1}\mathbf{u}$ (or $\mathbf{R}_A^{-1}\mathbf{u}$ in the semidiscretized formulation), i.e.

$$w_i^n = \lambda_i^n w_i^0; \quad w_i(t) = e^{\lambda_i t}w_i(0) \tag{166}$$

where λ_i stand for the eigenvalues of either $\mathbf{C}$ or $\mathbf{A}$.

If $\mathbf{C}$ or $\mathbf{A}$ do not admit a diagonal canonical form (i.e. they do not admit a complete set of independent eigenvectors), it is always possible to cast them in a block diagonal Jordan form by a similarity transformation (Gantmacher, 1960; Young, 1971) i.e.

$$\mathbf{C} = \mathbf{T}_C\mathbf{J}_C\mathbf{T}_C^{-1}, \quad \mathbf{A} = \mathbf{T}_A\mathbf{J}_A\mathbf{T}_A^{-1}$$

Hence, for the fully discretized formulation

$$\mathbf{u}^n = \mathbf{T}_C\mathbf{w}^n \quad \text{with } \mathbf{w}^n = \mathbf{J}_C^n\mathbf{w}^0$$

while for the semidiscretized formulation

$$\mathbf{u}(t) = \mathbf{T}_A\mathbf{w}(t) \quad \text{with } \mathbf{w}(t) = e^{\mathbf{J}_A t}\mathbf{w}^0$$

If the (block) diagonal of the matrix $\mathbf{J}_C$ is made out of q Jordan blocks $\{\mathbf{J}_{B_1}, \ldots, \mathbf{J}_{B_m}, \ldots, \mathbf{J}_{B_q}\}$, where the m-th block has order $r(m)$, $\mathbf{J}_C^n$ is $\{\mathbf{J}_{B_1}^n, \ldots, \mathbf{J}_{B_m}^n, \ldots, \mathbf{J}_{B_q}^n\}$. Then, one may rearrange the vector $\mathbf{w}$ in q subvectors (corresponding to the q Jordan blocks) $[\mathcal{W}_1, \ldots, \mathcal{W}_m, \ldots, \mathcal{W}_q]^T$, each subvector having $r(m)$ components $(W_m)_i$ $(i = 1, r(m))$, yielding

$$W_m^n = J_{B_m}^n W_m^0$$

with

$$J_{B_m}^n = (\lambda_m I + S)^n = \sum_{j=0,nd} \binom{n}{j} \lambda_m^{n-j} S^j$$

where S is a lower shift-type matrix of order $r(m)$ (having $S_{i,i-1} = 1$, and $S_{i,j} = 0$ for $j \leq i$ and $j \geq i - 1$), and $nd = \min(n, r(m) - 1)$. Each component of W_m then evolves according to

$$(W_m^n)_i = \sum_{j=0,i-1} \binom{n}{j} \lambda_m^{n-j} (W_m^0)_{j+1} \tag{167}$$

Analogously, for the semidiscretized formulation one has

$$\frac{d}{dt} W_m = (\lambda_m I + S) W_m$$

and

$$(W_m(t))_i = \exp(\lambda_m t) \sum_{j=0,i-1} \frac{t^j}{j!} (W_m(0))_{i-j} \tag{168}$$

The above expressions then show

$$\rho(C) < 1 \Longrightarrow w \text{ (and } u) \to 0 \qquad \text{for } n \to \infty$$

and

$$\rho(e^A) < 1, \text{ i.e. } \operatorname{Re}(\lambda_i)_A < 0 \Longrightarrow w \text{ (and } u) \to 0 \qquad \text{for } t \to \infty$$

where $\rho(\cdot)$ stands for the spectral radius.

Moreover, if the matrices are not diagonalizable, owing to the presence of polynomial terms in n (or t for the semidiscretized formulation), the growth of u can be of polynomial and/or exponential type. The consequence of this is that even if the condition on the spectral radius of C (or A) is verified, the polynomial law dependency may produce an initial growth. As an example, let us consider the function $t^j \exp(-k^2 t)$; it is easy to show that the function grows from zero up to a maximum value (equal to $(j/k^2)^j \exp(-j)$), when $0 \leq t \leq j/k^2$; it then decreases, approaching zero for $t \to \infty$. The example further shows that for $j \to \infty$ both j/k^2 and the attained maximum go to infinity. From a computational point of view the behaviour described with the simple example with j going to infinity may occur when the number of mesh points goes to infinity (generally a (theoretical) requirement in order to converge to the 'exact' solution!).

In conclusion, the linear evolution scenario that has been briefly outlined highlights the importance not only of the eigenvalues but also of the eigenvectors of the matrices $\mathbf{C}$ and $\mathbf{A}$. Hence, a stability analysis of an initial boundary value problem must identify the effects of the treatment of the boundary conditions on the spectra and on the structure of $\mathbf{C}$ and $\mathbf{A}$. Furthermore, the stability analysis should not only account for the temporal evolution of the perturbation state vector, but it should also take into account the effects of the variations of the computational parameters $(\Delta x, \Delta t)$. A variation of the dimensions of the mesh (e.g. a change in the mesh size Δx) produces a variation both in the number of components N of the vector $\mathbf{u}$ and in the dimensions of the related space and transition matrices. As a consequence, the tailoring of norms for spaces whose dimensions go to infinity (as $\Delta x \to 0$) is further complicated by the fact that the equivalence of norms does not exist for non-finite dimension spaces. A variation in the time step of a fully discretized formula requires an investigation of the evolution behaviour for n approaching infinity, even when the integration is carried out for a finite interval $(0, T)$ and with $\Delta t \to 0$.

Hence, for a fully discretized formulation, as a consequence of Δx and Δt approaching zero (and/or of the unboundedness of the space/time physical domain), a stability analysis may require an estimate of the limit behaviour of the evolution of the norm of the vector $\mathbf{u}$ by letting its dimensions and n go to infinity either separately (for the sake of simplicity) or simultaneously (as generally required in order to converge to the 'exact' solution). Analogously, the analysis of semidiscretized formulas may be carried out with fixed Δx either for a bounded or unbounded integration time interval; in practice one must also account (at a later stage) for the stability of the time integration algorithm (of the system of ordinary differential equations), and again the limit behaviour for $\Delta t \to \infty$ and/or $\Delta x \to 0$ must be considered.

The canonical forms of the matrices $\mathbf{C}$ and $\mathbf{A}$ are seldom available, and in general neither the similarity transformation, nor (167) and (168) are available. Hence, the (theoretical) need to formulate stability criteria and requirements arises in order to simplify the analysis.

2 Stability criteria

Zero stability

The simplest approach to the problem of numerical stability in a fixed dimension space considers the conditions under which the discretized solution (or a perturbation) approaches zero as n (or t) $\to \infty$, and relates the stability to the spectral radius of $\mathbf{C}$ (or $\mathbf{A}$). That is

$$\rho(\mathbf{C}) < 1, \ \mathrm{Re}(\lambda_{\mathrm{i}})_{\mathbf{A}} < 0 \ \forall \mathbf{u}^0 \tag{169}$$

In order to avoid a non-monotone decay of the perturbation state vector (for the reasons described in the linear evolution scenario), one may introduce a stronger requirement

$$\phi_\infty(\mathbf{u}^n) \leq k^n \phi_\infty(\mathbf{u}_0)$$

with $k < 1$, and ϕ_∞ being the maximum norm. This condition is also equivalent to imposing

$$\|\mathbf{C}\|_\infty < 1 \qquad (170)$$

where $\|\mathbf{C}\|_\infty = \max_i \sum_j |\mathbf{C}_{ij}|$. Analogously, for the semidiscretized case

$$\|e^{\mathbf{A}}\|_\infty < 1 \qquad (171)$$

which is ensured, for instance, if $\mathbf{A}$ has a strong diagonal dominance and negative diagonal coefficients.

Note that the zero stability requirements in the above form are rather strong. Indeed, if one considers a scalar advection problem with periodic boundary conditions, the consistency requirement would imply that a constant perturbation state is invariant in time, i.e. $\mathbf{C}$ and $\mathbf{A}$ have an eigenvalue equal to one. Then, the inequalities (170) and (171) can be weakened by allowing the L_∞-norms to be equal to one. Conversely, for an initial boundary value pure advection problem with homogeneous boundary conditions, the perturbation should vanish for $t \to \infty$, and the above requirements are physically acceptable.

Lax-Richtmyer stability

The (most) well-known stability definition is the so-called Lax–Richtmyer stability (LRS), which is a set property generally referred to the fully discretized formulation and which in some sense is weaker than the previous ones. The LRS refers to a set $(\mathcal{F})$ of matrices $\mathbf{C}$ associated with a parameter set $P = \{\Delta\mathbf{x}, \Delta t \mid \cdots\}$ that has $(\mathbf{0}, 0)$ as the limit point, and it consists of either requiring

$$\phi_2(\mathbf{u}^n) \leq k\phi_2(\mathbf{u}^0) \qquad \text{for } n\Delta t \leq T, \qquad \forall \Delta\mathbf{x}, \Delta t \in P \qquad (172)$$

or simply

$$\phi_2(\mathbf{u}^n) \leq k' \exp(\alpha n\Delta t)\phi_2(\mathbf{u}^0) \qquad \forall n\Delta t \qquad (173)$$

where k, k', and α do not depend on the choice of the parameters family P, xith, $k, k' \geq 1$, and ϕ_2 the L_2-norm.

In terms of the requirements for the matrix $\mathbf{C}$ (assumed to depend only on a single parameter, e.g. $\Delta x = \Delta x(\Delta t)$) one has

$$\|\mathbf{C}^n(\Delta t)\|_2 \leq k \qquad n = 1, 2, \ldots, T/\Delta t$$

or

$$\|\mathbf{C}^n(\Delta t)\|_2 \leq k' \exp(\alpha n\Delta t) \qquad \forall n\Delta t$$

and the following criteria follow

$$\|\mathbf{C}(\Delta t)\|_2 \leq 1 + O(\Delta t) \qquad \text{sufficient condition} \qquad (174)$$

$$\rho(\mathbf{C}(\Delta t)) \leq 1 + O(\Delta t) \qquad \text{necessary condition} \qquad (175)$$

where (175) is nothing but the well-known Von Neumann condition (see Chapter 1).

For all cases for which $\|\mathbf{C}\|_2 = \rho(\mathbf{C})$ the Von Neumann condition is necessary and sufficient for LRS. For example, for a linear hyperbolic problem with constant coefficients and periodic boundary conditions $\mathbf{C}$ is a finite circulant matrix (Hirsch, 1990) and LRS coincides with the Von Neumann condition. Indeed, the eigenvalues can easily be determined by a similarity transformation obtained by choosing a basis vector representation of $\mathbf{u}$ of Fourier type.

The application of the stability criteria discussed above requires an estimate of the eigenvalue spectrum of $\mathbf{C}$. For simple problems one might attempt an asymptotic estimate of the loci of the eigenvalues (Warming and Beam, 1990). The numerical evaluation of the eigenvalues is computationally expensive but still possible (Eberle *et al.*, 1992), even though the order of the matrix increases as the mesh spacing is reduced, thus making the procedure impractical.

Godunov–Ryabenkii stability

As pointed out in the previous section, the simple knowledge of the eigenvalues (sometimes referred as the proper values) of the matrix $\mathbf{C}$ can be a misleading guide for the stability when $\mathbf{C}$ is not diagonalizable. Godunov and Ryabenkii (1977) have used the more general definition of eigenvalue spectrum of a family $\mathcal{F}$ of transition matrices defined on a one-parameter family $P = \{\Delta x, \Delta t \mid \Delta t = \Delta t(\Delta x)\}$. In particular, they define as family eigenvalue any complex number λ such that

$$\forall \epsilon > 0, \Delta x > 0 \quad \exists \Delta x < \Delta x_0, \mathbf{u} \neq 0 \quad : \quad \|\mathbf{C}(\Delta x)\mathbf{u} - \lambda \mathbf{u}\|_{\Delta x} \leq \epsilon \|\mathbf{u}\|_{\Delta x}$$

where $\| \ \|_{\Delta x}$ stands for a norm that may depend on Δx; the Godunov–Ryabenkii spectrum $\mathcal{S}(\mathcal{F})$ is said to include ∞ if

$$\forall k > 0, \ \Delta x > 0 \quad \exists \Delta x < \Delta x_0, \mathbf{u} \neq 0 \quad : \quad \|\mathbf{C}(\Delta x)\mathbf{u}\|_{\Delta x} \leq k \|\mathbf{u}\|_{\Delta x}$$

Then, the Godunov-Ryabenkii stability amounts to requiring

$$\max \mathcal{S}(\mathcal{F}) = \rho(\mathcal{S}) \leq 1$$

The above condition is also a necessary condition for the Lax–Richtmyer stability, but it is a weaker condition. Indeed, there are situations where a scheme is GR stable but not LR stable (Richtmyer and Morton, 1967).

The usefulness of the GR criterium relies on the possibility of determining the eigenvalue spectrum $\mathcal{S}(\mathcal{F})$. First of all if the matrix $\mathbf{C}$ is symmetric, then the

Godunov-Ryabenkii spectrum coincides with the set $\mathcal{U}_S$ of the limit points of the proper values of $\mathbf{C}$.

In general, let $\mathbf{C}_{\lambda,\Delta x}$ be the matrix $\mathbf{C}(\Delta x) - \lambda \mathbf{I}$, the Godunov-Ryabenkii spectrum is the union of $\mathcal{U}_S$ and of the set of λ's for which the condition number of $\mathbf{C}_{\lambda,\Delta x}$ is unbounded (in the norm $\| \quad \|_{\Delta x}$), for sufficiently small Δx and $\|\mathbf{u}\|_{\Delta x} = 1$.

3 The energy method

The method introduces a suitable positive definite quadratic form $E(\mathbf{u})$ (the   -called energy) and finds a scalar inequality for the time evolution of $E($ $)$f the type:

- *For a fully discretized scheme:*

$$E(\mathbf{u}^n) \leq k_e \exp{(\alpha_e n \Delta t)} E(\mathbf{u}^0), \quad \forall \mathbf{u}^0 \tag{176}$$

- *For a semidiscretized scheme:*

$$E(\mathbf{u}(t)) \leq k_e \exp{(\alpha_e t)} E(\mathbf{u}^0), \quad \forall \mathbf{u}^0 \tag{177}$$

where $k_e \geq 1$.

For example, if one assumes for $E(\mathbf{u})$ a suitable scalar product (   ) that may depend on Δx, the norm $\phi_E(\mathbf{u}) = (\mathbf{u}, \mathbf{u})^{\frac{1}{2}}$ satisfies

$$\phi_E(\mathbf{u}^n) \leq \sqrt{k_e} \exp{(\alpha_e n \Delta t/2)} \phi_E(\mathbf{u}^0)$$

that is analogous to the LR condition given by (173) when (17   ensured for a family $\mathcal{F}(P)$; however, the equivalence between the ϕ_E and   norms must be investigated.

In practice, once the scalar product $(\cdot, \cdot)$ is defined,   nergy evolution equation is easily obtained by scalar multiplication o'   her the fully or semidiscretized equations ((162) and (163), respectively)   the vector $\mathbf{u}$, thus yielding:

- *For the fully discretized form:*

$$(\mathbf{u}^{n+1}, \mathbf{u}^{n+1}) = (\mathbf{u}^n, \mathbf{u}^n) + 2\Delta t(\mathbf{u}^n, \mathbf{D}\mathbf{u}^n) + \Delta t^2 (\mathbf{D}\mathbf{u}^n, \mathbf{D}\mathbf{u}^n)$$

i.e.

$$E(\mathbf{u}^{n+1}) = E(\mathbf{u}^n) + 2\Delta t(\mathbf{u}^n, \mathbf{D}'\mathbf{u}^n)$$

where $\mathbf{D}' = \mathbf{D} + \Delta t/2 \mathbf{D}^T \mathbf{D}$.

- *For the semidiscretized form:*

$$\frac{d}{dt} E(\mathbf{u}) = 2\left(\mathbf{u}, \frac{d}{dt}\mathbf{u}\right) = 2(\mathbf{u}, \mathbf{A}\mathbf{u})$$

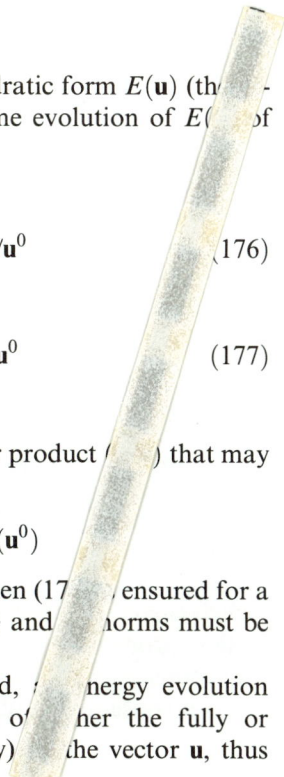

If the matrices $\mathbf{D}'$ and $\mathbf{A}$ of the family $\mathcal{F}(P)$ are semibounded, i.e.

$$(\mathbf{u}, \mathbf{D}'\mathbf{u}) \leq k_f(\mathbf{u}, \mathbf{u}); \quad (\mathbf{u}, \mathbf{A}\mathbf{u}) \leq k_s(\mathbf{u}, \mathbf{u}) \quad \forall \mathbf{u} \tag{178}$$

then, for the fully discretized formulation, it follows

$$E(\mathbf{u}^n) \leq (1 + 2k_f \Delta t)^n E(\mathbf{u}^0) \leq \exp\,(2k_f n \Delta t) E(\mathbf{u}^0) \tag{179}$$

while for the semidiscretized formulation

$$E(\mathbf{u}(t)) \leq \exp(2k_s t) E(\mathbf{u}^0) \tag{180}$$

and the inequalities (176), (177) are verified.

Note that the semiboundedness of $\mathbf{D}'$ can be ensured by requiring $(\mathbf{u}, \mathbf{D}\mathbf{u}) \leq k(\mathbf{u}, \mathbf{u})$. Indeed,

$$(\mathbf{u}, \mathbf{D}'\mathbf{u}) = (\mathbf{u}, \mathbf{D}\mathbf{u}) + \frac{\Delta t}{2}(\mathbf{D}\mathbf{u}, \mathbf{D}\mathbf{u}) \leq k + \frac{\Delta t}{2}\|\mathbf{D}\|_{\phi_E}^2(\mathbf{u}, \mathbf{u})$$

and (178) is conditionally verified (when $\|\mathbf{D}\|_{\phi_E}^2$ depends on Δx), provided the parameter family is restricted to the values for which $\Delta t \leq 2k_2/\|\mathbf{D}\|_{\phi_E}^2$, and $k_f = k + k_2$. To further illustrate the relation between the energy inequalities and the Lax–Richtmyer sufficient conditions, note that, if the geometric norm of the matrix $\mathbf{D}$ induced by the energy norm ϕ_E satisfies $\|\mathbf{D}\|_{\phi_E} \leq k$, then from (162) and (178) one has

$$\|\mathbf{C}\|_{\phi_E} = \|\mathbf{I} + \Delta t \mathbf{D}\|_{\phi_E} \leq 1 + \Delta t \|\mathbf{D}\|_{\phi_E}$$
$$\leq 1 + k\Delta t$$

which is equivalent to the Lax–Richtmyer condition when one assumes for the scalar product $(\cdot, \cdot)$ the usual one, i.e. $(\mathbf{a}, \mathbf{b}) = \mathbf{a}^T \mathbf{b}$ where T stands for transposed.

As a consequence of the Lax–Richtmyer-like sufficient condition, one has

$$(\mathbf{D}\mathbf{u}, \mathbf{D}\mathbf{u})^{\frac{1}{2}} \leq k(\mathbf{u}, \mathbf{u})^{\frac{1}{2}}$$

and, exploiting the Cauchy–Buniakowski inequality, the semiboundedness of $\mathbf{D}$ follows

$$(\mathbf{u}, \mathbf{D}\mathbf{u}) \leq (\mathbf{u}, \mathbf{u})^{\frac{1}{2}}(\mathbf{D}\mathbf{u}, \mathbf{D}\mathbf{u})^{\frac{1}{2}} \leq k(\mathbf{u}, \mathbf{u})$$

We now show the use of the energy method for analysing the stability of a semidiscretized formulation for an initial boundary value problem. Let us first consider the scalar one-dimensional problem

$$\begin{aligned} u_t &= -au_x \quad 0 \leq x \leq 1 \\ u(0, t) &= 0, \quad u(x, 0) = f(x) \end{aligned} \tag{181}$$

with $a > 0$ and f real. Assuming a central discretization formula with constant mesh spacing, the semidiscrete approximation is

$$\frac{du_j}{dt} = Lu_j$$

$$u_1(t) = 0, \quad u_j(0) = f(x)$$

where $L = -a(\delta_+ - \delta_-)/2\Delta x$, and $j = 1$ and $j = N$ represent, respectively, the left and right boundaries (with $N = 1/\Delta x + 1$).

The above equation can be applied everywhere but at the boundary points, where either the central operator needs to be modified or numerical boundary conditions must be prescribed. It is easy to show that the latter is ensured if a zero-th order extrapolation condition $u_{N+1} = u_N$ is assumed. Indeed, if one defines as scalar product $(\mathbf{u}, \mathbf{v}) = \Sigma_{j=2,N} u_j v_j \Delta x$, the following inequality holds

$$\frac{d}{dt} E(\mathbf{u}) = -au_N^2 \leq 0$$

thus showing that the energy inequality (180) is verified.

On the other hand, if one assumes at $x = 1$ a first-order extrapolation condition

$$u_{N+1} = 2u_N - u_{N-1} \tag{182}$$

then

$$(\mathbf{u}, L\mathbf{u}) = -a\left(u_N^2 - \frac{u_N u_{N-1}}{2}\right) \tag{183}$$

and no conclusions can be drawn on the energy stability. However, if one defines as scalar product

$$(\mathbf{u}, \mathbf{v}) = \sum_{j=2,N-1} u_j v_j \Delta x + u_N v_N \Delta x/2 \tag{184}$$

one obtains

$$\frac{d}{dt} E(\mathbf{u}) = -au_N^2 \leq 0 \tag{185}$$

thus, again verifying (180).

It is important to observe that the use of extrapolation conditions is equivalent to the (previously suggested) treatment of the numerical boundary conditions in terms of time evolution relations. Indeed, it is easy to verify that the use of a first-order upwind approximation at $N + 1$ amounts to solving

$$\frac{d}{dt} u_{N+1} = -a\frac{u_{N+1} - u_{N-1}}{2\Delta x} = \frac{d}{dt} u_N$$

i.e., $u_{N+1} = u_N$, which is the zero-th order extrapolation. Analogously, the use of a second-order upwind approximation at $N + 1$ is equivalent to solve

$$\frac{d}{dt} u_{N+1} = 2\left(-a \frac{u_{N+1} - u_{N-1}}{2\Delta x}\right) - \left(-a \frac{u_N - u_{N-2}}{2\Delta x}\right)$$

$$= 2 \frac{d}{dt} u_N - \frac{d}{dt} u_{N-1}$$

i.e., $u_{N+1} = 2u_N - u_{N-1}$, which is the first-order extrapolation.

To illustrate the application of the energy method to systems of conservation laws, we refer for simplicity to the linearized (with frozen coefficients) characteristic form of the one-dimensional Euler equations with subsonic inflow ($0 \le u \le c$) at $x = 0$. The use of central discretization yields

$$\frac{d}{dt} W_j^{(1)} = -(u - c) \frac{W_{j+1}^{(1)} - W_{j-1}^{(1)}}{2\Delta x}$$

$$\frac{d}{dt} W_j^{(2)} = -u \frac{W_{j+1}^{(2)} - W_{j-1}^{(2)}}{2\Delta x} \tag{186}$$

$$\frac{d}{dt} W_j^{(3)} = -(u + c) \frac{W_{j+1}^{(3)} - W_{j-1}^{(3)}}{2\Delta x}$$

and in vector form one has

$$\frac{d}{dt} \boldsymbol{W}_j = -\Lambda \frac{(\boldsymbol{W}_{j+1} - \boldsymbol{W}_{j-1})}{2\Delta x}$$

where $\boldsymbol{W}_j$ is the characteristic state vector at the j-th node. For subsonic flow (see Table 2) homogeneous boundary conditions can be imposed on the variables corresponding to the incoming waves

$$\begin{cases} W_0^{(2)} = 0 \\ W_0^{(3)} = 0 \\ W_{N+1}^{(1)} = 0 \end{cases} \tag{187}$$

where 0 and $N + 1$ indicate the exterior nodes. The set of boundary conditions can be completed by extrapolating the characteristic variables associated with the outgoing waves. If one assumes the following zero-th order extrapolation numerical boundary conditions

$$\begin{cases} W_0^{(1)} = W_1^{(1)} \\ W_{N+1}^{(2)} = W_N^{(2)} \\ W_{N+1}^{(3)} = W_N^{(3)} \end{cases} \tag{188}$$

the semiboundedness of the discrete spatial operator can be proved if one defines as scalar product

$$(\mathbf{a}, \mathbf{b}) = \sum_{j=1,N} [a_j^{(1)} b_j^{(1)} + a_j^{(2)} b_j^{(2)} + a_j^{(3)} b_j^{(3)}] \Delta x$$

Indeed, from (187) and (188) one gets

$$\frac{d}{dt}(\mathbf{W}_j, \mathbf{W}_j) = -\left[-(u-c) W_1^{(1)} W_0^{(1)} + u W_N^{(2)} W_{N+1}^{(2)} + (u+c) W_N^{(3)} W_{N+1}^{(3)} \right] / 2$$

$$= -\left[-(u-c) W_1^{(1)} W_1^{(1)} + u W_N^{(2)} W_N^{(2)} + (u+c) W_N^{(3)} W_N^{(3)} \right] / 2 \leq 0$$

which shows the energy stability of the scheme.

Conversely, if one assumes first-order extrapolation numerical boundary conditions

$$\begin{cases} W_0^{(1)} = 2 W_1^{(1)} - W_2^{(1)} \\ W_{N+1}^{(2)} = 2 W_N^{(2)} - W_{N-1}^{(2)} \\ W_{N+1}^{(3)} = 2 W_N^{(3)} - W_{N-1}^{(3)} \end{cases} \tag{189}$$

the energy stability of the scheme can be proved by defining as scalar product (Gustafsson, 1991)

$$(\mathbf{a}, \mathbf{b}) =$$

$$\left[\frac{a_1^{(1)} b_1^{(1)}}{2} + \sum_{2,N} a_j^{(1)} b_j^{(1)} + \sum_{1,N-1} a_j^{(2)} b_j^{(2)} + \frac{a_N^{(2)} b_N^{(2)}}{2} + \sum_{1,N-1} a_j^{(3)} b_j^{(3)} + \frac{a_N^{(3)} b_N^{(3)}}{2} \right] \Delta x$$

which yields

$$\frac{d}{dt}(\mathbf{W}_j, \mathbf{W}_j) = -\left\{ \frac{u-c}{2} \left[W_1^{(1)} \frac{W_2^{(1)} - W_0^{(1)}}{2} - W_2^{(1)} W_1^{(1)} \right] \right.$$

$$+ \frac{u}{2} \left[W_{N-1}^{(2)} W_N^{(2)} + W_N^{(2)} \frac{W_{N+1}^{(2)} - W_{N-1}^{(2)}}{2} \right]$$

$$\left. + \frac{u+c}{2} \left[W_{N-1}^{(3)} W_N^{(3)} + W_N^{(3)} \frac{W_{N+1}^{(3)} - W_{N-1}^{(3)}}{2} \right] \right\}$$

$$= -\left\{ -\frac{u-c}{2} W_1^{(1)} W_1^{(1)} + \frac{u}{2} W_N^{(2)} W_N^{(2)} + \frac{u+c}{2} W_N^{(3)} W_N^{(3)} \right\} \leq 0$$

and the energy inequality (180) is again verified.

It must be pointed out that for a state vector **u** satisfying arbitrary boundary conditions: (i) the tailoring of a suitable scalar product is a difficult question; and (ii) the proof of semiboundedness can be a critical issue especially for the multidimensional Navier–Stokes equations. In addition, we remark that the energy method only furnishes sufficient conditions for the stability of an initial boundary value problem.

4 Normal mode method

The normal mode method is a powerful method for the analysis of initial boundary value problems. Following the original idea of Babenko and Guelfand (see Godunov and Ryabenkii, 1977) the stability of an initial boundary value problem is reduced to the stability of three auxiliary problems: (i) an initial value problem; (ii) an initial left boundary value problem defined in the semibounded region whose right boundary is the left boundary of the original IBVP; and (iii) an initial right boundary value problem defined in the semibounded region whose left boundary is the right boundary of the original IBVP. For each of the auxiliary problems the interior and boundary schemes are those adopted for the IBVP that one intends to analyse.

The Babenko–Guelfand criterion was first applied to account for the influence of the boundaries in one-dimensional advection problems with variable coefficients. Godunov and Ryabenkii (1977) found the correlation between the Godunov–Ryabenkii (GR) stability of an IBVP and the Babenko–Guelfand criterion. Introducing the concept of the kernel of a family spectrum, they have shown that, for the scalar one-dimensional IBVP corresponding to (181), a spectrum S can be obtained as the union of the following sets:

(i)   the set of eigenvalues of the linear operator of the auxiliary initial value problem in $(-\infty, +\infty)$

(ii) the set of eigenvalues of the linear operator of the auxiliary initial left boundary value problem in $(-\infty, 1)$

(iii) the set of eigenvalues of the linear operator of the auxiliary initial right boundary value problem in $(0, +\infty)$

It is important to note that the state vectors and the discretized operators associated with the auxiliary problems are defined in dimensional spaces that are not finite, and a suitable norm must be defined for each problem.

The eigenvalues of the (auxiliary) initial value problem can be investigated by means of any classical analysis, e.g. Von Neumann analysis; obviously, if their moduli are greater than one, the interior scheme is unstable, and interior instability cannot be cured by any boundary treatment.

The normal mode analysis may be employed to obtain the eigenvalues and the eigenvectors associated to each auxiliary problem.

By means of separation of variables, the analysis consists in assuming that the discretized solution can be expressed as follows:

- *For fully discretized formulation:*

$$u_i^n = z(n)u_i^0$$

- *For semidiscretized formulation:*

$$u_i(t) = e^{st}u_i(0)$$

where $u_i^{(0)}$ (or $u_i(0)$) are the components of (a suitable) initial state.

From (162) and (163) one obtains

$$(z(n+1)\mathbf{I} - z(n)\mathbf{C})\mathbf{u}^0 = 0 \tag{190}$$

and

$$(s\mathbf{I} - \mathbf{A})\mathbf{u}(0) = 0 \tag{191}$$

from which it follows $z(n) = \lambda^n$, where λ and $\mathbf{u}^0$ must be, respectively, an eigenvalue and an eigenvector of $\mathbf{C}$; analogously, s and $\mathbf{u}(0)$ are, respectively, an eigenvalue and an eigenvector of $\mathbf{A}$. Moreover, $\mathbf{u}^0$ (and $\mathbf{u}(0)$) have to be bounded in the L_∞-norm for the initial value problem, and in the L_2-norm for the initial left and right boundary value problems (Godunov and Ryabenkii, 1977).

For each auxiliary problem one assumes as solution

$$u_i^n \sim \lambda^n \sum P_j(i)k_j^i; \quad u_i(t) \sim e^{st} \sum P_j(i)k_j^i \tag{192}$$

where $P_j(i)$ are polynomials of degree m_j in i, and λ (or s) and k_j are complex numbers such that:

(i)   the characteristic equation is verified, i.e.

$$F(\lambda, k) = 0 \qquad (\text{or } F(s, k) = 0)$$

where F is the algebraic equation obtained from the interior scheme substituting λ (or s) and k_j to the time shift (or time derivative) and spatial shift operators, respectively; for a given λ, $m_j + 1$ represents the multiplicity of the root k_j of the characteristic equation;

(ii) $\sum P_j k_j^i$ is the i-th component of the initial state vector that has to be bounded in the appropriate norm for each of the auxiliary problems (thus yielding restrictions on the admissible roots and on their moduli);

(iii) the appropriate boundary conditions for each auxiliary boundary value problem must be verified by u_i^n.

In conclusion, the normal mode analysis furnishes for each auxiliary problem the set of eigenvalues that must verify the GR stability condition.

To clarify the described methodology, let us consider again (181) and let us use for simplicity the CIR scheme with the fully discretized formulation of (51). The characteristic equation is $(1 - \lambda - \sigma)k + \sigma = 0$, i.e. $k = \sigma/(\lambda + \sigma - 1)$, yielding

$$u_i^0 \sim \left(\frac{\sigma}{\lambda + \sigma - 1}\right)^i$$

where $\sigma = a\Delta t/\Delta x$ and $a > 0$.

(i)   The boundedness of $\|\mathbf{u}^0\|_\infty$ of the initial value problem requires $|k| = 1$, i.e. $k = e^{I\phi}$ $(I = \sqrt{-1})$, and

$$\lambda = (1 - \sigma) + \sigma e^{-I\phi}$$

hence, the set of all possible eigenvalues is the circle of radius σ and centre $1 - \sigma$ on the real axis of the complex plane.

(ii) For the initial right boundary value problem, the boundedness of $\|\mathbf{u}^0\|_2$ implies $|k| < 1$, and all possible λ's are now exterior to the locus of the eigenvalues of the initial value problem. The satisfaction of the (homogeneous) left boundary condition (at $i = 1$) $u_i = \lambda^n u_i^0$ can be satisfied only with $\lambda = 0$ (excluding the trivial initial solution $u_i^0 = 0$).

(iii) The initial left boundary value problem does not require any numerical boundary condition to be imposed when CIR is employed. Hence, the boundedness of $\|\mathbf{u}^0\|_2$ is ensured with $|k| > 1$, and

$$\lambda = (1 - \sigma) + \sigma/|k|e^{-I\phi}$$

represents all points internal to the locus of the eigenvalues of the initial value problem.

Then, the Godunov–Ryabenkii spectrum is

$$\mathcal{S} = \{0\} \bigcup \{\lambda : |\lambda - (1 - \sigma)| \leq \sigma\}$$

and the GR stability reduces to the standard Courant–Friedrichs–Lewy (CFL) condition $\sigma \leq 1$.

It is easy to show that the use of central discretization yields an unstable initial value problem; indeed one obtains

$$\lambda = 1 - \frac{\sigma}{2}(k - 1/k)$$

which, with $|k| = 1$, becomes

$$\lambda = 1 - I \sin\phi$$

i.e. the locus of the eigenvalues is the segment that connects the points $(1, -I)$ and $(1, I)$, yielding $\max|\lambda| = \sqrt{2}$ and the GR stability condition is not verified.

Let us now consider the semidiscretized form with central discretization, whose characteristic equation is

$$s = -\frac{\beta}{2}\left(k - \frac{1}{k}\right) \tag{193}$$

where $\beta = a/\Delta x$; hence, if k_1 and k_2 are two distinct roots of the equation one has

$$u_i(t) = e^{st}(a_1 k_1^i + a_2 k_2^i)$$

or, for $k_1 = k_2 = k$, $u_i(t) = e^{st}(a_0 + a_1^i)k^i$.

(i)   The boundedness of $\|\mathbf{u}(0)\|_\infty$ (of the auxiliary initial value problem) again restricts the solution to the case $k_1 \neq k_2$ and $|k_1| = |k_2| = 1$; in addition from (193) $k_1 k_2 = 1$. Hence, the allowed values of s are

$$s = -I\beta \sin \phi$$

and the locus of the eigenvalues (e^s) coincides with the unit circle with the exception of the intercept with the positive real axis (that corresponds to $k_1 = k_2 = \pm 1$).

(ii) The boundedness of $\|\mathbf{u}(0)\|_2$ (of the right boundary value problem) restricts the solution to the case $|k| < 1$, thus excluding one of the roots of (193) (being $k_1 k_2 = 1$). In addition, the satisfaction of the homogeneous boundary condition at $x = 0$ gives

$$u_i(t) = e^{st}k = 0$$

where $|k| < 1$, i.e. $s = -\infty$ (excluding the trivial case $k = 0$) and the set of eigenvalues of the right boundary value problem is $\{e^{-\infty}\} = \{0\}$.

(iii) Analogously, for the left boundary value problem we have one admissible root of modulus $|k| > 1$. In addition, at the right boundary $x = 1$ $(i = M)$ the numerical boundary condition requires

$$\frac{d}{dt}u_M = -\frac{\beta}{2}(u_M - u_{M-1}) \quad \text{(if one assumes zero-th order extrapolation)}$$

or

$$\frac{d}{dt}u_M = -\beta(u_M - u_{M-1}) \quad \text{(if one assumes first-order extrapolation)}$$

which, recalling (193), gives $k = 1$ for both cases.

Hence, no normal modes with $|k| > 1$ exist and the set of eigenvalues associated with the left boundary value problem is empty (for the selected numerical boundary conditions).

Then, the Godunov–Ryabenkii spectrum is

$$S = \{e^s : |e^s| = 1; \; e^s \neq 1\} \bigcup \{0\}$$

and the GR stability is verified (as already shown by means of the energy method).

Note that the above conclusion is not inconsistent with the unstable behaviour of the fully discretized formulation based on central discretization (which is equivalent to the time integration of the semidiscretized form by an explicit Euler method). Indeed, the stability analysis described here for the semidiscretized form refers to the ideal case of the exact time evolution of the perturbation of the (ideal) semidiscretized solution. Obviously, any conclusion on the stability must also take into account the time integration procedure adopted for obtaining the numerical solution (this latter point will be further elaborated upon in the next section with reference to the Runge–Kutta-like time integration approaches for semidiscretized schemes).

It is important to realize that the use of the normal mode analysis is in general rather complex. Indeed, it requires the solution of algebraic problems that are often analytically intractable for the auxiliary left and right boundary value problems. To overcome these difficulties and to extend the analysis to hyperbolic systems of conservation laws, a significant theoretical work has been carried out. In particular, weaker stability requirements have been defined and criteria for solving the algebraic issues of the normal mode analysis have been proposed. The interested reader should refer for more details to Gustafsson *et al.* (1972), Strikwerda (1980) and Goldberg and Tadmor (1985, 1987).

VII NUMERICAL SOLUTION OF CONTROL VOLUME SCHEMES

In the previous sections attention has mainly been focused on the design of the numerical flux function both for the semi- and fully discretized approaches. In the former an approximation to the weak solution requires a time integration of a system of ordinary differential equations. In a fully discretized approach a non linear system of algebraic equations may have to be solved owing to the dependence of the numerical flux function upon time-updated unknowns. For both formulations two classes of methods can in general be devised: explicit and implicit. For the sake of brevity, in the present section we restrict our discussion to multistage explicit Runge–Kutta-type algorithms and implicit single-step approaches.

A Runge–Kutta algorithms

Referring to the semidiscretized form of the equations and recalling that in many instances we are mainly concerned with steady-state solutions, one constructs the integration in time as an iteration technique, thus simultaneously ensuring: (i) convergence toward a steady solution and, more importantly, (ii) the physical reliability of the solution of the non-linear set of algebraic equations associated with the steady conservation equations. Note that the uniqueness of the solution of non-linear algebraic equations is in general not guaranteed; moreover, the convergence toward the physically reliable solution may depend on the mesh dimensions (often fine meshes are required to cure the issue, with consequences on the stability and the computational costs).

One of the primary objectives is to devise algorithms that allow time steps as large as possible (compatible with the stiffness of the system of ordinary differential equations to be solved). Multistage Runge–Kutta-like explicit schemes satisfy such a requirement, and have been extensively and successfully employed for the solution of the Euler and/or Navier–Stokes equations (Jameson, 1983a,b; Martinelli et al., 1986; Venkatakrishnan and Jameson, 1988; Volpe and Jameson, 1990; Grasso et al., 1992; Volpe, 1993; Swanson and Turkel, 1993).

Let **R** be the right-hand-side of (33) with changed sign, the so-called residual of the steady equations. The semidiscretized form of the equations can be cast in a system of autonomous ordinary differential equations

$$V\frac{d\mathbf{W}}{dt} = -\mathbf{R}(\mathbf{W}) \tag{194}$$

where V is the cell volume.

The use of an m-stage Runge–Kutta-type algorithm to advance the solution from time t_n to t_{n+1} yields

$$W^{n+k/m} = \mathbf{W}^n - \lambda\alpha_k\mathbf{R}(\mathbf{W}^{n+(k-1)/m}); \quad k = 1, m \tag{195}$$

where $\lambda = \Delta t/V$ and the values of the coefficients α determine the order of accuracy of the scheme, its stability and damping properties. Jameson (1983b) has shown that for an m-stage algorithm, the maximum CFL number that can be realized is $m - 1$.

For the purpose of determining relationships for the α's, let us consider (for the sake of simplicity) a linear one-dimensional scalar conservation law corresponding to (39). Introducing a harmonic data $w = \hat{w}e^{I\vartheta x/\Delta x}$ (with $I = \sqrt{-1}$) the Fourier transform of the scalar equation corresponding to (194) is

$$\Delta t\frac{d\hat{w}}{dt} = -z\hat{w} \tag{196}$$

where z is the Fourier symbol and $\hat{w}$ is the amplitude.

Assuming that the approximate solution is piecewise constant within each cell, the Fourier symbol corresponding to a first-order upwind scheme is

$$z = \sigma(1 - e^{-I\vartheta})$$

where σ is the CFL number ($\sigma = a\Delta t/\Delta x$).

The Fourier symbol corresponding to a central differencing scheme is

$$z = I\sigma \sin \vartheta$$

Likewise, for a higher-order upwind scheme obtained by a MUSCL-type approach one has

$$z = \sigma(1 - e^{-I\vartheta})\left[1 + \frac{1 - \eta}{4}(1 - e^{-I\vartheta}) + \frac{1 + \eta}{4}(e^{I\vartheta} - 1)\right]$$

where different upwind discretization formulas are recovered depending on the values of η.

It is easy to show that the time integration of (196) by means of the m-stage Runge–Kutta algorithm corresponding to (195) yields

$$\hat{w}^{n+1} = g\hat{w}^n$$

where g is the so-called amplification factor

$$g = 1 - \alpha_m z(1 - \alpha_{m-1}z(\cdots(1 - \alpha_2 z(1 - \alpha_1 z))\cdots))$$

It is clear that the stability and damping properties depend on g which, in turn, is a complex polynomial of z (i.e. depends on the discretization scheme), and depends on the α's. If one only requires high accuracy (of order m) it is easy to show that the latter satisfy the relation

$$\alpha_k = \frac{(m - k)!}{(m - k + 1)!} = \frac{1}{m - k + 1}$$

On the other hand, as remarked by Powell and Van Leer (1990) and Van Leer et al. (1989), perfect damping of a disturbance of a given frequency will result when z coincides with a zero of g. This can be achieved by selecting optimal Runge–Kutta coefficients, where following Powell and Van Leer (1990) we interpret optimal as meaning damping the amplitudes of the high-frequency Fourier modes to a value not exceeding a selected threshold level. Then, the optimal α's depend on the discretization scheme (i.e. the form of z) and the maximum allowed level of $\|g\|$ in the high-frequency range $[\pi/2, \pi]$.

The details of the procedure for determining the α's can be found in Powell and Van Leer (1990) and Van Leer et al. (1989), who show that, for a given spatial discretization operator, the maximum allowed CFL number of the optimized m-stage scheme is always less than the maximum CFL number. The optimal Runge–Kutta coefficients for upwind schemes are given in Table 4.

The values of these coefficients strictly hold for the scalar linear one-dimensional case; a straightforward extension of the analysis to multidimensions may lead to poor convergence (Van Leer et al., 1989). It is important to note that the optimal behaviour may be meaningful only if used in conjunction with an acceleration technique such as multigrid, where the coefficients must ensure a selective damping of high-frequency error components.

Table 4
Runge–Kutta coefficients for upwind schemes

Order	α_1	α_2	α_3	α_4	α_5	m
First	0.1481	0.4	1.0	–	–	3
	0.0833	0.207	0.4265	1.0	–	4
	0.0533	0.1263	0.2375	0.4414	1.0	5
Second	0.1918	0.4929	1.0	–	–	3
	0.1084	0.2602	0.5052	1.0	–	4
	0.0695	0.1602	0.2898	0.5060	1.0	5
Third	0.2884	0.5010	1.0	–	–	3
	0.1666	0.3027	0.5275	1.0	–	4
	0.1067	0.1979	0.3232	0.5201	1.0	5

Table 5
Runge–Kutta coefficients for JST scheme

α_1	α_2	α_3	α_4	α_5	m
0.6	0.6	1	–	–	3
1/4	1/3	1/2	1	–	4
1/4	1/6	3/8	1/2	1	5

Jameson in his early works (Jameson, 1983a, b) on the semidiscretized approach, also determined optimal Runge–Kutta coefficients when using the Jameson–Schmidt–Turkel (JST) numerical flux function corresponding to (135), together with a multigrid acceleration technique. Values of the optimal Runge–Kutta coefficients corresponding to the JST scheme are reported in Table 5.

The Fourier symbol and the stability region (defined as the locus of z for which $g(z) \leq 1$) corresponding to first- and second-order upwind discretization, and JST scheme are shown in Figures 10–12. The figures clearly show the effects of the discretization scheme and the number of Runge–Kutta stages on the region of stability and its (maximum) extension along the real and imaginary axes.

B Implicit methods

Implicit schemes represent a class of methods that, in principle, are less constrained by stability restrictions and are well suited to compute steady-state via a time marching approach in very few iterations. In the following we briefly

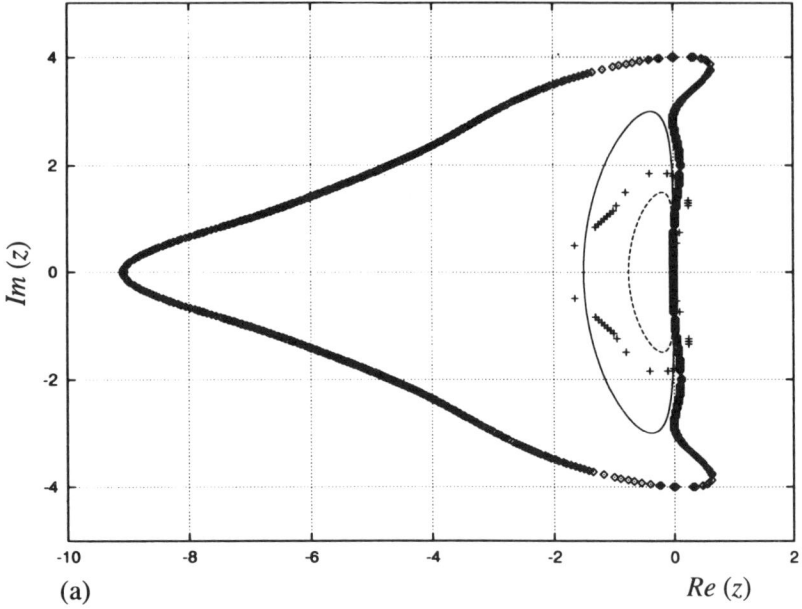

(a)

Figure 10 (a) Stability region and Fourier symbols for JST with optimal Runge–Kutta coefficients. Key: —, z (five stages); - - -, z (three stages); $\diamond\diamond\diamond$, $|g|=1$ (five stages); $+++$, $|g|=1$ (three stages).

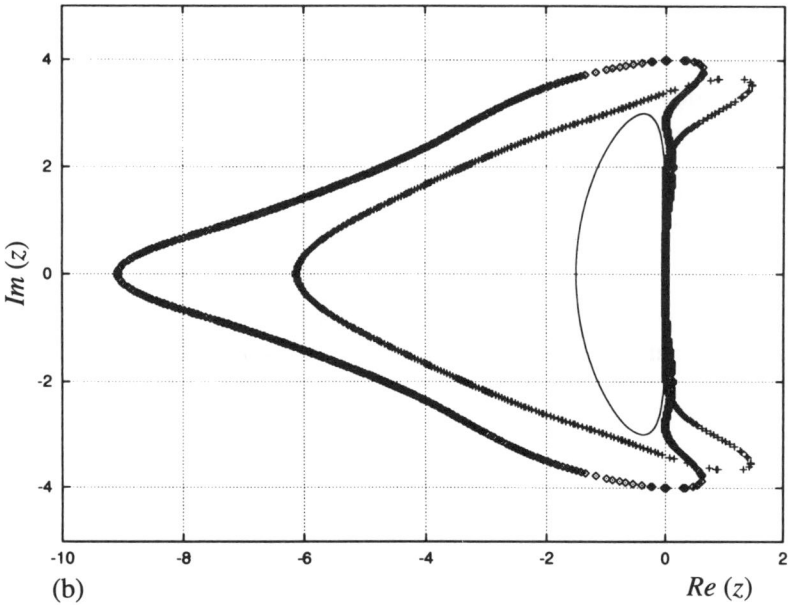

(b)

Figure 10 (b) JST scheme. Key: —, z (five stages); $\diamond\diamond\diamond$, $|g|=1$, optimal Runge–Kutta coefficients; $+++$, $|g|=1$, standard Runge–Kutta coefficients.

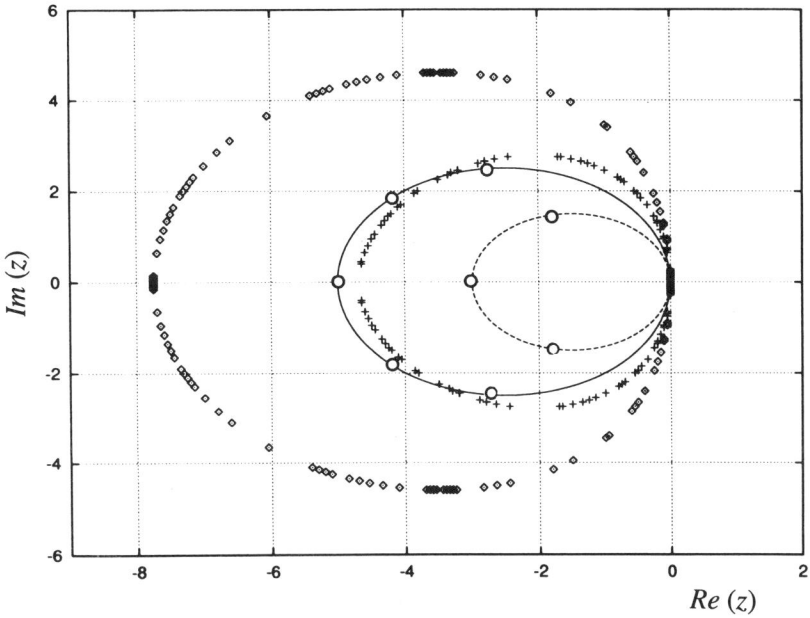

Figure 11 Stability region and Fourier symbols for first-order upwind with optimal Runge–Kutta coefficients. Key: —, z (five stages); - - -, z (three stages); $\diamond\diamond\diamond$, $|g| = 1$ (five stages); $+++$, $|g| = 1$ (three stages); $\bigcirc$, roots of $g(z) = 0$.

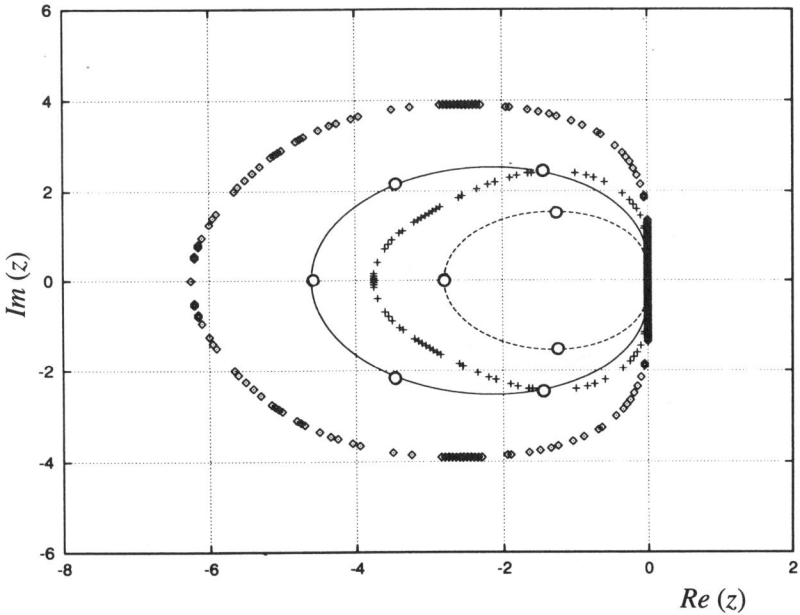

Figure 12 Stability region and Fourier symbols for second-order upwind with optimal Runge–Kutta coefficients. Key: —, z (five stages); - - -, z (three stages); $\diamond\diamond\diamond$, $|g| = 1$ (five stages); $+++$, $|g| = 1$ (three stages); $\bigcirc$, roots of $g(z) = 0$.

discuss the use of implicit methods for the solution of the Euler/Navier–Stokes equations.

Referring to the fully discretized formulation, (32) can be recast as

$$\Delta \mathbf{W} = \mathbf{W}^{n+1} - \mathbf{W}^n = -\lambda \mathbf{R}(\mathbf{W}^n, \mathbf{W}^{n+1}) \tag{197}$$

where $\mathbf{R}$ is the numerical approximation of the net flux rate, and $\lambda = \Delta t / V$. Equation (197) can be quasilinearized assuming that $\mathbf{R}$ is a differentiable function of $\mathbf{W}$ thus obtaining

$$\left[\mathbf{I} + \eta \lambda \frac{\partial \mathbf{R}}{\partial \mathbf{W}} \right] \Delta \mathbf{W} = -\lambda \mathbf{R}(\mathbf{W}^n) \tag{198}$$

where $\eta = 1$ corresponds to fully implicit schemes (reducing to the Newton method with $\Delta t \to \infty$), while $\eta = 0$ yields the fully explicit approach.

Let $\mathbf{P}$ be the matrix that premultiplies the (time) difference $\Delta \mathbf{W}$. Then (198) shows that a matrix inversion is required to update the solution. In general, for a three-dimensional problem $\mathbf{P}$ is an $N \times N \times N$ matrix of bandwidth $O(N^2)$, and the use of a direct method requires $O(N^7)$ floating point operations. This is somewhat impractical from a computational point of view. A reduction in the number of operation counts can be achieved by means of a factorization of $\mathbf{P}$.

In the following we describe two approximate factorization approaches that are usually employed in implicit methods: the alternating direction implicit scheme (ADI) and the lower–upper (LU) decomposition approach.

1 Alternating direction implicit method

Let us assume, for the sake of simplicity, that the structured grid is made out of regular control volumes (parallelepipeds in three-dimensions and rectangles in two-dimensions), and let $\mathbf{R} = \mathbf{R}_x + \mathbf{R}_y + \mathbf{R}_z$, where $\mathbf{R}_x (\mathbf{R}_y; \mathbf{R}_z)$ represent the contribution in the x (y; z) direction. The alternating direction implicit scheme is obtained by replacing the matrix $\mathbf{P}$ by the product of three matrices (Beam and Warming, 1976; Pulliam, 1985, 1993)

$$\left(\mathbf{I} + \lambda \frac{\partial \mathbf{R}_x}{\partial \mathbf{W}} \right) \left(\mathbf{I} + \lambda \frac{\partial \mathbf{R}_y}{\partial \mathbf{W}} \right) \left(\mathbf{I} + \lambda \frac{\partial \mathbf{R}_z}{\partial \mathbf{W}} \right) \Delta \mathbf{W} = -\lambda \mathbf{R}(\mathbf{W}^n) \tag{199}$$

where η has been set equal to 1. The solution of (199) may then proceed in three steps

$$\left(\mathbf{I} + \lambda \frac{\partial \mathbf{R}_x}{\partial \mathbf{W}} \right) \Delta \mathbf{W}^* = -\lambda \mathbf{R}(\mathbf{W}^n)$$

$$\left(\mathbf{I} + \lambda \frac{\partial \mathbf{R}_y}{\partial \mathbf{W}} \right) \Delta \mathbf{W}^{**} = \Delta \mathbf{W}^*$$

$$\left(\mathbf{I} + \lambda \frac{\partial \mathbf{R}_z}{\partial \mathbf{W}} \right) \Delta \mathbf{W} = \Delta \mathbf{W}^{**}$$

The method is computationally very efficient and it can be easily vectorized. However, according to a Von Neumann analysis, it is only conditionally stable when applied to the three-dimensional Euler equations (Jameson and Turkel, 1981; Peyret and Taylor, 1983; Yoon, 1985). The convergence properties of ADI methods are strongly affected by $O(\Delta t^2)$ factorization errors: as the time step increases these errors may dominate with a loss in convergence. Moreover, the method is strongly affected by the boundary conditions to be imposed on the intermediate variables $\Delta W^*, \Delta W^{**}$. For more details the reader is referred to Pulliam (1985).

2 Lower–Upper factorization

The LU method is generally based on a lower and upper decomposition of the matrix P that: (i) always produces two factors independently of the number of space dimensions; (ii) it only requires inversion of 5×5 (4×4 in two dimensions) matrices; (iii) can be made unconditionally stable; and (iv) furnishes steady-state solutions independent of the time step.

For the sake of deriving an LU formulation for a two-dimensional case (the extension to three-dimensions is straightforward), let A and K_x be, respectively, the x-components of the inviscid and viscous flux jacobians; B and K_y are the analogous components in the y-direction. Referring to Figure. 13, (199) becomes

$$\Delta W_{i,j} + \eta \lambda_{ij} \sum_{\beta=1,4} [(An_x + Bn_y)\Delta s]_\beta \Delta W_\beta$$

$$+ \eta \lambda \sum_{\beta=1,4} [(K_x n_x + K_y n_y)\Delta s]_\beta \Delta W_\beta = -\lambda_{ij} R_{i,j}^n \quad (200)$$

where Δs_β represents the length of the line segment of cell face β.

Let $\widehat{A}$ and $\widehat{B}$ represent, respectively, the normal flux jacobian at cell faces $i \pm \frac{1}{2}$ (i.e., $\beta = 1, 3$) and $j \pm \frac{1}{2}$ (i.e., $\beta = 2, 4$). To construct the lower and upper factors one can first split the inviscid jacobian matrices into positive and negative contributions

$$(\widehat{A}\Delta W)_{i+\frac{1}{2},j} = \widehat{A}_{i,j}^+ \Delta W_{i,j} + \widehat{A}_{i+1,j}^- \Delta W_{i+1,j}$$

$$(\widehat{A}\Delta W)_{i-\frac{1}{2},j} = \widehat{A}_{i-1,j}^+ \Delta W_{i-1,j} + \widehat{A}_{i,j}^- \Delta W_{i,j}$$

$$(\widehat{B}\Delta W)_{i,j+\frac{1}{2}} = \widehat{B}_{i,j}^+ \Delta W_{i,j} + \widehat{B}_{i,j+1}^- \Delta W_{i,j+1}$$

$$(\widehat{B}\Delta W)_{i,j-\frac{1}{2}} = \widehat{B}_{i,j-1}^+ \Delta W_{i,j-1} + \widehat{B}_{i,j}^- \Delta W_{i,j}$$

Diagonal dominance is generally required on the lower and upper factors (Yoon, 1985; Pulliam, 1985; Jameson and Yoon, 1987; Rieger and Jameson, 1988; Yoon and Kwak, 1991, 1994; Grasso and Marini, 1992) and consequently

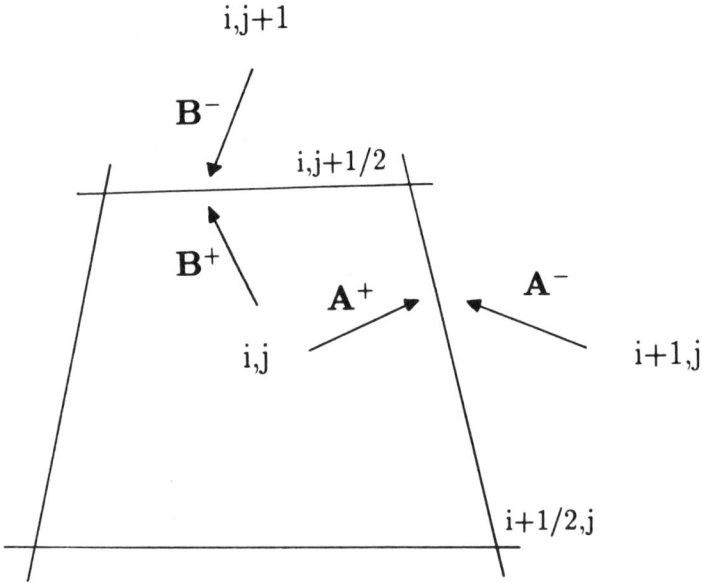

Figure 13 Two-dimensional cell for lower–upper (LU) factorization.

the $\widehat{\mathbf{A}}^{\pm}$ and $\widehat{\mathbf{B}}^{\pm}$ are constructed so that the positive (negative) matrices have non negative (positive) eigenvalues, i.e.

$$\widehat{\mathbf{A}}^{\pm} = \mathbf{L}_A^{-1}\boldsymbol{\Lambda}_A^{\pm}\mathbf{L}_A, \qquad \widehat{\mathbf{B}}^{\pm} = \mathbf{L}_B^{-1}\boldsymbol{\Lambda}_B^{\pm}\mathbf{L}_B$$

where

$$\boldsymbol{\Lambda}_A^{\pm} = (\boldsymbol{\Lambda}_A \pm g(\boldsymbol{\Lambda}_A))/2, \qquad \boldsymbol{\Lambda}_B^{\pm} = (\boldsymbol{\Lambda}_B \pm g(\boldsymbol{\Lambda}_B))/2$$

and

$$g(\boldsymbol{\Lambda}) \geq |\boldsymbol{\Lambda}|$$

The evaluation of the viscous flux jacobians is rather complicated; however, approximate jacobians can be used (Beam and Warming 1976, 1978; Yoon, 1985; Yoon and Jameson, 1988; Pulliam, 1985; Grasso and Marini, 1993a). For example, if one assumes for $\mathbf{K}_x$ and $\mathbf{K}_y$ the following diagonal forms

$$\mathbf{K}_x = k_x\mathbf{I}, \quad \mathbf{K}_y = k_y\mathbf{I}$$

where k_x and k_y are the spectral radii of the viscous jacobians, and further assuming that the grid is equally spaced in the x and y directions, one has

$$k_x = \frac{2\mu}{\rho\Delta x}, \quad k_y = \frac{2\mu}{\rho\Delta y}$$

Substitution of the above expressions in (200) yields the lower and upper factorization formulation

$$\mathbf{LU}\Delta\mathbf{W} = -\lambda\mathbf{R} \tag{201}$$

where $\lambda = \Delta t/\Delta x\Delta y$ and

$$
\begin{aligned}
\mathbf{L} = \mathbf{I} + \lambda\Big[& (\widetilde{\mathbf{A}}^+\Delta\mathbf{W})_{i,j}\Delta y - (\widehat{\mathbf{A}}^+\Delta\mathbf{W})_{i-1,j}\Delta y \\
& + (\widetilde{\mathbf{B}}^+\Delta\mathbf{W})_{ij}\Delta x - (\widehat{\mathbf{B}}^+\Delta\mathbf{W})_{i,j-1}\Delta x\Big]
\end{aligned}
\tag{202}
$$

$$
\begin{aligned}
\mathbf{U} = \mathbf{I} + \lambda\Big[& (\widehat{\mathbf{A}}^-\Delta\mathbf{W})_{i+1,j}\Delta y - (\widetilde{\mathbf{A}}^-\Delta\mathbf{W})_{ij}\Delta y \\
& + (\widehat{\mathbf{B}}^-\Delta\mathbf{W})_{i,j+1}\Delta x - (\widetilde{\mathbf{B}}^-\Delta\mathbf{W})_{ij}\Delta x\Big]
\end{aligned}
\tag{203}
$$

with

$$\widetilde{\mathbf{A}}^\pm = \widehat{\mathbf{A}}^\pm \pm k_x\mathbf{I}, \qquad \widetilde{\mathbf{B}}^\pm = \widehat{\mathbf{B}}^\pm \pm k_y\mathbf{I}$$

Relaxation parameters may also be introduced to further increase the diagonal dominance of the jacobian matrices, amounting to

$$\widehat{\mathbf{A}}^\pm = \omega_A^e(\widehat{\mathbf{A}} \pm g(\widehat{\mathbf{A}}))/2; \quad \widehat{\mathbf{B}}^\pm = \omega_B^e(\widehat{\mathbf{B}} \pm g(\widehat{\mathbf{B}}))/2$$
$$\mathbf{K}_x = \omega_x^v k_x\mathbf{I}; \qquad \mathbf{K}_y = \omega_y^v k_y\mathbf{I}$$

where ω^e and ω^v play the role of over and under relaxation parameters; typically $\omega^e = O(2)$ and $\omega^v = O(1/\omega^e)$.

The solution can be updated in two sweeps: the so-called 'forward' and 'backward' sweeps that require two 4×4 (5×5 in three-dimensions) matrix inversions

$$
\begin{aligned}
\mathbf{L}\Delta\mathbf{W}^* &= -\lambda\mathbf{R} \qquad \text{forward sweep} \\
\mathbf{U}\Delta\mathbf{W} &= -\Delta\mathbf{W}^* \qquad \text{backward sweep}
\end{aligned}
$$

It must be observed that if one uses an implicit algorithm one also faces the problem of reducing the operation counts. A more efficient LU decomposition can be obtained by means of a symmetric successive over-relaxation technique, thus eliminating the need for the 4×4 matrix inversion (Yoon, 1985; Rieger and Jameson, 1988; Yoon and Jameson, 1988; Yoon and Kwak, 1991; Grasso and Marini, 1993a). Let $\mathbf{N}, \mathbf{M}^+$ and $\mathbf{M}^-$ be the following matrices

$$\mathbf{N} = \mathbf{I} + \lambda[(\widetilde{\mathbf{A}}_{i,j}^+\Delta y - \widetilde{\mathbf{A}}_{i,j}^-\Delta y) + (\widetilde{\mathbf{B}}_{i,j}^+\Delta x - \widetilde{\mathbf{B}}_{i,j}^-\Delta x)]$$

$$\mathbf{M}^+\Delta\mathbf{W} = \lambda[-\widehat{\mathbf{A}}_{i-1,j}^+\Delta y\Delta\mathbf{W}_{i-1,j} - \widehat{\mathbf{B}}_{i,j-1}^+\Delta x\Delta\mathbf{W}_{i,j-1}]$$

$$\mathbf{M}^-\Delta\mathbf{W} = \lambda[\widehat{\mathbf{A}}_{i+1,j}^-\Delta y\Delta\mathbf{W}_{i+1,j} + \widehat{\mathbf{B}}_{i,j+1}^-\Delta x\Delta\mathbf{W}_{i,j+1}]$$

Equation (201) can then be recast as

$$(\mathbf{N} + \lambda\mathbf{M}^+)\mathbf{N}^{-1}(\mathbf{N} + \lambda\mathbf{M}^-)\Delta\mathbf{W} = -\lambda\mathbf{R}$$

where the matrix $\mathbf{N}$ is the scalar diagonal matrix

$$\mathbf{N} = [1 + \lambda(\rho_A\Delta y + \rho_B\Delta x)]\mathbf{I}$$

which is easily obtained if the matrices $g(\widehat{\mathbf{A}})$ and $g(\widehat{\mathbf{B}})$ are the diagonal matrices

$$g(\widehat{\mathbf{A}}) = \rho_A\mathbf{I}, \qquad g(\widehat{\mathbf{B}}) = \rho_B\mathbf{I}$$

where

$$\rho_A = r_A \max (|\lambda_{\hat{A}}|), \quad \rho_B = r_B \max (|\lambda_{\hat{B}}|)$$

and $\lambda_{\hat{A}}$ ($\lambda_{\hat{B}}$) are the eigenvalues of $\widehat{\mathbf{A}}$ ($\widehat{\mathbf{B}}$), and r_A (r_B) is a constant of $O(1)$. Hence, no block inversion is required and the solution is obtained as follows:

- *Forward sweep*

$$\Delta\mathbf{W}^*_{ij} = -\lambda\mathbf{R}_{ij} + \lambda\mathbf{N}^{-1}(\widehat{\mathbf{A}}^+_{i-1j}\Delta y\Delta\mathbf{W}^*_{i-1j} + \widehat{\mathbf{B}}^+_{ij-1}\Delta x\Delta\mathbf{W}^*_{ij-1})$$

Referring to Figure 14, boundary conditions on $\Delta\mathbf{W}^*$ are needed at the left ($i = 1$) and bottom ($j = 1$) boundaries, where usually one sets $\Delta\mathbf{W}^* = 0$.

- *Backward sweep*

$$\Delta\mathbf{W}_{ij} = \mathbf{N}^{-1}\left[\Delta\mathbf{W}^*_{ij} - \lambda(\widehat{\mathbf{A}}^-_{i+1j}\Delta y\Delta\mathbf{W}_{i+1j} + \widehat{\mathbf{B}}^-_{ij+1}\Delta x\Delta\mathbf{W}_{ij+1})\right]$$

Boundary conditions are needed at the top and right boundaries, where one usually imposes $\Delta\mathbf{W} = 0$.

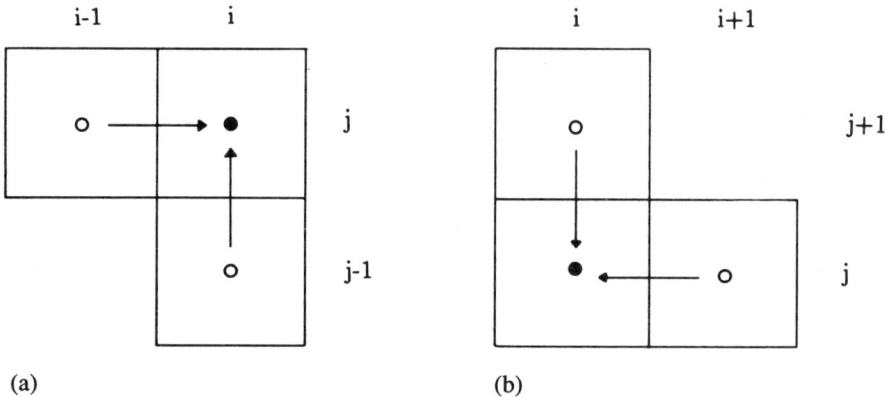

Figure 14 Schematic of sweeping modalities. (a) Forward sweep; (b) backward sweep.

Once $\Delta \mathbf{W}$ is computed the solution is updated as follows

$$\mathbf{W}^{n+1} = \mathbf{W}^n + \Delta \mathbf{W}$$

The three-dimensional extension of the method is straightforward.

C Multigrid techniques

The solution of either the Euler or the Navier–Stokes equations is generally dominated by local phenomena, thus requiring highly accurate schemes and very fine grids. However, high accuracy can in general be obtained only at the expense of computational efficiency. In the present section we briefly review the underlying concepts behind the multigrid, a technique that is generally employed to accelerate the convergence toward the steady state and to increase the computational efficiency of numerical schemes.

The multigrid technique was originally developed in the mid-1960s as a technique for solving linear elliptic equations (Federenko 1964). In the mid-1970s Brandt (1977) and Hackbush (1977) extended the methodology to the solution of non-linear partial differential equations, achieving a factor of 10 in convergence acceleration. The success in using multigrid in computational fluid dynamics is due to Jameson, who was a pioneer in extending the technique for computing the Euler equations (Jameson, 1983a, b, 1985; Jameson et al., 1986; Jameson and Yoon, 1987; Volpe and Jameson, 1990) and subsequently the Navier–Stokes equations for transonic and supersonic flows (Martinelli et al., 1986; Martinelli and Jameson, 1988; Yoon and Jameson, 1988). Recently, successful applications of the technique for hypersonic flows have also been reported (Koren and Hemker, 1991; Radespiel and Swanson, 1991; Vatsa et al., 1993; Leclercq and Stoufflet, 1993; Grasso and Marini, 1993b, 1995).

The main underlying idea of multigrid consists in transferring some of the tasks associated with the solution procedure to a sequence of successively coarser meshes (Brandt, 1977; Jameson, 1983b, 1985). As a consequence: (i) the time step can be successively increased on the coarse grids without violating the stability restriction; (ii) the number of floating point operations is reduced; and (iii) a faster convergence is achieved.

The multigrid can be interpreted as a general methodology applicable to different classes of flow problems as long as the transfer operators account correctly for their different characters. In particular, a multigrid technique should damp (optimally) all frequencies of the error spectra associated with the space discretization operator. High frequencies are generally damped in a few iterations on the given grid by exploiting the damping properties of the selected iteration technique (for example the Runge–Kutta), while the low frequencies are damped via the multigrid strategy on the coarser grids.

We now describe the basic principles of Full Approximation Scheme–Full Multigrid (FAS-FMG) approaches for solving the compressible Euler and/or

Navier–Stokes equations (for a broader overview, see the work of Brandt (1977) or Jameson (1985)). A FAS-FMG strategy amounts to introducing a sequence of successively coarser meshes, whereby a multigrid level is identified by a grid of the sequence. For each grid one introduces another sequence of successively coarser meshes where a full approximation scheme is applied.

Two approaches can be generally followed (or combinations of the two) for generating the coarse grid sequence: the so-called full- and semi-coarsening. The former amounts to generating coarser grids by doubling the mesh spacing in all coordinate directions (i.e. halving the number of grid points), thus generating, in two-dimensions, grids of (m,n), $(m/2,n/2)$, $(m/4,n/4),\ldots$, cells. In the semi-coarsening approach the sequence of coarser grids is obtained by halving the number of mesh points only in certain coordinate directions or selectively alternating the coarsening coordinate directions. Radespiel and Swanson (1991) have shown that, for two-dimensional applications, semicoarsening converges faster than full-coarsening. However, its extension to three-dimensional problems is not immediate with regard to the nesting of the grids.

The most commonly used grid sequences are the so-called V and W cycles. The former consists of a sequence of steps corresponding to: (i) evolution of the fine grid solution; (ii) fine grid solution residual restrictions; (iii) coarse grid evolution; and (iv) coarse grid correction prolongation. This sequence of operations is performed for a certain number of cycles; the solution is then transfered to the next finer grid (of the multigrid sequence) where the same steps are repeated until the finest multigrid level is reached. The W cycle consists of the same sequence of operations with additional V cycles on the intermediate coarse grids of any multigrid level (the sequence of typical operations is illustrated in Figure 15).

Let us now describe in some detail a V cycle multigrid strategy for two-dimensional problems (the extension to three dimensions is straightforward). If h and $2h$ indicate two successive grid levels, a V cycle strategy for a semidiscrete algorithm can then be formulated as follows:

- *Fine grid solution*

$$V_h \frac{\mathrm{d}}{\mathrm{d}t} \mathbf{W}_h + \mathbf{R}_h = 0 \tag{204}$$

where $\mathbf{R}_h$ represents the residual of the equation, and V_h is the cell volume. Let $\mathbf{W}'_h$ represent the approximation to the solution of the conservation equations obtained by integrating (204) from time t_n to t_{n+1}.

- *Restriction phase*
  The coarse-grid solution is initialized by restricting $\mathbf{W}'_h$ onto the $2h$ grid

$$\mathbf{W}_{2h} = \hat{I}_h^{2h} \mathbf{W}'_h$$

where, referring to Figure 16a, the restriction operator $\hat{I}_h^{2h}$ is usually defined

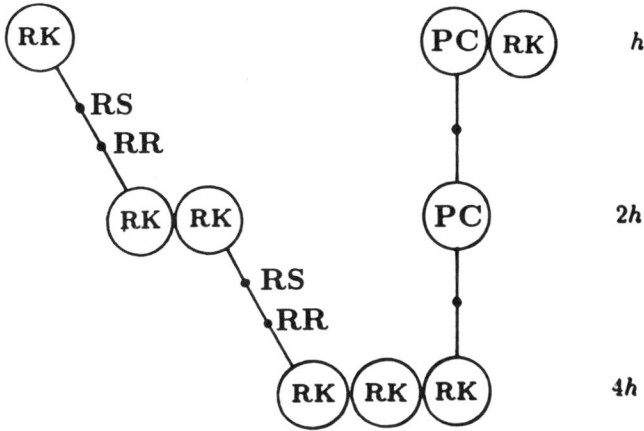

Figure 15 Multigrid V cycle strategy. Key: RR, residual restriction; RS, solution restriction; RK, Runge–Kutta iteration; PC, correction prolongation; PS, solution prolongation.

as

$$\hat{I}_h^{2h}(\cdot) = \frac{\sum_h V_h(\cdot)}{\sum_h V_h} \tag{205}$$

To maintain the accuracy of the h-grid its residual is restricted onto the $2h$-grid and a forcing function is defined according to

$$\mathbf{F}_{2h} = \bar{I}_h^{2h}\mathbf{R}_h(\mathbf{W}_h') - \mathbf{R}_{2h}(\hat{I}_h^{2h}\mathbf{W}_h') \tag{206}$$

where $\bar{I}_h^{2h} = \sum_h(\cdot)$.

- *Coarse-grid evolution*
  The coarse-grid evolves according to the following equation

$$V_{2h}\frac{\mathrm{d}}{\mathrm{d}t}\mathbf{W}_{2h} + \mathbf{R}_{2h}'' = 0 \tag{207}$$

where $\mathbf{R}_{2h}'' = \mathbf{R}_{2h} + \mathbf{F}_{2h}$.

The task demanded of the coarse grid is that of damping the low-frequency errors. Therefore, one simply requires the coarse-grid algorithm to be robust. The accuracy requirement on the space-discretization (i.e. on the numerical flux function) may then be relaxed to an order lower than the fine-grid one, the overall accuracy being maintained by the residual restriction. It is standard practice to employ for example: first-order TVD schemes on the coarse grid and second- or higher-order TVD schemes on the fine one; or first-order JST switch method (obtained by neglecting the third-order dissipation term) on the coarse grid and the standard JST scheme on the fine grid.

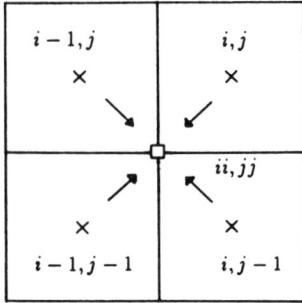

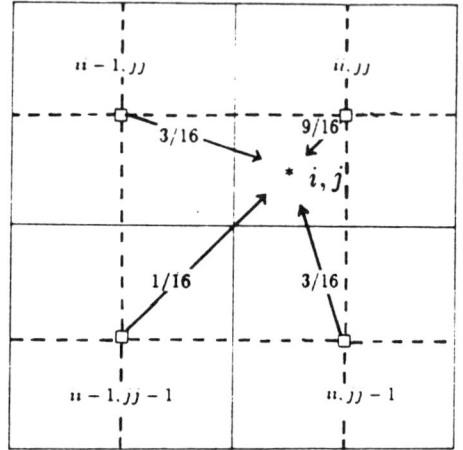

(a)   (b)

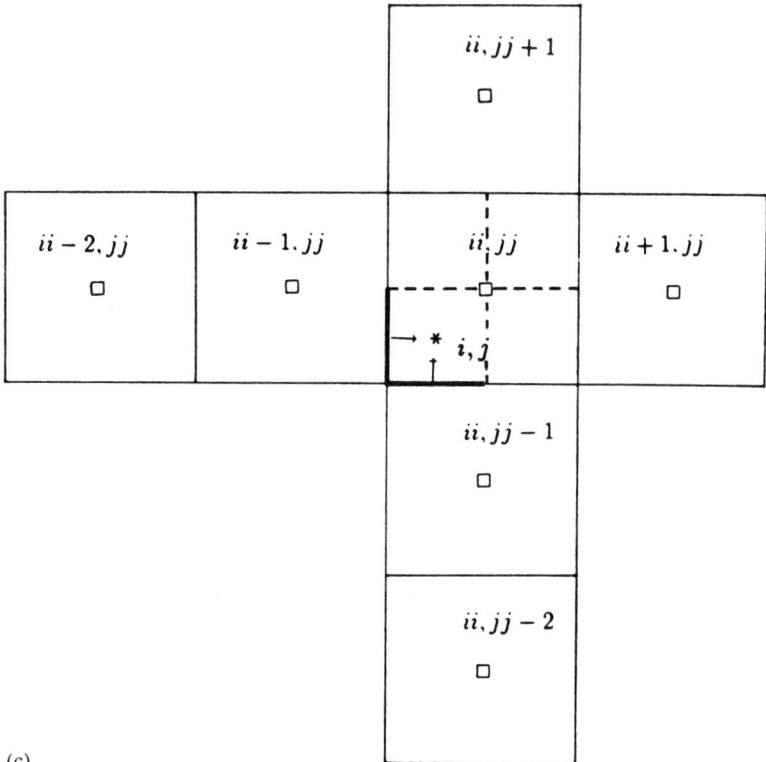

(c)

Figure 16 (a) Conservative (geometrical) restriction. ×, fine-grid centroid; □, coarse-grid centroid. Correction prolongation: (b) bilinear interpolation; (c) MUSCL interpolation.

It is important to point out that since the evolution on the coarse grids is driven by the residuals restricted from the next finer grid, the solution on the fine grid is independent of the choice of the coarse grid boundary conditions. Usually the wall condition is treated in the same way on all grids, while the other boundary conditions (e.g. inflow, outflow, etc.) are either recalculated or transferred from the next finer grid.

- *Coarse-grid correction prolongation*
Let $\mathbf{W}'_{2h}$ be the approximate solution of (207), the coarse grid correction $\Delta\mathbf{W}_{2h}$ is prolongated onto the fine grid

$$\mathbf{W}_h = \mathbf{W}'_h + \mathcal{P}^h_{2h}\Delta\mathbf{W}_{2h} = \mathbf{W}'_h + \mathcal{P}^h_{2h}(\mathbf{W}'_{2h} - \hat{I}^{2h}_h\mathbf{W}'_h) \qquad (208)$$

It must be pointed out that the prolongation step is a critical one, at least for supersonic (and even more so for hypersonic) flow simulations. Indeed, the domain of dependence should not be altered by the coarse-grid correction transfer.

For transonic and (low) supersonic flows the classical grid transfer operator defined in terms of the bilinear interpolation operator $\mathcal{B}^h_{2h}$ has been successfully used by several authors (Jameson, 1985; Yoon, 1985; Martinelli et al., 1986; Jameson and Yoon, 1987; Caughey, 1988, 1993; Martinelli and Jameson, 1988; Radespiel and Swanson, 1991; Swanson and Radespiel, 1991; Swanson and Turkel, 1992; Radespiel et al., 1993). Referring to Figure 16b, the operator is

$$\mathcal{B}^h_{2h}(\cdot) = \tfrac{9}{16}(\cdot)_{ii,jj} + \tfrac{3}{16}(\cdot)_{ii-1,jj} + \tfrac{3}{16}(\cdot)_{ii,jj-1} + \tfrac{1}{16}(\cdot)_{ii-1,jj-1} \qquad (209)$$

To damp high-frequency errors and make the scheme more robust, the prolongation operator can be modified by smoothing operations

$$\mathcal{P}^h_{2h} = \left(1 - \epsilon_c(a_j\delta^+_j - b_j\delta^-_j)\right)\left(1 - \epsilon_c(a_i\delta^+_i - b_i\delta^-_i)\right)\mathcal{B}^h_{2h} \qquad (210)$$

where

$$\delta^+_i = (\cdot)_{i+1,j} - (\cdot)_{i,j}; \quad \delta^+_j = (\cdot)_{i,j+1} - (\cdot)_{i,j}$$

and ϵ_c is a constant (usually 0.1–0.2).

The coefficients a and b may take any value between zero and one: the standard smoothing operator is recovered if $a_i = a_j = b_i = b_j = 1$; an upwind-type residual smoothing (Blazek et al., 1991; Grasso and Marini, 1995) is obtained for example if $a_i = a_j = 0, b_i = b_j = 1$; backward–forward smoothing (Zhu et al., 1993) is recovered with $a_i = \epsilon_{f_i}, b_i = \epsilon_{b_i}$, and likewise $a_j = \epsilon_{f_j}, b_j = \epsilon_{b_j}$.

For high Mach number flows, or even when using the multigrid technique in conjunction with an upwind scheme, the prolongation operator must have a directionality character (Blazek et al., 1991; Koren and Hemker, 1991;

Leclercq and Stoufflet, 1993; Grasso and Marini, 1993b, 1995). Under these conditions one may introduce a characteristic decomposition of the coarse grid correction with an 'upwind' smoothing operation. For example, in a one-dimensional flow this can be achieved by introducing a characteristic decomposition of the coarse grid correction, i.e.

$$
\mathcal{P}_{2h}^h(\mathbf{W}_{2h}' - \hat{I}_h^{2h}\mathbf{W}_h') = \mathbf{L}_h^{-1}\left[\mathbf{I} - \epsilon_c\left(\frac{\mathbf{I} - \mathrm{sign}(\Lambda_h)}{2}\delta^+ - \frac{\mathbf{I} + (\Lambda_h)}{2}\delta^-\right)\right]I_{2h}^h\alpha_{2h}
$$

$$(211)$$

where α_{2h} is the vector of the coarse grid characteristic correction and Λ is the diagonal jacobian of the inviscid flux. However, for vanishing eigenvalues the operator reduces to the standard one.

In two dimensions one may introduce an approximate factorization of the involved operators thus having

$$
\Delta\mathbf{W}_h^* = \left\{\mathbf{L}_h^{-1}\left[\mathbf{I} - \epsilon_c\left(\frac{\mathbf{I} - \mathrm{sign}(\Lambda_h)}{2}\delta^+ - \frac{\mathbf{I} + \mathrm{sign}(\Lambda_h)}{2}\delta^-\right)\right]I_{2h}^h\alpha_{2h}\right\}_i
$$

$$(212)$$

$$
\Delta\mathbf{W}_h = \left\{\mathbf{L}_h^{-1}\left[\mathbf{I} - \epsilon_c\left(\frac{\mathbf{I} - (\Lambda_h)}{2}\delta^+ - \frac{\mathbf{I} + (\Lambda_h)}{2}\delta^-\right)\right]\alpha_h^*\right\}_i
$$

where

$$
\alpha_h^* = \mathbf{L}_h\Delta\mathbf{W}_h^*
$$

and the i and j refer to the factorization directions. Hence

$$
\mathbf{W}_h = \mathbf{W}_h' + \Delta\mathbf{W}_h
$$

If one assumes for the transfer operator $I_{2h}^h = \mathcal{B}_{2h}^h$, then the directionality of the prolongation is obtained only through the use of the directional smoothing. On the other hand one can construct a directional transfer operator by means of an upstream monotone reconstruction of the coarse grid characteristic correction. Referring to Figure 16c, following the work of Koren and Hemker (1991) and Grasso and Marini (1995), an unsmoothed upstream monotone reconstruction of the coarse grid correction amounts to

$$
\mathbf{W}_{h_{i,j}} = \mathbf{W}_{h_{i,j}}' + \frac{1}{2}\left(\Delta^M\mathbf{W}_{2h_{ii-\frac{1}{2},jj}} + \Delta^M\mathbf{W}_{2h_{ii,jj-\frac{1}{2}}}\right)
$$

$$
= \mathbf{W}_{h_{i,j}}' + \mathbf{M}_{2h}^h\Delta\mathbf{W}_{2h}
$$

where $\mathbf{M}_{2h}^h$ represents the operator that yields a monotone upstream reconstruction and

$$\Delta^M \mathbf{W}_{2h_{ii-\frac{1}{2},jj}} = \frac{1}{2}\left(\Delta \mathbf{W}^R_{2h_{ii-\frac{1}{2},jj}} + \Delta \mathbf{W}^L_{2h_{ii-\frac{1}{2},jj}}\right)$$

$$\Delta^M \mathbf{W}_{2h_{ii,jj-\frac{1}{2}}} = \frac{1}{2}\left(\Delta \mathbf{W}^R_{2h_{ii,jj-\frac{1}{2}}} + \Delta \mathbf{W}^L_{2h_{ii,jj-\frac{1}{2}}}\right)$$

The left $(\Delta \mathbf{W}^L_{2h})$ and right $(\Delta \mathbf{W}^R_{2h})$ state corrections are, for example, evaluated by a limited third-order upwind monotone reconstruction (see 131), yielding

$$\Delta \mathbf{W}^L_{2h_{ii-\frac{1}{2},jj}} = \Delta \mathbf{W}_{2h_{ii-1,jj}} + \frac{1}{2}\left(\frac{1}{3}\overline{\delta}^- + \frac{2}{3}\hat{\delta}^+\right)\Delta \mathbf{W}_{2h_{ii-1,jj}}$$

$$\Delta \mathbf{W}^R_{2h_{ii-\frac{1}{2},jj-\frac{1}{2}}} = \Delta \mathbf{W}_{2h_{ii,jj}} - \frac{1}{2}\left(\frac{1}{3}\overline{\delta}^+ + \frac{2}{3}\hat{\delta}^-\right)\Delta \mathbf{W}_{2h_{ii,jj}}$$

In a compact form, introducing a smoothing operation the monotone upstream prolongation phase reduces to

$$\mathbf{W}_h = \mathbf{W}'_h + S_h M^h_{2h}\Delta \mathbf{W}_{2h} \tag{213}$$

where S_h is the directional smoothing operator (for example the one corresponding to (212)). The strategy can easily be extended to three-dimensional flows.

The damping properties of the grid transfer operators can be evaluated by means of the Fourier analysis. Details can be found for example in Jameson (1985), Radespiel and Swanson (1991), Grasso and Marini (1993b, 1995).

VIII APPLICATIONS

In the present section we list some applications of the methods that have been discussed in this chapter: the list is not exhaustive and is limited to recent applications for either inviscid or laminar flows.

Scott and Niu (1993) have presented a comparative study of the influence of limiters used in explicit MUSCL-type flux splitting algorithm to resolve accurately expansion fans, slip lines and shock waves.

Venkatakrishnan (1993) has addressed the issue of high accuracy and convergence requirements when using monotone upwind schemes.

Yoon and Kwak (1991) have presented an implicit multigrid method for the solution of the three-dimensional Euler equations by means of a lower–upper symmetric Gauss–Seidel relaxation scheme.

A diagonal alternating direction implicit multigrid algorithm based on a symmetric total variation diminishing discretization of the inviscid flux has been developed by Caughey (1993) for the solution of the inviscid transonic and supersonic flows around airfoils.

Zha and Bilgen (1993) have applied a flux vector splitting scheme to the solution of one-, two- and three-dimensional Euler equations. The scheme is based on a splitting that uses the velocity component normal to the volume interface as the characteristic speed and yields sharp monotonic shocks.

Vanden and Belk (1993) have studied the formation of asymmetric vortices over an ogive cylinder in supersonic flow by means of an implicit approximate factorization method and a second-order Roe's approximate Riemann solver.

Tysinger and Caughey (1991) have developed a multigrid (diagonal) alternating direction implicit method for the solution of the Navier–Stokes equations. They have shown that for laminar flows accounting for the viscous fluxes in the implicit factor improves the stability of the scheme.

Radespiel *et al.* (1990) have applied an explicit multigrid scheme to compute two- and three-dimensional viscous transonic flows, whereby the classical Jameson flux function is used for the inviscid flux.

Swanson and Turkel (1993) have addressed (some of) the aspects of high resolution for the Navier–Stokes equations. The issue of high-resolution schemes for the compressible Euler and Navier–Stokes equations is also addressed by Yamamoto and Daiguji (1993), who propose a fifth-order compact upwind scheme and a fourth-order compact MUSCL TVD scheme that increase the resolution of weak discontinuities, vortices, and shocks.

Simpson and Whitfield (1992) have developed an algorithm for the solution of the unsteady (thin-layer) Navier–Stokes equations whereby the numerical function corresponding to the inviscid flux is based on Roe's approximate Riemann solver, and have presented applications for oscillating airfoils and wings at transonic Mach numbers.

Volpe and Jameson (1990) have computed transonic and supersonic flows about realistic fighter configurations by means of an efficient method based on a structured multiblock approach, and have shown that the use of composite meshes is indeed appropriate for complex shapes, whereby grids are generated for each individual aircraft component such as wing, nacelle, tails, fuselage, etc. (Rai, 1986; Volpe, 1993).

To conclude the list of applications, in the last decade application codes, which exploit some of the principles and techniques discussed in this chapter, have been produced and widely applied for research and development.

IX CONCLUDING REMARKS

In the present chapter we have outlined (some of) the critical issues that arise in the numerical solution of the Euler and Navier–Stokes equations for compressible flows, and have analysed in some detail the design principle that should guide in: (i) designing the numerical flux function (i.e. the numerical approximation of inviscid/viscous fluxes); (ii) devising the computational strategy (e.g. explicit/implicit time integration, acceleration technique, etc.); and (iii) analysing the stability properties of a selected algorithm with due account of the boundary conditions.

It is important to point out that the numerical simulation of any fluid dynamic phenomena should always be accompanied (preceded) by a validation phase. Indeed, the aspect of validation should not be neglected, owing to its importance in the development of computational fluid dynamics tools.

Validation involves both numerical and physical aspects. The former requires the assessment of the sensitivity of the solution to the numerical algorithm, grid resolution and its geometry. Moreover, the sensitivity of the solutions to the parameters that affect the accuracy of the scheme should be investigated for an estimate of the error magnitude.

The physical aspect of validation implies that the experimental and theoretical knowledge of the phenomena must be taken into account when analysing the computational solution: one should always pose the question as to whether or not the computed solution is physically correct. To answer such a question one must guarantee that the computed results satisfy consistency constraints that have to come from theoretical and/or experimental knowledge: a successful simulation always requires a careful analysis of the physical problem.

REFERENCES

Abgrall, R. (1992). *INRIA Report 1584.*

Abgrall, R. (1994). *J. Comput. Phys.* **114**, 45–58.

Anderson, W.K., Thomas, J.L. and Van Leer, B. (1986). *AIAA J.* **24**, 1453–1460.

Atkins, H.L. (1991). *AIAA Paper 91-1557,* AIAA 10th Computational Fluid Dynamics Conference, Honolulu, Hawaii.

Barth, T.J. and Frederikson, P.O. (1990). *AIAA Paper 90-0013,* AIAA 28th Aerospace. Sciences Meeting, Reno, NV.

Beam, R.W. and Warming, R.F. (1976). *J. Comput. Phys.* **23**, 87–110.

Beam, R.W. and Warming, R.F. (1978). *AIAA J.* **16**, 393–402.

Blazek, J., Kroll, N., Radespiel, R. and Rossow C.C. (1991). *AIAA Paper 91-1533,* AIAA 10th Computational Dynamics Conference, Honolulu, Hawaii.

Boris, J.P. and Book, D.L. (1973). *J. Comput. Phys.* **11**, 38–69.

Brandt, A. (1977). *Math. Comput.* **31**, 333–390.

Cambier, L., Ghazzi, Veuillot, J. and Viviand, H. (1984). In *Computational Methods in Viscous Flows* (W.G. Habashi, ed.), pp. 513–539. Pineridge Press, Swansea.

Casper, J. (1992). *AIAA J.* **30**, 2829–2835.

Casper, J. and Atkins, H.L. (1993). *J. Comput. Phys.* **106**, 62–76.

Catalano, L.A., De Palma, P. and Pascazio, G. (1992). *Lecture Notes in Physics,* vol. 414, pp. 90–94. Springer Verlag, Berlin.

Caughey, D.A. (1988). *AIAA J.* **26**, 841–851.

Caughey, D.A. (1993). *AIAA Paper 93-3358,* AIAA 11th Computational Fluid Dynamics Conference, Orlando, FL.

Cavaliere, A., De Felice, G., Denaro, F.M. and Meola, C. (1993). *Proc. Comput. Methods Exp. Meas.* 1, 135–150.

Chakravarthy, S.R. (1990). *NASA CR-4285.*

Chakravarthy, S.R. and Osher, S. (1983). *AIAA Paper 86-1943,* AIAA 6th Computational Fluid Dynamics Conference, Honolulu, Hawaii.

Chorin, A.J. and Marsden, J.E. (1982). *A Mathematical Introduction to Fluid Mechanics.* Springer-Verlag, Berlin.

Courant, R., Isaacson, E. and Rees, M. (1952). *Comm. Pure Appl. Math.* 5, 243–255.

Crandall, M.G. and Majda, A. (1980). *Math. Comput.* 34, 285–314.

Crowley, W.P. (1968). *Mon. Weath. Rev.* 96, 1–11.

Dadone, A. and Grossman, B. (1992). *Lecture Notes in Physics,* vol. 414, pp. 95–99. Springer-Verlag, Berlin.

Davis, S.F. (1984). *J. Comput. Phys.,* 56, 65–92.

Deconinck, H., Hirsch, C. and Peuteman, J. (1986). *Lecture Notes in Physics,* vol. 264, pp. 216–221. Springer-Verlag, Berlin.

Deconinck, H., Strujis, R., Powell, K.G. and Roe, P. (1991). *AIAA Paper 91-1532,* AIAA 10th Computational Fluid Dynamics Conference, Honolulu, Hawaii.

De Felice, G., Denaro, F.M. and Meola, C. (1993a). *J. Wind Eng. Ind. Aero.* 50, 49–60.

De Felice, G., Denaro, F.M. and Meola, C. (1993b). *Numer. Heat Transfer* 23, 425–460.

De Zeeuw, D. and Powell K.G. (1991). *AIAA Paper 91-1542,* AIAA 10th Computational Fluid Dynamics Conference, Honolulu, Hawaii.

Dutt, P. (1988). *SIAM J. Numer. Anal.* 25, 245–267.

Eberle, A., Rizzi, A. and Hirschel, E.H. (1992). *Numerical Solutions of the Euler Equations for Steady Flow Problems,* Notes on Numerical Fluid Mechanics, vol. 34. Vieweg-Verlag, Braunschweig.

Engquist, B. and Majda, A. (1977). *Math. Comput.* 31, 629–651.

Engquist, B. and Majda, A. (1979). *Comm. Pure Appl. Math.* 32, 312–358.

Engquist, B. and Osher, S. (1980). *Math. Comput.* 34, 45–75.

Federenko, R.P. (1964). *USSR Comput Math. and Math. Phys.* 4, 227–235.

Fromm, J.E. (1968). *J. Comput. Phys* 3, 176–189.

Gantmacher, F.R. (1960). *The Theory of Matrices.* Chelsea Publishing Co., New York, NY.

Gnoffo, P.A. (1989). *AIAA Paper 89-1972,* AIAA 9th Computational Fluid Dynamics Conference, Buffalo, NY.

Godfrey, A.G., Mitchell, C.R. and Walters, R.W. (1993). *AIAA J.* 31, 1634–1642.

Godunov, S. (1959). *Math. Sbornik* 47, 271–306.

Godunov, S. and Ryabenkii (1977). *Schémas aux Différences.* Mir Ed., Moscow.

Godunov, S., Zabrodine, A., Ivanov, M., Kraiko, A. and Propokov, G. (1979). *Résolution Numérique des Problémes Multidimensionnels de la Dynamique des Gaz.* Mir Ed., Moscow.

Goldberg, M. and Tadmor, E. (1985). *Math. Comput.* 44, 361–377.

Goldberg, M. and Tadmor, E. (1987). *Math. Comput.,* 48, 503–520.

Goodman, J.B. and LeVeque, R.J. (1985). *Math. Comput.* 45, 15–27.

Grasso F. and Marini, M. (1992). *AIAA J.* 30, 2184–2185.

Grasso, F. and Marini, M. (1993a). *J. Prop. Power* 9, 255–262.

Grasso, F. and Marini, M. (1993b). *AIAA J.* **31**, 1729–1731.

Grasso, F. and Marini, M. (1995). *Computers and Fluids* **23**, 571–592.

Grasso, F., Marini, M. and Passalacqua, M. (1992). *AIAA J.* **30**, 1780–1788.

Gustafsson, B. (1975). *Math. Comput.* **29**, 396–406.

Gustafsson, B. (1981). *SIAM J. Numer. Anal.* **18**, 179–190.

Gustafsson, B. (1982) *J. Comput. Phys.* **48**, 270–283.

Gustafsson, B. (1991). *Von Karman Institute, Lecture Series, 1991-01.*

Gustafsson, B. and Kreiss, H.O. (1979). *J. Comput. Phys.* **90**, 333–351.

Gustafsson, B. and Sundström, A. (1978). *SIAM J. Appl. Math.* **35**, 343–357.

Gustafsson, B., Kreiss, H.O. and Sundström, A. (1972). *Math. Comput.* **26**, 649–682.

Hackbusch, W. (1978). *Computing,* **20**, 291–306.

Hänel, D., Schwane, R. and Seider, G. (1987). *AIAA Paper 87-1105,* AIAA 8th Computational Fluid Dynamics Conference Honolulu, Hawaii.

Harten, A. (1983). *J. Comput. Phys.* **49**, 357–393.

Harten, A. (1984). *SIAM J. Numer. Anal.* **21**, 1–23.

Harten, A. (1987). *J. Comput. Phys.* **83**, 148–184.

Harten, A. (1991). *ICASE Report, 91-8.*

Harten, A. and Hyman, J.M. (1983). *J. Comput. Phys.* **50**, 235–269.

Harten, A. Hyman, J.M. and Lax, P.D. (1976). *Comm. Pure Appl. Math.* **29**, 297–322.

Harten, A., Lax P.D. and Van Leer (1983). *SIAM Rev.* **25**, 35-61.

Harten, A., Osher. S., Engquist, B. and Chakravarty S. (1986). *Appl. Numer. Math.* **2**, 347–377.

Harten, A., Engquist, B., Osher, S. and Chakravarthy, S. (1987). *J. Comput. Phys.,* **71**, 231–323.

Hedström, G.W. (1979). *J. Comput. Phys.* **30**, 222–237.

Hirsch, C. (1990). *Numerical Computation of Internal and External Flows.* John Wiley & Sons, New York.

Hirsch, C. and Lacor, C. (1989). *AIAA Paper 89-1958,* AIAA 9th Computational Fluid Dynamics Conference, Buffalo, NY.

Hirsch, C., Lacor, C. and Deconinck, H. (1987). *AIAA Paper 87-1163,* AIAA 8th Computational Fluid Dynamics Conference, Honolulu, Hawaii.

Jameson, A. (1983a). *Appl. Math. Comput* **13**, 327–356.

Jameson, A. (1983b). *Mech. and Aero. Engrg Report 1651,* Princeton Univ.

Jameson, A. (1985). *Mech. and Aero. Engrg. Report 1743,* Princeton Univ.

Jameson, A. (1993). *AIAA Paper 93-3359,* AIAA 11th Computational Fluid Dynamics Conference, Orlando, FL.

Jameson, A. and Lax, P.D. (1984). *Mech. and Aero. Engrg Report 1650,* Princeton Univ.

Jameson, A. and Turkel, E. (1981). *Math. Comput.* **37**, 385–397.

Jameson, A. and Yoon, S. (1987) *AIAA J.* **25**, 929–935.

Jameson, A., Schmidt, W. and Turkel, E. (1981) *AIAA Paper 81-1259,* AIAA 5th Computational Fluid Dynamics Conference, Honolulu, Hawaii.

Jameson, A., Baker, T.J. and Weatherhill, N. (1986). AIAA Paper 86–0103, AIAA 24th Aerospace Sciences Meeting, Reno, NV.

Kordulla, W. (1987). *Von Karman Institute, Lecture Series 1987-04.*

Kordulla, W. and Vinokur, M. (1983). *AIAA J.* **21**, 917–918.

Koren, B. and Hemker, P.W. (1991). *Appl. Numer. Math.* **7**, 308–329.

Kreiss, H.O. (1989). *Initial Boundary Value Problems and the Navier–Stokes Equations,*

Academic Press, San Diego.

Lacor, C., Zhu, Z.W. and Hirsch, C. (1993). Internal Report, Vrije Univ. of Bruxelles, TN-9301.

Lax, P.D. (1972). *SIAM Regional Conference Series in Applied Mathematics*, vol. 11, Philadelphia.

Lax, P.D. and Wendroff, B. (1960). *Comm. Pure Appl. Math.* **13**, 217–237.

Lax, P.D. and Wendroff, B. (1964). *Comm. Pure Appl. Math.* **17**, 381–398.

Leclercq, M.P. and Stoufflet, B. (1993). *J. Comput. Phys.* **104**, 329–346.

Lerat, A. and Peyret, R. (1974). *Computers and Fluids* **2**, 35–52.

LeVeque, R.J. (1992). *Numerical Methods for Conservation Laws*, 2nd edn. Birkhäuser-Verlag, Basel.

Levy, D.W., Powell, K.G, and Van Leer, B. (1989). *AIAA Paper 89–1931*, AIAA 9th Computational Fluid Dynamics Conference, Buffalo, NY.

Liou, M.S. (1993). *Lecture Notes in Physics*, vol. 414, pp. 115–119. Springer-Verlag, Berlin.

Liou, M.S. and Steffen C.J. (1993). *J. Comput. Phys.* **107**, 23–39.

MacCormack, R.W. (1969). *AIAA Paper 69-354.*

Majda A. (1984). *Compressible Fluid Flow and Systems of Conservation Laws in Several Space Variables.* Springer-Verlag, New York.

Manna, M. (1992). PhD Thesis, Univ. Catholique de Louvain.

Martinelli, L., Jameson, A.J., and Grasso, F. (1986). *AIAA Paper 86-0208,* AIAA 24th Aerospace Sciences Meeting, Reno, NV.

Martinelli, L., and Jameson, A.J., (1988). AIAA Paper 88–0414, *AIAA 26th Aerosp. Sci. Meeting,* Reno, NV.

Oden, J.T. and Reddy, J.N. (1976). *An Introduction to the Mathematical Theory of Finite Elements.* J. Wiley & Sons, New York.

Oliger, J. and Sundström, A. (1978). *SIAM J. Appl. Math.* **35**, 419–446.

Orszag, S.A. (1971) *J. Fluid Mech.* **49**, 75–112.

Osher, S. (1984). *SIAM J. Numer. Anal.* **21**, 217–235.

Osher, S. (1985). *SIAM J. Numer. Anal.* **22**, 947–961.

Osher, S. and Chakravarthy S.R. (1984). *SIAM J. Numer. Anal.* **21**, 995–984.

Osher, S. and Solomon F. (1982). *Math. Comput.* **22**, 339–374.

Pandolfi, M. (1989). *Notes on Numerical Fluid Mechanics,* vol. 24, pp. 462–481. Vieweg-Verlag, Braunschweig.

Peyret R. and Taylor T.D. (1983). *Computational Methods for Fluid Flow.* Springer-Verlag, Berlin.

Poinsot, T.J. and Lele, S.K. (1992). *J. Comput. Phys.* **101**, 104–129.

Powell, K.G. and Van Leer, B. (1990), *Von Karman Institute, Lecture Series 1990-03.*

Pulliam, T.H. (1985). *Von Karman Institute, Lecture Series 1985-03.*

Pulliam, T.H. (1993) *AIAA Paper 93-3360,* AIAA 11th Computational Fluid Dynamics Conference, Orlando, FL.

Quirk, J.J. (1994). *Int. J. Numer. Meth. Fluids* **18**, 555–574.

Radespiel, R. and Swanson R.C. (1991). *ICASE Report 91-89.*

Radespiel, R., Rossow, C. and Swanson R.C. (1990). *AIAA J.* **28**, 1464–1472.

Rai M.M. (1986). *J. Comput. Phys.* **66**, 99–131.

Richtmyer, R.D. and Morton, K.W. (1967). *Difference Methods for Initial-value Problems.* John Wiley & Sons, New York.

Rieger, H. and Jameson, A.J. (1988). *AIAA Paper 88-0619,* AIAA 26th Aerospace Sciences Meeting, Reno, NV.

Roe, P.L. (1981a). *J. Comput. Phys.* **43**, 357–372.

Roe, P.L. (1981b). *Royal Aircraft Establishment Technical Report 81407.*

Roe, P.L. (1984). *ICASE Report 84-53.*

Roe, P.L. (1985). *Lectures in Applied Mathematics,* vol. 22, pp. 163–193.

Roe, P.L. (1986). *J. Comput. Phys.* **63**, 458–476.

Rudy, D.H. and Strikwerda, J.C. (1981). *Computers and Fluids* **9**, 327–338.

Scott, J.N. Niu, Y.-Y. (1993). *AIAA Paper 93-0068,* AIAA 31st Aerospace Sciences Meeting, Reno, NV.

Shu, C. and Osher, S. (1988). *J. Comput. Phys.* **77**, 439–471.

Simpson, L.B. and Whitfield, D.L. (1992). *AIAA J.,* **30**, 914–922.

Sod, G.A. (1978). *J. Comput. Phys.,* **27**, 1–31.

Steger, J.L. and Warming, R.F. (1981). *J. Comput. Phys.* **40**, 263–293.

Strang, G. (1968). *SIAM J. Numer. Anal.* **5**, 506–517.

Strikwerda, J.C. (1977). *Comm. Pure Appl. Math.* **30**, 797–805.

Strikwerda, J.C. (1980). *J. Comput. Phys.* **34**, 94–107.

Strujis, R., Deconinck, H., DePalma, P., Roe, P. and Powell, K.G. (1991). *AIAA Paper 91-1550,* AIAA 10th Computational Fluid Dynamics Conference, Honolulu, Hawaii.

Swanson, R.C. and Radespiel, R. (1991). *AIAA J.* **29**, 697–703.

Swanson, R.C. and Turkel, E. (1992). *J. Comput. Phys.* **101**, 297–306.

Swanson, R.C. and Turkel, E. (1993). *AIAA Paper 93-3372,* AIAA 11th Computational Fluid Dynamics Conference, Orlando, FL.

Sweby, P.K. (1984). *SIAM J. Numer. Anal.* **21**, 995–1011.

Tatsumi, S., Martinelli, L. and Jameson, A. (1994). *AIAA Paper 94-0647,* AIAA 32nd Aerospace Sciences Meeting, Reno, NV.

Thompson, K.W. (1987). *J. Comput. Phys.* **68**, 1–24.

Thompson, K.W. (1990). *J. Comput. Phys.* **89**, 439–461.

Tysinger, T.L. and Caughey, D.A. (1991). *AIAA Paper 91-0242.* AIAA 29th Aerospace Sciences Meeting, Reno, NV.

Van Albada, G.D, Van Leer, B. and Roberts, W.W. (1982). *Astron. Astrophysics,* **108**, 76–84.

Van Leer, B. (1977). *J. Comput. Phys.* **23**, 276–299.

Van Leer, B. (1979). *J. Comput. Phys.* **32**, 101–136.

Van Leer, B. (1982). *Lecture Notes in Physics,* vol 170, pp. 507–512. Springer-Verlag, Berlin.

Van Leer, B. (1984a). *J. Comput. Phys.* **14**, 361–370.

Van Leer, B. (1984b). *SIAM J. Sci. Statist Comp.* **5**, 1–20.

Van Leer, B. (1985). *Lectures in Applied Mathematics, vol 22, pp.* 327–336.

Van Leer, B. (1992). *ICASE Report 92-43.*

Van Leer, B., Tai, C.H. and Powell, K.G. (1989). *AIAA Paper 89-1933,* AIAA 9th Computational Fluid Dynamics Conference, Buffalo, NY.

Vanden, K.J and Belk, D.M. (1993). *AIAA J.* **31**, 1377–1391.

Van Ransbeeck, P. and Hirsch, C. (1993). *AIAA Paper 93-3304,* AIAA 11th Computational Fluid Dynamics Conference, Orlando, FL.

Vatsa, V.N., Turkel, E. and Abalhassani, J.S. (1993). *Int. J. Numer. Methods Fluids* **17**, 115–144.

Venkatakrishnan V. (1993). *AIAA Paper 93-0880*, AIAA 31st Aerospace Sciences Meeting, Reno, NV.

Venkatakrishnan V. and Jameson A. (1988). *AIAA J.* **26**, 974–981.

Volpe, G. (1993). *AIAA Paper 93-2390*, AIAA 24th Fluid Dynamics Conference, Orlando, FL.

Volpe, G. and Jameson, A. (1990). *J. Aircraft* **27**, 223–231.

Wada, Y. and Liou, M.S. (1994). *AIAA Paper 94–0083*, AIAA 32nd Aerospace Sciences Meeting, Reno, NV.

Warming, R.F. and Beam, R.W. (1976). *AIAA J.* **14**, 1241–1249.

Warming, R.F. and Beam, R.W. (1990). *Notes in Numerical Fluid Mechanics* vol. 29, pp. 564–573. Wieweg-Verlag, Braunschweig.

Yamamoto, S. and Daiguji, H. (1993). *Computers and Fluids* **22**, 259–270.

Yanenko, N.N. (1971). *The Method of Fractional Step.* Springer-Verlag, Berlin.

Yee, H. (1989). *NASA Technical Memorandum* no. 101088.

Yoon, S. (1985). PhD Thesis no. 1720-T, Princeton University.

Yoon, S. and Jameson, A. (1988). *AIAA J.* **26**, 1025–1026.

Yoon, S. and Kwak, D. (1991). *AIAA Paper 91-1555*, AIAA 10th Computational Fluid Dynamics Conference, Honolulu, Hawaii.

Yoon, S. and Kwak, D. (1994). *AIAA J.* **32**, 950–955.

Young, D. M. (1971). *Iterative Solutions of Large Linear Systems.* Academic Press, New York.

Zha, G.C., and Bilgen, E. (1993). *Int. J. Numer. Methods, Fluids* **17**, 115-144.

Zhong, X. (1994). *AIAA J.* **32**, 1606-1616.

Zhu, Z.W. Lacor, C. and Hirsch, C. (1993). *AIAA Paper 93-3356*, AIAA 11th Computational Fluid Dynamics Conference, Orlando, FL.

5 Turbulent flows: direct numerical simulation and large-eddy simulation

Carlos Härtel

I INTRODUCTION

The precise description of turbulent flow phenomena and their reliable prediction are matters of primary concern in the geophysical and engineering sciences. The reason is that turbulence prevails in most flows of practical interest and may strongly affect important global features of these flows such as the wall friction, the heat transfer, the dispersion of additives or pollutants and the course of combustion processes, to name but a few. So far no predictive theory has been found which might be applied with success to the vast diversity of turbulent flows encountered in practice and consequently the treatment of most of these flows is based on theoretical and numerical approximations.

HANDBOOK OF COMPUTATIONAL FLUID MECHANICS
ISBN 0-12-532200-3

flows. Section III reviews the fundamentals and some recent developments in subgrid modelling. The question of suitable numerical schemes for application in turbulence simulations will then be addressed in Section IV. At the end of the chapter a survey is given of typical DNS and LES applications for fundamental studies and more engineering-type problems in order to illustrate the historical development and the state-of-the-art. The literature cited in the last section represents typical examples, but is not meant to furnish a complete survey of all past contributions to the fields. Further reviews of works related to DNS and LES can be found in the articles of Ferziger (1983), Grötzbach (1987), Reynolds (1990), Rogallo and Moin (1984), Schumann and Friedrich (1987) and Voke and Collins (1982). Comprehensive discussions of many aspects relevant to LES are given in Galperin and Orszag (1993) and in the book of Lesieur (1990). The volumes edited by AGARD (1994), Hussaini *et al.* (1992, 1994), Ragab and Piomelli (1993), Kral and Zang (1993) and Voke *et al.* (1994) compile DNS and LES studies which were presented at more recent conferences and workshops. In Hussaini *et al.* (1992, 1994) and Kral and Zang (1993) works are also presented on the numerical simulation of transition to turbulence. Though transition simulation is closely related to turbulence simulation we will not address this issue here. Reviews about DNS of transition are given by Kleiser and Zang (1991) and Reed (1993). More recent work on DNS of transition in compressible flows is reviewed in Guo *et al.* (1994b).

II FUNDAMENTALS OF DNS AND LES

A Basic equations

In the numerical simulation of turbulent fluid flows the Navier–Stokes equations, supplemented by appropriate initial and boundary conditions, are discretized and integrated in time. In the subsequent two sections these equations will be laid out for both incompressible and compressible flows. For convenience the equations will be given in common cartesian tensor notation only. Unlike in most CFD applications where body-fitted and, more recently, unstructured meshes are routinely applied, in DNS and LES cartesian meshes are still most widely used. This is due partly to the fact that so far both techniques have primarily been applied to flows in rather simple geometries where cartesian grids are the natural choice. Moreover, such grids are required by many of the highly accurate and efficient numerical schemes that were devised for direct and large-eddy simulations. A detailed description of the basic equations in more general coordinate systems can be found in Chapter 2 of this book and in standard textbooks on computational methods in fluid dynamics (e.g. Anderson *et al.*, 1984).

1 Incompressible flow

In the absence of external body forces, the governing equations for an incompressible newtonian fluid of constant density and constant viscosity read

$$\frac{\partial u_k}{\partial x_k} = 0 \tag{1}$$

$$\frac{\partial u_i}{\partial t} + \frac{\partial (u_i u_k)}{\partial x_k} = -\frac{\partial p}{\partial x_i} + \frac{1}{Re} \frac{\partial^2 u_i}{\partial x_k \partial x_k}. \tag{2}$$

In the above equations, where the Einstein summation convention applies to repeated indices, (u_1, u_2, u_3) is the velocity vector and p denotes the pressure. Note that in (2) the density does not explicitly occur, since it has been absorbed in the pressure. In (1, 2) all variables are non-dimensionalized by a reference velocity U_{ref} and a reference length L_{ref}. Consequently the Reynolds number Re is defined as $Re = U_{ref} L_{ref} / \nu$, where ν designates the kinematic viscosity of the fluid.

Frequently the dispersion of passive scalar quantities is of interest and the above equations need to be supplied with the appropriate transport equations. Such a scalar quantity might be the temperature, which can be treated as dynamically passive unless temperature differences in the flow become sufficiently large to induce appreciable density gradients. Approximating the temperature as a passive scalar, however, rules out any buoyancy effects which play a key role in many geophysical flows and for numerous engineering applications as well. Therefore the investigation of buoyant flows usually is based on the well-known Boussinesq equations which are an extension of the equations given above. Within this chapter I will not address the problem of buoyant flows any further, but detailed discussions of many of the pertinent computational and physical aspects can be found e.g. in Grötzbach (1987), Lesieur (1990) and Galperin and Orszag (1993).

2 Compressible flow

In the case of compressible flows, we want to consider a thermally and calorically perfect gas here. If body forces are absent or neglected as in the above case, the flow of a perfect gas is governed by the following equations which can be derived from the fundamental conservation laws for mass, momentum and energy

$$\frac{\partial \rho}{\partial t} + \frac{\partial (\rho u_k)}{\partial x_k} = 0 \tag{3}$$

$$\frac{\partial (\rho u_i)}{\partial t} + \frac{\partial (\rho u_i u_k)}{\partial x_k} = -\frac{\partial p}{\partial x_i} + \frac{1}{Re} \frac{\partial \sigma_{ik}}{\partial x_k} \tag{4}$$

$$\frac{\partial p}{\partial t} + u_k \frac{\partial p}{\partial x_k} + \gamma p \frac{\partial u_k}{\partial x_k} = \frac{1}{M^2 \, Pr \, Re} \frac{\partial q_k}{\partial x_k} + \frac{\gamma - 1}{Re} \Phi \tag{5}$$

$$\gamma M^2 p = \rho T \tag{6}$$

where

$$\sigma_{ij} = -\tfrac{2}{3} \mu \frac{\partial u_k}{\partial x_k} \delta_{ij} + \mu \left(\frac{\partial u_i}{\partial x_j} + \frac{\partial u_j}{\partial x_i} \right) \tag{7}$$

is the viscous stress tensor,

$$q_i = \kappa \frac{\partial T}{\partial x_i} \tag{8}$$

the heat flux, and

$$\Phi = \frac{\partial u_k}{\partial x_\ell} \sigma_{kl} \tag{9}$$

the viscous dissipation. In the above equations ρ signifies the density and T and $\gamma = c_p/c_v$ are the temperature and the ratio of specific heats, respectively. M, Re and Pr denote the Mach number, the Reynolds number and the Prandtl number of the flow which are defined by the reference quantities U_{ref}, L_{ref} and the reference values for the temperature, density, viscosity and conductivity T_{ref}, ρ_{ref}, μ_{ref} and κ_{ref}, respectively. The system consisting of the equations (3)–(9) still has to be supplemented by the laws of dependence of the viscosity μ and the thermal conductivity κ on the fluid state. To this end Sutherland's law for the viscosity is frequently applied which has the following dimensionless form

$$\mu(T) = T^{\frac{3}{2}} \frac{1 + S}{T + S} \tag{10}$$

In (10) S designates the Sutherland parameter which, e.g. for air, takes a value of $S = 110.4 \, K/T_{ref}$. The thermal conductivity can then be computed from the viscosity assuming a constant Prandtl number Pr which for air is commonly set to the value $Pr = 0.7$.

B Direct numerical simulation

As noted above, the precise numerical treatment of turbulence in principle requires that the entire band of scales which ranges from the energy-carrying to the dissipative motions is resolved in time and space. Figure 1 gives a sketch of typical spectra of the kinetic energy and the viscous dissipation for a free turbulent flow at a Reynolds number of the order 10^5. It is seen that the energy spectrum takes its maximum within the low-wavenumber regime while the

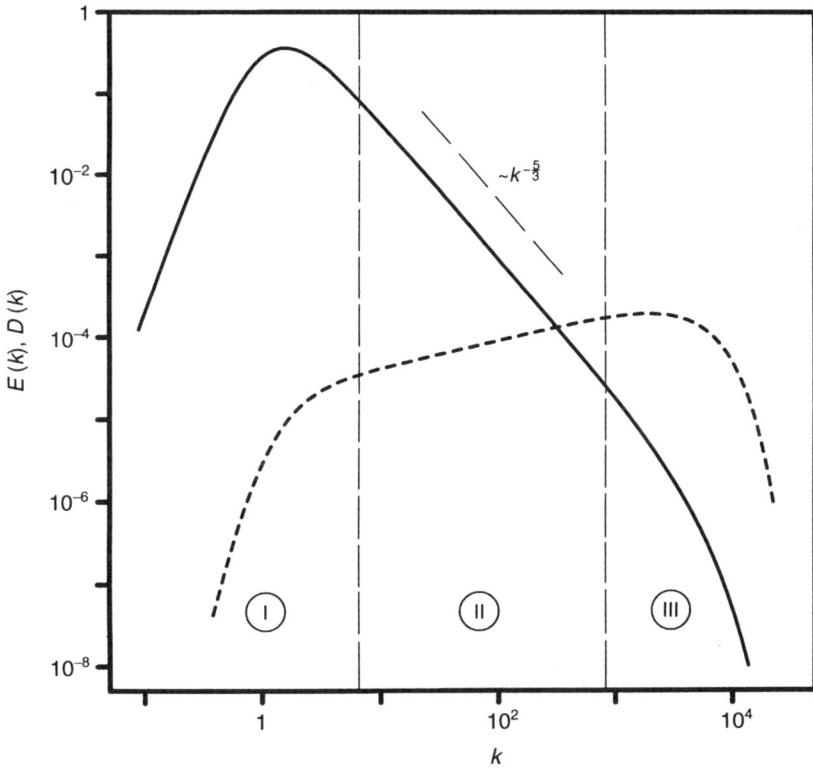

Figure 1 Typical energy spectrum $E(k)$ (—) and dissipation spectrum $D(k)$ (- - -) of a free turbulent flow at a Reynolds number (based on integral length and velocity scale L and U, respectively) of the order 10^5. All quantities are normalized by L and U. (I) Energy-containing range; (II) inertial subrange, exhibiting the characteristic $E(k) \propto k^{-\frac{5}{3}}$ behaviour; (III) dissipation range.

maximum dissipation can be found in the high-wavenumber part. In the direct numerical simulation approach a full resolution of all these scales is performed. Since no *ad hoc* models are needed to account for unresolved motions, DNS results, unlike results from statistical simulations or LES, essentially are free of errors due to empirical assumptions about the intrinsic turbulence physics. There remain, however, some external sources of errors due to uncertainties with respect to the initial and boundary conditions and, of course, due to the discrete (and finite) numerical approximation of the problem.

The application of DNS presently is confined to moderate Reynolds numbers owing to the fact that the bandwidth of the turbulence spectrum grows rapidly with *Re*. From a practical point of view this constraint is rather severe, since turbulence primarily prevails at high Reynolds numbers which are common in geophysical and engineering problems. In fact, DNS is not likely to become a predictive tool for engineering purposes in the foreseeable future and its principal field of application will be fundamental turbulence-physics studies. The insight which is gained from DNS studies at fairly low Reynolds numbers, however, may frequently be generalized for higher Re. Numerous features of turbulent flows can already be well studied at moderate Reynolds numbers where only a limited range of excited scales is present. The mean-flow properties of turbulent jets, wakes or mixing layers, for example, are known to be largely independent of Reynolds number, since in these flows viscosity mainly determines the scale of the dissipative structures leaving the larger scales almost unaffected. As long as the behaviour of the larger eddies of such flows is of primary concern, DNS may be employed as an accurate and efficient means of investigation.

For wall-bounded flows, the situation is generally less favourable than for free flows. In principle features of the near-wall turbulence are known to become Reynolds-number independent if properly scaled in viscous wall units, but this universal scaling of integral flow quantities can only be expected for Reynolds numbers considerably higher then those which are currently amenable to DNS (for an analysis of low-Re effects in channel flow see e.g. Antonia and Kim, 1994). Consequently an accurate quantitative prediction of near-wall turbulence features for the more relevant case of high-Re flows cannot be accomplished by DNS. However, in practice this weakness might not be very severe, since turbulence studies often concentrate on fundamental aspects and qualitative insight. If no distinct Reynolds-number effects prevail such insight can well be gained from the examination of currently available DNS databases.

Concerning the analysis of the strong non-linear dynamics which are characteristic for all turbulent flows, the interactions between various scales of motion are usually of more interest than the low-order mean-field statistics of the flow. Clearly, non-linear interactions of scales which differ by many orders of magnitude in size cannot be examined using DNS, but fortunately such interactions are often less complicated than those between neighbouring scales and are therefore more amenable to theoretical analyses. On the other hand, interactions of scales which are comparable in size reflect the full complexity of the turbulence problem and little is yet known about their governing mechanisms. Since these interactions are significant within a rather narrow band of neighbouring scales only, DNS can successfully be employed to study them.

1 Resolution requirements for DNS

In this section I want to outline the resolution requirements which apply to the DNS of simple free and wall-bounded turbulent flows (see also Saffmann, 1978; Reynolds, 1990). First of all, these resolution requirements are determined by the physical properties of the flow and depend on the ratio of the largest and smallest scales. They also depend on the accuracy of the numerical methods which are employed in the simulation, but this to a much lesser extent. Since I aim at providing estimates of the respective orders of magnitude only, I will not address the latter aspect here.

For incompressible free turbulence, the ratio of the largest and smallest spatial scales can rather easily be assessed: while the size of the largest energy-containing eddies is set by the integral length scale L of the flow (for example the diameter of a turbulent jet), the dissipation is known to peak at about ten times the Kolmogorov scale $\eta = (\nu^3/\varepsilon)^{\frac{1}{4}}$, where ε signifies the average rate of energy dissipation per unit mass in the flow. Since the smallest scales which have to be resolved are of the order η, the number N of grid points required in each spatial direction scales with the ratio L/η and this is well known to vary with the integral Reynolds number Re as $Re^{\frac{3}{4}}$ (Tennekes and Lumley, 1972). From this follows that the number of mesh cells required for full three-dimensional resolution scales like $Re^{\frac{9}{4}}$.

To illustrate the above relation more clearly one may consider homogeneous isotropic turbulence which has repeatedly been studied using DNS in the past. The first extensive calculations for isotropic turbulence were reported by Orszag and Patterson (1972) where the authors employed numerical grids with slightly more than 32 000 nodes. The Reynolds numbers Re_λ of their simulations, based on Taylor microscale λ and root-mean-square velocity, were limited to around 40. About two decades later Vincent and Meneguzzi (1991) performed a DNS of this flow at a Reynolds number of about $Re_\lambda = 150$ and in this study a mesh with almost 14 million grid points had to be employed to achieve full resolution. Hence, increasing the Reynolds number Re_λ by about a factor of four requires a number of grid points larger by roughly a factor of 500. The increase in the total computing effort is even much more dramatic and amounts to more than four orders of magnitude.

A thorough resolution of all relevant scales of motion is particularly costly in wall-bounded turbulent flows. The reason is that in these flows the size of the smallest eddies is determined by the thickness of the viscous wall layer, which becomes very small at higher Reynolds numbers. For a turbulent channel flow or simple turbulent boundary layers this thickness is known to scale with ν/u_τ, where u_τ denotes the common friction velocity. The ratio of the integral length scale L (e.g. the width of the channel or the displacement thickness of the boundary layer) to the size of the smallest eddies hence grows about linearly with the wall Reynolds number Re_τ. Using the empirical relation $Re_\tau \propto Re^{\frac{7}{8}}$ (Dean, 1978) between Re_τ and the integral Reynolds number Re (based on L and bulk velocity), it follows that the total number of grid points approximately varies with Re as $Re^{\frac{21}{8}}$. Clearly, these resolution requirements are much more stringent than those for free turbulent flows as discussed above. With currently available supercomputers discretizations of the order of 10^7 to 10^8 grid points are feasible, which allows for direct simulations e.g. of turbulent channel flow up to bulk Reynolds numbers of about $Re = 20\,000$. Simple boundary layer flows may be simulated up to momentum-thickness Reynolds numbers Re_θ of the order of 2000.

An important issue with respect to the computational needs of a time-accurate numerical simulation is that a high spatial resolution necessitates a high temporal resolution. Consequently more individual time steps have to be computed for a given physical span of time when the discretization is refined. For the case of simple free turbulent flows, one has to integrate the governing equations over order N time steps (as above, N denotes the number of grid points required in each spatial direction) in order to simulate the flow evolution for a period L/U where U designates the characteristic velocity of the energy-carrying eddies (thus L/U corresponds to the characteristic large-eddy turnover time). In this case the total computing expense for a simulation will at least vary like N^4, an estimate which even holds only under idealized conditions (Schumann, 1991). If a statistically steady state is to be reached and evaluated, numerical simulations frequently have to be conducted over a span of time T_{sim}, which is much longer than a single large-eddy turnover period. In turbulent channel flow, for example, T_{sim} typically is of the order of $T_{sim} \approx 100L/U$.

The key problem with the analysis of statistically steady states is that the various scales of motion are not evenly sampled when statistics are taken from the individual realizations, i.e. the discrete time steps of the simulation. In homogeneous isotropic turbulence a time period L/U allows sample sizes to be obtained for the smallest scales which are of the order $\sqrt{Re}$ (Tennekes and Lumley, 1972), but only a few 'large-eddy events' will occur during this period. Therefore care has to be taken that the total time of a simulation is large enough to provide sufficient samples for all scales. This can be very expensive for genuinely three-dimensional flows, where no averaging in homogeneous directions can be exploited in the statistical evaluation. In this case sufficiently large samples must be acquired at each location in physical space individually. Further difficulties arise when turbulent flows exhibit low-frequency fluctuations which are due to global instabilities and have timescales much larger than L/U. If the statistically steady state depends considerably on such disturbances, a particularly long integration time will be necessary.

In general, low-order statistical moments can be evaluated quite accurately based on samples of the order of 10^3, but much larger samples are required for higher-order statistics. This is of particular importance with respect to turbulence modelling where frequently higher-order moments are of primary interest. Many of these moments are difficult to measure experimentally, but may in principle be computed from DNS databases.

The resolution requirements for the DNS of compressible flows are essentially quite similar to those for incompressible flows. Like their incompressible counterparts, compressible DNS are restricted to rather low Reynolds numbers because only in this case is a thorough numerical resolution of all scales feasible. Strictly speaking, the Reynolds number range is even more restricted in the compressible case owing to several additional factors. One of them is that in compressible simulations more variables have to be stored and advanced in time, which increases both the storage needs and the computing time. Moreover the increased complexity of the equations requires that more arithmetic operations have to be made per time step. Another important factor is the presence of acoustic waves which, in flows at low Mach numbers, propagate at a speed much higher than the characteristic velocities of the turbulence field and hence restrict the time step of the simulation severely. In highly compressible flows at higher Mach numbers, where the flow velocities are comparable to the speed of sound, these acoustic waves tend to steepen, which then demands for a considerably refined spatial resolution (Zang *et al.*, 1992).

With respect to computational requirements, geophysical flows clearly represent an exceptional case: dynamically important large-scale vortices may extend over hundreds or even thousands of kilometres while the dissipative structures are only centimetres or millimetres in size. The number of grid points for a (purely hypothetical) direct numerical simulation of such flows was estimated e.g. by Voke and Collins (1982) to be of the order of 10^{24}. This almost certainly is far in excess of the capabilities of any computer ever available. For flow problems of aerodynamic interest the resolution requirements are less extreme, but they are still far beyond of what is feasible today. An estimate of the computational needs for DNS applications in aerodynamics is given in Figure 2, which is taken from Peterson *et al.* (1989). Direct numerical simulations are termed 'Full Navier–Stokes' in this figure. The estimates given in Figure 2 were made for various turbulence-research applications.

C Large-eddy simulation

The LES technique was developed, and for the first time applied, for three-dimensional and time-dependent simulations of atmospheric flow phenomena. Today LES is still most widely used in this field and has essentially become an operational tool for applications like weather forecasting. The starting point for the application of LES to engineering-type flows is marked by the study of Deardorff (1970), who simulated incompressible turbulence in a plane channel. In the engineering field LES primarily was employed for fundamental studies in the past, but has recently started to become more extensively utilized for realistic flow problems.

The LES approach occupies an intermediate position between the extremes of DNS, where all fluctuations are resolved, and the statistical simulation based on the Reynolds-averaged Navier–Stokes equations. In LES only the large-scale (grid-scale, GS) part of the turbulent fluctuations is computed explicitly while the small-scale (subgrid-scale, SGS) motions are modelled. Computationally LES clearly is less costly than DNS, but in general much more expensive than statistical simulations of the same flow. The reason is that independent of the

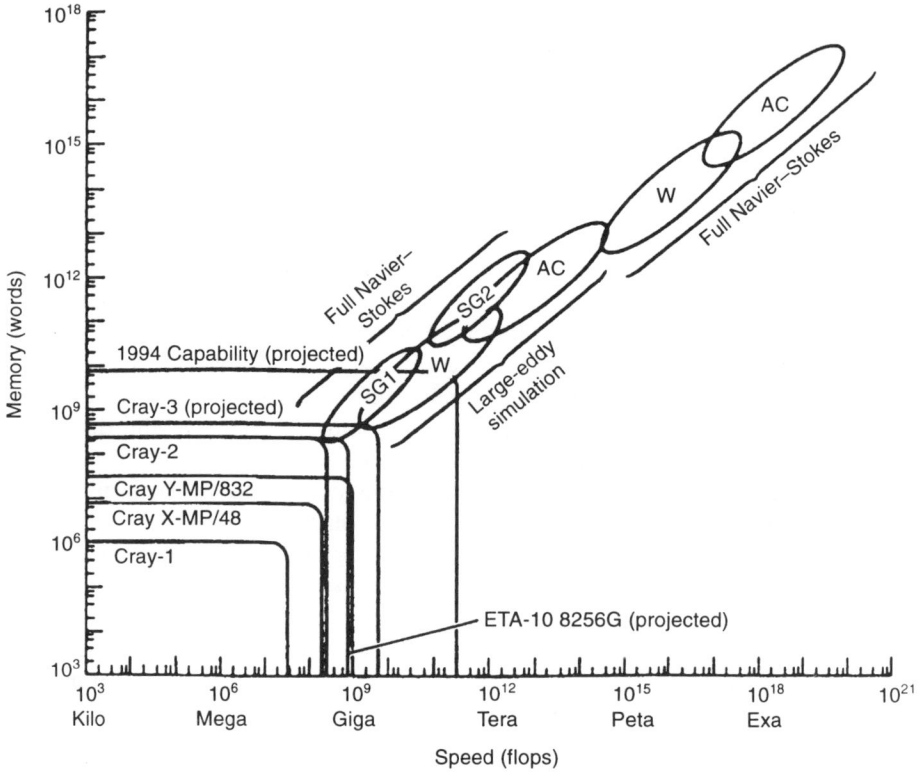

Figure 2 Computer-speed and memory requirements for DNS ('Full Navier–Stokes') and LES for turbulence research applications. Included in the figure are the capabilities of various computers. The estimates are based on 200 CPU-hour runs using typical 1988 algorithms. SG1: plates, channels; Reynolds number 10^6 to 5×10^6. SG2: plates, channels; Reynolds number 10^7 to 5×10^7. W: wing; Reynolds number 5×10^6 to 3×10^7. AC: aircraft; Reynolds number 5×10^7 to 3×10^8. From Peterson et al. (1989).

problem at hand LES always requires fully three-dimensional and time-dependent calculations even for flows which are two- or one-dimensional in the mean. Moreover, large-eddy simulations, like DNS, need to be conducted over long periods of time to obtain stable and significant statistics. The application of LES should prove most fruitful for the analysis of complex three-dimensional flows and time-dependent problems where statistical simulations frequently fail. Among those problems are separated flows and flows in and around complicated geometries. A rough estimate of the computational expenses of LES for turbulence-research applications in aerodynamics is provided in Figure 2.

The concept of LES mainly rests on two presumptions which appear plausible in view of both practical experience and theoretical considerations. The first of these is that most global features of turbulent flows, like averaged mixing rates or averaged losses, are governed by the dynamics of the largest scales and depend only little on the small-scale turbulence. From the practical point of view such global features are of primary interest and hence a reliable simulation of the largest scales will usually suffice. The other presumption is that the small-scale turbulence, especially at high Reynolds numbers, becomes independent of the strong inhomogeneities which are typical for the energy-containing eddies and thus tends to local isotropy (this issue will be addressed further in Section II.C.1). It is reasonable to suppose that in this case models for the small-scale turbulence can be much simpler than statistical turbulence models and will be more universally applicable, because statistical models have to account for the inhomogeneous eddies as well.

To illustrate the role of the SGS model it is useful to consider possible consequences if turbulence simulations are performed with insufficient resolution. An obvious implication of a too coarse resolution is that the viscous dissipation in the flow cannot properly be accounted for. This deficiency will typically result in an accumulation of energy at the high-wavenumber end of the spectrum which reflects a distorted equilibrium state between the production and dissipation of turbulent kinetic energy. For sufficiently high Reynolds numbers (or conversely sufficiently coarse grids) the discrete representation of the flow even becomes essentially inviscid and the non-linear transfer of energy can lead to an unbounded growth of turbulence intensities and eventually to numerical instability of the computation. In cases where no production of kinetic energy occurs, an equipartition of the existing energy among the finite degrees of freedom may be attained when energy-conserving algorithms are employed (see Rogallo and Moin, 1984). Such results have little to do with the real flow physics and they elucidate the need for an additional mechanism which accounts for the effect of the missing small-scale motions. This mechanism should mimic as realistically as possible the interaction between resolved and unresolved turbulent scales. In many flows this interaction is characterized by a non-linear transfer of energy to successively higher wavenumbers which is known as the turbulence cascade.

Before the problem of suitable subgrid models for large-eddy simulations can be tackled, several important issues have to be clarified. Among those which deserve particular attention are a clear definition of what one wants to consider the resolved and unresolved motions, and the derivation of the according governing equations. These issues will be addressed in Sections II.C.2 and II.C.3, respectively.

1 GS and SGS turbulence

As noted above, the concept of LES is strongly supported by the observation that frequently the characteristics of large and small turbulent scales are qualitatively different: while the energy-carrying large-scale structures mainly

Table 1
Qualitative differences between GS turbulence and SGS turbulence under idealized conditions (after Schumann and Friedrich, 1987)

GS turbulence	SGS turbulence
Produced by mean flow	Produced by larger eddies
Depends on boundaries	Universal
Ordered	Chaotic
Requires deterministic description	Can be modelled statistically
Inhomogeneous	Homogeneous
Anisotropic	Isotropic
Long-lived	Short-lived
Diffusive	Dissipative
Difficult to model	Easier to model

provide for the turbulent transport, the dissipative small-scale motions carry most of the vorticity and act as a sink of turbulent kinetic energy. For free turbulent flows at sufficiently high Reynolds numbers the dissipative part of the spectrum becomes clearly separated from the energy-containing short-wavenumber range (see Figure 1). Between these two bands the so-called inertial subrange forms where energy is transferred non-linearly to successively higher wavenumbers, at a rate which is prescribed by the turbulent kinetic energy production of the large scales (Tennekes and Lumley, 1972). Since the occurrence of an inertial subrange may give rise to (approximate) local isotropy in the flow, the modelling problem will be facilitated considerably if the numerical discretization is fine enough to ensure that the smallest resolved motions fall within that part of the spectrum. Some of the significant differences between GS and SGS turbulence which can be observed under these idealized conditions have been given e.g. by Schumann and Friedrich (1987) and are summarized in Table 1.

In many practical applications the flow may exhibit regions where the local Reynolds numbers are too low to give rise to a developed inertial subrange, and where at the same time no clear separation of energy-carrying and dissipative scales can be made. An illustrative example is the flow adjacent to a solid wall, where the dominant flow structures that account for most of the production of turbulent kinetic energy become very small. There are further factors, for example strong inhomogeneities or intermittency, which may have a considerable impact on the small-scale turbulence structure and may render a reliable subgrid modelling more difficult.

2 Scale separation and filtering

In large-eddy simulation any dependent variable f of the flow is split into a GS part $\bar{f}$ and a SGS part f', which will hereafter be indicated by an overbar and a single prime, respectively

$$f(x_1, x_2, x_3, t) - \bar{f}(x_1, x_2, x_3, t) + f'(x_1, x_2, x_3, t) \qquad (11)$$

Generally speaking, the GS component $\bar{f}$ represents that part of the turbulent fluctuation which remains after some smoothing has been applied to the flow field. This smoothing should be designed such that an accurate numerical treatment of the GS field is possible on the computational mesh which will be employed in the simulation. A clear definition of how the splitting in (11) has to be performed is an important conceptual issue in LES and deserves special care, since it has implications for the development of an appropriate numerical method. An obvious approach might be to define the GS fields as the averages of the corresponding flow quantities over the mesh cells of the numerical grid, as was done in the early studies on LES (see Smagorinsky, 1963; Lilly, 1967; Deardorff, 1970). However, a consistent numerical method based on this concept was developed for the first time by Schumann (1973, 1975) and has become known as the 'volume-balance method'. In Schumann's approach the governing equations are integrated by parts to obtain discrete budget equations for the individual mesh cells. Particularly advantageous with this method is that anisotropies and inhomogeneities can readily be incorporated in the scheme.

In Schumann's method a rather intimate relationship exists between the numerical solution scheme and the smoothing of the flow variables. From a theoretical standpoint, however, it is desirable to distinguish the effects due to the smoothing from those which are due to the application of a particular numerical scheme. A concept which in principle allows the SGS turbulence to be examined without reference to a certain numerical method was suggested by Leonard (1974). In this approach the GS turbulence is defined by applying explicitly a spatial filtering operation which is based on a convolution integral

$$\bar{f}(x_1, x_2, x_3, t) = \int_D \prod_{i=1}^{3} h_i(x_i - x_i', \Delta_i) f(x_1', x_2', x_3', t) \, dx_1' \, dx_2' \, dx_3' \qquad (12)$$

where the integration extends over the whole domain D. In (12) h_i denotes the filter function in the i-th direction and Δ_i is the width of the filter which selects the size of the smallest resolved eddies. Using an appropriate filter function, the classical volume average can also be represented by this ansatz. A useful illustration of the effect of such filtering can be gained in spectral space. According to the convolution theorem one finds for the Fourier transforms (symbolized by a hat)

$$\hat{\bar{f}} = \prod_{i=1}^{3} \hat{h}_i \cdot \hat{f} \qquad (13)$$

which shows that the filtering corresponds to a weighted selection of certain scales (wavenumbers) in the flow. To ensure that the filtering operation will reproduce any spatially uniform and constant quantity, it is necessary to impose the following normalization condition

$$\int_D \prod_{i=1}^{3} h_i(x_i) \, \mathrm{d}x_1 \, \mathrm{d}x_2 \, \mathrm{d}x_3 = 1 \tag{14}$$

Once the flow field has been smoothed by means of the filter, a high-order numerical discretization may be employed with grid sizes significantly smaller than the filter width, thus keeping further discretization errors negligibly small. Though this decoupling of physics and numerics looks attractive from a theoretical point of view, it must be emphasized that it will hardly be feasible in practice. In fact, in LES the filter width usually cannot be chosen much larger than the size of the computational grid in order to ensure that as many turbulent scales as possible are resolved with the available computer resources (see Rogallo and Moin, 1984; Boris, 1990; Schumann, 1991). Such a maximum resolution is essential for LES because the physical truncation errors associated with the unresolved scales generally are much larger than the truncation errors of the numerical scheme.

In the literature a number of filter functions have been suggested for application in LES, the most prominent among them being the gaussian filter, the cut-off filter in spectral space and the top-hat filter in real space. A common feature of all these filter functions is that the filtering operation and the partial differentiation of a variable f commute, i.e. it holds that

$$\frac{\overline{\partial f}}{\partial x_i} = \frac{\partial \bar{f}}{\partial x_i} \tag{15}$$

This commutivity gains its central importance from the fact that it allows the required evolution equations for the filtered fields to be obtained by applying the filtering operation to the Navier–Stokes equations.

An important difference between the filtering in LES and the common Reynolds averaging applied in statistical simulations is that the filtered product of a GS and an SGS quantity does not vanish identically, i.e.

$$\overline{f g'} \neq 0 \tag{16}$$

where f and g denote arbitrary flow variables. Furthermore, in general

$$\bar{\bar{f}} \neq \bar{f} \quad \text{and} \quad \overline{f'} \neq 0 \tag{17}$$

The inequalities in (17), however, have to be replaced by equal signs for the cut-off filter where single-filtered and double-filtered quantities are identical. The reason is that for the cut-off filter the grid and subgrid scales do not overlap in spectral space as will subsequently be discussed.

From the three previously mentioned filters only the cut-off filter h_i^c allows a clear distinction between large and small scales, since in wavenumber space they are represented by disjoint bands. This is most conveniently illustrated by looking at the Fourier transform $\hat{h}_i^c$ of the cut-off filter

$$\hat{h}_i^c(k_i) = \begin{cases} 1 & \text{for } |k_i| \leq K_i^c \equiv 2\pi/\Delta_i \\ 0 & \text{otherwise} \end{cases} \tag{18}$$

In (18) K_i^c denotes the so-called 'cut-off wavenumber'. The width of the filter Δ_i is usually related to the mesh size Δx_i of the computational grid by

$$\Delta_i = 2\Delta x_i \tag{19}$$

The shape of the cut-off filter in spectral and real space is depicted in non-dimensional form in Figure 3. Generally speaking, a cut-off filtering is implicitly applied whenever a continuous problem is treated by means of discrete approximations. In simulations where spectral methods are employed that are based on a Fourier representation of the flow variables, the cut-off filter arises naturally. Since in the past spectral methods have been extensively utilized for DNS and LES of a number of canonical flows (e.g. homogeneous isotropic turbulence or turbulent channel flow), the cut-off filter has received particular attention. However, the details of the interactions between grid-scale and subgrid-scale turbulence depend only little on the filter function applied (see Härtel, 1994).

At this point it should be noted that for more practical applications where homogeneous directions are missing and where orthogonal grids cannot be used, the only feasible way of filtering might be some type of averaging over the computational mesh cells. In this case the filtering concept can still prove useful when the characteristics of the numerical scheme are related to the characteristics of some filter functions. This may be exploited in the analysis of such schemes with respect to their application in LES. For example, the top-hat filter in real space is related to the application of a central finite-difference approximation (Rogallo and Moin, 1984), whereas the cut-off filter, according to the Nyquist theorem, is implicitly involved in any discrete approximation as noted above. Most numerical methods, however, will imply some filtering with properties which are not known in detail.

For the LES of compressible flows besides the common filtering a Favre-type, i.e. density-weighted, filtering may be employed which is defined analogously to the common Favre averaging applied in the statistical theory of compressible turbulence (see e.g. Hinze, 1975). To distinguish commonly filtered quantities from Favre-filtered ones, the latter will subsequently be indicated by a tilde for the Favre-filtered GS part and a double prime for the corresponding SGS part, respectively, i.e.

$$f = \tilde{f} + f'' \tag{20}$$

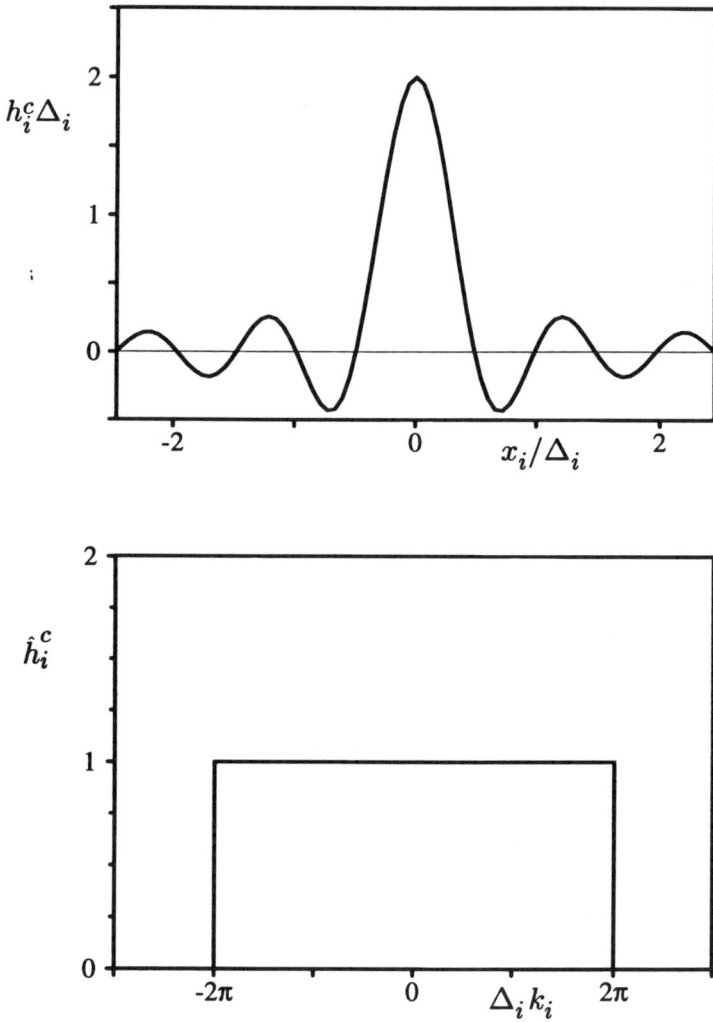

Figure 3 Cut-off filter h_i^c in real space (above) and in spectral space (see (18)).

where the Favre filter is given by

$$\tilde{f} = \frac{\overline{\rho f}}{\overline{\rho}}$$ (21)

For the Favre filtering some inequalities hold that correspond to those which were noted in (16, 17) for the conventional filtering. If f and g denote arbitrary flow variables one finds for the Favre filtering in general

$$\widetilde{\tilde{f}} \neq \tilde{f}, \quad \widetilde{f''} \neq 0, \quad \widetilde{\tilde{f}g''} \neq 0 \tag{22}$$

In this section the filtering concept has been laid out for spatial filters only, but it can likewise be applied in time which was suggested e.g. by Dakhoul and Bedford (1986a, b) and Aldama (1990). An additional explicit time filtering, however, turns out to be necessary in exceptional cases only since the high-frequency components in a turbulent flow are usually highly correlated with the high-wavenumber components. Thus a spatial low-pass filtering simultaneously eliminates high-frequency turbulence structures. A comprehensive discussion of this issue and a detailed survey of many important aspects of the filtering concept can be found in Aldama (1990).

3 Filtered equations of motion

Applying the filtering operation to the Navier–Stokes equations yields the equations of motion for the large-scale flow field. Like in the statistical averaging, the filtering of the non-linearities is of particular interest since it gives rise to additional unknown terms.

Incompressible flow
Employing the decomposition of the flow variables introduced in (11), one finds for the filtered convective term of the incompressible Navier–Stokes equations

$$\frac{\partial}{\partial x_k}(\overline{u_i u_k}) = \frac{\partial}{\partial x_k}(\bar{u}_i \bar{u}_k) + \frac{\partial}{\partial x_k}(L_{ik} + C_{ik} + R_{ik}) \tag{23}$$

where

$$L_{ij} = \overline{\bar{u}_i \bar{u}_j} - \bar{u}_i \bar{u}_j \tag{24}$$

$$C_{ij} = \overline{\bar{u}_i u_j'} + \overline{u_i' \bar{u}_j} \tag{25}$$

$$R_{ij} = \overline{u_i' u_j'} \tag{26}$$

The terms in equations (24)–(26) are usually denoted as 'Leonard stresses' (L_{ij}), 'cross stresses' (C_{ij}) and 'SGS Reynolds stresses' (R_{ij}), respectively. The cross stresses and the Reynolds stresses reflect directly the decomposition of the velocity fields into a GS and SGS part, and invariably have to be modelled. The term L_{ij}, on the other hand, is defined by the GS velocities only and can in principle either be modelled or computed.

The special character of the stresses L_{ij} was emphasized for the first time by Leonard (1974). An illustrative interpretation of these terms can be given if a cut-off filter is applied. In this case the Leonard terms represent those higher wavenumbers which are non-linearly excited by the grid scales, but which cannot adequately be resolved by the available discrete spectrum. The misinterpretation of these modes may contaminate the well-resolved lower wavenumbers, a problem which is known as aliasing (see Canuto *et al.*, 1988). If a fully spectral method is employed in the simulation aliasing errors do not occur. In pseudo-spectral computations, where non-linear products are evaluated in physical rather than in spectral space, such aliasing errors can efficiently be eliminated. If filters other than the cut-off filter are applied the Leonard stresses are no longer identical with the aliasing errors, but still should be computed explicitly whenever possible. However, for certain second-order finite-difference methods the evaluation of the Leonard terms may be of the same form and order on the numerical truncation error and their evaluation will be unnecessary in this case (Ferziger, 1977).

The SGS stresses C_{ij} and R_{ij}, which require modelling, can be decomposed into their isotropic and anisotropic parts. The isotropic part $\delta_{ij}(C_{kk} + R_{kk})/3$ of the tensor $C_{ij} + R_{ij}$, which is related to the subgrid-scale kinetic energy, does not need particular attention since it can be absorbed in the pressure. The sum of the pressure and the isotropic part of the SGS stresses is frequently referred to as pseudopressure q.

$$q = \bar{p} + \tfrac{1}{3} Q_{kk}, \quad \text{where} \quad Q_{ij} = C_{ij} + R_{ij} \tag{27}$$

The anisotropic part of the SGS stress tensor will subsequently be denoted τ_{ij}, i.e.

$$\tau_{ij} = Q_{ij} - \tfrac{1}{3} \delta_{ij} Q_{kk} \tag{28}$$

It must be emphasized that the decomposition discussed above is necessary for consistency reasons if an eddy-viscosity ansatz is employed for the subgrid modelling. In eddy-viscosity models the SGS stresses are set proportional to the GS rate-of-strain tensor $\bar{S}_{ij}$ and this tensor is traceless in incompressible flows. The trace of the SGS stress tensor, on the other hand, vanishes identically only if there is no SGS turbulence at all.

Using the definitions given in equations (24)–(28) the filtered equations of motion for an incompressible fluid can be written in the following form

$$\frac{\partial \bar{u}_k}{\partial x_k} = 0 \tag{29}$$

$$\frac{\partial \bar{u}_i}{\partial t} + \frac{\partial}{\partial x_k}(\bar{u}_i \bar{u}_k) = \frac{\partial q}{\partial x_i} - \frac{\partial \tau_{ik}}{\partial x_k} - \frac{\partial L_{ik}}{\partial x_k} + \frac{1}{Re} \frac{\partial^2 \bar{u}_i}{\partial x_k \partial x_k} \tag{30}$$

Compressible flow

For the LES of compressible flows consider the filtered form of the equations of motion for a thermally and calorically perfect gas which were given in (3)–(6)

$$\frac{\partial \bar{\rho}}{\partial t} + \frac{\partial (\bar{\rho} \tilde{u}_k)}{\partial x_k} = 0 \tag{31}$$

$$\frac{\partial (\bar{\rho} \tilde{u}_i)}{\partial t} + \frac{\partial (\bar{\rho} \tilde{u}_i \tilde{u}_k)}{\partial x_k} = - \frac{\partial \bar{p}}{\partial x_i} + \frac{1}{Re} \frac{\partial \bar{\sigma}_{ik}}{\partial x_k} + \frac{\partial t_{ik}}{\partial x_k} \tag{32}$$

$$\frac{\partial \bar{p}}{\partial t} + \tilde{u}_k \frac{\partial \bar{p}}{\partial x_k} + \gamma \bar{p} \frac{\partial \tilde{u}_k}{\partial x_k} = \frac{1}{M^2 \, Pr \, Re} \frac{\partial \bar{q}_k}{\partial x_k} + \frac{\gamma - 1}{Re} \bar{\Phi} + \frac{1}{M^2} \frac{\partial Q_k}{\partial x_k}$$

$$+ (\gamma - 1) \left(u_k \frac{\partial p}{\partial x_k} - \tilde{u}_k \frac{\partial \bar{p}}{\partial x_k} \right) \tag{33}$$

$$\gamma M^2 \bar{p} = \bar{\rho} \tilde{T} \tag{34}$$

The stress tensor σ_{ij}, which appears in (32), and the viscous dissipation Φ and the heat flux q_i in (33) are given by (7)–(9). The dependence of the molecular diffusion coefficients μ and κ on the fluid state are assumed to be given by Sutherland's law together with a constant Prandtl number (see Section II.A.2). Note that in the above equations both Favre-type filtering and conventional filtering have been applied. The SGS stress tensor t_{ij} and the SGS heat flux Q_i are defined by

$$t_{ij} = -\bar{\rho} \left(\widetilde{\tilde{u}_i \tilde{u}_j} - \tilde{u}_i \tilde{u}_j + \widetilde{\tilde{u}_i u_j''} + \widetilde{u_i'' \tilde{u}_j} + \widetilde{u_i'' u_j''} \right) \tag{35}$$

$$Q_i = -\bar{\rho} \left(\widetilde{\tilde{u}_i \tilde{T}} - \tilde{u}_i \tilde{T} + \widetilde{u_i'' \tilde{T}} + \widetilde{\tilde{u}_i T''} + \widetilde{u_i'' T''} \right) \tag{36}$$

Analogous to the decomposition performed in (23), the subgrid-scale stress tensor and the SGS heat flux can be decomposed into their respective Leonard, cross, and Reynolds components. For the SGS stresses this gives

$$L_{ij} = -\bar{\rho} \left(\widetilde{\tilde{u}_i \tilde{u}_j} - \tilde{u}_i \tilde{u}_j \right) \tag{37}$$

$$C_{ij} = -\bar{\rho} \left(\widetilde{\tilde{u}_i u_j''} + \widetilde{u_i'' \tilde{u}_j} \right) \tag{38}$$

$$R_{ij} = -\bar{\rho} \left(\widetilde{u_i'' u_j''} \right) \tag{39}$$

and decomposing the SGS heat flux in a similar fashion yields

$$Q_i^L = -\bar{\rho} \left(\widetilde{\tilde{u}_i \tilde{T}} - \tilde{u}_i \tilde{T} \right) \tag{40}$$

$$Q_i^C = -\bar{\rho} \left(\widetilde{u_i'' \tilde{T}} + \widetilde{\tilde{u}_i T''} \right) \tag{41}$$

$$Q_i^R = -\bar{\rho} \left(\widetilde{u_i'' T''} \right) \tag{42}$$

For the individual components of the subgrid-scale stress tensor and the subgrid-scale heat flux similar arguments apply as for the SGS stresses in incompressible flow which were discussed in the preceding section. While C_{ij}, R_{ij}, Q_i^C and Q_i^R can only be modelled, the Leonard components L_{ij} and Q_i^L can be computed directly. However, an important difference is that, unlike in incompressible flows, the isotropic part $\delta_{ij} Q_{kk}/3$ of the tensor $Q_{ij} = C_{ij} + R_{ij}$ (see (27)) cannot be absorbed into the pressure, but also needs to be modelled.

4 Resolution requirements for LES

The SGS effects in a turbulent flow depend qualitatively and quantitatively on the width of the filter function applied, i.e. on the size of the smallest resolved eddies. Generally, the filter width should be as large as possible in order to minimize the computational needs of a simulation, but larger filter widths give rise to a more complex SGS turbulence. Obviously, SGS effects vanish completely if the filter width is vanishing and this should formally be the case in direct numerical simulations. As long as discrete methods are applied, however, one cannot rigorously comply with this requirement in DNS, but in practice has to choose grid sizes of the order of the Kolmogorov microscale η (see Section II.B.1). For LES the matter is less clear than for DNS because the resolution does not only depend on the properties of the flow but also on the accuracy of the subgrid model. Frequently the required resolution cannot readily be determined beforehand and preliminary studies have to be conducted in order to assess the adequate grid sizes of the computational mesh.

For free flows at sufficiently high Reynolds numbers the energy-carrying scales and the dissipative structures become disparate, as was emphasized before (see Section II.C.1). Between the according spectral bands an inertial subrange develops which may give rise to approximate local isotropy in the flow. Whenever the energy spectrum exhibits a developed inertial subrange the filter width (and hence the grid size) of a LES should be chosen such that the smallest resolved eddies fall within that part of the spectrum. Consequently, the dissipative structures will be much smaller than the numerical grid size and hence viscous effects are essentially absent in the simulation. In this case an LES, unlike a DNS, will be free of any Reynolds number limitation. In fact, the simulation will then be performed for the limiting case of an infinite Reynolds number.

In geophysical flows the situation outlined above is quite common, which is one of the reasons why LES is applied with particular success in that field. In many engineering applications, however, it is frequently not possible to obey the requirement that the smallest resolved scales exhibit local isotropy. Examples are flows where the formation of an inertial subrange is ruled out by a too low Reynolds number or by pronounced inhomogeneities which are present in the flow. From a practical point of view it is therefore desirable to develop resolution requirements which are weaker, but still suffice to ensure reliable LES results. For example, one may require that the SGS turbulence should be in a state of local equilibrium between production and dissipation of kinetic energy – a prerequisite for the applicability of most eddy-viscosity models as will be discussed in Section III. Another requirement which has frequently been suggested is that the large scales should contain about 80–90% of the kinetic energy, while the small unresolved scales should account for about 80–90% of the viscous dissipation (Ferziger, 1977; Schumann, 1991).

To illustrate the requirements given above, it may be useful to derive the according discretizations for a prototype spectrum (see also Schumann, 1991). A typical energy spectrum $E(k)$ of the production type, which was considered e.g. by Leslie and Quarini (1979), is given by

$$E(k) = E_P \frac{2(k/k_P)}{1 + (k/k_P)^{\frac{8}{3}}}, \quad E_P = \frac{1}{2}\alpha\varepsilon^{\frac{2}{3}}k_P^{-\frac{5}{3}}, \tag{43}$$

where k_P is proportional to the wavenumber at which the spectrum is a maximum. In (43) α denotes the Kolmogorov constant ($\alpha \approx 1.5$) and ε is the viscous energy dissipation. The shape of this spectrum is depicted in non-dimensional form in Figure 4 where $E(k)/E_P$ is given as a function of the normalized wavenumber k/k_P. In order to capture as many of the large scales as possible, a reasonable choice for a computational box size L might be $L = 4\pi/k_P$, which is sufficient to accommodate all but the largest eddies with wavenumbers below $k_{\min} = k_P/4$. Together these largest eddies account for no more than about 2% of the kinetic energy. If all remaining scales up to $k_{\max} = 20\,k_P$ are resolved, another 12% of the energy spectrum is cut off at the high-wavenumber end, thus achieving that in total more than 85% of the kinetic energy is contained within the resolved band of scales. Note that $k_{\max}$ lies within the inertial subrange that quickly forms for wavenumbers above $k/k_P > 5$. The number N of required grid points in each direction can then be computed to be $N = k_{\max}/k_{\min} = 80$.

It might be of interest to compare this estimate of the resolution requirement for an LES with the according estimate for a DNS of the same flow. However, although the viscous dissipation ε appears in (43), the kinematic viscosity ν does not explicitly enter the spectrum and hence no information is available about the size $\eta = (\nu^3/\varepsilon)^{\frac{1}{4}}$ of the smallest scales (see Section II.B.1). The spectrum in (43) rather approximates the high-wavenumber end with an infinitely extended inertial subrange which corresponds to an infinite Reynolds number. A DNS of such a flow is impossible in principle since it would require an infinite numerical resolution.

In LES, like in DNS, the situation generally is more demanding for wall-bounded flows than for free flows. Adjacent to solid walls there always exists a region where viscosity plays a major role and where locally significant turbulence structures become very small. For the treatment of the near-wall flow in an LES essentially two different approaches exist. The first approach, which in practice is the more common one, attempts to circumvent a very costly resolution of near-wall structures by bridging the wall region with the aid of

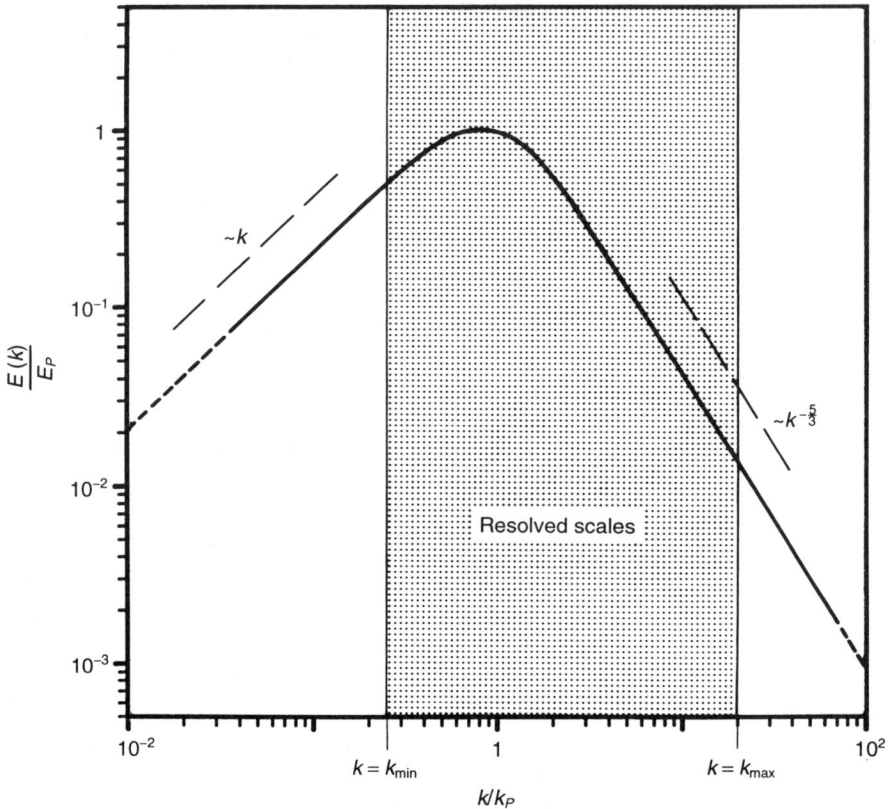

Figure 4 Production-type energy spectrum according to (43) in non-dimensional form. For low wavenumbers the spectrum is proportional to k. For high wavenumbers the characteristic $E(k) \propto k^{-\frac{5}{3}}$ inertial-subrange behaviour can be observed. The wavenumber band between k_{min} and k_{max} contains more than 85% of the turbulent kinetic energy of the flow.

empirical boundary conditions for the outer layer. These empirical boundary conditions, which will briefly be addressed in Section III.C, correspond to the so-called wall functions frequently used in statistical turbulence simulations. For high-Reynolds-number flows this approach appears to be the only feasible one. It should be stressed, however, that the particular dynamics of near-wall turbulence may strongly influence the entire flow field and that it is desirable to treat them more accurately whenever possible. This path is pursued in the second approach where a refined mesh is employed near the solid boundaries in order to resolve the dominant near-wall structures.

What makes LES of the near-wall flow particularly difficult is the fact that in this flow regime no clear separation between energy-carrying and dissipative scales can be made. Resolving 80–90% of the turbulent kinetic energy requires numerical grids which differ only marginally from those necessary for full resolution, i.e. for a DNS. On the other hand, if a considerable fraction of the kinetic energy resides in the subgrid scales the SGS turbulence will contain structures which play an important role for the turbulence dynamics and do not merely dissipate energy. It is therefore important that judicious guidelines for the spatial resolution are developed which ensure that important features like the evolution of near-wall streaks or bursting events can be captured. As minimum requirement Zang (1991) suggested grid spacings of about $\Delta x_1^+ = 80$ in the streamwise and $\Delta x_2^+ = 30$ in the perpendicular wall-parallel direction. The superscript '+' indicates that these grid spacings are given in common wall units, i.e. non-dimensionalized with ν/u_τ (u_τ being the friction velocity). According to current practice in LES (see e.g. Härtel et al., 1994; Piomelli, 1993) a resolution of $\Delta x_2^+ = 30$ appears to be the upper limit, while in the streamwise direction a much coarser resolution might still be sufficient. Compared with the required resolution for a DNS, such grid spacings result in a number of grid points which is lower by about two orders of magnitude in each wall-parallel plane, corresponding to considerable savings in terms of both storage needs and computing time. In the wall-normal direction x_3, however, not much can be saved in an LES compared with a DNS. At least three grid points between the wall and $x_3^+ = 10$ are considered to be necessary in LES (Zang, 1991) which differs by no more than a factor of 2 to 3 from what is required for full resolution (see Härtel et al., 1994).

The crucial point with the explicit treatment of the wall layer in LES is that the resolution requirements are set in wall units. Consequently, the number of grid points required will depend on the Reynolds number of the flow. To gain an idea of this dependence one may consider simulations of simple wall-bounded flows for various increasing Reynolds numbers: at sufficiently high Reynolds numbers the outer flow can be expected to exhibit an inertial subrange, and hence the resolution requirements in this flow regime will approach those for infinite Reynolds number. The inner layer, on the other hand, will retain its Reynolds-number dependence as was stated before. Since the thickness of the wall layer is inversely proportional to the wall Reynolds number Re_τ, the number of grid points required to resolve the near-wall zone adequately will increase with Re_τ^2. Employing the empirical relation introduced in Section II.B.1 this corresponds to an increase like $Re^{\frac{14}{8}}$ with the integral Reynolds number of the flow. Clearly, this increase with Re is much smaller than that found for a DNS, but it still may be steep enough to restrict LES to Reynolds numbers considerably lower than those of practical interest. Therefore in most engineering LES applications an explicit resolution of the near-wall dynamics will hardly be affordable.

III SGS MODELLING

In this section the fundamentals of currently applied SGS models are addressed for both incompressible and compressible flows. According to the philosophy of LES, SGS modelling is usually based on a simple gradient-diffusion ansatz.

An SGS model should mimic as realistically as possible the interaction between those scales that are filtered out and those which are resolved. Whether a simple model can be applied with success will depend on both the magnitude and the nature of these interactions. In general one distinguishes between 'local' interactions (local in wavenumber space) which are confined to a small band of neighbouring scales, typically one octave in width, and 'nonlocal' interactions which occur between widely separated scales. The local interactions retain the full complexity of the turbulence problem, and this renders their modelling very difficult. The non-local interactions, on the other hand, can more easily be analysed and modelled. Eddy-viscosity models are essentially based on the assumption that the energy transfer between the GS and SGS scales is governed by non-local interactions. This, however, contradicts the classical idea of a cascade process which implies that the transfer is governed by local processes (see Tennekes and Lumley, 1972). The analysis of DNS data revealed that both types of interactions play a significant role, but their relative importance depends on the flow and the Reynolds number (see e.g. Domaradzki and Rogallo, 1990; Domaradzki et al., 1994). At least for high-Reynolds-number flows the modelling of the energy transfer between large and small scales appears to be rather well represented by an eddy-viscosity concept (Domaradzki and Rogallo, 1990).

Usually SGS models are not supposed to provide directly information required for the flow statistics of interest. Those statistics should be available from the resolved scales alone and the key role of the SGS model is to ensure that the simulation of the large-scale features of the flow is not strongly affected by the absence of small-scale motions (Rogallo and Moin, 1984). One can expect that rather simple closures work well for SGS modelling if the unresolved scales exhibit local isotropy and provide mainly for the viscous dissipation (see Section II.C.1). In fact, this presumption has been confirmed by the experiences gained from numerous LES applications in the past. However, if the subgrid motions contribute considerably to certain large-scale features of the flow and if the important prerequisite of approximate local isotropy is not met, the application of simple closures can be precluded. Typically this may be the case for near-wall turbulence, turbulence in the presence of strong shear and intermittent, transitional or relaminarizing flows in general. In order to extend the applicability of LES to more demanding situations, considerable research effort is currently spent for the development and improvement of SGS models. It should be noted, however, that SGS models for such complicated flows will probably not evolve in a fashion similar to statistical models which have become much more complex and computationally more costly in the past. In LES an obvious improvement is always given by a refined resolution and consequently more elaborate subgrid models with increased computational costs will have to compete with standard models employed at refined resolution.

A Models for incompressible flows

The first subgrid model for an LES of incompressible flow was introduced by Smagorinsky (1963) more than 30 years ago. Today this model and its variants are still widely applied. The Smagorinsky model is of the eddy-viscosity type where the SGS stresses are set proportional to the strain rates of the resolved field

$$\tau_{ij} = -2\nu_t \bar{S}_{ij} \tag{44}$$

In (44) ν_t denotes the SGS eddy viscosity and $\bar{S}_{ij}$ the rate-of-strain tensor of the GS velocity field

$$\bar{S}_{ij} = \frac{1}{2}\left(\frac{\partial \bar{u}_i}{\partial x_j} + \frac{\partial \bar{u}_j}{\partial x_i}\right) \tag{45}$$

Assuming that the SGS turbulence is locally in a state of equilibrium between production and dissipation, Smagorinsky (1963) derived the following expression for the eddy viscosity

$$\nu_t = L^2 \|\bar{S}\|, \quad \|\bar{S}\| = \sqrt{2\bar{S}_{kl}\bar{S}_{kl}} \tag{46}$$

The SGS length scale L which appears in (46) can usually be related to the filter width Δ or the numerical mesh size Δx

$$L = C_S \Delta x \tag{47}$$

where C_S designates a dimensionless constant, known as the Smagorinsky constant. Since very few LES calculations of interest are conducted with an equal mesh spacing in all three spatial directions, Δx in (47) in practice has to be replaced by an average Δx_{av} of the individual grid sizes Δx_i. For meshes with moderate anisotropies the proper average is the geometric mean (Scotti et al., 1993)

$$\Delta x_{av} = (\Delta x_1 \Delta x_2 \Delta x_3)^{\frac{1}{3}} \tag{48}$$

which usually works well up to aspect ratios of about 20 : 1 (Reynolds, 1990). A major contribution to the theoretical support of the Smagorinsky model was made by Lilly (1967), who was able to show that, under idealized conditions, the Smagorinsky model is consistent with an infinitely extended inertial subrange. He derived the proportionality relation (47) for the subgrid length scale with the constant C_S given by

$$C_S = \frac{1}{\pi}\left(\frac{2}{3\alpha}\right)^{\frac{3}{4}} \tag{49}$$

Assuming a Kolmogorov constant of $\alpha = 1.5$, one finds $C_S \approx 0.17$. Lilly conducted his analysis only for an infinitely extended inertial subrange and a cut-off filter, but refined theoretical studies for more realistic spectra and other filter functions did not reveal a considerable sensitivity of the value of the Smagorinsky constant C_S (Leslie and Quarini, 1979). From practical experience, however, it was found that the optimum value for C_S may vary from flow to flow, ranging between about 0.07 and 0.24 (Rogallo and Moin, 1984; Schumann, 1991). A substantial improvement of the Smagorinsky model was later suggested by Schumann (1973, 1975), who introduced an additional term to account explicitly for the strong inhomogeneities in the turbulent flow near solid walls.

A qualitatively different approach to SGS modelling was proposed by Bardina *et al.* (1980). Starting from the filtering concept, these authors employed a similarity assumption to derive a model for the SGS stresses, which is defined by single and double filtered quantities

$$\tau_{ij} = C_B(\bar{u}_i\bar{u}_j - \bar{\bar{u}}_i\bar{\bar{u}}_j) \tag{50}$$

The constant C_B in (50) has to be unity in order to preserve galilean invariance of the equations of motion, as was shown by Speziale (1985). This ansatz which became known as the 'scale-similarity model', cannot be applied without an additional eddy-viscosity term since it does not dissipate sufficient energy from the resolved scales. An interesting feature of this model, however, is that it basically allows for both the transfer of energy from·large to small scales and the inverse transfer from small- to large-scale turbulent motions. This feature is very attractivè since the flux of energy is well known to exhibit a strongly intermittent nature, giving frequent changes of sign in physical space. The associated backward flux of energy, sometimes termed 'backscatter', has only recently received closer attention, although it was suggested theoretically for some time (Leslie and Quarini, 1979). It must be emphasized that such backscatter cannot be modelled with usual SGS closures like the Smagorinsky model which are absolutely dissipative in nature, meaning that they take energy from the resolved scales throughout the flow domain.

In numerical simulations a backward flux of energy has frequently been observed in various types of flow such as incompressible and compressible isotropic turbulence (see e.g. Domaradzki *et al.*, 1990) or turbulent channel flow (Horiuti, 1989; Piomelli *et al.*, 1991; Härtel and Kleiser, 1992). In order to account for this backscatter explicitly, a number of special models were developed and investigated in recent years, among others by Leith (1990), Chasnov (1991), Mason and Thomson (1992) and Carati *et al.* (1995). In all of these studies a usual eddy-viscosity model was supplemented by a parameter-ization for the backscatter based on a stochastic forcing function. In the context of wall-bounded turbulence the study of Mason and Thomson (1992), where boundary layer flows are considered, is of particular interest. Although the authors did not resolve the viscous wall layer in their computations, they found that the LES results were improved, especially near the wall when a backscatter term was included.

As mentioned above, the model of Bardina *et al.* (1980) was based on the assumption that a similarity relation exists between the various scales of motion present in a turbulent flow. Another model based on this concept was recently developed by Germano *et al.* (1991) and has become known as the 'dynamic model'. The procedure employed for the derivation of this model will briefly be outlined below. In recent years this dynamic model and its variants have been applied to a number of free and wall-bounded flows with considerable success (see e.g. Cabot and Moin, 1991; Germano *et al.*, 1991; Moin *et al.*, 1991; Akselvoll and Moin, 1993; Squires, 1993; Zang *et al.*, 1993).

For the application of the dynamic modelling procedure it is necessary to introduce a second filter H in addition to the usual filter h. In order to distinguish these two different filters, the former has frequently been termed the test filter while the latter is usually referred to as the grid filter. The test filter has a width Δ^H which is larger than the width Δ^h of the grid filter. In the following the overbars $—H$ and $—h$ are used to denote the filtering performed with the test filter H and the grid filter h, respectively. The corresponding SGS stresses t_{ij} and T_{ij} which arise from the application of h and $H \cdot h$ to the equation of motion are given by

$$t_{ij} = \left(\overline{u_i u_j}^h - \overline{u_i}^h \overline{u_j}^h \right) - \tfrac{1}{3} \delta_{ij} \left(\overline{u_k u_k}^h - \overline{u_k}^h \overline{u_k}^h \right) \tag{51}$$

$$T_{ij} = \left(\overline{\overline{u_i u_j}^h}^H - \overline{\overline{u_i}^h}^H \overline{\overline{u_j}^h}^H \right) - \tfrac{1}{3} \delta_{ij} \left(\overline{\overline{u_k u_k}^h}^H - \overline{\overline{u_k}^h}^H \overline{\overline{u_k}^h}^H \right). \tag{52}$$

Note that t_{ij} and T_{ij} in (51, 52) contain the associated Leonard terms which is crucial for the formalism. Germano (1992) derived the following algebraic relation for t_{ij} and T_{ij}

$$\mathcal{L}_{ij} = T_{ij} - \overline{t_{ij}}^H = \left(\overline{\overline{u_i}^h \overline{u_j}^h}^H - \overline{\overline{u_i}^h}^H \overline{\overline{u_j}^h}^H \right) - \tfrac{1}{3} \delta_{ij} \left(\overline{\overline{u_k}^h \overline{u_k}^h}^H - \overline{\overline{u_k}^h}^H \overline{\overline{u_k}^h}^H \right) \tag{53}$$

which relates a term $\mathcal{L}_{ij}$, that can be computed, to the unknown stresses T_{ij} and t_{ij}. From (53) it follows that the difference $T_{ij} - \overline{t_{ij}}^H$ can be evaluated, but note that neither T_{ij} nor t_{ij} can be obtained from this equation, and that consequently the closure problem has not been eliminated. However, although the identity in (53) does not remove the need for a SGS model, it may be utilized in order to adjust a given model to the actual state of the flow. To this end, the fundamental assumption must be introduced that the stresses due to both the grid filter and the test filter are similar in the sense that they can be parameterized by the identical model. In Germano *et al.* (1991) a Smagorinsky model was applied for this purpose but other models can be used as well. For the Smagorinsky model

this concept was developed by Deardorff (1973), who employed it for the simulation of a convective boundary layer. The results showed some improvement over conventional modelling, but the high price which had to be paid was a doubling in both the computing time and the storage for the simulation. Such a complex SGS modelling has not been well received by other authors yet. A general weakness of the method is that modelling assumptions for the unknown third-order moments have to be devised, but so far no reliable reference data for these terms are available. Deardorff therefore employed closure assumptions which were essentially analogous to those used in statistical simulations. However, the characteristics and the relative importance of the respective higher-order moments are probably not the same in LES and statistical simulations and this approach might therefore not be appropriate (Rogallo and Moin, 1984).

In the past various authors employed statistical theories of homogeneous isotropic turbulence for the development and analysis of SGS models. In these studies typically a spectral-space representation of the eddy viscosity is used. The concept of a spectral eddy viscosity was introduced by Kraichnan (1976), who examined the energy transfer in a turbulent flow using the test field model (TFM). The analysis suggested that the eddy viscosity should exhibit a steep increase in the proximity of the cut-off wavenumber (the so-called 'cusp') while being almost constant away from the cut-off. This result was later confirmed by the analysis of DNS databases of homogeneous isotropic turbulence (see e.g. Domaradzki et al., 1987). Other frequently employed theories are the direct interaction approximation (DIA) and the EDQNM theory. Studies based on DIA were conducted among others by Yoshizawa (1982, 1989), Love and Leslie (1979) and Leslie and Quarini (1979). The EDQNM theory was applied e.g. by Chollet and Lesieur (1981), who utilized it to derive a spectral eddy viscosity of the following form

$$\nu_t(k/K^c, t) = \nu_t^+(k/K^c)\sqrt{\frac{E(K^c, t)}{K^c}} \qquad (60)$$

where K^c denotes the spectral cut-off and $E(K^c, t)$ is the kinetic-energy density at $k = K^c$. In (60) $\nu_t^+(k/K^c)$ is a dimensionless function of the wavenumber k which can be evaluated explicitly assuming that K^c falls within the inertial subrange. If the dependence of ν_t^+ on k is neglected, as in Comte et al. (1990a) or Métais and Lesieur (1992), ν_t^+ can be approximated as a constant which, for an infinitely extended inertial spectrum, assumes a value of

$$\nu_t^+ = \tfrac{2}{3}\alpha^{-\frac{2}{3}} \approx 0.4 \qquad (61)$$

The eddy viscosity given in (60) was extensively used in large-eddy simulations performed by Cambon et al. (1981), Chollet and Lesieur (1981), Bertoglio (1982), Dang (1985) and Aupoix (1986), to cite a few. Since a spectral representation of the eddy-viscosity frequently is inappropriate, Métais and

Lesieur (1992) developed a variant of this model where the spectral eddy viscosity is evaluated in real space by the aid of the second-order velocity structure function.

In the context of these rather formal and complex theories, I finally wish to cite some studies where the renormalization-group theory (RNG) was applied for the purpose of SGS modelling. Examples are the works of Rose (1977), Yakhot and Orszag (1985), McComb and Shanmugasundaram (1986) and Yakhot *et al.* (1989). Models based on RNG were implemented and tested e.g. by Yakhot *et al.* (1989), Karniadakis *et al.* (1990) and Piomelli *et al.* (1990). Details of the RNG methodology can be found in Yakhot and Orszag (1986) and Smith and Reynolds (1992).

B Models for compressible flows

The governing equations for the LES of compressible flows were discussed in Section II.C.3 (equations (31)–(34)). In the compressible case further non-linearities exist in addition to the convective term, which give rise to further unknown correlations when the filtering is applied. According to current practice in compressible LES, only the SGS stresses and the SGS heat flux as given in equations (37)–(42) are explicitly accounted for while the small-scale components in the other unknown correlations are neglected (Yoshizawa, 1986; Speziale *et al.*, 1988; Lesieur *et al.*, 1991; Moin *et al.*, 1991; Erlebacher *et al.*, 1992; Kral and Zang, 1992; Zang *et al.*, 1992). For example, Zang *et al.* (1992) use the following approximations in their LES of compressible isotropic turbulence ($\mathbf{u}$ denotes the velocity vector u_i)

$$\overline{\sigma_{ij}(\mathbf{u})} \approx \sigma_{ij}(\tilde{\mathbf{u}}), \quad \overline{q_i(T)} \approx q_i(\tilde{T}), \quad \overline{\Phi(\mathbf{u})} \approx \Phi(\tilde{\mathbf{u}}), \quad \left(\overline{u_k \frac{\partial p}{\partial x_k}} - \tilde{u}_k \frac{\partial \bar{p}}{\partial x_k} \right) \approx 0$$

$$(62)$$

The difficulty with the modelling of the subgrid components of the pressure gradient-velocity correlation can be illustrated by rewriting this term in the following alternative form (see Erlebacher *et al.*, 1992)

$$\overline{u_k \frac{\partial p}{\partial x_k}} = \frac{1}{\gamma M^2} \left(\frac{\partial}{\partial x_k} (\bar{\rho} \tilde{u}_k \tilde{T} + Q_k) - \bar{\rho} \tilde{T} \frac{\partial \tilde{u}_k}{\partial x_k} - \overline{\rho T' \frac{\partial \tilde{u}_k}{\partial x_k}} - \bar{\rho} \tilde{T} \frac{\partial \tilde{u}'_k}{\partial x_k} - \overline{\rho T' \frac{\partial \tilde{u}'_k}{\partial x_k}} \right)$$

$$(63)$$

From the above equation it becomes clear that the unknown subgrid part can be represented as a temperature-dilatation correlation, a quantity which is very difficult to model. However, based on DNS results of Sarkar *et al.* (1991), Erlebacher *et al.* (1992) conjectured that in compressible isotropic turbulence at low Mach numbers this term should be very small and might therefore be neglected.

In the LES conducted by Erlebacher *et al.* (1992) a gaussian filter function was applied. For the modelling of the SGS stresses the authors employed the scale-similarity model of Bardina *et al.* (1980) for the cross stresses together with an eddy-viscosity ansatz for the deviatoric part of the subgrid-scale Reynolds stresses, which is a compressible generalization of the Smagorinsky model (Speziale *et al.*, 1988)

$$C_{ij} = -\rho \left(\widetilde{u_i u_j} - \tilde{u}_i \tilde{u}_j \right) \tag{64}$$

$$R_{ij}^d = 2 \, C_R \, \bar{\rho} \, \Delta^2 \, ||\widetilde{S}|| \, (\widetilde{S}_{ij} - \widetilde{S}_{kk} \delta_{ij}/3) \tag{65}$$

where

$$R_{ij} = R_{ij}^d + R_{ij}^i, \quad \text{and} \quad R_{ij}^i = -\bar{\rho} \, \widetilde{u_k'' u_k''} \delta_{ij}/3 \tag{66}$$

In (65) C_R is a dimensionless constant and $\widetilde{S}_{ij}$ and $||\widetilde{S}||$ denote the Favre-filtered rate-of-strain tensor and its norm, respectively,

$$\widetilde{S}_{ij} = \frac{1}{2} \left(\frac{\partial \tilde{u}_i}{\partial x_j} + \frac{\partial \tilde{u}_j}{\partial x_i} \right), \quad ||\widetilde{S}|| = \sqrt{\widetilde{S}_{kl} \widetilde{S}_{kl}} \tag{67}$$

Note that the norm $|| \cdot ||$ in (67) differs from that introduced in (46) by a factor of $1/\sqrt{2}$. If a cut-off filter is applied rather than a gaussian filter, the scale-similarity part of the model given above vanishes and the eddy-viscosity part has to account for both the SGS Reynolds stresses and the cross stresses. Compressible large-eddy simulations using a cut-off filter were conducted e.g. by Kral and Zang (1992).

Unlike in incompressible flows, the isotropic part of the subgrid-scale Reynolds stress tensor cannot be absorbed into the pressure in compressible turbulence and possibly requires separate modelling. An eddy-viscosity model for this isotropic part was proposed by Yoshizawa (1986), who analysed the various SGS terms occurring in compressible turbulence by means of a two-scale DIA method. Since Yoshizawa's analysis was founded on a small-compressibility expansion around an incompressible state, the applicability of his model is restricted to flows with small density variations. However, results of direct numerical simulations for isotropic turbulence revealed that for weakly-compressible flows the isotropic part of the SGS Reynolds stresses may be neglected without introducing considerable errors (Erlebacher *et al.*, 1992). Zang *et al.* (1992) studied isotropic turbulence exhibiting moderate compressibility and accounted for R_{ij}^i by the following model which was suggested by Speziale *et al.* (1988)

$$R_{ij}^i = -\tfrac{2}{3} C_I \bar{\rho} \Delta^2 ||\widetilde{S}|| \delta_{ij} \tag{68}$$

where C_I is another dimensionless constant.

Similar to the modelling of the SGS stresses, the modelling of the unknown components Q_i^R and Q_i^C of the SGS heat flux Q_i can be based on a gradient-diffusion ansatz where the subgrid-scale heat flux is assumed to be proportional to the gradient of the filtered temperature $\widetilde{T}$. This ansatz may be augmented with a scale-similarity part, depending on the filter function employed. Speziale *et al.* (1988) proposed the following model

$$Q_i^R + Q_i^C = -\frac{C_R}{Pr_t}\Delta^2\|\widetilde{S}\|\frac{\partial \widetilde{T}}{\partial x_i} - \bar{\rho}(\widetilde{u_i T} - \widetilde{\tilde{u}_i}\widetilde{T}) \tag{69}$$

Pr_t being the SGS turbulent Prandtl number.

The above model constants C_R, C_I and Pr_t were evaluated by Speziale *et al.* (1988) and Erlebacher *et al.* (1992) based on the analysis of compressible DNS and LES results. The authors recommended values of $C_R = 0.012$, $C_I = 0.0066$ together with a turbulent Prandtl number Pr_t in the range of 0.4 to 0.5. In these studies it was found that the values were rather insensitive to the Mach number of the flow which was varied between 0 and 0.6. However, owing to the particular choice of initial conditions compressibility effects remained very small throughout the simulations. Results for moderately compressible flows were later furnished by Zang *et al.* (1992) and this study revealed that the model constants are far from universal if significant compressibility effects are present. An attempt to adjust the constants to the actual state of the flow was made by Moin *et al.* (1991), who extended the dynamic procedure of Germano *et al.* (1991) to the SGS modelling in compressible turbulence.

The closure models currently employed in LES of compressible flows are essentially analogous to those used in incompressible simulations. The applicability of these models requires that strong compressibility effects are absent in the turbulence field or do not need particular attention. For compressible isotropic turbulence this will be the case if the Mach number and the initial gradients are not very high. Indeed, for initial Mach numbers up to 0.6 very good results were obtained with such models as reported for example by Comte *et al.* (1990a), Erlebacher *et al.* (1992) and Zang *et al.* (1992). However, for higher initial Mach numbers these models may fail, as was observed by Comte *et al.* (1990a). The authors conjectured that the failure can be attributed to the formation of eddy shocklets in the flow which increase the energy dissipation considerably and should therefore be taken into account (see Zeman, 1990, for the modelling of such effects in statistical simulations). For the case of boundary-layer flows, which are of more practical interest, one may expect that small-scale turbulence is only slightly affected by compressibility as long as the free-stream Mach number of the flow is below 5. This corresponds to the so-called 'Morkovin hypothesis' which assumes that in this range of Mach numbers no feedback from the acoustic modes and the entropy modes onto the turbulent velocity field occurs (see Bradshaw, 1977). SGS models which were developed for incompressible flows should therefore be applicable.

C Near-wall modifications

As in statistical simulations, the proper treatment of near-wall flows is a particularly difficult task in LES. In this context one should distinguish between rough and smooth walls, since for the latter it is possible to resolve the near-wall flow explicitly while for the former it generally is not. This distinction is most significant for flows at moderate Reynolds numbers where the computational costs for the explicit treatment of the near-wall layer may be acceptable. For high-Reynolds-number flows, on the other hand, a prohibitively large number of grid points will be required if an adequate resolution of the near-wall turbulence structures is attempted (see Section II.C.4). As for rough walls, some approximate treatment of the wall-layer will be inevitable in this case too. In simple wall-bounded flows such as attached turbulent boundary layers, the near-wall flow exhibits well-established features which considerably facilitate an approximate treatment. The most important among them is the existence of the logarithmic regime where turbulence is in a state of equilibrium. For more complicated flows, however, not much is known about the near-wall physics at the present time.

The approximate treatment of the wall layer in LES is usually based on boundary conditions for the outer flow which have to represent the dominant features of the interaction of outer and inner layer. These boundary conditions are imposed at the first mesh plane off the wall assuming that the near-wall flow exhibits a universal behaviour and can be described by a logarithmic profile or similar laws. In essence, this method is analogous to the way in which the near-wall flow is treated in statistical simulations of high-Reynolds-number flows where so-called wall functions are applied rather than physical no-slip boundary conditions.

In his simulations of turbulent channel flow, Schumann (1975) employed a model where the instantaneous shear stress $\tau_{13,w}$ at the wall is set proportional to the instantaneous velocity at the first grid point directly above it

$$\tau_{13,w}(x_1, x_2) = \frac{\bar{u}_1(x_1, x_2, x_3^*)}{\langle \bar{u}_1(x_1, x_2, x_3^*) \rangle} \langle \tau_{13,w}(x_1, x_2) \rangle \tag{70}$$

In (70) x_3^* denotes the height of the first grid point and x_1 and x_2 designate the streamwise and spanwise coordinate directions, respectively. The operator $\langle \cdot \rangle$ symbolizes a statistical mean value. The averaged velocity $\langle \bar{u}_1(x_1, x_2, x_3^*) \rangle$ can be computed from the logarithmic law-of-the-wall, while the average shear stress $\langle \tau_{13,w}(x_1, x_3) \rangle$ is derived from a global momentum balance and hence is prescribed. For the modelling of the stress component $\tau_{23,w}$ Schumann assumed a linear distribution of $\bar{u}_2$ within the wall-adjacent mesh cell

$$\tau_{23,w}(x_1, x_2) = \frac{1}{Re} \frac{\bar{u}_2(x_1, x_2, x_3^*)}{x_3^*} \tag{71}$$

Refined versions of Schumann's model were later devised e.g. by Grötzbach (1987), who allowed for variations in the average wall stress, and by Piomelli *et al.* (1989), who tried to incorporate explicitly the dynamics of important near-wall coherent structures. All of these models have in common that they require the logarithmic law-of-the-wall to be satisfied on the average. In contrast, the wall model proposed by Mason and Callen (1986) assumes that the logarithmic profile also holds instantaneously and locally in the flow. Recently, Balaras and Benocci (1994) developed a more sophisticated method for the near-wall treatment. In their approach a simplified set of equations, derived from the two-dimensional boundary layer equations, is solved between the wall and the first mesh point. For the solution of these equations a refined grid has to be embedded in the mesh cells adjacent to the wall.

Approximate boundary conditions of the kind discussed above work well in flows where the coupling between the wall layer and the outer flow is weak. When near-wall effects play a dominant role for the dynamics of the outer flow such models may not be appropriate and an improved treatment of the wall layer is required. This may typically be the case for flows including viscous separation. As pointed out above, a resolution of the wall layer will generally be feasible for flows at moderate Reynolds numbers and over smooth surfaces only, but this includes a number of interesting engineering applications. For example, cascade flows in turbomachinery typically exhibit Reynolds numbers such that a coarse resolution of the wall layer appears feasible.

If the near-wall flow is resolved in an LES then special care has to be taken that the reduction of the turbulent scales close to the wall is properly accounted for in the model. For example, the Smagorinsky model yields a non-vanishing eddy viscosity even on the wall and consequently needs to be augmented with an additional correction. To accomplish this, an exponential Van Driest (1956) damping function may be applied to the SGS length scale given in (47), which leads to (Moin and Kim, 1982)

$$L = C_S[1 - \exp(-y^+/A^+)]^{\frac{1}{2}}(\Delta x_1 \Delta x_2 \Delta x_3)^{\frac{1}{3}} \tag{72}$$

In (72) y^+ denotes the wall distance in wall units and A^+ is a dimensionless constant usually set to the value $A^+ = 25$. The above formula may still be modified to achieve an improved asymptotic behaviour of the SGS eddy viscosity ν_t near the wall (Piomelli *et al.*, 1987). An alternative formulation for the near-wall modification of L was furnished by Mason and Callen (1986), who matched the length scale given by (47) to a near-wall Prandtl mixing length. This modification, however, was employed to improve the modelling within the logarithmic regime rather than within the viscosity-dominated inner layer for which the damping in (72) is devised.

Strictly speaking, empirical corrections like the one given in (72) are applicable to some simple flows only where important characteristics of the near-wall turbulence are known in advance. This is well illustrated by the fact that wall units are employed in the definition of the damping function. Other means like the dynamic procedure discussed in Section III.A are superior in this respect since the reduction of ν_t is achieved without introducing *ad hoc* knowledge about the flow. However, it must be emphasized that the proper asymptotic near-wall behaviour of the eddy viscosity is only a smaller problem for the SGS modelling in wall turbulence. The major difficulties stem from the fact that flows adjacent to solid boundaries exhibit pronounced inhomogeneities throughout all the scales, from which it follows that important prerequisites for the applicability of the eddy-viscosity concept are missing. In fact, a detailed analysis of the according interactions in wall turbulence reveals that simple subgrid models fail to account for important energy-transfer mechanisms (see Härtel, 1994; Härtel *et al.*, 1994). This deficiency may lead to considerable errors in the simulation results and can usually be remedied only by a greatly increased numerical resolution. It is likely that more sophisticated subgrid models will be needed if LES is applied to flows where near-wall features are of central importance.

D   Validation of SGS models

Like models used for statistical simulations, subgrid models for LES also need to be checked and validated carefully. Any comprehensive validation will include the application of the model in an actual LES and the subsequent comparison of the obtained results with data from a reliable source. Since LES is a means to provide efficiently and rather precisely low-order statistics of a turbulent flow, this comparison will focus on mean values and fluctuation intensities of the flow quantities of interest. This type of validation is usually called an *a posteriori* test. Another, qualitatively different way, to assess a subgrid model is to compare the exact values of the modelled quantities directly with the SGS model itself by analysing reference data from theory, experiment, or direct simulations. Since this analysis can be performed prior to the application of a model in a LES, it is termed an *a priori* test. In the past two decades *a priori* tests were primarily based on results obtained from direct simulations. Clark *et al.* (1979), McMillan and Ferziger (1979) and McMillan *et al.* (1980) were probably the first to conduct such studies when they analysed DNS results for homogeneous isotropic turbulence and turbulence in the presence of irrotational strain. Meanwhile DNS databases became available for a number of more complex flows and virtually all of them have been employed for *a priori* tests. Among those flows which received most attention is the incompressible channel flow (see e.g. Piomelli *et al.*, 1988; Horiuti,1993; Härtel *et al.*, 1994). In addition to DNS data approximate turbulent flow fields were occasionally employed for *a priori* tests. For example, Kaneda and Leslie (1983) analysed simulation results from a so-called 2.5-dimensional flow model for near-wall turbulence. In recent years turbulence measurements with a high spatial resolution became feasible by the use of techniques like particle displacement velocimetry (PDV). This paved the way for *a priori* tests based on two-dimensional instantaneous flow fields obtained from experiments. Such a study was conducted for the far field of a turbulent jet by Liu *et al.* (1994). Meneveau (1994) performed *a priori* tests of various subgrid models and various filters using one-dimensional data obtained from single-probe measurements in grid turbulence.

The *a priori* test and *a posteriori* test should be considered as complementary rather then alternative ways to assess the performance of SGS models, and therefore should be combined whenever possible. Clearly, the final word on the success or failure of a subgrid model can only come from simulations in which the model is employed, i.e. from *a posteriori* tests. However, the *a priori* test can make the specific merits and weaknesses of the models more evident and may therefore provide both support for the interpretation of LES results and suggestions for model improvements.

IV NUMERICAL METHODS

The numerical simulation of turbulent flows requires very accurate numerical methods because turbulence exhibits a broad range of eddies and the evolution of even the smallest of them should be treated as accurately as possible. Numerical approximations with significant artificial viscosity, such as low-order upwind methods, are in general unsuitable for use in such applications. In DNS and LES the numerical discretization is typically based on higher-order finite-difference methods or spectral schemes. The advantage of spectral methods is that they allow for a very accurate numerical differentiation and achieve a higher accuracy with given number of grid points (or discrete modes) than finite-difference methods. On the other hand, a spectral method often requires more floating-point operations per mesh point than does a finite-difference scheme and consequently a finer resolution can be chosen with the latter for a given computing time. This aspect may be more important for large-eddy simulations where it is necessary to resolve as much of the turbulent scales as possible. However, an important requirement for the numerical method in LES is that the discretization error remains small compared with the SGS effects which are modelled.

Since the additional subgrid terms in an LES do not pose severe additional problems for the numerical scheme, virtually the same numerical methods are used for direct and large-eddy simulations. The basic concepts employed for the discretization will be outlined in this section together with a brief discussion of proper initial and boundary conditions. The focus will be on spectral methods since the fundamentals of finite-difference and finite-volume schemes for incompressible and compressible flows are laid out in Chapters 2 and 4 of this volume. Very extensive discussions of many relevant theoretical and practical aspects of spectral methods can be found in the monographs of Gottlieb and Orszag (1977) and Canuto et al. (1988). A review of these methods was also given by Hussaini and Zang (1987).

Recent progress in spectral and other high-order methods is reported in VKI (1994) and in the proceedings of the International Conferences on Spectral and High Order Methods (ICOSAHOM), edited by Canuto and Quarteroni (1989) and Bernardi and Maday (1994).

A Spatial discretization

If finite-difference schemes are used for the spatial discretization, the difference formulas for the numerical differentiation can either be defined in an explicit or in an implicit manner. In the latter case the derivatives at the mesh points are related to each other and to the function values while in the former case only function values, are included. Essential advantages of implicit difference approximations over explicit ones are that they require many fewer stencil points to achieve the same order of accuracy and that they may exhibit very favourable (in fact spectral-like) resolution properties. In DNS and LES applications traditionally explicit methods based on second- or fourth-order accurate central difference formulas have been more common. However, more recently implicit methods, in the form of compact differencing approximations (see Lele, 1992; also Chapter 1 of this book), have received closer attention in particular for the DNS of transitional flows (see e.g. Sandham and Reynolds, 1991; Adams, 1993; Adams and Shariff, 1995; Pruett et al., 1995). Applications of such methods for DNS and LES of turbulence were reported, for example, by Kral and Zang (1992), Guo and Adams (1994) and Schiestel and Viazzo (1995). Detailed discussions of high-order implicit methods and their application to fluid dynamics problems can be found in Hirsh (1975), Aubert and Deville (1983), Peyret and Taylor (1983) and Lele (1992).

Spectral methods

For the application of DNS and LES to transitional and turbulent flows spectral methods have become a widely used numerical tool. In spectral methods the flow variables are expanded in terms of smooth, global and, usually, orthogonal functions $\{\phi_k\}$, the so-called trial functions. The term 'global' is to denote that the trial functions are defined throughout the computational domain while 'smooth' means that they are infinitely differentiable. Let $u(x)$ be an arbitrary function, then the approximation of $u(x)$ by a finite series reads

$$u_N(x) = \sum_{k=0}^{N} a_k \phi_k(x) \tag{73}$$

where a_k denotes the unknown expansion coefficients. To evaluate derivatives of the function $u(x)$, the expansion in (73) can be differentiated analytically and may be re-expanded in terms of the trial functions, i.e.

$$\frac{d^p}{dx^p} u_N(x) = \sum_{k=0}^{N} a_k \frac{d^p}{dx^p} \phi_k(x) = \sum_{k=0}^{N'} a_k^{(p)} \phi_k(x) \tag{74}$$

where $a_k^{(p)}$ designates the expansion coefficients of the p-th derivative.

The determination of the coefficients a_k of the expansion in (73) may be based on the Galerkin, tau, or collocation method, which all belong to the general class of the 'methods of weighted residuals' (see Canuto *et al.*, 1988). In the Galerkin and tau methods, the solution is sought in terms of these expansion coefficients directly, while in the collocation method, which is concentrated upon here, the values at the nodal points are usually taken directly as the unknowns. Since in the collocation approach the expansion coefficients are only used to perform the analytical differentiation according to (74), this method is often referred to as the pseudospectral method in the literature (Orszag, 1972). The collocation approach generally is the simplest of the three methods and is computationally less expensive than the others if non-linearities have to be evaluated. Another attractive feature of this method is that boundary conditions can be more easily implemented owing to the fact that the nodal unknowns are explicitly used.

The proper choice of the expansion functions is a crucial aspect of spectral methods and will depend on the characteristics of the flow domain, the nature of the solution and the respective boundary conditions. For the case of a bounded cartesian domain where periodic boundary conditions can be applied, Fourier series are most efficient. Taking for simplicity the periodicity length equal to 2π, the expansion in (73) reads for (complex) trigonometric polynomials

$$u_N(x) = \sum_{|k|<N/2} \hat{u}_k \exp(ikx) \tag{75}$$

$\hat{u}_k$ being the complex Fourier coefficients. If u_N is real, one finds that

$$\hat{u}_{-k} = \hat{u}_k^* \tag{76}$$

where the asterisk symbolizes a complex conjugate. For the collocation method the Fourier coefficients are given by the common formula for trigonometric interpolation, i.e.

$$\hat{u}_k = \frac{1}{N} \sum_{j=0}^{N-1} u(x_j) \exp(-ikx_j) \tag{77}$$

using the equidistant collocation points $x_j = 2\pi j/N, j = 0, 1, ..., N - 1$. The transformations between the 'spectral space' $\{\hat{u}_k\}$ and 'physical space' (the nodal values) can be efficiently accomplished using a fast Fourier transform (FFT). For trigonometric polynomials the differentiation of u_N is particularly simple and (74) may be written in this case

$$\frac{d^p}{dx^p} u_N(x) = \sum_{|k|<N/2} (ik)^p \hat{u}_k \exp(ikx) \tag{78}$$

If the domain is bounded but periodic boundary conditions cannot be chosen, Jacobi polynomials are the appropriate class of ansatz functions (Hussaini and Zang, 1987). The application of Jacobi polynomials requires the computational domain to be mapped onto the interval $[-1, 1]$. Among the Jacobi polynomials Chebyshev and Legendre polynomials are those most frequently used in spectral schemes. The Chebyshev polynomials $T_k(x)$ are given by

$$T_k(x) = \cos(k \arccos x) \tag{79}$$

and satisfy the following recursion relation

$$T_{k+1}(x) + T_{k-1}(x) = 2xT_k(x) \tag{80}$$

where

$$T_0(x) = 1, \quad T_1(x) = x \tag{81}$$

Given the coefficients a_k of the Chebyshev expansion, the coefficients $a_k^{(p)}$ of the p-th derivative (74) can be obtained from simple recursion formulas (see Canuto et al., 1988). If the collocation points are given by the cosine distribution $x_j = \cos(\pi j/N), j = 0, ..., N$, which in practice is the most common choice, the coefficients a_k for the Chebyshev collocation method can be computed as (Hussaini and Zang, 1987)

$$a_k = \frac{2}{N\bar{c}_k} \sum_{j=0}^{N} \frac{1}{\bar{c}_j} u(x_j) T_k(x_j) \tag{82}$$

where $\bar{c}_0 = \bar{c}_N = 2$, and $\bar{c}_k = 1$ otherwise. With the above choice of collocation points the Chebyshev transform reduces to a cosine transform which allows FFT algorithms to be employed for the transformation between spectral and real space.

If the domain is unbounded, neither Fourier nor Chebyshev series can be directly utilized, but a mapping of the domain onto a proper finite interval may allow for the application of these spectral expansions. An important point is that the favourable exponential convergence of spectral methods can frequently be retained, provided the mapping is carefully chosen and the function $u(x)$ decays sufficiently fast at infinity (see Canuto et al., 1988).

In collocation methods the differentiation can be represented by multiplying the vector $\{u(x_j)\}$ of the nodal values with a matrix D. A differentiation of order p is therefore given by $d^p/dx^p u(x_i) = D^p u(x_i)$. Since the matrix D is known in closed form for Fourier and Chebyshev collocation (see Canuto *et al.*, 1988), derivatives may in principle be evaluated either in spectral or in real space. However, unlike the matrices typically applied in finite-difference methods, the matrix D is full and therefore requires $2N^2$ operations, which for larger values of N becomes considerably more expensive than a pair of fast transforms plus the differentiation in spectral space. On modern super-computers this will typically be the case if N exceeds 60.

The numerical stability properties of spectral collocation methods depend on the scheme which is applied for the time discretization (see following section) and on the properties of the discrete spatial differentiation operators, i.e. the differentiation matrices (for detailed discussions see Gottlieb and Orszag, 1977; Canuto *et al.*, 1988). For a Fourier method the spectral radius of the spatial operator of the first derivative grows linearly with the number N of grid points, while that of the second derivative varies as N^2. If explicit time-integration schemes are employed, the stability limit for the time step will consequently scale as $1/N$ and $1/N^2$, respectively, and this is similar to what holds for the spatial operators of finite-difference methods. The time-step restriction is much more severe for the case of Chebyshev collocation where the spectral radius grows with N as N^2 for the first derivative and as N^4 for the second derivative owing to the clustering of the collocation points near the boundaries. If implicit time-integration schemes are used, this severe time-step restriction disappears, but the efficient solution of the resulting implicit equations may pose difficulties. In general cases an implicit time discretization will require iterative solution methods (see e.g. Zang *et al.*, 1984; Deville and Mund, 1985; Malik *et al.*, 1985). However, for a number of important special cases fast direct solution methods are known (see e.g. Moin and Kim, 1980; Kleiser and Schumann, 1984; Le Quéré and de Roquefort, 1985; Ehrenstein and Peyret, 1989). A detailed discussion of iterative and direct solution methods is given in Chapters 5 and 7 of Canuto *et al.* (1988).

In many practical flow problems the expansion of the variables in terms of global expansion functions is not appropriate. This may either be due to a more complicated geometry of the flow domain or to the physical properties of the flow, for example if large variations of the numerical resolution over the computational domain are required but cannot be achieved by simple global mappings. A more recent development to tackle these problems with spectral methods is the spectral domain-decomposition technique (see Chapter 3 of this volume and recent reviews by Karniadakis *et al.*, 1993; Rønquist, 1994; Timmermans, 1994). In this approach, the computational domain is partitioned into a number of subdomains on each of which a spectral scheme is used. The individual domains have then to be patched together by the aid of additional interface conditions (see Hussaini and Zang, 1987). In recent years spectral domain-decomposition methods have been used for DNS of turbulent flows in slightly more complex geometries. Examples are the flow over surface-mounted riblets which was simulated by Chu *et al.* (1992) and the grooved-channel flow studied by Amon (1993).

B Time advancement

The time advancement in the numerical scheme may either be done explicitly or implicitly. The advantage of explicit time integration schemes is that they usually are more easily implemented, but stability considerations frequently suggest the use of at least partially implicit methods. For example, in simulations of wall-bounded flows an implicit treatment of the viscous term can be advantageous to overcome the severe time-step restrictions which arise from the very fine grid spacings in wall-normal direction. In DNS such fine grids are necessary to resolve the steep near-wall gradients.

For compressible simulations a generally applicable and effective semi-implicit time discretization is still missing and most compressible simulations have utilized fully explicit schemes. In incompressible simulations, on the other hand, typically mixed explicit/implicit methods are chosen with an implicit treatment of the viscous term, the pressure term, and the continuity constraint together with an explicit discretization of the convective term. The implicit time discretization is traditionally done using the second-order accurate Crank–Nicolson scheme, but second-order backward-differentiation formulas have also been used successfully (see e.g. Vanel *et al.*, 1986). Explicit time discretizations are mostly based on second-, third-, or fourth-order accurate Runge–Kutta methods which replaced the second-order accurate Adams–Bashforth and Leapfrog schemes within recent years. While the latter two methods require fewer computations per time step, the Runge–Kutta schemes exhibit a larger stability domain and therefore allow for larger time steps. This profit can more than balance the additional computational work due to the multiple evaluation of the convective terms. In many turbulent flows, however, the time step imposed by the convective stability limit may still be much smaller than the characteristic time-scale of the smallest resolved eddies. In some cases this time-step restriction can be ameliorated by utilizing a coordinate system which moves at constant speed. As a result the convection of the flow relative to the mesh will be reduced which then allows for larger time steps (see e.g. Deardorff, 1970).

C Initial and boundary conditions

The proper choice of boundary conditions is a crucial part of the numerical simulation of turbulent flows. This refers to both the specification of physical conditions and the selection of numerical boundary closures. Inflow and outflow boundaries pose particularly severe difficulties when a fully turbulent flow prevails, since detailed knowledge is required about the state of the flow outside the computational domain. This can usually be accomplished by *ad hoc* assumptions only. Because the influence of upstream conditions can persist far downstream, the inflow problem appears to be the most difficult one (Rogallo and Moin, 1984). If the flow exhibits one or more directions of statistical homogeneity, typically periodic boundary conditions are applied in these directions. For this type of boundary conditions the flow at the corresponding inflow and outflow positions is identical. Care has then to be taken that the box size is sufficiently large to allow all flow quantities to decorrelate completely within half the periodicity lengths of the computational domain.

In some cases periodic boundary conditions can be applied to flows where the required spatial homogeneity is missing. This can be done if the spatially inhomogeneous flow is homogeneous in time and may be approximated by a corresponding flow which develops in time but is homogeneous in space. Typical examples for flows where such an approximation may work are free turbulent shear flows like jets, wakes or mixing layers and turbulent boundary layers where the streamwise growth is sufficiently small (see e.g. Moser and Rogers, 1994, for a DNS of a plane wake; Leith, 1990, for the LES of a compressible mixing layer, and Kral and Zang, 1992, for the LES of a supersonic boundary layer). The core of this approximation is to assume that the flow is strictly parallel in the mean which, however, usually result in some changes in the flow characteristics. In recent years several improvements of this approach were developed to take into account some of the non-parallel effects in spatially evolving boundary layers. For example, Spalart (1988) applied a multiple-scale analysis to obtain a system of equations which can be solved with periodic boundary conditions to provide an approximation of the flow in a slowly growing boundary layer. Guo *et al.* (1995) used a spatially periodic simulation to compute the turbulence quantities, but coupled this with an additional simulation of the mean flow based on the Reynolds-averaged Navier–Stokes equations. To allow for a marching procedure in streamwise direction, the authors employed a parabolization of the mean-flow equations. Another recent and possibly promising development is to provide a short transition regime at the end of the computational domain where the outflowing disturbances are damped or rescaled in order to allow for the application of periodic boundary conditions (and possibly to provide realistic turbulent inflow data). This approach has been followed by several groups and has successfully been applied in the study of transitional and turbulent boundary layers in both incompressible and compressible flows (Spalart, 1989; Spalart and Watmuff, 1993; Guo *et al.*, 1994a).

In cases where three-dimensional and time-dependent inflow boundary conditions have to be provided explicitly, results from separate computations can often be utilized. For example, Friedrich and Arnal (1990), who studied the flow over a backward-facing step, and Werner and Wengle (1993), who simulated the flow around a cube in a channel, employed data from large-eddy simulations of developed turbulent channel flow to obtain the desired inflow conditions. This approach is particularly efficient if the supplementary simulation can be performed with periodic boundary conditions.

The relative importance of initial conditions for turbulence simulations depends strongly on the type of flow. For flows like decaying homogeneous isotropic turbulence one finds that the time history to a large extent depends on the initial state of the flow and, hence, realistic initial conditions play a central role. Inhomogeneous flows, on the other hand, like turbulent channel flow or turbulent boundary layers tend to establish their own characteristic statistically steady state and become completely independent from the initially prescribed flow field in the end. However, in an actual simulation the length of the transient phase which has to be computed before the desired steady state is achieved may depend strongly on the initial conditions. Consequently a realistic initial flow field is desirable in this case as well, in order to minimize the computer time required to obtain a statistically steady flow.

Particularly well-suited as initial conditions are instantaneous flow fields from previously performed direct or large-eddy simulations of similar flow problems. If such flow fields can be adapted to the actual problem without significant alterations, only a short transient may be necessary to redevelop a statistical steady state (Moin and Kim, 1982; Grötzbach, 1987). If no data from previous simulations are available, the turbulence fields may be initialized by random fluctuations which should match the boundary conditions and possibly the incompressibility constraint. These random fluctuations can be modified to obtain a realistic spatial distributions of the fluctuating intensities or realistic energy spectra (see e.g. Schumann, 1973). Initial conditions for the mean-field quantities, like the mean velocities or the mean temperature, may be obtained from experimental data or empirical relations like the logarithmic law-of-the-wall (Grötzbach, 1987). In principle one can think of matching the initial conditions not only to first- and second-order moments, but also to moments of third or even higher order. However, a turbulence field cannot uniquely be determined with a finite set of statistical moments and, therefore, the initial conditions will not necessarily become more realistic if further statistical moments are prescribed. An adjustment period will be indispensable in any case to attain a physically meaningful turbulent state.

V APPLICATIONS OF DNS AND LES

In this section I want to present a survey of DNS and LES applications for fundamental research and more engineering-type problems during the past two decades. This overview is not meant to be a complete summary of all contributions to the fields, but represents a selection of typical works in order to illustrate the development of these methods and their current state-of-the-art. Many further applications, also to geophysical problems, can be found in the books and reviews which were cited in Section I.

Fundamental studies using DNS and LES have frequently concerned homogeneous turbulence fields. Large-eddy simulations of incompressible isotropic turbulence were performed e.g. by Bardina *et al.* (1980), Dang (1985), Lesieur and Rogallo (1989), Chasnov (1991), Moin *et al.* (1991) and Métais and Lesieur (1992). Direct simulations of this flow were performed, among others, by Orszag and Patterson (1972), Clark *et al.* (1979), Brachet *et al.* (1983), Kerr (1985) and Vincent and Meneguzzi (1991). DNS results for compressible isotropic turbulence were reported by Erlebacher *et al.* (1992), Zang *et al.* (1992) and Blaisdell *et al.* (1993). Hannappel and Friedrich (1994) used DNS to investigate the interaction of compressible isotropic turbulence with a shock wave. Homogeneous incompressible turbulence suddenly exposed to rotation was studied using LES by Bardina *et al.* (1985), Dang and Roy (1985b) and Squires (1993). Bardina *et al.* (1985) and Dang and Roy (1985b) also present DNS results for this type of flow. Another homogeneous flow that was extensively studied by direct and large-eddy simulations is incompressible turbulence subject to homogeneous shear. Large-eddy simulations of this problem were made by Cambon *et al.* (1981), Dang (1985), Dang and Roy (1985a), Aupoix (1986) and Laurence (1986); DNS results can be found in Rogallo (1981), Dang and Roy (1985a) and Rogers *et al.* (1986). Direct simulations of compressible turbulence in the presence of homogeneous shear were conducted by Blaisdell *et al.* (1993).

An unbounded but inhomogeneous turbulent flow that has repeatedly been investigated is the turbulent mixing layer. Large-eddy simulations of mixing layers were conducted by Mansour *et al.* (1978), Marayuma (1988) and Comte *et al.* (1990b) for incompressible flow, and by Leith (1990), Ragab and Sheen (1991) and Sheen *et al.* (1993) for compressible flow. DNS results for incompressible mixing layers can be found in Moser and Rogers (1992). Further direct and large-eddy simulations of compressible mixing layers were performed by Vreman *et al.* (1994). Closely related to the turbulent mixing layer is the turbulent jet which was studied using LES by Baron and Laurence (1983), Pourquie and Eggels (1993) and Voke and Gao (1993).

The majority of the more recent investigations have been concerned with wall-bounded turbulent flows in various geometries. For example, Biringen and Reynolds (1981) used large-eddy simulation to study incompressible decaying turbulence in the presence of a solid wall, the so-called 'shear-free boundary layer'. Direct simulation results of this flow were recently reported by Perot and Moin (1993). Large-eddy simulations of 'regular' incompressible boundary layers were conducted by Schmitt and Friedrich (1984), Esmaili and Piomelli (1992) and Mason and Thomson (1992). Direct numerical simulations of turbulent boundary layers are reported by Spalart (1988) and Spalart and Watmuff (1993) for incompressible flow, and by Guo and Adams (1994) for supersonic flow. Large-eddy simulations of a supersonic boundary layer were conducted by Kral and Zang (1992). Further studies on turbulent boundary layers concerned the flow over curved walls. For example, Friedrich and Su (1982) and Moin et al. (1994) performed large-eddy simulations of an incompressible boundary layer with longitudinal curvature. A direct simulation of a transversely curved boundary layer was presented by Neves et al. (1992).

A very extensive application of DNS and LES is the simulation of turbulent flow in a plane channel. Large-eddy simulations of this flow were conducted, among others, by Deardorff (1970), Schumann (1973, 1975), Grötzbach (1977), Moin and Kim (1982), Horiuti (1985), Mason and Callen (1986), Piomelli et al. (1989), Yakhot et al. (1989), Germano et al. (1991), Härtel and Kleiser (1992) and Härtel (1994). In the study of Piomelli et al. (1989) the case of a channel flow including transpiration is also investigated. Direct simulations of turbulent channel flow were reported by Kim et al. (1987), Gilbert and Kleiser (1991) and Härtel (1994). Sendstad and Moin (1993) presented a DNS study of a turbulent channel flow which is suddenly exposed to a spanwise pressure gradient.

Several flow problems which are closely related to plane channel flow have occasionally been studied in the past. Examples are the flow in a straight pipe, which was investigated by Unger and Friedrich (1993) using LES, and by Eggels et al. (1994) and Zhang et al. (1994) using DNS, and the flow in an annular tube (see e.g. Schumann, 1975; Grötzbach, 1987). Another related problem is the flow in a rotating channel which was investigated using LES by Kim (1983) and Piomelli and Liu (1995). Direct simulations of rotating-channel flow were reported by Kristoffersen and Andersson (1993) and Piomelli and Liu (1995). The turbulent Couette flow was also concerned in several DNS and LES studies. Kobayashi and Kano (1986) investigated this flow by large-eddy simulation while Lee and Kim (1991) and Bech and Andersson (1994) performed direct numerical simulations. Large-eddy simulations of the turbulent flow in a square duct were reported by Madabhushi and Vanka (1991), Balaras and Benocci (1994) and Su and Friedrich (1994); DNS results for this flow were presented by Gavrilakis (1992) and Huser and Biringen (1993). A flow problem considerably different with respect to the physics, but with a similar flow geometry is the turbulent Rayleigh–Bénard convection. Large-eddy and direct simulations of this flow were conducted, among others, by Grötzbach (1987) and Grötzbach and Wörner (1994), respectively.

A more complex flow problem which was studied by a number of authors is the flow over a backward-facing step. This flow exhibits several complicated features such as solid walls, separation and reattachment. Moreover, for this problem a careful choice of suitable inflow and outflow boundary conditions is particularly important. Large-eddy simulations of the backward-facing step flow were performed by Friedrich and Arnal (1990), Karniadakis et al. (1990), Morinishi and Kobayashi (1990), Silveira Neto et al. (1991) and Akselvoll and Moin (1993). DNS studies of this flow were performed by Le and Moin (see Akselvoll and Moin, 1993). Related problems are the sudden pipe expansion, which was studied using DNS by Wagner and Friedrich (1994), and the flow of confined coannular jets, which was investigated by Akselvoll and Moin (1995) using LES. Another type of flow where separation plays a key role, but where boundary conditions can be chosen in a natural way, is the driven cavity flow. LES studies of this problem were conducted e.g. by Jordan and Ragab (1993), who also performed a direct simulation, and by Zang et al. (1993). Further examples of the applications of LES to more complex problems are the flow over square ribs, which was studied by Werner and Wengle (1989) and Yang and Ferziger (1993), and the flow over and around a cube in a plane channel which was studied by Kobayashi et al. (1984), He and Song (1993) and Werner and Wengle (1993). In addition to the simulation of the flow around a cube, He and Song (1993) present LES results for several other interesting applications including diffuser flows or flows in bifurcation pipes and draft tubes which are used in power engineering. Another more complex application of LES is the flow around turbulence-control devices (LEBU devices) which was studied by Klein and Friedrich (1990). Typical examples of recent DNS applications for more complex problems are the flow over surface-mounted riblets, which was simulated by Chu et al. (1992) and Choi et al. (1993), and the grooved-channel flow studied by Amon (1993).

Many more applications could have been listed, and their number certainly will grow rapidly in the future. Readers interested in the future progress should, in addition to the relevant archival journals, check the biannual conference series 'Symposium on Turbulent Shear Flows', 'European Turbulence Conference', and 'ERCOFTAC Workshop on Direct and Large-Eddy Simulation'.

ACKNOWLEDGEMENT

The author wishes to thank Professor L. Kleiser for many helpful suggestions and for reviewing a previous version of this paper.

REFERENCES

Adams, N.A. (1993). Doctoral Dissertation, Technical Univ. of Munich, Germany. Also: DLR-FB 93-29, Deutsche Forschungsanstalt für Luft- und Raumfahrt, Germany.

Adams, N.A. and Shariff, K. (1995). *J. Comp. Phys.* (submitted).

AGARD (1994). *Application of Direct and Large Eddy Simulation to Transition and Turbulence*, AGARD-CP-551. AGARD, Neuilly-sur-Seine.

Akselvoll, K. and Moin, P. (1993). In *Engineering Applications of Large-Eddy Simulations* (S.A. Ragab and U. Piomelli, eds), FED vol. 162, pp. 1–5, ASME, New York.

Akselvoll, K. and Moin, P. (1995). Report No. TF-63, Dept. Mech. Engng., Stanford Univ.

Aldama, A.A. (1990). *Filtering Techniques for Turbulent Flow Simulation*, Lecture Notes in Engineering, vol. 56. Springer-Verlag, Berlin.

Amon, C.H. (1993). *AIAA J.* **31**, 42–48.

Anderson, D.A., Tannehill, J.C. and Pletcher, R.H. (1984). *Computational Fluid Mechanics and Heat Transfer*. Hemisphere Publishing Co., New York.

Antonia, R.A. and Kim, J. (1994). *J. Fluid Mech.* **276**, 61–80.

Aubert, X.–L. and Deville, M. (1983). *J. Comput. Phys.* **49**, 490–522.

Aupoix, B. (1986). In *Direct and Large Eddy Simulation of Turbulence* (U. Schumann and R. Friedrich, eds), pp. 37–66, Notes on Numerical Fluid Mechanics, vol. 15. Vieweg-Verlag, Braunschweig.

Balaras, E. and Benocci, C. (1994). In *Application of Direct and Large Eddy Simulation to Transition and Turbulence*, AGARD-CP-551. AGARD, Neuilly-sur-Seine.

Bardina, J., Ferziger, J.H. and Reynolds, W.C. (1980). *AIAA Paper 80-1357*, AIAA 13th Fluid and Plasma Dynamics Conference, Snowmass, CO.

Bardina, J., Ferziger, J.H. and Rogallo, R.S. (1985). *J. Fluid Mech.* **154**, 321–336.

Baron, F. and Laurence, D. (1983). In *Proc. 4th Symposium on Turbulent Shear Flows*, Karlsruhe, 12–14 September, 1983.

Bech, K.H. and Andersson, H.I. (1994). In *Direct and Large-Eddy Simulation I* (P.R. Voke *et al.*, eds), pp. 13–24. Kluwer Academic Publishers, Dordrecht.

Bernardi, C. and Maday, Y. (eds) (1994). *Analysis, Algorithms and Applications of Spectral and High Order Methods for Partial Differential Equations*. North-Holland, Amsterdam.

Bertoglio, J.-P. (1982). In *Turbulent Shear Flow 3* (L.J.S. Bradbury *et al.*, eds), pp. 253–261. Springer-Verlag, Berlin.

Biringen, S. and Reynolds, W.C. (1981). *J. Fluid Mech.* **103**, 53–63.

Blaisdell, G.A., Mansour, N.N. and Reynolds, W.C. (1993). *J. Fluid Mech.* **256**, 443–485.

Boris, J.P. (1990). In *Whither Turbulence? Turbulence at the Crossroads* (J.L. Lumley, ed.), Lecture Notes in Physics, vol. 357, pp. 344–353, . Springer–Verlag, Berlin.

Brachet, M.E., Meiron, D.I., Orszag, S.A., Nickel, B.G., Morf, R.H. and Frisch, U. (1983). *J. Fluid Mech.* **130**, 411–452.

Bradshaw, P. (1977). In *Annual Review of Fluid Mechanics*, vol. 9, pp. 33–54. Annual Reviews Inc., Palo Alto, CA.

Cabot, W. and Moin, P. (1991). In *Large Eddy Simulation of Complex Engineering and Geophysical Flows* (B. Galperin and S.A. Orszag, eds), pp. 141–158. Cambridge Univ. Press, Cambridge.

Cambon, C., Jeandel, D. and Mathieu, J. (1981). *J. Fluid Mech.* **104**, 247–262.

Canuto, C. and Quarteroni, A. (eds) (1989). *Spectral and High Order Methods for Partial Differential Equations*. North-Holland, Amsterdam.

Canuto, C., Hussaini, M.Y., Quarteroni, A. and Zang, T.A. (1988). *Spectral Methods in Fluid Dynamics*. Springer–Verlag, New York.

Carati, D., Ghosal, S. and Moin, P. (1995). *Phys. Fluids A* **7**, 606–616.

Chasnov, J.R. (1991). *Phys. Fluids A* **3**, 188–200.

Choi, H., Moin, P. and Kim, J. (1993). *J. Fluid Mech.* **255**, 503–539.

Chollet, J.-P. and Lesieur, M. (1981). *J. Atmos. Sci.* **38**, 2747–2757.

Chu, D., Henderson, R. and Karniadakis, G.E. (1992). *Theoret. Comput. Fluid Dynamics* **3**, 219–229.

Clark, R.A., Ferziger, J.H. and Reynolds, W.C. (1979). *J. Fluid Mech.* **91**, 1–16.

Comte, P., Lee, S. and Cabot, W.H. (1990a). In *Proc. 1990 CTR Summer Program*, pp. 31–45. NASA Ames/Stanford Univ., Stanford, CA.

Comte, P., Lesieur, M. and Fouillet, Y. (1990b). In *Topological Fluid Mechanics* (H.K. Moffatt and A. Tsinober, eds), pp. 649–658. Cambridge Univ. Press, Cambridge.

Dang, K.T. (1985). *AIAA J.* **23**, 221–227.

Dang, K.T. and Roy, P. (1985a). In *Macroscopic Modelling of Turbulent Flows* (U. Frisch *et al.*, eds), pp. 134–147. Springer-Verlag, Berlin.

Dang, K.T. and Roy, P. (1985b). *Proc. 5th Symposium on Turbulent Shear Flows.* Cornell Univ., Ithaca, NY, 7–9 August, 1985.

Dakhoul, Y.M. and Bedford, K.W. (1986a). *Int. J. Numer. Methods Fluids* **6**, 49–64.

Dakhoul, Y.M. and Bedford, K.W. (1986b). *Int. J. Numer. Methods Fluids* **6**, 65–82.

Dean, R.B. (1978). *J. Fluids Engng.* **100**, 215–223.

Deardorff, J.W. (1970). *J. Fluid Mech.* **41**, 453–480.

Deardorff, J.W. (1973). *J. Fluids Engng.* **95**, 429–438.

Deville, M. and Mund, E. (1985). *J. Comput. Phys.* **60**, 517–533.

Domaradzki, J.A. and Rogallo, R.S. (1990). *Phys. Fluids A* **2**, 412–426.

Domaradzki, J.A., Metcalfe, R.W., Rogallo, R.S., and Riley, J.J. (1987). *Phys. Rev. Lett.* **58**, 547–550.

Domaradzki, J.A., Rogallo, R.S. and Wray, A.A. (1990). In *Proc. 1990 CTR Summer Program*, pp. 319–329. NASA Ames/Stanford Univ., Stanford, CA.

Domaradzki, J.A., Liu, W., Härtel, C. and Kleiser, L. (1994). *Phys. Fluids A* **6**, 1583–1599.

Eggels, J.G.M., Unger, F., Weiss, M.H., Westerweel, J., Adrian, R.J., Friedrich, R. and Nieuwstadt, F.T.M. (1994). *J. Fluid Mech.* **268**, 175–209.

Ehrenstein, U. and Peyret, R. (1989). *Int. J. Numer. Methods Fluids* **9**, 427–452.

Erlebacher, G., Hussaini, M.Y., Speziale, C.G. and Zang, T.A. (1992). *J. Fluid Mech.* **238**, 155–185.

Esmaili, H. and Piomelli, U. (1992). *Theoret. Comput. Fluid Dynamics* **6**, 369.

Ferziger, J.H. (1977). *AIAA J.* **15**, 1261–1267.

Ferziger, J.H. (1983). In *Computational Methods for Turbulent, Transonic and Viscous Flows* (J.A. Essers, ed.), pp. 93–182. Hemisphere Publ. Co., Washington.

Friedrich, R. and Arnal, M. (1990). *J. Wind Eng. and Ind. Aerodyn.* **35**, 101–128.

Friedrich, R. and Su, M.D. (1982). In *Proc. 8th International Conference on Numerical Methods in Fluid Dynamics* (E. Krause, ed.), Lecture Notes in Physics, vol. 170, pp. 196–202. Springer–Verlag, Berlin.

Galperin, B. and Orszag, S.A. (eds) (1993). *Large Eddy Simulation of Complex Engineering and Geophysical Flows.* Cambridge Univ. Press, Cambridge.

Gavrilakis, S. (1992). *J. Fluid Mech.* **244**, 101–129.

Germano, M. (1992). *J. Fluid Mech.* **238**, 325–336.

Germano, M., Piomelli, U., Moin, P. and Cabot, W.H. (1991). *Phys. Fluids A* **3**, 1760–1765.

Ghosal, S., Lund, T.S., Moin, P. and Akselvoll, K. (1995). *J. Fluid Mech.* **286**, 229–255.

Gilbert, N. and Kleiser, L. (1991). In *Proc. 8th Symposium on Turbulent Shear Flows*, Munich, 9–11 September, 1991.

Gottlieb, D. and Orszag, S.A. (1977). *Numerical Analysis of Spectral Methods: Theory and Applications.* SIAM, Regional Conference Series in Applied Mathematics, vol. 26.

Grötzbach, G. (1977). Doctoral Dissertation, Technical Univ. of Karlsruhe, Germany. Also: KFK–Bericht 2426, Kernforschungszentrum Karlsruhe.

Grötzbach, G. (1987). In *Encyclopedia of Fluid Mechanics* (N.R. Cheremisinoff, ed.), vol. 6, pp. 1337–1391. Gulf Publishers, Houston, TX.

Grötzbach, G. and Wörner, M. (1994). In *Direct and Large-Eddy Simulation I.* (P.R. Voke *et al.*, eds), pp. 387–397. Kluwer Academic Publishers, Dordrecht.

Guo, Y. and Adams, N.A. (1994). In *Proc. 1994 CTR Summer Program*, pp. 245–267. NASA Ames/Stanford Univ., Stanford, CA.

Guo, Y., Adams, N.A. and Kleiser, L. (1994a). In *Direct and Large-Eddy Simulation I* (P.R. Voke *et al.*, eds), pp. 249–259. Kluwer Academic Publishers, Dordrecht.

Guo, Y., Adams, N.A., Sandham, N.D. and Kleiser, L. (1994b). In *Application of Direct and Large Eddy Simulation to Transition and Turbulence*, AGARD-CP-551. AGARD, Neuilly-sur-Seine.

Guo, Y., Adams, N. A. and Kleiser, L. (1995). *Theoret. Comput. Fluid Dynamics* **7**, 141–157.

Hannappel, R. and Friedrich, R. (1994). In *Direct and Large-Eddy Simulation I* (P.R. Voke *et al.*, eds), pp. 359–373. Kluwer Academic Publishers, Dordrecht.

Härtel, C. (1994). Doctoral Dissertation, Technical Univ. of Munich, Germany. Also: DLR-FB 94-22, Deutsche Forschungsanstalt für Luft- und Raumfahrt, Germany.

Härtel, C. and Kleiser, L. (1992). In *Computational Fluid Dynamics '92* (Ch. Hirsch *et al.*, eds), vol. 1, pp. 215–222. Elsevier Science Publishers, Amsterdam.

Härtel, C., Kleiser, L., Unger, F. and Friedrich, R. (1994). *Phys. Fluids A* **6**, 3130–3143.

He, J. and Song, C.C.S. (1993). In *Engineering Applications of Large-Eddy Simulations* (S.A. Ragab and U. Piomelli, eds), FED vol. 162, pp. 147–154. ASME, New York.

Hinze, J.O. (1975). *Turbulence*. McGraw-Hill, New York.

Hirsh, R.S. (1975). *J. Comp. Phys.* **19**, 90–109.

Horiuti, K. (1985). *J. Phys. Soc. Jap.* **54**, 2855–2865.

Horiuti, K. (1989). *Phys. Fluids A* **1**, 426–428.

Horiuti, K. (1993). *Phys. Fluids A* **5**, 146–157.

Horiuti, K. and Yoshizawa, A. (1986). In *Direct and Large Eddy Simulation of Turbulence* (U. Schumann and R. Friedrich, eds), pp. 119–134, Notes on Numerical Fluid Mechanics, vol. 15. Vieweg-Verlag, Braunschweig.

Huser, A. and Biringen, S. (1993). *J. Fluid Mech.* **257**, 65–95.

Hussaini, M.Y. and Zang, T.A. (1987). In *Annual Review of Fluid Mechanics*, vol. 19, pp. 339–367. Annual Reviews Inc., Palo Alto, CA.

Hussaini, M.Y., Kumar, A. and Streett, C.L. (eds) (1992). *Instability, Transition and Turbulence*. Springer–Verlag, New York.

Hussaini, M.Y., Gatski, T.B. and Jackson, T.L. (eds) (1994). *Transition, Turbulence and Combustion*. Kluwer Academic Publishers, Dordrecht.

Jordan, S.A. and Ragab, S.A. (1993). In *Engineering Applications of Large-Eddy Simulations* (S.A. Ragab and U. Piomelli, eds), FED vol. 162, pp. 127–146. ASME, New York.

Kaneda, Y. and Leslie, D.C. (1983) *J. Fluid Mech.* **132**, 349–373.

Karniadakis, G.E., Orszag, S.A. and Yakhot, V. (1990). In *Engineering Turbulence Modelling and Experiments* (W. Rodi and E.N. Ganić, eds), pp. 269–278. Elsevier Science Publishing Co., New York.

Karniadakis, G.E., Orszag, S.A., Rønquist, E.M. and Patera, A.T. (1993). In *Incompressible Computational Fluid Dynamics* (M.D. Gunzburger and R.A. Nicolaides, eds), pp. 203–266. Cambridge Univ. Press, Cambridge.

Kerr, R.M. (1985). *J. Fluid Mech.* **153**, 31–58.

Kim, J. (1983). In *Proc. 4th Symposium on Turbulent Shear Flows*, Karlsruhe, 12–14 September, 1983.

Kim, J., Moin, P. and Moser, R. (1987). *J. Fluid Mech.* **177**, 133–166.

Klein, H. and Friedrich, R. (1990). In *Turbulence Control by Passive Means* (E. Coustols, ed.), pp. 41–65. Kluwer Academic Publishers, Amsterdam.

Kleiser, L. and Schumann, U. (1984). In *Spectral Methods for Partial Differential Equations* (R. G. Voigt *et al.*, eds), pp. 141–163. SIAM, Philadelphia.

Kleiser, L. and Zang, T.A. (1991). In *Annual Review of Fluid Mechanics*, vol. 23, pp. 495–537. Annual Review Inc., Palo Alto, CA.

Kobayashi, T. and Kano, M. (1986). In *Direct and Large Eddy Simulation of Turbulence* (U. Schumann and R. Friedrich, eds), pp. 135–146, Notes on Numerical Fluid Mechanics, vol. 15., Vieweg-Verlag, Braunschweig.

Kobayashi, T., Kano, M. and Ishihara, T. (1984). *Bull. JSME* **27**, 1893–1898.

Kraichnan, R.M. (1976). *J. Atmos. Sci.* **33**, 1521–1536.

Kral, L.D. and Zang, T.A. (1992). In *Instability, Transition and Turbulence* (M.Y. Hussaini *et al.*, eds), pp. 589–599. Springer-Verlag, New York.

Kral, L.D. and Zang, T.A. (eds) (1993). *Transitional and Turbulent Compressible Flows*, FED vol. 151. ASME, New York.

Kristoffersen, R. and Andersson, H.I. (1993). *J. Fluid Mech.* **256**, 163–197.

Laurence, D. (1986). In *Direct and Large Eddy Simulation of Turbulence* (U. Schumann and R. Friedrich, eds), pp. 147–160, Notes on Numerical Fluid Mechanics, vol. 15. Vieweg-Verlag, Braunschweig.

Lee, M.J. and Kim, J. (1991). In *Proc. 8th Symposium on Turbulent Shear Flows*, Munich, 9–11 September, 1991.

Leith, C.E. (1990). *Phys. Fluids A* **2**, 297–299.

Lele, S.K. (1992). *J. Comput. Phys.* **103**, 16–42.

Leonard, A. (1974). *Adv. Geophys. A* **18**, 237–248.

LeQuere, P. and de Roquefort, T.A. (1985). *J. Comput. Phys.* **57**, 210–228.

Lesieur, M. (1990). *Turbulence in Fluids*. Kluwer Academic Publishers, Amsterdam.

Lesieur, M. and Rogallo, R. (1989). *Phys. Fluids A* **1**, 718–722.

Lesieur, M., Comte, P. and Normand, X. (1991). *AIAA Paper 91-0335.*

Leslie, D.C. and Quarini, G.L. (1979). *J. Fluid Mech.* **91**, 65–91.

Lilly, D.K. (1967). In *Proc. IBM Scientific Computing Symposium on Environmental Sciences*, pp. 195–210. Thomas Watson Research Center, Yorktown Heights, N.Y.

Lilly, D.K. (1992). *Phys. Fluids A* **4**, 633–635.

Liu, S., Meneveau, C. and Katz, J. (1994). *J. Fluid Mech.* **275**, 83–119.

Love, M.D. and Leslie, D.C. (1979). In *Turbulent Shear Flows I* (F. Durst *et al.*, eds), pp. 353–369. Springer-Verlag, Berlin.

Madabhushi, R.K. and Vanka, S.P. (1991). *Phys. Fluids A* **3**, 2734–2744.

Malik, M.R., Zang, T.A., and Hussaini, M.Y. (1985). *J. Comput. Phys.* **61**, 64–88.

Manhart, M. and Wengle, H. (1994). In *Direct and Large-Eddy Simulation I* (P.R. Voke *et al.*, eds), pp. 299–310. Kluwer Academic Publishers, Dordrecht.

Mansour, N., Ferziger, J.H. and Reynolds, W.C. (1978). Report No. TF–11, Dept. Mech. Engng., Stanford Univ.

Maruyama, Y. (1988). *Trans. Jap. Soc. Aeronaut. Space Sci.* **31**, 79–93.

Mason, P.J. and Callen, N.S. (1986). *J. Fluid Mech.* **162**, 439–462.

Mason, P.J, and Thomson, D.J. (1992). *J. Fluid Mech.* **242**, 51–78.

McComb, W.D. and Shanmugasundaram, V. (1986). *J. Phys.* **18 A**, 2191–2198.

McMillan, O.J. and Ferziger, J.H. (1979). *AIAA J.* **17**, 1340–1346.

McMillan, O.J., Ferziger, J.H. and Rogallo, R.S. (1980). *AIAA Paper 80-1339*, AIAA 13th Fluid and Plasma Dynamics Conference, Snowmass, CO.

Meneveau, C. (1994). *Phys. Fluids* **6**, 815–833.

Métais, O. and Lesieur, M. (1992). *J. Fluid Mech.* **239**, 157–194.

Moin, P. and Kim, J. (1980). *J. Comput. Phys.* **35**, 381–392.

Moin, P. and Kim, J. (1982). *J. Fluid Mech.* **118**, 341–377.

Moin, P., Squires, K., Cabot, W. and Lee, S. (1991). *Phys. Fluids A* **3**, 2746–2757.

Moin, P., Carati, D., Lund, T., Ghosal, S. and Akselvoll, K. (1994). In *Application of Direct and Large Eddy Simulation to Transition and Turbulence*, AGARD-CP-551. AGARD, Neuilly-sur-Seine.

Morinishi, Y. and Kobayashi, T. (1990). In *Engineering Turbulence Modelling and Experiments* (W. Rodi and E.N. Ganić, eds), pp. 279–286. Elsevier Science Publishing Co., New York.

Moser, M.D. and Rogers, M.M. (1992). *J. Fluid Mech.* **247**, 275–320.

Moser, M.D. and Rogers, M.M. (1994). In *Application of Direct and Large Eddy Simulation to Transition and Turbulence*, AGARD-CP-551. AGARD, Neuilly-sur-Seine.

Neves, J.C., Moin, P. and Moser, R.D. (1992). Report No. TF–54, Dept. Mech. Engng., Stanford Univ.

Orszag, S.A., (1972). *Stud. Appl. Math.* **51**, 253–259.

Orszag, S.A. and Patterson, M.R. (1972). *Phys. Rev. Lett.* **28**, 76–79.

Perot, J.B. and Moin, P. (1993). Report No. TF–60, Dept. Mech. Engng., Stanford Univ.

Peterson, V.L., Kim, J., Holst, T.L., Deiwert, G.S., Cooper, D.M., Watson, A.B. and Bailey, F.R. (1989). *Proc. IEEE* **77**, 1038–1055.

Peyret, R. and Taylor, T.D. (1983). *Computational Methods for Fluid Flow*. Springer–Verlag, New York.

Piomelli, U. (1993). *Phys. Fluids A* **5**, 1484–1490.

Piomelli, U. and Liu, J. (1995). *Phys. Fluids A* **7**, 839–848.

Piomelli, U., Ferziger, J.H. and Moin, P. (1987). Report No. TF-32, Dept. Mech. Engng., Stanford Univ.

Piomelli, U., Moin, P. and Ferziger, J.H. (1988). *Phys. Fluids* **31**, 1884–1891.

Piomelli, U., Ferziger, J.H. and Moin, P. (1989). *Phys. Fluids A* **1**, 1061–1068.

Piomelli, U., Zang, T.A., Speziale, C.G. and Lund, T.S. (1990). In *Instability and Transition* (M.Y. Hussaini and R.G. Voigt, eds), pp. 480–496. Springer–Verlag, Berlin.

Piomelli, U., Cabot, W.H., Moin, P. and Lee, S. (1991). *Phys. Fluids A* **3**, 1766–1771.

Pourquie, M. and Eggels, J.G.M. (1993). In *Engineering Applications of Large–Eddy Simulations* (S.A. Ragab and U. Piomelli, eds), FED vol. 162, pp. 37–44. ASME, New York.

Pruett, C.D., Zang, T.A., Chang, C.-L. and Carpenter, M.H. (1995). *Theoret. Comput. Fluid Dynamics* **7**, 49–76.

Ragab, S.A. and Piomelli, U. (eds) (1993). *Engineering Applications of Large-Eddy Simulations*, FED vol. 162. ASME, New York.

Ragab, S.A. and Sheen, S.-C. (1991). AIAA paper 91-0233, AIAA 29th Aerospace Sciences Meeting, Reno, NV.

Reynolds, W.C. (1990). In *Whither Turbulence? Turbulence at the Crossroads* (J.L.

Lumley, ed.), pp. 313–342, Lecture Notes in Physics, vol. 357. Springer–Verlag, Berlin.

Reed, H. (1993). In *Progress in Transition Modeling*, AGARD Report No. 793. AGARD, Neuilly-sur Seine.

Rogallo, R.S. (1981). *NASA TM-81315*.

Rogallo, R.S. and Moin, P. (1984). In *Annual Review Fluid Mechanics*, vol. 16, pp. 99–137. Annual Reviews Inc., Palo Alto, CA.

Rogers, M.M., Moin, P. and Reynolds, W.C. (1986). Report No. TF-25, Dept. Mech. Engng., Stanford Univ.

Rønquist, E.M. (1994). In *Spectral Methods for Flow Simulation*. Von Karman Institute for Fluid Dynamics, Rhode–Saint–Genèse, Belgium.

Rose, H.A. (1977). *J. Fluid Mech.* **81**, 719–734.

Saffman, P.G. (1978). In *Structure and Mechanisms of Turbulence II* (H. Fiedler, ed.), pp. 273–306, Lecture Notes in Physics, vol. 76. Springer–Verlag, Berlin.

Sandham, N.D. and Reynolds, W.C. (1991). *J. Fluid Mech.* **224**, 133–158.

Sarkar, S., Erlebacher, G., Hussaini, M.Y. and Kreiss, H.O. (1991). *J. Fluid Mech.* **227**, 473–493.

Schiestel, R. and Viazzo, S. (1995). *Computers and Fluids* **24**, 739–752.

Schmidt, H. and Schumann, U. (1989). *J. Fluid Mech.* **200**, 511–562.

Schmitt, L. and Friedrich, R. (1984). In *Proc. 5th GAMM–Conference on Numerical Methods in Fluid Mechanics* (M. Pandolfi and R. Riva, eds), pp. 299–306, Notes on Numerical Fluid Mechanics, vol. 7. Vieweg-Verlag, Braunschweig.

Schumann, U. (1973). Doctoral Dissertation, Technical Univ. of Karlsruhe, Germany. Also: KFK–Bericht 1854, Kernforschungszentrum Karlsruhe.

chumann, U. (1975). *J. Comput. Phys.* **18**, 376–404.

Schumann, U. (1991). In *Introduction to the Modelling of Turbulence*, VKI lecture series 1991-02. Von Karman Institute for Fluid Mechanics, Rhode–Saint–Genèse, Belgium.

Schumann, U. and Friedrich, R. (1987). In *Advances in Turbulence* (G. Comte–Bellot and J. Mathieu, eds), pp. 88–104. Springer-Verlag, Berlin.

Scotti, A., Menereau, C. and Lilly, D.K. (1993). *Phys. Fluids* **A5**, 2306–2308.

Sendstad, O. and Moin, P. (1993). Report No. TF–57, Dept. Mech. Engng., Stanford Univ.

Sheen, S., Sreedhar, M. and Ragab, S.A. (1993). In *Engineering Applications of Large-Eddy Simulations* (S.A. Ragab and U. Piomelli, eds), FED vol. 162, pp. 53–64, ASME, New York.

Silveira Neto, A., Grand, D., Métais, O. and Lesieur, M. (1991). *Phys. Rev. Lett.* **66**, 2320–2323.

Smagorinsky, J. (1963). *Monthly Weather Rev.* **91**, 99–164.

Smith, L.M. and Reynolds, W.C. (1992). *Phys. Fluids A* **4**, 364–390.

Spalart, P.R. (1988). *J. Fluid Mech.* **187**, 61–98.

Spalart, P.R. (1989). In *Laminar–Turbulent Transition* (D. Arnal and R. Michel, eds), pp. 621–630. Springer-Verlag, Berlin.

Spalart, C.G. and Watmuff, U.H. (1993). *J. Fluid Mech.* **249**, 337–371.

Speziale, C.G. (1985). *J. Fluid Mech.* **156**, 55–62.

Speziale, C.G., Erlebacher, G., Zang, T.A. and Hussaini, M.Y. (1988). *Phys. Fluids* **31**, 940–942.

Squires, K.D. (1993). In *Engineering Applications of Large-Eddy Simulations* (S.A. Ragab and U. Piomelli, eds), FED vol. 162, pp. 65–71. ASME, New York.

Su, M.D. and Friedrich, R. (1994). *J. Fluids Eng.* **116**, 677–684.

Tennekes, H. and Lumley, J.L. (1972). *A First Course in Turbulence.* The MIT Press, Cambridge, Massachusetts.

Timmermans, L. (1994). Doctoral Dissertation, Technical Univ. of Eindhoven, The Netherlands.

Unger, F. and Friedrich, R. (1993). In *Flow Simulation With High–Performance Computers I* (E.H. Hirschel, ed.), pp. 201–215, Notes on Numerical Fluid Mechanics, vol. 38. Vieweg-Verlag, Braunschweig.

Van Driest, E.R. (1956). *J. Aero. Sci.* **23**, 1007–1011.

Vanel, J.M., Peyret, R. and Bontoux, P. (1986). In *Numerical Methods for Fluid Dynamics II* (K.W. Morton and M.J. Baines, eds), pp. 463–475, Clarendon Press, Oxford.

Vincent, A. and Meneguzzi, M. (1991). *J. Fluid Mech.* **225**, 1–20.

VKI (1994). *Spectral Methods for Flow Simulation.* Von Karman Institute for Fluid Dynamics, Rhode-Saint-Genèse, Belgium.

Voke, P.R. and Collins, M.W. (1982). *PhysicoChem. Hydrodyn.* **4**, 119–161.

Voke, P.R. and Gao, S. (1993). In *Engineering Applications of Large-Eddy Simulations* (S.A. Ragab and U. Piomelli, eds), FED vol. 162, pp. 29–35. ASME, New York.

Voke, P.R., Chollet, J.–P. and Kleiser, L. (eds) (1994). *Direct and Large-Eddy Simulation I*, Kluwer Academic Publishers, Dordrecht.

Vreman, A.W., Geurts, B.J. and Kuerten, J.G.M. (1994). In *Direct and Large-Eddy Simulation I* (P.R. Voke *et al.*, eds), pp. 133–144. Kluwer Academic Publishers, Dordrecht.

Wagner, C. and Friedrich, R. (1994). In *Application of Direct and Large Eddy Simulation to Transition and Turbulence*, AGARD-CP-551. AGARD, Neuilly-sur-Seine.

Werner, H. and Wengle, H. (1989). In *Advances in Turbulence 2* (H.H. Fernholz and H.E. Fiedler, eds), pp. 418–423. Springer-Verlag, Berlin.

Werner, H. and Wengle, H. (1993). In *Turbulent Shear Flows 8* (F. Durst *et al.*, eds), pp. 155–168. Springer-Verlag, Berlin.

Yakhot, V. and Orszag, S.A. (1985). In *Nonlinear Dynamics of Transcritical Flows* (H.L. Jordan *et al.*, eds), pp. 155–174, Lecture Notes in Engineering, vol. 13. Springer-Verlag, Berlin.

Yakhot, V. and Orszag, S.A. (1986). *J. Sci. Comput.* **1**, 3–51.

Yakhot, A., Orszag, S.A., Yakhot, V. and Israeli, M. (1989). *J. Sci. Comput.* **4**, 139–158.

Yang, K.-S. and Ferziger, J. H. (1993). *AIAA Paper 93-0542*, AIAA 31st Aerospace Sciences Meeting and Exhibit, Reno, NV.

Yoshizawa, A. (1982). *Phys. Fluids* **25**, 1532–1538.

Yoshizawa, A. (1986). *Phys. Fluids* **29**, 2152–2164.

Yoshizawa, A. (1989). *Phys. Fluids A* **1**, 1293–1295.

Zang, T.A. (1991). *Phil. Trans. Roy. Soc. Lond. A* **336**, 95–102.

Zang, T.A., Wong, Y.-S. and Hussaini, M.Y. (1984). *J. Comput. Phys.* **54**, 489–507.

Zang, T.A., Dahlburg, R.B. and Dahlburg, J.P. (1992). *Phys. Fluids A* **4**, 127–140.

Zang, Y., Street, R.L. and Koseff, J.R. (1993). *Phys. Fluids A* **12**, 3186–3196.

Zeman, O. (1990). *Phys. Fluids A* **2**, 178–188.

Zhang, Y., Gandhi, A., Tomboulides, A.G. and Orszag, S.A. (1994). In *Application of Direct and Large Eddy Simulation to Transition and Turbulence*, AGARD-CP-551. AGARD, Neuilly-sur-Seine.

6 Turbulent flows: model equations and solution methodology

Thomas B. Gatski

I INTRODUCTION

Most of the practical aerodynamic and hydrodynamic fluid flows of interest are turbulent. Attempts to understand the physics of such flows, as well as efforts to predict and control them, have been ongoing for 100 years. These efforts constitute the so-called 'turbulence problem', whose complete solution still eludes scientists and engineers. In the earlier part of this century, both theoretical and experimental efforts were made by many scientists; however, their understanding of the problem was incomplete, and the analyses constrained by the techniques available at the time.

The dawn of the supercomputer age brought hope that scientists would be

HANDBOOK OF COMPUTATIONAL FLUID MECHANICS   1996 Academic Press
ISBN 0-12-532200-3

able to compute directly such flows from the Navier–Stokes equations and ultimately solve the turbulence problem. Unfortunately, even with the advent of the new generation of vector and parallel processors, the direct numerical simulation (DNS) of complex turbulent flows from the Navier–Stokes equations will probably not be possible for the foreseeable future. Because the problem is the inability to resolve all the component scales within the flow, an obvious approximation is to model the smaller scales and only simulate the larger scales. Such an alternative has been found in the large-eddy simulation (LES) technique. The Navier–Stokes equations are filtered so that the flow variables are split into resolvable and subgrid-scale motions. Suitable models are required for the subgrid-scale motions. The significant strides made in this area have been outlined in Chapter 5.

Although the simulation techniques mentioned above may ultimately become common engineering tools sometime in the next century, a void currently exists for computing complex engineering turbulent flows. The LES technique models only the unresolvable scales within the flow; the next level models the entire flow through suitable averaged quantities for both the mean and turbulent motion. This averaging procedure, known as Reynolds averaging, decomposes the motion into both mean and fluctuating components. Although the procedure eliminates the need to completely resolve the turbulent motion, its drawback is that unknown higher-order correlations appear in both the mean and turbulent equations. The need to model these correlations is the 'closure problem' associated with this technique. Nevertheless, the engineering tool of choice in computing turbulent flows is the Reynolds-averaged Navier–Stokes equations for the mean motion, coupled with a closure scheme for the unknown turbulent Reynolds stresses. From a physical standpoint, the task at hand is to somehow characterize the turbulence. One obvious characterization is by adequately describing the evolution of representative turbulent velocity and length scales, an idea that originated over 50 years ago (Kolmogorov, 1942). Normally, the velocity scale is characterized by the turbulent Reynolds stresses or turbulent kinetic energy, and the length scale is characterized by coupling the turbulent energy dissipation rate with this velocity scale. Although this specific representation is not unique, the physical cornerstone behind the development of turbulent closure models is this ability to model correctly the characteristic scales associated with the turbulent flow.

The focus of this chapter is to outline the development of the turbulent closure schemes and to discuss their implementation into numerical algorithms. An unambiguous evaluation of the performance of the various second-moment or two-equation models is difficult, if not impossible, to provide. The main problems are how to isolate the behaviour of the individual correlations and how to perform the calculations so that the results are not dependent on the numerical test bed. Alternative attempts at model evaluation require very detailed experiments or simulations. The experiments require hard to measure correlations which, even if obtained, may not be sufficiently accurate, and the direct numerical simulations are of limited use because the Reynolds numbers are low and, as such, are inconsistent with the evaluation of the high-Reynolds-number models. Thus, the number of such experimental test flows is rather small. Nevertheless, attempts have been made to evaluate models, and these studies are referenced throughout the chapter. The reader is encouraged to review this material before embarking on any extensive calculations of turbulent flows. The intent of this chapter is to provide the reader with a knowledge of the current trends in turbulent modelling research, as well as an up-to-date outline of the capabilities available to the researcher in solving turbulent flow problems. Additional topics that relate to multiple-species turbulent flows or reacting turbulent flows are outside the scope of this chapter.

In the first part of the chapter, the problem of solving turbulent flow fields, is formulated mathematically. The mean and turbulent equations derived are applicable to both incompressible and compressible flows. The turbulent closure problem is identified, and a hierarchy of closure models is discussed applicable to high-Reynolds number flows. Also discussed is the wall-function approach, and the alternative near-wall modifications to the high-Reynolds-number forms of the models.

In the second part of the chapter, the numerical aspects of solving the Reynolds-averaged Navier–Stokes equations and the turbulent closure equations are presented. Lower order models such as the algebraic and half-equation models simply impact the Navier–Stokes equations through a variable viscosity, and the half-equation model only adds an ordinary differential equation as an extension to the algebraic model. Thus, the viscous Navier–Stokes solution techniques, described in the previous chapters, suffice as solution methodologies for these types of turbulent closure schemes.

II INCOMPRESSIBLE TURBULENT MODELLING

In this section, the appropriate governing equations for the solution of incompressible turbulent flows are developed. The basis of the formulation is the continuity equation (mass conservation) and the Navier–Stokes equations given by

$$\frac{\partial u_j}{\partial x_j} = 0 \tag{1}$$

$$\frac{\partial u_i}{\partial t} + u_j \frac{\partial u_i}{\partial x_j} = -\frac{1}{\rho}\frac{\partial p}{\partial x_i} + \frac{1}{\rho}\frac{\partial \sigma_{ij}}{\partial x_j} \tag{2}$$

Above, σ_{ij} is the deviatoric part of the viscous stress tensor and is given by the newtonian relationship

$$\sigma_{ij} = 2\mu S_{ij}, \tag{2a}$$

where ρ is the density, u_i is the fluid velocity, p is the pressure, μ is the molecular viscosity, and $S_{ij} = \frac{1}{2}(\partial u_i/\partial x_j + \partial u_j/\partial x_i)$ is the strain rate tensor. For the present purposes, the viscosity μ is assumed to be constant throughout the discussion on incompressible flows. Throughout this chapter, cartesian tensor notation, along with the Einstein summation convention, is adopted in the presentation of the governing equations and models. Readers not familiar with this notation will need to review one of the many texts that contain introductory tensor analysis (e.g. Aris, 1962; Kundu, 1990).

In a Reynolds decomposition, any flow variable can be decomposed into mean and fluctuating components as

$$f = \bar{f} + f', \tag{3}$$

where the mean variable $\bar{f}$ can be extracted if a statistically steady or a statistically homogeneous turbulence is assumed. In the former case, a long-time average of f is taken; in the latter case, a volume average is taken. If the flow is neither *stationary* nor *homogeneous*, and the average characteristics *vary with time or space*, then an *ensemble mean* over all realizations (samples) is required. For example, if the turbulence is stationary, then the appropriate average is

$$\overline{f(\mathbf{x})} = \lim_{\tau \to \infty} \frac{1}{\tau} \int_{t_0}^{t_0 + \tau} f(\mathbf{x}, t)\, \mathrm{d}t, \tag{3a}$$

whereas for a homogeneous turbulence the formula is similar but with the average taken over a large spatial volume. Whichever procedure is used, the average of a fluctuating quantity is zero ($\bar{f'} = 0$), and the average of the product of two quantities is the equality $\overline{fg} = \bar{f}\bar{g} + \overline{f'g'}$.

A Reynolds-averaged Navier–Stokes equations

If the procedure just outlined is applied to the continuity and Navier–Stokes equations for a constant property fluid, then

$$\frac{\partial \bar{u}_j}{\partial x_j} = 0 \tag{4}$$

$$\frac{\partial \bar{u}_i}{\partial t} + \bar{u}_j \frac{\partial \bar{u}_i}{\partial x_j} = -\frac{1}{\rho}\frac{\partial \bar{p}}{\partial x_i} + \frac{1}{\rho}\frac{\partial \bar{\sigma}_{ij}}{\partial x_j} - \frac{\partial \overline{u_i' u_j'}}{\partial x_j} \tag{5}$$

Equation (5) is the Reynolds-averaged Navier–Stokes equation (RANS). It has the same form as the instantaneous equation (2), with the exception of the second-moment or correlation tensor $\tau_{ij} \equiv \overline{u_i' u_j'}$. The Reynolds stress tensor is then given by $-\rho\tau_{ij}$, which is the basis for reference to Reynolds stress modelling. For consistency throughout this chapter and with most of the cited literature, the quantity τ_{ij} is preferred; however, the terminology Reynolds stress or Reynolds stress modelling will be used freely because of the implied constant-density formulation throughout the incompressible discussion.

As is readily seen, (5) is not closed because the turbulent correlation τ_{ij} is not known. This closure problem is well known in association with the solution of the RANS. Transport equations for τ_{ij} can be derived by taking the second moment of the fluctuating momentum equation. These equations then contain triple moments, which also must be determined. All higher-order moment equations generate still unknown correlations, and the process must be truncated at some level. This discussion leads to the topic of phenomenological turbulence modelling in which a variety of closure models, which range in complexity from very simple algebraic models to fully coupled sets of partial differential equations that represent the higher-order moments, are used to model τ_{ij}.

B Turbulent correlation closure models

In this section, the hierarchy of closure models for the second moment or Reynolds stress tensor will be discussed. These closure models are generally presented in order of increasing complexity, starting with the algebraic or zero-equation models, and ending with the full Reynolds stress model. Although this development makes sense from the standpoint of introducing models of increasing difficulty, it masks common theoretical threads that underlie the models. For this reason, this chapter starts with the full Reynolds stress closure, which is easily derivable from the fluctuating momentum equation. This closure has always been viewed as the most difficult to implement because it leads to six partial differential equations for the independent stress components plus a differential scale equation that must be solved. However, with the exception of the algebraic or zero-equation models to be discussed, the other closure models are simplifications, or truncations, of the full Reynolds stress closure. This linkage needs to be more explicitly presented to show both the theoretical connection and the commonality of the assumptions that underlie most of the models.

1 Reynolds stress models

A straightforward approach to finding the appropriate closure for τ_{ij} is to simply form the second-moment equation and attempt to find models for the higher-order correlations in the equation. The starting point for developing the second moment or the Reynolds stress transport equations is the fluctuating momentum equation, which is obtained by subtracting the mean momentum equation (5) from (2) and using the Reynolds decomposition (3). If we write the resulting fluctuating momentum equation for u_i'

$$\mathcal{L}u_i' = \frac{\partial u_i'}{\partial t} + \bar{u}_j \frac{\partial u_i'}{\partial x_j} + u_j' \frac{\partial u_i'}{\partial x_j} + u_j' \frac{\partial \bar{u}_i}{\partial x_j} + \frac{1}{\rho} \frac{\partial p'}{\partial x_i} - \frac{1}{\rho} \frac{\partial \sigma_{ij}'}{\partial x_j} - \frac{\partial \tau_{ij}}{\partial x_j} = 0 \qquad (6)$$

and take the second moment (Speziale, 1991; Wilcox, 1993)

$$\overline{u_i' \mathcal{L}u_j'} + \overline{u_j' \mathcal{L}u_i'} = 0 \qquad (7)$$

we obtain the transport equation

$$\frac{\partial \tau_{ij}}{\partial t} + \bar{u}_k \frac{\partial \tau_{ij}}{\partial x_k} = \bar{P}_{ij} + \Pi_{ij} + \frac{\partial \bar{D}_{ijk}^t}{\partial x_k} - \varepsilon_{ij} + \nu \frac{\partial^2 \tau_{ij}}{\partial x_k \partial x_k} \qquad (8)$$

where the right-hand side represents the rate of change of τ_{ij} produced by the turbulent production $\bar{P}_{ij}$, the pressure-strain rate correlation Π_{ij}, the turbulent diffusion $\bar{D}_{ijk}^t$, the turbulent dissipation rate ε_{ij}, and the molecular diffusion. These terms are given by

$$\bar{P}_{ij} = -\tau_{ik} \frac{\partial \bar{u}_j}{\partial x_k} - \tau_{jk} \frac{\partial \bar{u}_i}{\partial x_k} \qquad (8a)$$

$$\Pi_{ij} = \overline{\frac{p'}{\rho} \left(\frac{\partial u_i'}{\partial x_j} + \frac{\partial u_j'}{\partial x_i} \right)} \qquad (8b)$$

$$\bar{D}_{ijk}^t = - \left[\overline{u_i' u_j' u_k'} + \overline{\frac{p'}{\rho} \left(u_i' \delta_{jk} + u_j' \delta_{ik} \right)} \right] \qquad (8c)$$

$$\varepsilon_{ij} = 2\nu \overline{\frac{\partial u_i'}{\partial x_k} \frac{\partial u_j'}{\partial x_k}} \qquad (8b)$$

These equations represent six independent tensor components that constitute an incompressible Reynolds stress transport equation. Unfortunately, these equations are not closed and require modelling of the third-order correlations, the pressure–velocity and pressure–strain rate correlations, and the turbulent dissipation rate. Because in the modelling of the turbulent diffusion term $\bar{D}_{ijk}^t$ in incompressible flows the pressure–velocity correlation is commonly dropped and only the turbulent transport (triple-moment) term is modelled, the above transport equation requires three terms to be modelled: the pressure–strain correlation Π_{ij}, the turbulent diffusion $\bar{D}_{ijk}^t$, and the dissipation rate ε_{ij}.

For the second-moment closures, a transport equation for the dissipation rate has been almost exclusively coupled with (8), although Wilcox (1988b) has proposed a multiscale model that couples the Reynolds stress transport equation to the specific dissipation rate ω (*turbulent dissipation rate/turbulent kinetic energy*). Nevertheless, the focus here will be exclusively on the dissipation rate closures. To best interpret the closure modelling for the turbulent stress transport equations, partition the tensor dissipation rate into its isotropic and deviatoric parts

$$\varepsilon_{ij} = \tfrac{2}{3}\varepsilon\delta_{ij} + {}_D\varepsilon_{ij} \tag{9}$$

All attempts at modelling the deviatoric part of (9) to date have focused on a coupling with the Reynolds stress anisotropy (e.g. Hanjalić and Launder, 1976; Hallbäck et al., 1990). In general, however, the deviatoric dissipation has been assimilated into the pressure–strain correlation Π_{ij} with the composite term

$$\Pi_{ij} = \Pi_{ij} - {}_D\varepsilon_{ij} \tag{10}$$

to be modelled. The isotropic dissipation rate is then explicitly retained in the τ_{ij} transport equation (8) and must be modelled. In both the Reynolds stress equation formulation and the two-equation formulation to follow, a transport equation for the isotropic dissipation rate is used. This transport equation for ε can be derived from the fluctuating momentum equation by taking the moment (Speziale, 1991)

$$\nu\,\overline{\frac{\partial u_i'}{\partial x_j}\frac{\partial(\mathcal{L}u_i')}{\partial x_j}} = 0 \tag{11}$$

A complex equation is obtained in which higher-order moments appear that all have to be modelled. The form of these individual terms can be found in Speziale (1991) or Wilcox (1993) (cf. Launder et al., 1975) and can be grouped into production, destruction, and diffusion of dissipation contributions to the transport equation balance. The lack of rigour associated with the modelling of the terms in the dissipation rate equation is an obvious deficiency in the overall modelling of turbulent flows. Nevertheless, even with the associated problems, the form of the isotropic dissipation rate equation that is used is relatively standard. In the present context the high-Reynolds-number form of the modelled isotropic dissipation rate equation is simply written as

$$\frac{\partial \varepsilon}{\partial t} + \bar{u}_k \frac{\partial \varepsilon}{\partial x_k} = \bar{\mathcal{P}}_\varepsilon - \bar{\mathcal{D}}_\varepsilon + \bar{\mathcal{D}}_\varepsilon^t + \nu \frac{\partial^2 \varepsilon}{\partial x_k \partial x_k} \tag{12}$$

with

$$\bar{\mathcal{P}}_\varepsilon = -C_{\varepsilon 1}\frac{\varepsilon}{K}\tau_{kl}\frac{\partial \bar{u}_k}{\partial x_l} \tag{12a}$$

$$\bar{\mathcal{D}}_\varepsilon = C_{\varepsilon 2}\frac{\varepsilon^2}{K} \tag{12b}$$

$$\bar{\mathcal{D}}_\varepsilon^t = C_\varepsilon \frac{\partial}{\partial x_k}\left(\frac{K}{\varepsilon}\tau_{kl}\frac{\partial \varepsilon}{\partial x_l}\right) \tag{12c}$$

where the right-hand-side represents the production, destruction, turbulent diffusion, and viscous diffusion, respectively, and the coefficients $C_{\varepsilon 1}$, $C_{\varepsilon 2}$, and C_ε are closure constants. The values for $C_{\varepsilon 1}$ and $C_{\varepsilon 2}$ are set in conjunction with the pressure–strain closure, although the value of $C_{\varepsilon 2}$ is chosen close to 1.90 to match the decay rate of isotropic turbulence. Depending on the stress closure model and the values of the coefficients $C_{\varepsilon 1}$ and $C_{\varepsilon 2}$, the closure coefficient C_ε is generally set between the values 0.15 and 0.18 to have the correct log-law slope in the equilibrium boundary layer (see Launder *et al.*, 1975; Abid and Speziale, 1993).

Pressure–strain rate correlation models

One of the most active areas of research in Reynolds stress modelling is focused on the development of pressure–strain correlation models. This term is significant because it is of the same order as the production term, and it acts as a redistribution term between the Reynolds stress components. It acts on the anisotropy of the stress field in such a way as to diminish the difference between the normal stress components. While the number of such closures is rather plentiful, recent research has focused on a few such models that characterize those in use. This characterization has focused on the classification of the various models by their dependency on the Reynolds stress anisotropy tensor defined as

$$b_{ij} = \frac{\tau_{ij}}{2K} - \tfrac{1}{3}\delta_{ij} \tag{13}$$

In all of the commonly used second-moment closure models, Π_{ij} is modelled in the general form (e.g. Lumley, 1978; Reynolds, 1987) as

$$\Pi_{ij} = \varepsilon \mathcal{A}_{ij}(\mathbf{b}) + K \mathcal{M}_{ijkl}(\mathbf{b}) \frac{\partial \bar{u}_k}{\partial x_l} \tag{14}$$

where $\mathcal{A}_{ij}(\mathbf{b})$ and $\mathcal{M}_{ijkl}(\mathbf{b})$ are tensor functions of the anisotropy tensor and are related to integrals over the flow volume derived from the pressure Poisson equation. As has been recently pointed out by Speziale (1992b), the interpretation of each of these terms is exact in homogeneous turbulent flows but ambiguous in inhomogeneous turbulent flows. The $\mathcal{A}_{ij}(\mathbf{b})$ term is usually associated with the 'slow' relaxation of the turbulence toward isotropy, and the $\mathcal{M}_{ijkl}(\mathbf{b})$ term is usually associated with the 'rapid' response of the turbulence to imposed mean velocity gradients. This partitioning has its origins in the splitting of the turbulent pressure field p' into slow $p'^{(S)}$ and rapid $p'^{(R)}$ parts. The $p'^{(S)}$ part is the solution of a Poisson equation that involves gradients of the turbulent velocity field, and the $p'^{(R)}$ part is the solution of a Poisson equation involving the mean velocity gradients. The terminology originates from the observation that the slow part will only adjust as the turbulence itself adjusts, but the rapid part will adjust instantly through the mean velocity gradient. The reader is encouraged to review the references just cited for details of the theoretical foundation of such models. Representative models that are linear, quadratic, and cubic in the anisotropy tensor will be presented in order to acquaint the reader with their form.

In the past 25 years, Launder and colleagues at Imperial College and UMIST have extended the formulation of Naot *et al.* (1970) to develop a variety of pressure–strain correlation models. The linear Launder, Reece and Rodi (1975) LRR model in its high-Reynolds-number form has been quite popular. In terms of the anisotropy tensor b_{ij}, it can be written as:

Launder, Reece and Rodi model

$$\Pi_{ij} = - C_1 \varepsilon b_{ij} + C_2 K \overline{S}_{ij} + C_3 K \left(b_{ik} \overline{S}_{jk} + b_{jk} \overline{S}_{ik} - \tfrac{2}{3} b_{mn} \overline{S}_{mn} \delta_{ij} \right)$$

$$+ C_4 K (b_{ik} \overline{W}_{jk} + b_{jk} \overline{W}_{ik}) \tag{15}$$

where

$$C_1 = 3.0, \quad C_2 = 0.8, \quad C_3 = 1.75, \quad C_4 = 1.31 \tag{16}$$

with

$$C_{\varepsilon 1} = 1.44, \quad C_{\varepsilon 2} = 1.90 \tag{16a}$$

and $\overline{W}_{ij} = \tfrac{1}{2}(\partial \bar{u}_i / \partial x_j - \partial \bar{u}_j / \partial x_i)$ is the rotation rate tensor (i.e. the antisymmetric part of the mean velocity gradient tensor). Note that all of the models to be discussed can be generalized to non-inertial frames by adding $\epsilon_{mji} \Omega_m$ (where Ω_m is the rotation rate of the non inertial frame relative to the inertial frame) to $\overline{W}_{ij}$. In addition to the LRR model, the isotropization-of-production (IP) model (Launder *et al.*, 1975; Gibson and Launder, 1978) also fits into this form with the following coefficients:

Isotropization-of-production model

$$C_1 = 3.6, \quad C_2 = 0.8, \quad C_3 = 1.2, \quad C_4 = 1.2 \tag{17}$$

with

$$C_{\varepsilon 1} = 1.44, \quad C_{\varepsilon 2} = 1.90 \tag{17a}$$

A point should be made about the closure constants C_1–C_4. Of the four coefficients, only two are independent. These added constants appear due to the writing of the pressure–strain model in terms of b_{ij}, $\overline{S}_{ij}$, and $\overline{W}_{ij}$ rather than the τ_{ij} tensor and the mean velocity gradient tensor, as was the case when the models were originally derived. However, the present notation is preferred because it shows the interrelationship between the various pressure–strain models. The values of the coefficient of the slow term C_1 were optimized by comparisons with homogeneous shear flow (LRR model) and further modified for the IP model to better match free-shear flow data. In both the LRR and IP models, the coefficients (recall that $C_2 - C_4$ are not independent) are calibrated from homogeneous shear flow data, and rapid distortion of homogeneously strained initially isotropic turbulence data. As indicated earlier, the values for the closure coefficients in the dissipation rate equation (12) are optimized in conjunction with the respective pressure–strain–rate models. The value of $C_{\varepsilon 1}$ was originally determined (Hanjalić and Launder, 1972) from the properties of a constant stress layer adjacent to a wall at which a production equal dissipation balance occurs, but was later refined (Launder *et al.*, 1975) by computer optimization. The coefficient $C_{\varepsilon 2}$ was determined from the decay of grid-generated turbulence. Note that in comparisons with large-eddy simulations of homogeneous shear flow, as well as with experimental results, the *IP* model gives slightly better predictions of both the temporal evolution of the turbulent kinetic energy and the equilibrium values of the components of the stress anisotropy b_{ij} than the more complete model.

It should be noted that both the LRR and IP models which have been presented do not include the 'wall reflection' term which is necessary to insure the correct log-law behaviour in wall-bounded flows. It depends on distance from the wall, and as such does not fit into the classification structure presently being outlined. It needs to be included in wall-bounded calculations using the LRR and IP models and its effect is felt over a wide range of the boundary layer. The other models to be presented do not have this restriction which can be troublesome in actual calculations.

At the next level of complexity, the pressure–strain correlation models are *quadratic* in the stress anisotropy b_{ij}. These closures also have coefficients C_1–C_4 that are functions of the invariants of b_{ij}, and depend as well on $\bar{\mathcal{P}}/\varepsilon$ ($\bar{\mathcal{P}} \equiv \bar{P}/2$). One such model is the Speziale, Sarkar and Gatski (1991) SSG model. In the SSG model, the quadratic behaviour was originally included as a higher-order correction to the slow part of the pressure–strain correlation, and the rapid part was quasilinear in the stress anisotropy. As such, the composite pressure–strain Π_{ij} model is:

Speziale, Sarkar and Gatski model

$$\Pi_{ij} = -\,C_1\varepsilon b_{ij} + C_1'\varepsilon\left(b_{ik}b_{kj} - \tfrac{1}{3}b_{mn}b_{nm}\delta_{ij}\right) + C_2 K \overline{S}_{ij}$$

$$+ C_3 K\left(b_{ik}\overline{S}_{jk} + b_{jk}\overline{S}_{ik} - \tfrac{2}{3}b_{mn}\overline{S}_{mn}\delta_{ij}\right) + C_4 K\left(b_{ik}\overline{W}_{jk} + b_{jk}\overline{W}_{ik}\right) \qquad (18)$$

where

$$C_1 = 3.4 + 1.8\frac{\mathcal{P}}{\epsilon}, \quad C_1' = 4.2, \quad C_2 = 0.8 - 1.30(b_{ij}b_{ij})^{\frac{1}{2}}, \quad C_3 = 1.25, \quad C_4 = 0.40$$

$$\qquad (19)$$

with

$$C_{\varepsilon 1} = 1.44, \quad C_{\varepsilon 2} = 1.83 \tag{19a}$$

These values are obtained by the calibration with homogeneous shear flows (with and without rotation), the decay of isotropic turbulence and return to isotropy of initially anisotropic turbulence, and the rapid distortion of homogeneously strained turbulence that is initially isotropic. The dependence on $\mathcal{P}/\epsilon$ is introduced into the coefficient C_1 in order to achieve an asymptotically consistent truncation of the Taylor series expansion for Π_{ij}. Consistent with the behaviour of the anisotropy tensor b_{ij}, this ratio also attains an equilibrium value in flows such as plane homogeneous shear (≈ 1.5) and the log layer of a channel flow ($= 1$).

The final level of pressure–strain closures to be discussed are models that are *cubic* in the stress anisotropy tensor. A popular representative is the Fu, Launder and Tselepidakis (1987) FLT model. In the previous section, the deviatoric part of the tensor dissipation was assimilated into the pressure–strain correlation, and the composite was modelled. This procedure was done for the two models just discussed. In the FLT model, however, the deviatoric part of the dissipation rate $_D\varepsilon_{ij}$ is explicitly represented in the formulation and a model for Π_{ij} rather than Π_{ij} is formed.

Fu, Launder and Tselepidakis model

$$\Pi_{ij} = - C_1 \varepsilon b_{ij} + C_1' \varepsilon \left(b_{ik} b_{kj} - \tfrac{1}{3} b_{kl} b_{kl} \delta_{ij} \right)$$

$$+ C_2 K \overline{S}_{ij} + 1.2 K \left(b_{ik} \overline{S}_{jk} + b_{jk} \overline{S}_{ik} - \tfrac{2}{3} b_{kl} \overline{S}_{kl} \delta_{ij} \right)$$

$$+ \tfrac{26}{15} K \left(b_{ik} \overline{W}_{jk} + b_{jk} \overline{W}_{ik} \right) - \tfrac{112}{5} K \left[II \left(b_{ik} \overline{W}_{jk} + b_{jk} \overline{W}_{ik} \right) \right]$$

$$+ \tfrac{4}{5} K \left(b_{ik} b_{kl} \overline{S}_{jl} + b_{jk} b_{kl} \overline{S}_{il} - 2 b_{ik} \overline{S}_{kl} b_{lj} - 3 b_{kl} \overline{S}_{kl} b_{ij} \right)$$

$$+ \tfrac{4}{5} K \left(b_{ik} b_{kl} \overline{W}_{jl} + b_{jk} b_{kl} \overline{W}_{il} \right) - \tfrac{168}{5} K \left(b_{ik} b_{kl} \overline{W}_{lm} b_{mj} + b_{jk} b_{kl} \overline{W}_{lm} b_{mi} \right) \tag{20}$$

and

$$\varepsilon_{ij} = \tfrac{2}{3} \varepsilon \delta_{ij} +_D \varepsilon_{ij} = \tfrac{2}{3} \varepsilon \delta_{ij} + 2(1 - \sqrt{F}) b_{ij} \varepsilon \tag{21}$$

where

$$F = 1 + 9II + 27III \tag{22}$$

$$II = -\tfrac{1}{2}b_{ij}b_{ij}, \quad III = \tfrac{1}{3}b_{ij}b_{jk}b_{ki} \tag{23}$$

$$C_1 = -120II\sqrt{F} - 2\sqrt{F} + 2, \quad C_1' = 144II\sqrt{F}, \quad C_2 = 0.8 \tag{24}$$

with

$$C_{\varepsilon 1} = 1.45, \quad C_{\varepsilon 2} = 1.90 \tag{25}$$

While this model is tensorially more complex than either of the other two, its calibration is similar and is based on comparisons with the same benchmark flows used in the previous two calibrations. In addition, it also satisfies the constraint of realizability (Schumann, 1977; Lumley, 1978), which requires that a Reynolds stress model yield non-negative component energies. Because insufficient space is available here to discuss the details of the formulation, the interested reader is referred to the paper by Fu et al. (1987) for more details of this model, and to the paper of Speziale et al. (1994) for a discussion of the realizability aspects of the model.

Recently, Ristorcelli et al. (1994) developed a 2D MFI model that is similar in structure to the models just described, keeping terms cubic in the anisotropy tensor. However, two notable differences are associated with this model: (i) the coefficients are functions of the state of the flow, with the functional form deduced from mathematical constraints satisfied by the rapid pressure term and (ii) the model is also frame indifferent in the two-dimensional limit. This latter constraint insures that the model is consistent with the Taylor–Proudman theorem (Speziale, 1989); as such, the model is expected to perform well in situations where the flow is two dimensionalized by various body forces, such as buoyancy or rotation, or in situations where streamline curvature is important. Whether or not building more physics and mathematics into the model will actually produce a representation capable of successfully computing these more complex flows remains to be seen.

Some coordinated attempts have been made to assess the performance of the various pressure–strain models. One approach has been to simply solve the full Reynolds stress closure equations for a specific flow and compare the predictive capability of the models tested. The other approach has been to take either experimental or direct simulation data and calculate the pressure–strain correlation from the data and compare it with the various model representations using the same database.

In the former approach, practical engineering flows that are sufficiently simple to compute, and in which the results are not tainted by either numerical or model form ambiguities when applied, are required. Flows which fit these criteria are two-dimensional free shear flows such as jets, wakes and mixing layers, as well as two-dimensional flat-plate boundary–layer flows. More complex flows, such as separated wall-bounded flows, flows with curvature, and flows with imposed pressure gradients, are a few examples where numerical or model form ambiguities can produce misleading results. Nevertheless, organized attempts at such tests are over 25 years old and date back to the Stanford 1968 competition on computation of two-dimensional boundary layers (Kline *et al.*, 1969); these earlier attempts were followed by the Stanford 1980/ 1981 conference on complex turbulent flows (Kline *et al.*, 1982). A similar endeavour on collaborative testing of turbulent models (Bradshaw *et al.*, 1994) was initiated in 1991.

Simpler, more generic flows, such as homogeneous shear both with and without rotation, can be calculated in an unambiguous way because the turbulent transport is zero and the mean velocity gradient is constant. Such numerically generated flows, where the turbulence is statistically homogeneous in the streamwise and transverse directions, are not simply lagrangian descriptions of a corresponding laboratory experiment. Experiments in uniformly sheared turbulence have been performed for some time; one of the most recent is that of Tavoularis and Karnik (1989). This and previous studies have shown that the uniformly sheared turbulence can be a close, although not exact, representation of the homogeneous turbulence generated by the numerical experiments, and can display the same energetic and dissipation growth rates.

These homogeneous shear flows in equilibrium have a common theoretical connection to the log layer in channel flow (Abid and Speziale, 1993) in that a second-order closure model will yield the same equilibrium values for the anisotropy b_{ij} in both flows. As such, they are important indicators of the performance level of the models when applied to practical wall-bounded flows. Direct numerical simulations and large-eddy simulations of the homogeneous shear flows have been performed to compare the predictive performance of the pressure–strain models directly. Recently, Speziale *et al.*, (1990, 1991) have tested the LRR, SSG, and FLT models in homogeneous shear flows with and without rotation. The results showed that the SSG model outperforms the linear LRR and cubic FLT models in predicting the evolution of turbulent kinetic energy; however, none of the models was able to predict accurately the evolution of turbulent kinetic energy for the most energetic (unstable) homogeneous shear flow case with rotation.

In the latter approach, in which experimental and/or DNS data are used, the results are less ambiguous but require extensive data to compute the various pressure-strain correlations and the model representation. Schwarz and Bradshaw (1994) have used experimental data from a three-dimensional boundary layer flow to compute the behaviour of a variety of pressure–strain models. They found that the quadratic SSG model predictions were generally closest to the data but differed little from the linear LRR model. The cubic FLT model predicted the normal stresses well, but did not predict the τ_{12} shear stress as well as the LRR model. Some of their results are shown in Figures 1 and 2 for the streamwise normal stress component and the τ_{12} shear stress component. In these figures, the behaviour of the Π_{11} and Π_{12} correlations are compared with

(Term non-dimensionalized by δ_{99}/U_{ref}^3)

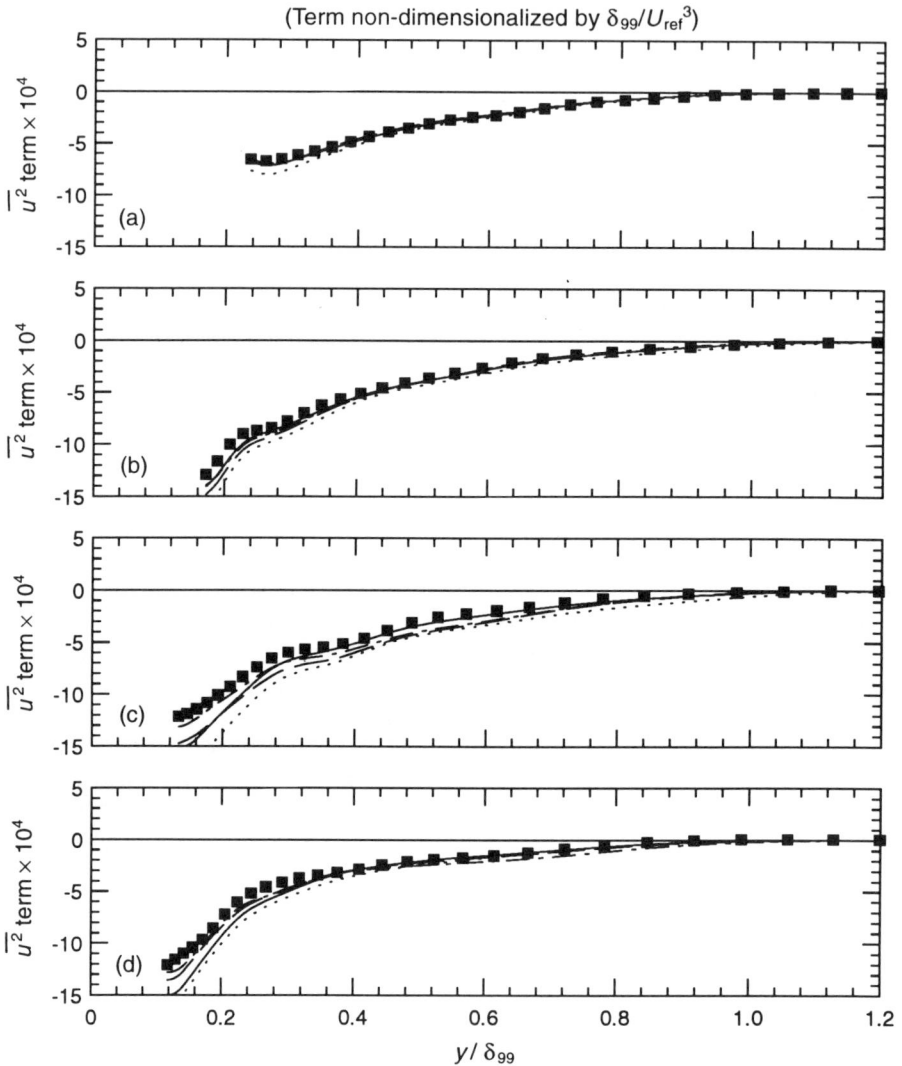

Figure 1 Comparisons of models for the Π_{11} pressure–strain term (Schwarz and Bradshaw, 1994). (a) Station 1: $x = 978$ mm. (b) Station 7: $x = 1775$ mm. (c) Station 13: $x = 2415$ mm. (d) Station 18: $x = 2948$ mm. Key: ■, data; ⋯, LRR; - - -, SSG; —, SL/CL; ---, FLT.

experimental results at various downstream locations in the three-dimensional flow. Also included in the comparison is the non-linear Shih and Lumley (1985) SL rapid model, coupled with the Choi and Lumley (1984) slow model. The model is not given here because the details of the formulation are complex; however, the authors attempt to utilize a strong form of the realizability

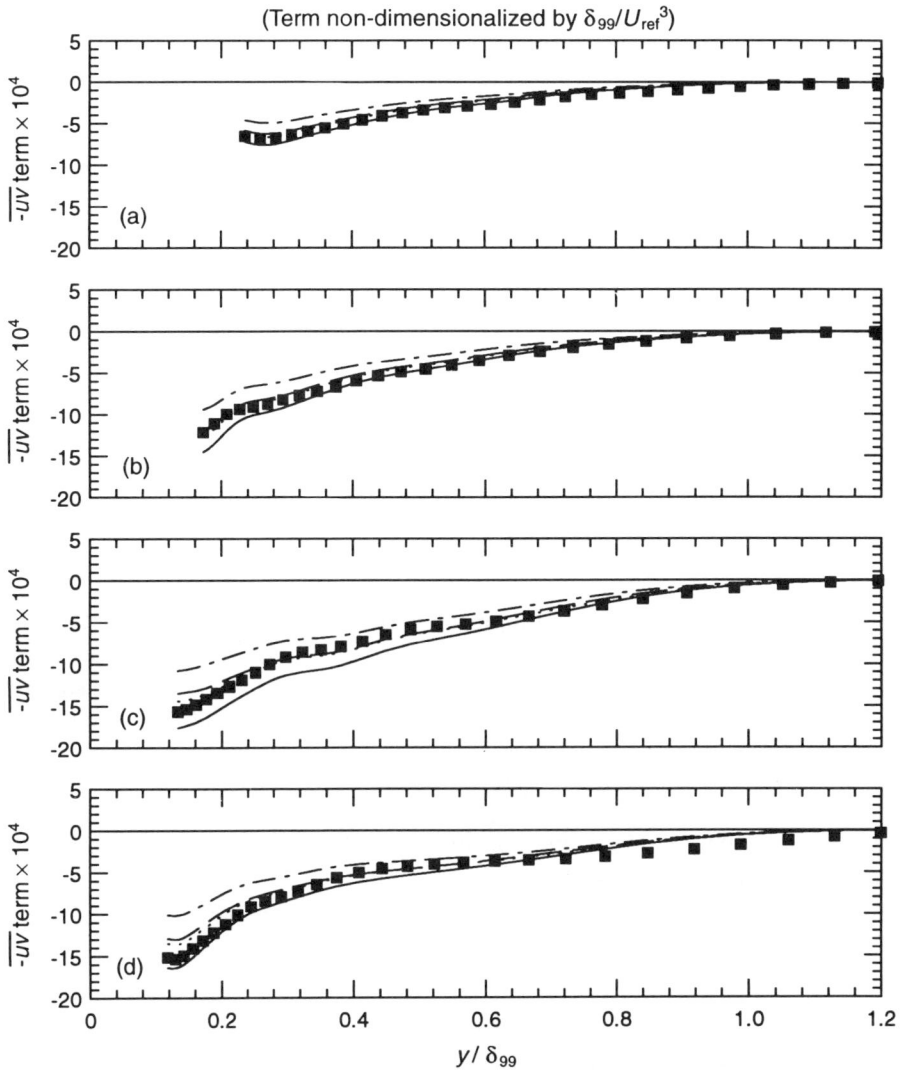

Figure 2 Comparisons of models for the Π_{12} pressure–strain term (Schwarz and Bradshaw, 1994). (a) Station 1: $x = 978\,mm$. (b) Station 7: $x = 1775\,mm$. (c) Station 13: $x = 2415\,mm$. (d) Station 18: $x = 2948\,mm$. Key: ■, data; $\cdots$, LRR; - - -, SSG; —, SL/CL; ---, FLT.

constraint. This constraint requires that when a principal Reynolds stress component vanishes, its time rate of change must vanish and its second derivative must be positive. Recently, this model has come under closer scrutiny by Speziale *et al.* (1994), who found that the model is not realizable and who questioned the use of the realizability constraint on closure models based on near-equilibrium conditions. Speziale *et al.* (1992b) have performed a comparative study using DNS data of homogeneous turbulence to assess the performance of the pressure–strain models. The models tested did not include the linear LRR model; however, the quadratic SSG model and the cubic FLT model were tested. They concluded that both the SSG and FLT models performed equally well in predicting the various pressure–strain correlations. Another recent study by Younis *et al.* (1994) has tested the IP and SSG models in three complex turbulent flows: rotating channel, boundary layer over convex curved surface, and boundary layer over an infinite swept wing. In these flows the SSG model performed as well as or better than the IP model.

The correlation models presented are not meant to be all inclusive; however, they are put forward as typical of the functional form encountered for present day pressure–strain models.

Turbulent diffusion models

A model for the turbulent diffusion term $\bar{D}^t_{ijk}$ in (8c), with the pressure–velocity correlation neglected, has traditionally been developed, although Lumley (1978) has made a theoretical estimate that the pressure–velocity correlation is about 20% percent of its triple-velocity correlation counterpart. Nevertheless, this term has received very little scrutiny in numerical simulations and experiments and has not been explicitly modelled in incompressible flows. The gradient transport hypothesis assumes that turbulent diffusion behaves analogously to molecular diffusion and, thus, a distinct scale separation exists between the two diffusion processes.

The full second-moment closures use the gradient transport hypothesis to model the turbulent diffusion term $\bar{D}^t_{ijk}$. The forms are more complex, however, because the individual stress components are utilized in the resulting models. The three most common models used in conjunction with the second-moment closures fall into the following general functional form (Speziale, 1991)

$$\bar{D}^t_{ijk} = d_{ijklmn} \frac{\partial \tau_{lm}}{\partial x_n} \tag{26}$$

where d_{ijklm} can be a function of τ_{ij}.

Daly and Harlow (1970) (DH) proposed the rather simple model

$$\bar{D}^t_{ijk} = C_s \frac{K}{\varepsilon} \tau_{kl} \frac{\partial \tau_{ij}}{\partial x_l} \tag{27}$$

where $C_s = 0.22$ is a diffusion constant determined from model optimization. While (27) is straightforward, it suffers from the obvious deficiency that it is not invariant under the permutation of indices. A second, less serious deficiency is the fact that in two-dimensional thin shear flows where the dominant diffusion is in the cross stream direction, the model gives an isotropic effective diffusion coefficient; although in more complex flows the model is anisotropic. It is important to recognize that these closure constants for the turbulent diffusion models are determined in the calibration process of the entire second-moment model. For example, the value 0.22 is the value recommended by Daly and Harlow (1970), but Launder *et al.* (1975) in the LRR model found an optimized value of 0.25 for C_s in their calibration studies. Therefore, it is necessary to use the value of C_s associated with each model for $\bar{D}^t_{ijk}$ as an initial estimate; for each second-moment closure model (fully modelled stress transport equation plus dissipation rate model), a unique set of closure coefficients is determined. An example of this modelling *in toto* is the LRR model, where the modelled terms in the stress transport equation and the dissipation rate equation have been optimized together.

A much more complicated expression has been proposed by Hanjalić and Launder (1972) (HL), who modelled the transport equation for the third-order moments and then extracted an algebraic expression from the modelled equation by neglecting the transport terms and assuming a gaussian relation to link the fourth-order correlations to the second-order correlations. Their invariant expression for $\bar{D}^t_{ijk}$ is

$$\bar{D}^t_{ijk} = C_s \frac{K}{\varepsilon} \left(\tau_{il} \frac{\partial \tau_{jk}}{\partial x_l} + \tau_{jl} \frac{\partial \tau_{ki}}{\partial x_l} + \tau_{kl} \frac{\partial \tau_{ij}}{\partial x_l} \right) \tag{28}$$

The value for C_s was originally chosen by Hanjalić and Launder (1972) (HL) as 0.08; however, later LRR chose a value of 0.11. In both cases, the value was chosen to optimize the predictive capability of the *entire* closure model.

The final model to be presented is the Mellor and Herring (1973) (MH) model. This model is an isotropized version of the HL model and is given by

$$\bar{D}^t_{ijk} = C_s \frac{K^2}{\epsilon} \left(\frac{\partial \tau_{jk}}{\partial x_i} + \frac{\partial \tau_{ki}}{\partial x_j} + \frac{\partial \tau_{ij}}{\partial x_k} \right) \tag{29}$$

where the model constant C_s is taken as two-thirds of the value chosen for the HL model. This model also retains the proper coordinate invariance properties.

Other more complicated models have also been proposed. Lumley (1978) proposed a model that contains the HL model as a tensorial basis. He also proposed an explicit model for the pressure–velocity diffusion to be used in conjunction with the triple-moment model. Recently, Magnaudet (1993) has proposed a model that is consistent with the two-dimensional turbulence encountered near a wall and at a free surface. This model is also quite complex and has not been tested for a variety of engineering flows. The interested reader is also referred to the recent review of Hanjalić (1994) for further commentary.

Recently, Demuren and Sarkar (1993) tested the performance of the DH, HL, and MH models in plane channel flow and found that the MH model gives the best performance because of its ability to properly model the relaxation to isotropy in the outer part of the layer (Figure 3). Schwarz and Bradshaw (1994) have compared the DH and HL models, as well as the Lumley model, in a three-dimensional boundary layer flow. They concluded that all three models

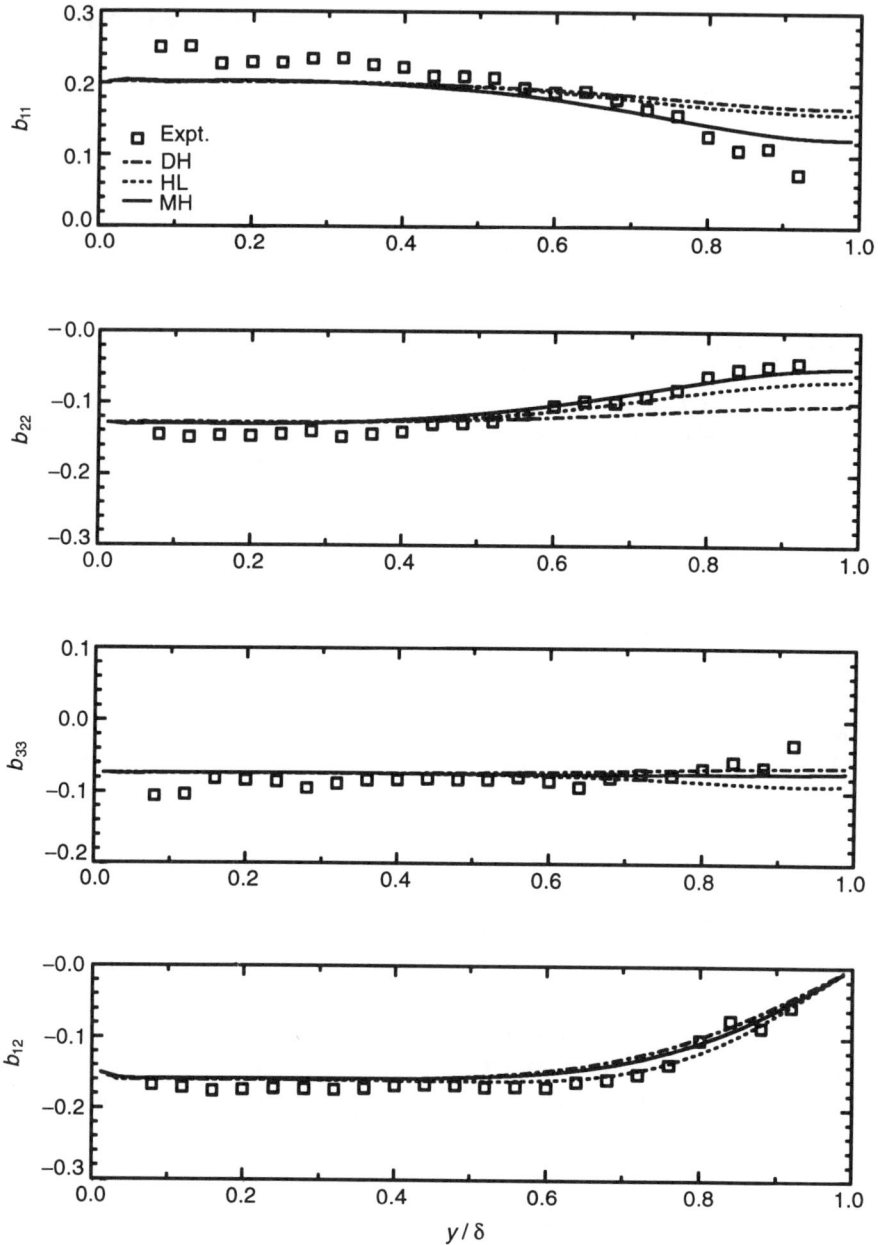

Figure 3 Comparison of components of the Reynolds stress anisotropy tensor, computed with SSG pressure–strain model and three diffusion models, with experimental data of Laufer (1951).

performed reasonably well; the Lumley model outperformed the other models in some component predictions.

2 Two-equation models

As the previous subsection has shown, the Reynolds stress formulation is capable of yielding the individual components of the second-moment or Reynolds stress tensor at the expense of rather complex differential equations and the need to model higher-order correlations. This complexity is eased somewhat by only looking at the trace of the τ_{ij} equations. To continue with the terminology introduced in Section II.B.1, the trace of τ_{ij} is designated by $\tau_{ii}(\equiv 2K)$, and the turbulent kinetic energy is given by ρK. For brevity and consistency, because the density ρ is constant, the quantity K will sometimes be termed the turbulent kinetic energy. This equation is constructed by taking the contraction of (8). Because the pressure–strain redistribution term is traceless ($\Pi_{ii} = 0$), the equation is less difficult to close because only diffusion models must be found. The disadvantage of this formulation is that the individual stress components cannot be determined directly from the solved equation set. The turbulent kinetic energy equation is

$$\frac{\partial K}{\partial t} + \bar{u}_j \frac{\partial K}{\partial x_j} = \bar{\mathcal{P}} + \frac{\partial \bar{\mathcal{D}}_j^t}{\partial x_j} - \epsilon + \nu \frac{\partial^2 K}{\partial x_j \partial x_j} \tag{30}$$

where the right-hand-side represents the transport of K by the turbulent production $\bar{\mathcal{P}} \equiv \bar{P}_{ii}/2$, the turbulent diffusion $\bar{\mathcal{D}}_j^t \equiv \bar{D}_{ijj}^t/2$, the isotropic turbulent dissipation rate, $\varepsilon \equiv \varepsilon_{ii}/2$, and the viscous diffusion. Consistent with the definitions used in the Reynolds stress formulation, the production, diffusion, and dissipation terms are given by

$$\bar{\mathcal{P}} = -\tau_{ij} \frac{\partial \bar{u}_i}{\partial x_j} \tag{30a}$$

$$\bar{\mathcal{D}}_j^t = -\left[\frac{1}{2} \overline{u_i' u_i' u_j'} + \frac{\overline{p' u_i'}}{\rho} \delta_{ij} \right] \tag{30b}$$

$$\varepsilon = \nu \overline{\frac{\partial u_i'}{\partial x_j} \frac{\partial u_i'}{\partial x_j}} \tag{30c}$$

respectively. Even with this formulation, the individual stress components, which are required in the production term of the kinetic energy equation, still must be known. Depending on the formulation of the mean-flow problem, the individual stress components appear explicitly in the Reynolds-averaged Navier–Stokes equations. In the two-equation formulation, the explicit appearance of the individual stress components is generally not the case because the effect of the turbulence on the mean flow is usually accounted for through an eddy viscosity.

Dimensional analysis considerations dictate that the eddy viscosity ν_t be given by the product of a turbulent velocity scale and a turbulent length scale. In a two-equation K–ϵ formulation, these are proportional to $K^{\frac{1}{2}}$ and $K^{\frac{3}{2}}/\varepsilon$, respectively. Thus, the eddy viscosity ν_t is given by the relation

$$\nu_t = C_\mu \frac{K^2}{\varepsilon} \tag{31}$$

where $C_\mu = 0.09$ is the usual value assumed.

Consistent with the simplification that results from the two-equation formulation, a simple form for the turbulent transport $\bar{D}^t_{ijk}$ is usually used in the kinetic energy equation. Although the models required for the Reynolds stress transport equation are more complex, the underlying gradient transport hypothesis that is used is also applied to the simpler turbulent diffusion model for the kinetic energy turbulent transport term. The turbulent diffusion model is given by

$$\bar{D}^t_j = \frac{\nu_t}{\sigma_K} \frac{\partial K}{\partial x_j} \tag{32}$$

where σ_K is a constant that acts like an effective Prandtl number for diffusion. In most cases, σ_K is assumed to be unity. With this model for the turbulent diffusivity, both the turbulent and molecular diffusion terms can be combined into a gradient transport model with effective viscosity $\nu + \nu_t/\sigma_K$. The resulting simple form of the modelled turbulent kinetic energy equation is the obvious appeal of the formulation.

Note that in both the kinetic energy and Reynolds stress equations, the turbulent dissipation rate is chosen as the fundamental scale variable, and, as was just shown, can be combined with the turbulent stresses to form both turbulent length and time scales. Other scale variables, which will be discussed shortly, can be derived from this quantity; however, for the most part, the dissipation rate is the scale variable used.

The modelled transport equation for the isotropic dissipation rate is once again given by (12), but now the turbulent diffusion term $\bar{D}^t_\varepsilon$ in the equation can be simplified relative to the form used in the Reynolds stress transport formulation. The analogue to the turbulent diffusion term that appears in (12) is the simpler

$$\bar{D}^t_\varepsilon = \frac{\partial}{\partial x_k} \left(\frac{\nu_t}{\sigma_\varepsilon} \frac{\partial \varepsilon}{\partial x_k} \right) \tag{33}$$

where σ_ε is a constant that acts like an effective Prandtl number for dissipation diffusion. In most cases, σ_ε is taken as 1.3 in order to ensure the correct log-law behaviour in boundary layer flows.

Although the $K-\varepsilon$ formulation is the most popular two-equation model, it would be remiss not to mention that several other two-equation models have been formulated. Most of these other formulations have used alternatives to the dissipation-rate-scale equation. Kolmogorov (1942) was the first to propose the dissipation rate per unit kinetic energy ω or specific dissipation rate as a scale variable. Saffman (1970) offered an improved alternative version, and Wilcox (see Wilcox (1993) for details and associated references) has refined and applied the model in the last 20 years to such a wide variety of incompressible and compressible flows that it is the most widely used alternative to the dissipation rate equation. Speziale *et al.* (1992a) have offered a $K-\tau$ model that has the correct asymptotic behaviour near the wall, and have also proposed the use of a transport equation for the time scale variable $\tau = K/\varepsilon$, which also has the correct near-wall asymptotics and is not singular at the wall. Two other alternatives have been the $K-Kl$ (Wolfshtein, 1970; Mellor and Herring, 1973), and the $K-K\tau$ model (Zeierman and Wolfshtein, 1986). Both of these models use the two-point velocity correlation tensor (in space and time, respectively) and the associated transport equations to extract the appropriate scale equations.

In the last few years, renormalization group (RNG or RG) methods have been employed (Yakhot and Orszag, 1986) to develop alternate two-equation $K-\varepsilon$ models for turbulent flows. The approach has since been further developed, and a variety of validation calculations performed (e.g. Yakhot *et al.*, 1992). Although the details of the formulation are outside the scope of this chapter, note that the Yakhot and Orszag approach is formally based on an ϵ-RNG methodology, in which a small parameter ϵ is introduced into the exponent of a forcing correlation function, which is in turn introduced into the momentum equation. The theory takes into account the non-local (spectral) interactions that occur in the flow. Recently, an alternative *Recursion*-RNG approach has been used to develop a modification to the usual Boussinesq formulation used in conjunction with the two-equation $K-\varepsilon$ formulation (Zhou *et al.*, 1994a, b). The formulation does not rely on an ϵ expansion and accounts for local (spectral) interactions that occur in the flow. The technique has not been extended to developing a corresponding *Recursion*-RNG two-equation model. Both approaches may provide advantages over the more traditional two-equation $K-\varepsilon$ formulations, but more widespread use and validation are required to assess their general applicability.

Much more could be discussed about the two-equation models and, in particular, the $K-\varepsilon$ model because of the ease with which they can be applied and the widespread use they have enjoyed. Space limitations prevent such a discussion, but the interested reader is referred to the recent book by Mohammadi and Pironneau (1994), which is devoted entirely to the $K-\varepsilon$ turbulence model.

Boussinesq approximation

Normally, in all of the two-equation formulations that have been mentioned, τ_{ij} components are *approximated* by analogy with a newtonian type of (linear) constitutive relationship between the turbulent stress and the mean strain-rate tensor $\overline{S}_{ij}$. This Boussinesq type of eddy-viscosity approximation is given by

$$\tau_{ij} = \tfrac{2}{3}K\delta_{ij} - 2\nu_t \overline{S}_{ij} \tag{34}$$

Equation (34) can also be obtained formally from a simple continuum mechanics approach (Speziale, 1991) by assuming that the stress tensor τ_{ij} is a tensor function of the mean velocity gradient and by assuming single turbulent length and time scales that are spectrally distinct from the larger mean scales. This single-scale assumption is the underlying basis contained in all standard turbulent closure models that are discussed. Attempts have been made at multiscale closures (e.g. Hanjalić *et al.* (1979) and Wilcox (1986) with moderate success); however, their usage has been generally confined to the originators.

Although the two-equation formulation with a Boussinesq constitutive relation is appealing, its omission of any anisotropic eddy-viscosity effects can lead to serious predictive deficiencies. A simple square-duct flow is an ideal example. The two-equation formulation with an isotropic eddy viscosity predicts no secondary streamline pattern; however, as Figure 4 shows, the qualitatively correct pattern is clearly predicted by the Reynolds stress model because of the proper accounting of the normal stress differences in the flow.

Explicit algebraic stress models

The two-equation formulation is appealing because it provides the transport equations for the turbulence that help account for some nonlocal and history effects but, at the same time, is not overly computationally intensive. One drawback, however, is the Boussinesq constitutive relationship (34), which assumes an isotropic eddy viscosity that is unable to account for strong anisotropies in the turbulence and results in poor predictions in flows with secondary motions. In addition, body forces such as an imposed rotational field are also not properly accounted for with the Boussinesq relationship. Furthermore, as noted, the pressure–strain correlation term has disappeared in the two-equation formulation, which precludes the effects of intercomponent transfer between τ_{ij} components. An effective compromise between the full second-moment closure and the two-equation model is the algebraic stress model (ASM). This approach extracts an algebraic relationship between the turbulent Reynolds stress and mean velocity field from the Reynolds stress closures discussed in Section II.B.1.

Algebraic stress models are derivable from equilibrium hypotheses imposed on both convective and diffusive effects. The convective equilibrium hypothesis

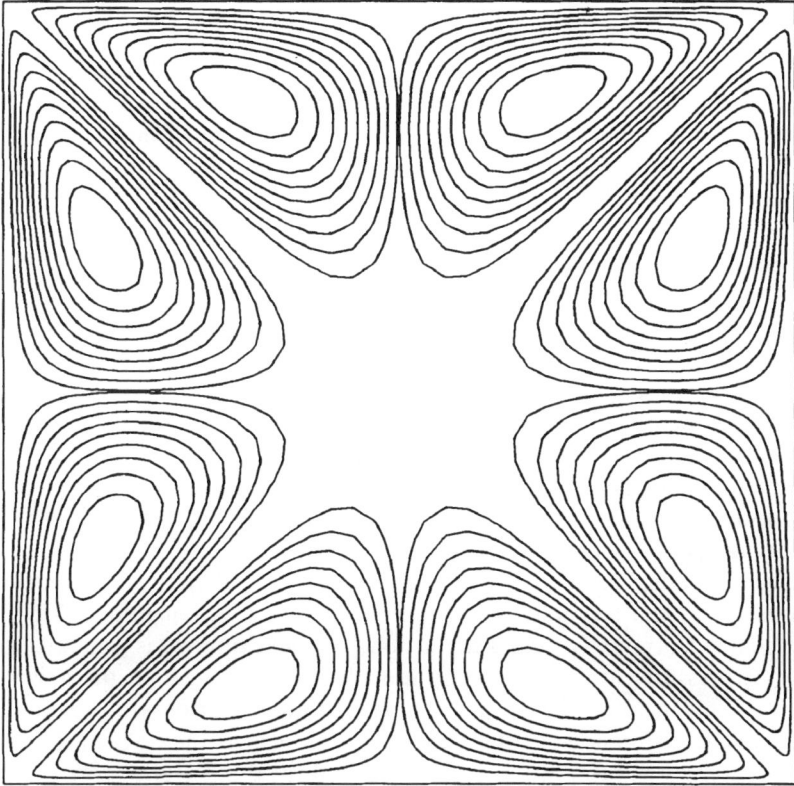

Figure 4 Secondary flow streamlines in a square duct with SSG Reynolds stress model (So, private communication).

is represented through the stress anisotropy tensor b_{ij} given by

$$\frac{Db_{ij}}{Dt} \equiv \frac{D\tau_{ij}}{Dt} - \frac{\tau_{ij}}{K}\frac{DK}{Dt} = 0 \tag{35}$$

and the equilibrium hypothesis for the diffusive terms is given by

$$\frac{\partial}{\partial x_k}\left(\bar{D}^t_{ijk} + \nu\frac{\partial \tau_{ij}}{\partial x_k}\right) = 0 \tag{36}$$

If we use these constraints as well as the definition of the stress anisotropy given in (13), and consider the fact that the mean velocity gradient is the sum of the strain- and rotation-rate tensors, then (35) can be expanded as

$$(\bar{P} - \epsilon)b_{ij} = -\tfrac{2}{3}K\bar{S}_{ij} - K\left(b_{ik}\bar{S}_{jk} + b_{jk}\bar{S}_{ik} - \tfrac{2}{3}b_{mn}\bar{S}_{mn}\delta_{ij}\right)$$

$$-K\left(b_{ik}\bar{W}_{jk} + b_{jk}\bar{W}_{ik}\right) + \tfrac{1}{2}\Pi_{ij} \tag{37}$$

This relation has been the traditional starting point for the algebraic stress models that have been developed.

Rodi (1976) was the first to implement an *implicit* algebraic relationship with the LRR model. It led to a *non-constant* coefficient C_μ in the eddy-viscosity definition (31), which was a function of the turbulent production and dissipation rate. Unfortunately, the potential of the formulation did not fully materialize because of the implicit nature of the algebraic relationship. At each iteration (time) step, an additional inner iteration is required to solve for the appropriate τ_{ij} component. This iterative process can cause solution divergence, and increased numerical overhead because of the extensive matrix inversions required at the different iteration levels. With the ASM formulated in this way, it is more appropriately viewed as a subset of the Reynolds stress formulation rather than a subset of the two-equation formulation.

Recently, interest has been renewed in algebraic stress models, as well as in the related area of non-linear two-equation models; the intent has been development of higher-order constitutive relations between the Reynolds stress tensor and the mean velocity gradient field. In this approach, the emphasis shifts from an alternative Reynolds stress formulation to an improved two-equation formulation. Gatski and Speziale (1993) have developed an *explicit* constitutive relationship for the Reynolds stress tensor with the ideas of Pope (1975), who proposed a tensorial polynomial expansion. The development is generalizable to a class of pressure–strain correlation models that are linear in the anisotropy tensor b_{ij}. This includes the linear LRR and IP models discussed previously, as well as linearized versions of such non-linear models as the SSG model. A 10-term polynomial representation, applicable to three-dimensional flows, which is quartic in products of the strain-rate and rotation-rate tensors, is found. For a two-dimensional flow, a general quadratic constitutive relationship can be written as

$$\tau_{ij} = \tfrac{2}{3}K\delta_{ij} - \nu_t^*\left(S_{ij}^* + \frac{K}{\varepsilon}[(S_{ik}^* W_{kj}^* + S_{jk}^* W_{ki}^*) - 2(S_{ik}^* S_{kj}^* - \tfrac{1}{3}S_{kl}^* S_{kl}^* \delta_{ij})]\right) \quad (38)$$

where

$$\nu_t^* = \frac{6(1+\eta^2)\alpha_1}{3 + \eta^2 + 6\zeta^2\eta^2 + 6\zeta^2}\frac{K^2}{\varepsilon} \quad (39)$$

$$\eta = \frac{K}{\varepsilon}(S_{ij}^* S_{ij}^*)^{\frac{1}{2}}, \quad \zeta = \frac{K}{\varepsilon}(W_{ij}^* W_{ij}^*)^{\frac{1}{2}} \quad (39a)$$

$$S_{ij}^* = \tfrac{1}{2}g(2 - C_3)\overline{S}_{ij} \quad (39b)$$

$$W_{ij}^* = \tfrac{1}{2}g(2 - C_4)\overline{W}_{ij} \quad (39c)$$

with $g = (C_1/2 - 1 + \mathcal{P}/\epsilon)^{-1}$, and $\alpha_1 = (C_2 - \frac{4}{3})/(C_3 - 2)$. The coefficients are obtained from the general linear pressure–strain model (15).

Gatski and Speziale (1993) have compared the present formulation with other nonlinear two-equation models such as those by Yoshizawa (1984), Speziale (1987), and Rubinstein and Barton (1990). This comparison has shown that these models have a similar basis and should represent the same essential physical behaviour when implemented. Taulbee *et al.* (1994) has also derived algebraic stress models but has modified the coefficients proposed in the original LRR model to be able to develop a truncated series expansion that requires no regularization as the more general formulation just described. Unfortunately, this recalibration results in b_{33} anisotropy levels for homogeneous shear flow that are essentially zero, which contradicts both the experimental and numerical simulation results.

3 One-equation models

The next level in this hierarchy of turbulent closure models results in only the solution of a single transport equation. As outlined in Section I, the physical motivation behind the development of the closure schemes is to be able to describe the turbulence through both characteristic velocity and length scales. Both the Reynolds stress and two-equation models just described provide this information through combinations of the Reynolds stress (kinetic energy) field and the dissipation rate field. At the one-equation level of closure and below, insufficient information from the transport equation to be solved is available to determine both quantities.

The earliest one-equation models date back to Prandtl and involve the turbulent kinetic energy equation (velocity scale) and a prescription for the turbulent length scale. Modern-day approaches have evolved beyond this formulation to the solution of transport equations for the turbulent Reynolds number or the turbulent eddy viscosity (velocity scale × length scale). These more recent formulations will be discussed here, and the interested reader can refer to the text by Wilcox (1993) for a more complete historical record.

Baldwin and Barth (1991) devised a one-equation model that has its origins in the K–ε two-equation formulation. A turbulent eddy viscosity $\nu \bar{R}_t = K^2/\varepsilon$ is defined, as well as its transport equation, which is derivable from the modelled K–ε equations (30) and (12) (with (33)). This equation is further modified by simplifying the form of the turbulent diffusion term and neglecting the two coupling terms. The result is a self-consistent one-equation model for $\nu \bar{R}_t$. The neglect of the coupling terms is formally valid in equilibrium flows where turbulent energy production equals dissipation. For the present introductory purposes, only a relatively simple form of the model will be presented; such factors as near-wall damping and intermittency effects on the form of the model are excluded but can be found in the original paper. The resulting equation is

$$\frac{\partial(\nu\bar{R}_t)}{\partial t} + \bar{u}_k \frac{\partial(\nu\bar{R}_t)}{\partial x_k} = (C_{\varepsilon 2} - C_{\varepsilon 1})\sqrt{(\nu\bar{R}_t)\mathcal{P}} + \left(\nu + \frac{\nu_t}{\sigma_\varepsilon}\right)\frac{\partial^2(\nu\bar{R}_t)}{\partial x_k \partial x_k} - \frac{1}{\sigma_\varepsilon}\frac{\partial\nu_t}{\partial x_k}\frac{\partial(\nu\bar{R}_t)}{\partial x_k}$$

$$(40)$$

where $\nu_t = C_\mu(\nu\bar{R}_t)$, and

$$C_{\varepsilon 1} = 1.2, \quad C_{\varepsilon 2} = 2.0, \quad C_\mu = 0.09, \quad \sigma_\varepsilon = \frac{(C_{\varepsilon 2} - C_{\varepsilon 1})\sqrt{C_\mu}}{\kappa^2} \qquad (41)$$

with the Von Karman constant $\kappa = 0.41$. Both $C_{\varepsilon 1}$ and $C_{\varepsilon 2}$ have values somewhat different from their values in the two-equation models discussed previously. The decay coefficient $C_{\varepsilon 2}$ is set to 2 to correspond to early experimental data on the decay of grid turbulence that showed a decay inversely proportional to downstream distance. Later experiments showed a more rapid decay that is the basis for the decay coefficient utilized in most of the two-equation models. With the decay coefficient fixed, the coefficient $C_{\varepsilon 1}$ is calibrated to match the skin-friction coefficient for incompressible flow over a flat plate.

In this one-equation formulation, the turbulence field is coupled to the mean field only through the turbulent eddy viscosity, which appears as part of an effective viscosity $(\nu + \nu_t)$ in the diffusion term of the Reynolds-averaged Navier–Stokes equation.

Other contemporary one-equation models using the eddy viscosity have been proposed. One is the Gulyaev et al. (1993) model which is an improved version of the model developed by Sekundov (1971). It has been shown in the Russian literature to solve a variety of incompressible and compressible flow problems (see Gulyaev et al. (1993) for selected references). Another is the model by Spalart and Allmaras (1994), which was motivated by the Baldwin–Barth model, and has become a popular model among industrial users owing to its ease of implementation and relatively inexpensive cost. Even though this one-equation level of closure is heavily based on empiricism and dimensional analysis, with characterizing flow features usually accounted for on a term by term basis using phenomenological based models, it has tended to perform well on the type of flows the models have been sensitized to, and as the recent study by Shur et al. (1995) has shown, can even outperform some two-equation models in separating and reattaching flows.

4 Zero- and half-equation models

To this point in the presentation of the closure models, the format has been to start from a general closure (the Reynolds stress model) and proceed to the simpler closure schemes. In this section, the two least complex closure models will be discussed. However, in the presentation of these two models, the simpler zero-equation model will be presented first and then the more complex half-equation model. The reversal in order of presentation is because the underlying assumptions associated with the half-equation model are closely associated with the algebraic model or the zero-equation model.

Throughout the earlier parts of this chapter, the presentation of the models has been confined to the high-Reynolds-number forms in order to minimize the complications associated with the presence of solid boundaries. As will be shown shortly, the historical motivation behind the development of the algebraic model was the need to account properly for the presence of a wall in the solution of turbulent flows. This fact necessitates the introduction of some of the fundamental modelling ideas associated with the presence of solid boundaries which will also serve as an introduction to the next two sections in which the ideas behind both wall functions and near-wall closure models will be discussed.

Both the zero- and half-equation models discussed are *two-layer models*, which divide the boundary layer into an inner and outer layer, and they both use mixing-length ideas to specify the length and velocity scales that are needed. This is the basis for the derivation of the mixing-length theory and serves as the best means of presenting the key ideas of the formulation. In contrast to the previous sections where the formulations were given in a general mathematical form, the discussion in this section will be confined to simple planar flows. Extension to more complex geometries is certainly possible, but each case would have to be evaluated on its own merits to determine the appropriate measures for the relevant turbulent scales.

The zero-equation model derives its name from the fact that the eddy viscosity required in the turbulent stress–strain relationship is defined from an algebraic relationship rather than a differential one. The earliest example of such a closure is Prandtl's mixing-length theory (Prandtl, 1925). In analogy with the kinetic theory of gases, Prandtl assumed the form for the turbulent eddy viscosity in a plane shear flow with unidirectional mean flow $\bar{u}_1(x_2) = \bar{u}(y)$ and shear stress $\tau_{12} = \tau_{xy} = -\nu_t \, d\bar{u}/dy$. The eddy viscosity was assumed to have the form

$$\nu_t = l^2 \left| \frac{d\bar{u}}{dy} \right| \tag{42}$$

where l is the mixing length which requires specification for each flow under consideration. In a free shear flow, the mixing length would be a characteristic measure of the width of the shear layer. In a planar wall-bounded flow, the mixing length l in the near-wall region would be proportional to the distance from the wall. These relationships, though simple, give rise to significant results about the structure of turbulent flows. In the case of wall-bounded flows, the law of the wall and the structure of the outer layer of the boundary–layer flow can be deduced. Several texts and reviews in the literature provide an insightful description of the physical basis for this type of modelling. These include

Tennekes and Lumley (1972), Reynolds (1976), Speziale (1991), and Wilcox (1993).

Two of the most popular and versatile algebraic models are the Cebeci–Smith (see Cebeci and Smith, 1974) and the Baldwin–Lomax (see Baldwin and Lomax, 1978) models. These modern-day mixing-length models are two-layer eddy-viscosity models that have an inner layer eddy viscosity given by

$$\text{Cebeci-Smith:} \qquad \nu_{ti} = l^2 \left(\frac{\partial \bar{u}_i}{\partial x_j} \frac{\partial \bar{u}_i}{\partial x_j} \right)^{\frac{1}{2}} \qquad (43)$$

$$\text{Baldwin-Lomax:} \qquad \nu_{ti} = l^2 (\bar{\omega}_i \bar{\omega}_i)^{\frac{1}{2}} \qquad (44)$$

where $\omega_i = \epsilon_{ijk} \bar{u}_{k,j}$ is the vorticity vector, and an outer layer eddy viscosity given by

$$\text{Cebeci-Smith:} \qquad \nu_{to} = 0.0168 \bar{u}_e \delta^* F_K(y; \delta) \qquad (45)$$

$$\text{Baldwin-Lomax:} \qquad \nu_{to} = 0.0168 F_{wk} F_K(y; y_m/0.3) \qquad (46)$$

The mixing length is defined similarly in both models for zero-pressure-gradient flows. That is,

$$l = \kappa y \left[1 - e^{-y^+/A^+} \right] \qquad (47)$$

where $\kappa = 0.41$ is the von Karman constant, A^+ is the Van Driest damping coefficient, and y^+ is the distance from the wall in wall units $(u_\tau y/\nu)$. In the expressions for the outer layer eddy viscosity, δ is the boundary layer thickness, δ^* is the displacement thickness, and $\bar{u}_e$ is the edge velocity. In general, the damping coefficient A^+ can be a function of the pressure gradient, but for the present purposes it will be assumed to be constant. Throughout this subsection, attention will be focused on the form of the models for zero-pressure-gradient flows; extensions that include pressure gradient effects can be found in the references cited for the particular algebraic models. The functions F_K and F_{wk} are an intermittency and a wake function, respectively. The Klebanoff intermittency function F_K is given by

$$F_K(y; \Delta) = \left[1 + 5.5 \left(\frac{y}{\Delta} \right)^6 \right]^{-1} \qquad (48)$$

and the wake function F_{wk} is given by

$$F_{wk} = \min \left(y_m F_m; y_m \bar{u}_{dif}^2/F_m \right) \qquad (49)$$

with

$$F_m = \frac{1}{\kappa} \left[\max_y (l|\bar{\omega}_i|) \right] \qquad (49a)$$

Where y_m is the location where F_m occurs, Δ is the boundary layer thickness δ in the Cebeci–Smith model, and Δ is $y_m/0.3$ in the Baldwin–Lomax model. The quantity $\bar{u}_{dif}$ is the difference between the maximum and minimum total velocity in the profile. Unlike the Cebeci–Smith model, the Baldwin–Lomax model does not need to know the location of the boundary layer edge. As (49a) suggests, the Baldwin–Lomax model bases the outer layer length scale on the vorticity in the layer rather than the displacement thickness, as in the Cebeci–Smith model. Extensions and generalizations to more complex flows can be found in Wilcox (1993) and Cebeci and Smith (1974).

The next level of complexity in closure modelling, which is closely related to the algebraic models just discussed, is the half-equation model. This level of closure derives its name somewhat subjectively from the fact that an ordinary differential equation is solved rather than a partial differential equation. Nevertheless, this level of closure does generalize the algebraic model by specifying a smooth functional behaviour for the eddy viscosity across the boundary layer, and by accounting in a limited way for history effects by solving a 'transport equation' for the maximum shear stress. Johnson and King (1985) devised this model, and since its inception it has undergone modifications (Johnson, 1987; Johnson and Coakley, 1990) to improve its predictive capabilities for a wider class of flows, in particular, compressible flows. For the present purpose, only the simpler incompressible formulation will be outlined so not to unnecessarily complicate the explanation of the model formulation.

The Johnson–King model is also a two-layer eddy-viscosity type model; however, in this model, the eddy viscosity transitions in a prescribed functional manner from the inner layer form to the outer layer form. This functional form is given by (Johnson and King, 1985)

$$\nu_t = \nu_{to}[1 - \exp(\nu_{ti}/\nu_{to})] \tag{50}$$

In the later form of the model (Johnson and Coakley, 1990), which was used in the solution of transonic flow problems, this functional dependency is based on a hyperbolic tangent function. The inner layer eddy viscosity is given by

$$\nu_{ti} = l^2 \frac{\sqrt{\tau_{xy}}}{\kappa y}\bigg|_m \tag{51}$$

where l is the mixing length defined in (47) with $A^+ = 15$, and the subscript m denotes maximum value. In zero-pressure-gradient two-dimensional flows where the law of the wall holds, this expression for ν_{ti} corresponds to the Cebeci–Smith inner layer eddy viscosity given in (43). The outer layer eddy viscosity is given by

$$\nu_{to} = 0.0168\bar{u}_e\delta^* F_K(y; \delta)\sigma(x) \tag{52}$$

which is the Cebeci–Smith form (45) with the addition of the factor $\sigma(x)$ that accounts for nonequilibrium streamwise evolution of the flow. At each streamwise station, $\sigma(x)$ is adjusted so that the relation

$$\nu_t|_m = \frac{-\tau_{xy}|_m}{\partial \bar{u}/\partial y|_m} \tag{53}$$

is satisfied. The remaining quantity that is needed is $\tau_{xy}|_m \equiv \tau_m$, and this is determined from a transport equation for the shear stress τ_{xy}. Unlike the Reynolds stress closures where the transport equation for the turbulent shear stress contains modelled pressure–strain correlations and turbulent transport terms, this turbulent shear stress equation is extracted from the turbulent kinetic energy equation (cf. equation (30)) by assuming that the shear stress anisotropy $a_{12} = -\tau_{xy}/K = 0.25$ is constant at the point of maximum shear. If the viscous diffusion effects are neglected, the evolution equation for τ_m is

$$\bar{u}_m \frac{d\tau_m}{dx} = a_{12}\left(\sqrt{\tau_{meq}} - \sqrt{\tau_m}\right)\frac{\tau_m}{L_m} - C_{dif}\frac{\tau_m^{\frac{3}{2}}}{(0.7\delta - y_m)}\left[1 - \sigma^{\frac{1}{2}}(x)\right] \tag{54}$$

where τ_{meq} is the equilibrium value $(\sigma(x) = 1)$ for the shear stress, $C_{dif} = 0.5$ for $\sigma(x) \geq 1$ and zero otherwise, and L_m is the dissipation length scale given by

$$L_m = \kappa y, \qquad y_m/\delta \leq 0.09/\kappa \tag{55}$$

$$L_m = 0.09\delta, \qquad y_m/\delta > 0.09/\kappa \tag{56}$$

Because $\sigma(x)$ is not known *a priori* at each streamwise station, it is necessary to iterate on the equation set at each station in order to determine its value.

Once again, the reader should be reminded that the Johnson–King model that has been discussed here has evolved since its inception to a slightly different form than has been presented. The alterations are straightforward and are intended to account for compressibility effects or unique features of separated flows in order to improve its predictive capability. Nevertheless, the intent here has been to present the underlying ideas behind the formulation, with the opportunity for the interested reader to refer to the suggested references for the latest modifications and applications at this time.

C Wall functions for incompressible flows

In the previous sections in which the second-moment, two-equation and one-equation models were presented, the focus was centred on the high-Reynolds-number forms. Thus, modifications due to the presence of solid boundaries or other low-Reynolds-number effects were not presented. In the case of a solid boundary, these additional effects require modifications to the models themselves or necessitate the introduction of an intermediate layer where the turbulence quantities can be specified. In this section, the construction of such a layer with the appropriate *wall functions* will be presented. In the next section, an alternative approach is presented where the high-Reynolds-number form of the models are themselves modified.

The wall-function approach is the most common method employed in the solution of incompressible wall-bounded flows. Although their range of applicability is formally limited to flows with constant stress layers and log layers, they have been applied to more complex flows with some success. While the formulation is somewhat standard, many authors omit the details of the formulation, especially the underlying assumptions that are used. It is instructive to review these assumptions to better ascertain the range of applicability of such methods, and to set the stage for the compressible formulation to be discussed in Section III.D.

In the constant stress layer near the solid surface, where the convective terms are small in comparison with the diffusion terms, the steady form of the streamwise component of the momentum equation (5) reduces to

$$\bar{\sigma}_{xy} - \rho \tau_{xy} = \bar{\sigma}_{xy}\big|_w \tag{57}$$

where in the present discussion the streamwise pressure gradient is neglected and the discussion is confined to a simple planar geometry with x and y as the streamwise and normal coordinates, respectively. In applying the wall functions along more complex surfaces, the appropriate spatial coordinate is aligned along the normal to the surface. Equation (57) simply states that the total stress $\bar{\sigma}_{tot} = \bar{\sigma}_{xy} - \rho \tau_{xy}$ in the layer is equivalent to the wall shear stress $\bar{\sigma}_w$.

The incompressible law of the wall is easily obtained from either dimensional analysis or mixing-length arguments (e.g. Tennekes and Lumley, 1972) and yields

$$\frac{\partial \bar{u}}{\partial y} = \frac{(\bar{\sigma}_w/\rho)^{\frac{1}{2}}}{\kappa y} \tag{58}$$

or

$$\bar{u}^+ \equiv \frac{\bar{u}}{\bar{u}_\tau} = \frac{1}{\kappa} \ln y^+ + B \tag{59}$$

where $\bar{u}_\tau = (\bar{\sigma}/\rho)^{\frac{1}{2}}_w$, $y^+ = \bar{u}_\tau y/\nu$, and (from experiments) $B \approx 5$. As shown in the discussion of the algebraic closure models, the length scale in the constant stress layer varies linearly with distance from the wall, so that the eddy viscosity ν_t is given by (31) and

$$\nu_t = \kappa y \bar{u}_\tau \tag{60}$$

Consistent with the fact that the log layer is a constant stress layer, a structural equilibrium for the turbulence is assumed and can be characterized by the fact that the Reynolds shear stress anisotropy $\tau_{xy}/K = -\sqrt{C_\mu}$ achieves an equilibrium value that is independent of the upstream conditions and dissipation rate values. This constraint, coupled with (31) and (60), leads to the following scaling of the dissipation rate in the inner layer

$$\varepsilon = \frac{K^{\frac{3}{2}}}{C_l y} \tag{61}$$

where $C_{l_1} = \kappa / C_\mu^{\frac{3}{4}} \approx 2.5$. Note also that in the log layer $\varepsilon^+ = 1/\kappa y^+$ and $K^+ = C_\mu^{-\frac{1}{2}}$. Equation (61) is an inertial estimate of the dissipation rate and, as such, is applicable in turbulent regions close to the wall. If the wall function interface is too close to the wall, then such an estimate would cause significant errors. In flows in which separation occurs, this formulation is unacceptable because the wall shear stress would vanish. Even though the applicability of the wall function approach is suspect in such flows, the question of computability does arise. In such cases, Launder and Spalding (1974) have replaced the velocity scale $\bar{u}_\tau$ with $K^{\frac{1}{2}}$, which does not vanish for separated flows and is consistent with the $\bar{u}_\tau$ approach in constant stress flows.

It is worthwhile to outline the implementation aspects of the wall-function formulation. Although it is not possible to outline this in complete generality because of the differing numerical grid structures and geometries that can be considered, a simple framework can serve as a basis for implementation. Consider a flat-plate boundary layer flow with the computational domain spanned by a cartesian grid.

Because the mean-flow equations are evaluated at the first interior point $y_{(2)}$ away from the wall, the wall shear $\bar{\sigma}_w$ and the surface pressure $\bar{p}_w$ must be specified for these equations. The wall shear is related directly to the friction velocity which is extracted from the log-law (59). The value of the surface pressure is extracted from the normal momentum equation, with the assumption that the normal gradient of pressure at the surface is zero and that the mean pressure is constant across the area bounded by the first grid cell. This leads to the simple relation $\bar{p}_w = \bar{p}_{(2)}$.

At the same point, the turbulent kinetic energy or stresses as well as the turbulent dissipation rate need to be specified. For the two-equation formulation, application of the wall functions is straightforward and simply involves specifying both K and ε at the first interior point $y_{(2)}$. The kinetic energy is determined from $K = \bar{u}_\tau^2 / \sqrt{C_\mu}$ and the dissipation rate is determined from (61). An alternative two-layer formulation has been used by Launder and Spalding (1974) and Chieng and Launder (1980) (see also Johnson and Launder, 1982), who solved the turbulent kinetic energy equation at point (2). Since this formulation explicitly accounts for the fact that the viscous sublayer is located within the first grid cell, it is necessary to treat consistently the source terms in the kinetic energy equation since the behaviour of the dependent variables in the sublayer differs significantly from that in the fully turbulent region.

For the Reynolds stress formulation, application of the wall functions is more complex since the energy redistribution between the component stresses is unknown. This requires solving the component stress equations at the first interior point and applying appropriate boundary conditions at the wall. For the single-layer model presented here, these boundary conditions are zero convective and diffusive flux at the wall.

The turbulent dissipation rate can be specified at the first interior point based on the local equilibrium assumption given by (61), or by a more general expression extracted from the turbulent dissipation rate equation itself. Younis (1984) extracts the value of the dissipation rate at the first interior point from the dissipation rate equation (12) by assuming streamwise convection is negligible and that ε varies inversely with distance from the wall in the logarithmic region. In equilibrium conditions this is consistent with (61).

The main steps for implementing the wall function approach can be outlined as follows:

(a) Set an initial guess for the wall shear stress with either $\bar{\sigma}_w = 0.3\rho K_{(2)}$ or $\bar{\sigma}_w$ from the previous iteration.
(b) Given the velocity profile from the previous iteration (time) step, iteratively calculate $\bar{\sigma}_w$ from the implicit relationship (59).
(c) Update the turbulent kinetic energy $K_{(2)}$ from $\bar{u}_\tau^2/\sqrt{C_\mu}$ or the turbulent stresses $\tau_{ij(2)}$ from their respective transport equations.
(d) Update the dissipation rate $\epsilon_{(2)}$ from (61).
(e) Return to (b) for the next iteration (time) step.

Since the various wall function distributions are developed from log-law considerations, it is necessary that the first grid point be located in this region. However, the dissipation rate is rapidly varying in this region which necessitates the use of relatively small spacing with the next interior grid point. Such issues require that the results presented in this section be used as a general guide to the basis and implementation of incompressible wall functions; however, the individual flows with their unique geometries and computational meshes may require some alterations to the above description.

D Incompressible near-wall modelling

An alternative to the wall-function method described in the previous section is to develop closure models that are capable of being integrated directly to the wall. In this section, attention is focused on modifications of the second-moment and two-equation models. Due to the relatively limited use of the one-equation models, their near-wall extensions will not be discussed in this section. The interested reader is referred to Section II.B.3 for the appropriate references. In the context of the higher order closures, the task is to account for the presence of the wall in the various closure models. Although this effort has sometimes been called low-Reynolds-number modelling, it is more properly called near-wall modelling to distinguish it from the separate problem of transition prediction. While these near-wall models may be applicable in the transition region, their theoretical foundation is based on asymptotic consistency in the fully turbulent region. For this reason, there will be no discussion of the extension into the transition region.

The Reynolds stress formulation presents the biggest challenge to the development of near-wall closures because of the complex behaviour of the modelled correlations in the presence of the wall. Both the pressure redistribution term and the anisotropic dissipation rate, which appear in the τ_{ij} transport equation, require modification. In addition, some researchers maintain that the pressure–velocity diffusion term, which as shown in Section II.B.1 is usually neglected, also needs to be taken into account in the near-wall region. Coupled with these issues, which focus on the stress transport equations, is the behaviour of the dissipation rate equation in the vicinity of the wall. The issue of near-wall modelling is at least as complex as the issue of developing the high-Reynolds-number models. Somewhat of a reprieve is attained when only a two-equation model is used. Because the turbulent kinetic energy equation contains an isotropic dissipation rate and no pressure redistribution term, the overall task is greatly simplified. In the remainder of this section, both a near-wall second-moment closure and near-wall two-equation model will be presented in order to get a perspective of the underlying rationale behind the near-wall corrections.

Several attempts have been made at developing near-wall closure corrections to the Reynolds stress transport equations and the dissipation rate equation. For the turbulent stress transport equations, the attempts have almost exclusively focused on extending the high-Reynolds-number pressure–strain models developed by Launder and colleagues. For the dissipation rate equation, additional terms and modified coefficients have been proposed to account for the anisotropic near-wall effects and the correct limiting behaviour at the wall. Although it is not possible to detail the modifications required for the second-moment stress closures, it is beneficial to give an overview of the types of corrections made to the various modelled correlations appearing in the equations. So *et al.* (1991) have reviewed the performance of eight different near-wall Reynolds stress models. The models were tested for asymptotic consistency with the exact transport equations and against DNS and experimental data of fully developed two-dimensional channel flow. These comparisons showed that the models of Shima (1988), Launder and Shima (1989), and Lai and So (1990) yielded the best overall performance. Only a new composite model that consists of the near-wall Lai and So (1990) model and the near-wall dissipation rate equation model of So *et al.* (1991) yielded agreement over the range of variables tested. Figure 5 shows the distribution of the various turbulent correlations in the near-wall region of fully developed channel flow for the Lai and So (1990) model. The results show the level of agreement that can be expected from near-wall second-moment closures when properly calibrated.

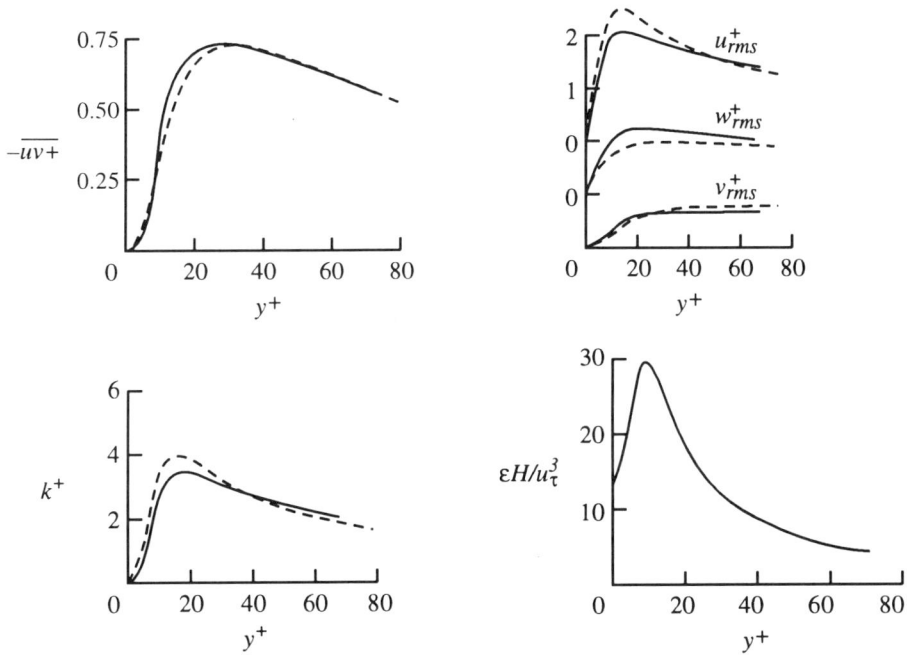

Figure 5 Near-wall distribution of turbulent correlation in fully developed channel flow (So *et al.*, 1991). Key: —, model calculations; - - -, DNS (Kim *et al.*, 1987).

It is worthwhile discussing in some detail one of the Reynolds stress near-wall models alluded to in order to gain a better perspective of the modifications required for a near-wall closure. For this purpose, the Lai and So (1990) model, which was one of the more successful models tested, will be used. Several additions are incorporated into the pressure–strain and dissipation rate equations that will be outlined. In the near-wall model, the deviatoric part of the dissipation rate is not assimilated into the pressure–strain correlation Π_{ij} but is modelled separately. The pressure–strain model is then composed of the high-Reynolds-number LRR model (15) plus a near-wall correction $\Pi_{ij}^w f_1^w$ given by

$$\Pi_{ij}^w = C_1 \varepsilon b_{ij} - \frac{\varepsilon}{K}\left(\tau_{ik}n_k n_j + \tau_{jk}n_k n_i\right) + \alpha^*\left(P_{ij} - \tfrac{1}{3}P_{kk}\right) \tag{62}$$

with

$$f_1^w = \exp\left[-(R_t/150)^2\right] \tag{62a}$$

where $\alpha^* = 0.45$, $R_t = K^2/\nu\varepsilon$ is the usual high-Reynolds-number definition for the turbulent Reynolds number, and n_k is the unit surface normal. The form of (62) is based on an optimized balance with (15) to ensure the same asymptotic behaviour in the near-wall region as in the exact pressure–strain correlation Π_{ij} term that appears in the τ_{ij} transport equation (8). By following the ideas of Hanjalić and Launder (1976), and Kebede *et al.* (1985) (cf. Launder and Reynolds, 1983; Launder and Tselepidakis, 1991), Lai and So (1990) used a dissipation rate model in which the anisotropy is attributed solely to the presence of the wall. The deviatoric part of the dissipation rate is then given by

$$\varepsilon_{ij} = \varepsilon_{ij}^w = -f_1^w \left[\tfrac{2}{3}\varepsilon\delta_{ij} - \frac{\varepsilon}{K}\frac{\left(\tau_{ij} + \tau_{ik}n_k n_j + \tau_{jk}n_k n_i + n_i n_j \tau_{kl}n_k n_l\right)}{(1 + 3\tau_{kl}n_k n_l/2K)} \right] \tag{63}$$

These two modifications to the pressure–strain correlation and the tensor dissipation rate are applied to the Reynolds stress transport equation.

In the isotropic dissipation rate equation, both the production and destruction terms in (12) are modified, and an additional term is introduced to satisfy the coincidence of $\partial\varepsilon/\partial t$ and $\partial(\nu\partial^2 K/\partial x_j\partial x_j)/\partial t$ at the wall. The isotropic dissipation rate equation now becomes

$$\frac{\partial\varepsilon}{\partial t} + \frac{\partial}{\partial x_j}\left(\bar{u}_j\varepsilon\right) = \left(1 + \sigma^w f_2^w\right)\bar{\mathcal{P}}_\varepsilon - C_{\varepsilon 2}f_\varepsilon\frac{\epsilon\epsilon_i^w}{K} + \bar{\mathcal{D}}_\varepsilon^t + \nu\frac{\partial^2\varepsilon}{\partial x_j\partial x_j} + \xi^w \tag{64}$$

with

$$\xi^w = f_2^w\left[\left(\tfrac{7}{9}C_{\varepsilon 2} - 2\right)\frac{\varepsilon\varepsilon_1^w}{K} - \frac{1}{2}\frac{(\varepsilon_2^w)^2}{K}\right] \tag{65}$$

where $\varepsilon_1^w = \varepsilon - 2\nu\left(\partial\sqrt{K}/\partial x_2\right)^2$, $\varepsilon_2^w = \varepsilon - 2\nu K/x_2^2$, and $\sigma^w = 1 - 0.6\exp[-Re/10^4]$. The damping function f_2^w is given by $\exp\left[-(Re_t/64)^2\right]$, and the modelling coefficients $C_{\varepsilon 1}, C_{\varepsilon 2}, C_\varepsilon$ (from (12c)) take the values $1.35, 1.80$, and 0.15, respectively. The variables ε_1^w and ε_2^w appear in (65) to ensure that the respective terms remain finite at the wall. The production $\bar{\mathcal{P}}_\varepsilon$ and diffusion $\bar{\mathcal{D}}_\varepsilon^t$ terms are as given in (12a) and (12c), respectively. The form adopted here for the near-wall isotropic dissipation rate equation is similar to that adopted by Shima (1988); however, changes to the form of ε_1^w and ε_2^w have been made to increase computational efficiency. The effect of these near-wall corrections diminishes sufficiently far from the wall, so that the suitably calibrated high-Reynolds-number form of the models holds.

It would be remiss not to point out the recent work by Durbin (1993), who developed a near-wall model based on the concept of kinematic wall blocking. This concept is based on the fact that the normal component of velocity to the wall must be brought to zero inviscidly. This inviscid phenomena affects the normal velocity component at considerably greater distances from the wall than the viscous effects. Durbin (1991, 1993) accounts for this (elliptic) effect through a Poisson equation for the pressure. Both a Reynolds stress model and a simpler $K - \varepsilon - \overline{v^2}$ model (in the notation of Durbin (1991), v is the normal fluctuating velocity) are developed.

Development of a near-wall closure for a two-equation model is obviously simpler than the Reynolds stress counterpart because no pressure–strain correlation terms and anisotropic dissipation terms exist. In effect, only the requisite modifications to the isotropic dissipation rate equation are invoked, along with a suitable damping coefficient applied to the definition of the eddy viscosity (31). This modification simply alters the eddy-viscosity definition so that

$$\nu_t = C_\mu \, f_\mu \frac{K^2}{\varepsilon} \tag{66}$$

with the damping function f_μ given as a function of normalized distance from the wall (wall units) and also as a function of turbulent Reynolds number Re_t. With the introduction of the near-wall damping into the eddy-viscosity definition, the turbulent kinetic energy equation is then left unaltered from its high-Reynolds-number form.

Several attempts have been made to develop near-wall two-equation closures. So et al. (1991) have compared the performance of four different near-wall models with both DNS simulations and measurements of channel and boundary layer flows for comparison. Wilcox (1993) has compared the performance of other earlier two-equation models. This evaluation of earlier models was somewhat limited because it did not focus heavily on the performance of the turbulent quantities such as shear stress or dissipation rate. In the So et al. (1991) study which focused on more recent models, the models of Myong and Kasagi (1990) and Speziale et al. (1990) gave reasonable predictions of the kinetic energy and the shear stress in the wall region but poorly predicted the dissipation rate below x_2^+ values of 30. The two-equation K–ω model of Wilcox (1988a) was also tested, but in a more limited way because it was not formulated to be asymptotically consistent near the wall. The model proposed by So et al. (1991) gave good predictions in the $x_2^+ > 30$ region, and also gave the correct qualitative behaviour below the $x_2^+ = 30$ level. In the So et al. model, the turbulent kinetic energy equation (30) is used with the turbulent diffusion term given by (32) ($\sigma_K = 0.75$) and the eddy viscosity defined in (66) ($C_\mu = 0.096$). The damping function f_μ is the same as that used by Speziale et al. (1990) and given by

$$f_\mu = \left[1 + \frac{3.45}{\sqrt{Re_t}} \right] \tanh \left(\frac{x_2^+}{115} \right) \tag{67}$$

Once again, the form of the damping function f_μ is such that its effect diminishes away from the wall in regions where the high-Reynolds-number form of the models holds.

The isotropic dissipation rate equation used is given by (64) with $\sigma^w = 0$ and $\bar{\mathcal{D}}_\varepsilon^t$ given by (33) ($\sigma_\varepsilon = 1.45$). The coefficients $C_{\varepsilon 1}$ and $C_{\varepsilon 2}$ are assigned values of 1.5 and 1.83, respectively. The form of the additional term ξ^w is simplified and is now given by

$$\xi^w = f_2^w \left[-2 \frac{\varepsilon \varepsilon_1^w}{K} + 1.5 \frac{\varepsilon_2^{w2}}{K} \right] \tag{68}$$

where f_2^w is the same as that used in (65). Additional insight into the behaviour of two-equation models can be extracted from a perturbation analysis of the equations across the boundary layer. Wilcox (1993) has analysed several two-equation models asymptotically to assess their ability to predict the correct flow behaviour in the viscous sublayer, the log layer, and the defect layer of the turbulent boundary layer.

As evident from the discussion in this section, near-wall models can and have been developed which, at the very least, give good qualitative predictions of both mean and turbulent flow quantities in simple planar wall-bounded flows. For some flow variables, the predictions are quantitatively accurate as well. The remaining issue with any of the models developed for this technologically important class of flows is their ability to predict the relevant complex engineering flows. Insufficient tests have been conducted to answer this question. However, if the models that are employed are not sufficiently mathematically consistent with the exact equation set, then the near-wall equations clearly introduce a stiffness into the overall equation set and make it difficult to solve the entire system efficiently over the entire domain.

III COMPRESSIBLE TURBULENT MODELLING

Once again, the starting point for the formulation of the compressible analogue to the Reynolds-averaged Navier–Stokes equations is the mass and momentum conservation equations, now coupled to the conservation of energy equation. The mass and momentum conservation equations are now best written in the form (cf. equations (1) and (2))

$$\frac{\partial \rho}{\partial t} + \frac{\partial}{\partial x_j} \left(\rho u_j \right) = 0 \tag{69}$$

$$\frac{\partial (\rho u_i)}{\partial t} + \frac{\partial (\rho u_i u_j)}{\partial x_j} = -\frac{\partial p}{\partial x_i} + \frac{\partial \sigma_{ij}}{\partial x_j} \tag{70}$$

where σ_{ij} is the deviatoric part of the viscous stress tensor and is given by the relationship

$$\sigma_{ij} = 2\mu \left(S_{ij} - \tfrac{1}{3} S_{kk} \delta_{ij} \right) \tag{70a}$$

and p is now a thermodynamic variable. In (70a), the bulk viscosity is assumed to be small, which is the case for dilute monatomic gases.

For a perfect gas, the total energy is the sum of the internal energy and the kinetic energy, and the corresponding conservation equation is

$$\frac{\partial}{\partial t}\left[\rho\left(e + \frac{u_i u_i}{2}\right)\right] + \frac{\partial}{\partial x_j}\left[u_j\rho\left(e + \frac{u_i u_i}{2} + \frac{p}{\rho}\right)\right] = \frac{\partial}{\partial x_j}\left(u_i\sigma_{ij}\right) - \frac{\partial q_j}{\partial x_j} \quad (71)$$

where $e = c_v T$, and p is given by ρRT (R is the gas constant). The heat flux vector q_j is given by the Fourier heat conduction law $q_j = -k_T \partial T/\partial x_j$, where k_T is the thermal conductivity and c_v is the specific heat at constant volume (which is constant because the fluid is assumed to be calorically perfect). Note that the second term on the left in (71) can be written in terms of the total enthalpy $H = \rho(c_p T + u_i u_i/2 + p/\rho)$, where c_p is the (constant) specific heat at constant pressure.

The optimal way, both mathematically and physically, to formulate both the mean conservation equations and the transport equations for any turbulence quantities is to employ Favre or mass-weighted averages (Favre, 1965, 1991). The resulting similarity between the incompressible and compressible form of the equations suggests straightforward variable density extensions of some closure models used in the incompressible formulation and also clearly identifies new correlations that are unique to the compressible formulation. For a dependent variable f, the Favre average is defined as

$$\tilde{f} = \frac{\overline{\rho f}}{\overline{\rho}}, \quad (72)$$

with the associated mean and fluctuating decomposition given by

$$f = \tilde{f} + f'' \quad (73)$$

and the time average defined in (3a). Useful guides to the properties associated with Favre averaging can be found in Rubesin and Rose (1973) and Cousteix and Aupoix (1990). If (73) is substituted into the conservation equations just discussed, the appropriate average taken, and the second moments formed, the compressible analogues to the Reynolds averaged Navier–Stokes equations and turbulent transport equations found in the incompressible formulation are obtained.

A Mean conservation equations

In the compressible formulation, additional equations for the mean conservation of mass (which replace the simpler continuity equation) and the mean conservation of (total) energy are needed. The forms of these equations and the momentum equation are intended to parallel the incompressible forms as closely as possible to highlight both the similarities and differences between the two formulations.

1 Mass and momentum

If we decompose the dependent variables according to (73) and follow the procedure outlined in the incompressible section, then the mean conservation equations for mass and momentum are given by

$$\frac{\partial \bar{\rho}}{\partial t} + \frac{\partial}{\partial x_j}(\bar{\rho}\tilde{u}_j) = 0 \tag{74}$$

and

$$\frac{\partial(\bar{\rho}\tilde{u}_i)}{\partial t} + \frac{\partial}{\partial x_j}(\tilde{u}_j\bar{\rho}\tilde{u}_i) = -\frac{\partial \bar{p}}{\partial x_i} + \frac{\partial \bar{\sigma}_{ij}}{\partial x_j} - \frac{\partial(\overline{\rho u_i'' u_j''})}{\partial x_j} \tag{75}$$

where

$$\bar{\sigma}_{ij} = \overline{2\mu\left(S_{ij} - \tfrac{1}{3}S_{kk}\delta_{ij}\right)} \simeq 2\bar{\mu}\left(\tilde{S}_{ij} - \tfrac{1}{3}\tilde{S}_{kk}\delta_{ij}\right) \tag{75a}$$

is the viscous stress tensor, $\bar{\mu}$ the mean molecular viscosity, and $\tau_{ij} \equiv \widetilde{u_i'' u_j''}$ is the Favre-averaged correlation tensor. Equation (75a) neglects contributions from the fluctuating viscosity μ' and assumes that the Favre-averaged mean velocity and Reynolds-averaged mean velocity are approximately equal, although Ristorcelli (1993) has recently shown some evidence to the contrary.

2 Total energy

Unlike the incompressible case, the compressible formulation requires that the conservation of energy be explicitly used. In the formulation to be discussed here, a mean total energy equation with specific total energy

$$\tilde{E} = c_v\tilde{T} + \frac{\tilde{u}_i\tilde{u}_i}{2} + \frac{\widetilde{u_i'' u_i''}}{2} \tag{76}$$

is derived. That is,

$$\frac{\partial(\bar{\rho}\tilde{E})}{\partial t} + \frac{\partial}{\partial x_j}\left(\tilde{u}_j\bar{\rho}\tilde{H}\right) = \frac{\partial}{\partial x_j}\bar{\Sigma}_j - \frac{\partial}{\partial x_j}\left(\bar{q}_j + \overline{\rho E'' u_j''}\right) \tag{77}$$

with

$$\tilde{H} = \tilde{E} + \frac{\bar{p}}{\bar{\rho}} \tag{77a}$$

$$\bar{\Sigma}_j = \bar{\sigma}_{ij}\tilde{u}_i + \bar{\sigma}_{ij}\overline{u_i''} + \overline{\sigma_{ij}' u_i'} \tag{77b}$$

$$\bar{q}_j = -\overline{k_T T_{,j}} \simeq -\bar{k}_T\tilde{T}_{,j} \tag{77c}$$

$$\overline{\rho E'' u_j''} = c_p\overline{\rho u_j'' T''} + \bar{\rho}\tilde{u}_i\tau_{ij} + \frac{\overline{\rho u_i'' u_i'' u_j''}}{2} \tag{77d}$$

where $\bar{k}_T$ is the mean thermal conductivity. In (77c), fluctuations in the thermal conductivity are neglected, and the Favre-averaged mean temperature and Reynolds-averaged mean temperature are taken as approximately equal. The mass flux, heat flux, and turbulent transport are additional correlations that must be modelled. An equation of state is also required for the specification of the mean pressure and for a perfect gas is given by $\bar{p} = \bar{\rho}R\tilde{T}$. In terms of the total energy, $\bar{p}$ is given by the equivalent form

$$\bar{p} = (\gamma - 1)\left[\bar{\rho}\tilde{E} - \tfrac{1}{2}\bar{\rho}(\tilde{u}^2 + \tilde{v}^2 + \tilde{w}^2) - \bar{\rho}K\right] \tag{78}$$

where γ is the ratio of specific heats (c_p/c_c). The presence of the turbulent kinetic energy term in the equation of state suggests a strong coupling between the mean equations and either the normal Reynolds stress components for the Reynolds stress models or the turbulent kinetic energy equation for the two-equation models.

The formulation of the mean conservation equations has yielded unknown correlations that require closure. These include the Favre-averaged correlation τ_{ij}, the turbulent heat flux $\bar{\rho}c_p\widetilde{u_j''T''}$, and the turbulent transport or diffusion $\bar{\rho}\widetilde{u_i''u_i''u_j''}$. In the incompressible formulation, a hierarchy of models for the closure of the turbulent stresses was presented as well as closures for the turbulent transport. Even at the mean equation level due to the need for an energy equation, an additional term for the turbulent heat flux has appeared, and, indirectly, for the mass flux through the equating of the Reynolds-averaged and Favre-averaged mean velocity. These additional terms, as well as the stress tensor τ_{ij}, require modelling.

B Compressible Reynolds stress and two-equation models

Even though closures for the heat and mass flux appeared in the energy conservation equation, the major impediment to future progress in treating the compressible closure problem lies with the choice of a suitable model for the turbulent stress tensor $\bar{\rho}\tau_{ij}$. The hierarchy of closure models outlined in the incompressible formulation still applies in the compressible case; however, the lower-order models as formulated do not introduce any new correlations when applied to compressible flows. Thus, in the discussion to follow, only the Reynolds stress formulation will be presented in some detail. The two-equation model will only be discussed briefly because its form is easily extracted from the full second-moment equations. The low-order models such as the zero-, half- and one-equation models will not be discussed because their extension to compressible flows is done empirically and can be found in the references that define the methods given in Sections II.B.3 and II.B.4.

Since significant similarities exist between the incompressible and compressible formulations, simple variable density extensions for some models can be justified. In other cases, terms that arise solely due to the compressibility of the flow can be easily identified, and suitable models can be developed. As in the incompressible case, the starting point for the development of the Favre-averaged second-moment equation is (7) with the differential operator $\mathcal{L}$ defined from the conservation of momentum equation. The resulting Favre-averaged equation for the Reynolds stress tensor $\bar{\rho}\tau_{ij}$ is then given by

$$\frac{\partial \bar{\rho}\tau_{ij}}{\partial t} + \frac{\partial}{\partial x_k}\left(\tilde{u}_k\bar{\rho}\tau_{ij}\right) = \bar{\rho}\widetilde{P}_{ij} + \bar{\rho}\Pi^d_{ij} + \bar{\rho}\Pi^{dl}_{ij} + M_{ij} - \bar{\rho}\epsilon_{ij} + \frac{\partial \bar{\rho}\widetilde{D}^t_{ijk}}{\partial x_k} + \frac{\partial D^v_{ijk}}{\partial x_k} \quad (79)$$

where the right-hand-side represents the rate-of-change of $\bar{\rho}\tau_{ij}$ produced by the turbulent production $\bar{\rho}\widetilde{P}_{ij}$, the deviatoric part of the pressure strain-rate correlation Π^d_{ij}, the pressure dilatation Π^{dl}_{ij}, the mass flux variation M_{ij}, the turbulent diffusion $\widetilde{D}^t_{ijk}$, the viscous diffusion D^v_{ijk}, and the turbulent dissipation rate ϵ_{ij}. These terms are given by

$$\bar{\rho}\widetilde{P}_{ij} = -\bar{\rho}\tau_{ik}\frac{\partial \tilde{u}_j}{\partial x_k} - \bar{\rho}\tau_{jk}\frac{\partial \tilde{u}_i}{\partial x_k} \quad (79a)$$

$$\bar{\rho}\Pi^d_{ij} = \overline{p'\left(\frac{\partial u'_i}{\partial x_j} + \frac{\partial u'_j}{\partial x_i}\right)} - \tfrac{2}{3}\overline{p'\frac{\partial u'_k}{\partial x_k}}\delta_{ij} \quad (79b)$$

$$\bar{\rho}\Pi^{dl}_{ij} = \tfrac{2}{3}\overline{p'\frac{\partial u'_k}{\partial x_k}}\delta_{ij} \quad (79c)$$

$$M_{ij} = \overline{u''_i}\left(\frac{\partial \bar{\sigma}_{jk}}{\partial x_k} - \frac{\partial \bar{p}}{\partial x_j}\right) + \overline{u''_j}\left(\frac{\partial \bar{\sigma}_{ik}}{\partial x_k} - \frac{\partial \bar{p}}{\partial x_i}\right) \quad (79d)$$

$$\bar{\rho}\widetilde{D}^t_{ijk} = -[\overline{\bar{\rho}u''_iu''_ju''_k} + \overline{p'(u'_i\delta_{jk} + u'_j\delta_{ik})}] \quad (79e)$$

$$D^v_{ijk} = (\overline{\sigma'_{ik}u'_j + \sigma'_{jk}u'_i}) \quad (79f)$$

$$\bar{\rho}\epsilon_{ij} = \overline{\sigma'_{ik}\frac{\partial u'_j}{\partial x_k}} + \overline{\sigma'_{jk}\frac{\partial u'_i}{\partial x_k}} \quad (79g)$$

The form of this compressible Reynolds stress transport equation should be compared with the incompressible form in (8).

A few comments should be made about the structuring of the pressure–strain correlation and dissipation rate terms that appear in (79). As was outlined in Section II.B.1, the deviatoric part of the dissipation rate is generally absorbed into the pressure–strain correlation to account for any anisotropic dissipation effects. The same holds for the compressible case, where the quantity

$\Pi_{ij}^d = \Pi_{ij}^d - {}_D\epsilon_{ij}$ is modelled. Note that in the definition of the pressure–strain correlation (79b), the trace (the pressure dilatation) was subtracted out. This approach assumes that the compressibility effects associated with the pressure–strain correlation are confined to the modelling of the pressure dilatation term Π_{ij}^{dl} and that the deviatoric part Π_{ij}^d can be modelled by a simple variable mean density extension of the incompressible models presented in Section II.B.1. A recent study by Speziale *et al.* (1995) has shown, however, that in compressible homogeneous shear flow simple variable density extensions of the deviatoric part of the pressure–strain correlation leads to incorrect predictions of the Reynolds stress anisotropies. Their study indicates that such variable density extensions are only applicable in flows where the turbulent Mach number $(M_t = \sqrt{(2K/\gamma RT)})$ is about 0.3 or smaller; otherwise, deviatoric pressure–strain models must be developed that explicitly account for compressibility effects. As will be shown shortly, a similar approach can be taken for the turbulent dissipation rate.

Recall that as in the incompressible case, the deviatoric part of the tensor dissipation rate was assimilated into the pressure–strain correlation, which leaves the isotropic dissipation rate ϵ in the compressible Reynolds stress or turbulent kinetic energy formulations to be obtained from a modelled transport equation. However, unlike the incompressible case, a partitioning of the isotropic dissipation rate is generally invoked (Zeman, 1990; Sarkar *et al.*, 1991) to explicitly account for compressibility effects. The form of the partitioning is

$$\epsilon = \varepsilon + \varepsilon^d \tag{80}$$

where ε is the solenoidal (incompressible) dissipation and ε^d is the dilatation (compressible) dissipation. The dilatation dissipation that is unique to the compressible case will be discussed in Section III.C.4. Because the compressibility effects are represented as turbulent Mach number corrections to the incompressible (solenoidal) dissipation rate, this allows for the direct extension of the incompressible form of the dissipation rate transport equation.

This structuring is not unique, however, and an obvious alternative is to develop compressible models for the pressure–strain correlation and the dissipation rate directly. This approach is taken by El Baz and Launder (1993) in their study of compressible mixing layers. They incorporated the effects of compressibility directly into their rapid pressure–strain model through a turbulent Mach number. Because they were not only dealing with the deviatoric part of the correlation, the trace of their rapid term did not vanish in the compressible flow. Space constraints preclude the writing of the full pressure–strain model here, but its form is similar to the incompressible FLT pressure–strain correlation model developed at UMIST, which is cubic in the turbulent stress anisotropy. A similar approach is taken for the turbulent dissipation rate equation, where modifications are made to the decay term in the equation rather than explicitly partitioning the dissipation rate to account for compressibility.

The viscous diffusion D_{ijk}^v is modelled as in the incompressible case, but accounts for the variation of viscosity. Thus, the viscous diffusion term is given by (Speziale and Sarkar, 1991)

$$D_{ijk}^v \approx \bar{\mu}\left(\frac{\partial \tau_{jk}}{\partial x_i} + \frac{\partial \tau_{ki}}{\partial x_j} + \frac{\partial \tau_{ij}}{\partial x_k}\right) \tag{81}$$

Unlike the incompressible turbulent diffusion term where the pressure–velocity correlation is absorbed into the triple-correlation term and the combination modelled through gradient diffusion, a separate model for $\overline{p'u_i'}$ needs to be developed in order to be consistent with a truly compressible formulation. This need is evident by noting that $\overline{p'u_i'} \equiv \overline{p'u_i''}$ with $p' = R(\rho T'' + \rho' \widetilde{T})$, which leads to the expression

$$\overline{p'u_i''} = R(\overline{\rho u_i'' T''} - \bar{\rho}\widetilde{T}\overline{u_i''}) \tag{82}$$

Equation (82) also requires models for the mass and heat flux. At this time no attempt has been made to incorporate models for either of these fluxes into the pressure–velocity diffusion term because of the inaccuracy of the models and the significance of the contribution in governing the dynamics of the flow. The correlation has simply been neglected in the modelling of the turbulent diffusion, and variable density extensions to the models given in Section II.B.1 have been used. Unfortunately, with the lack of experimental and simulation data, no *a priori* justification exists for neglecting this contribution.

However, before discussing any aspects of the dissipation rate closure, the compressible turbulent kinetic energy equation will be derived. It is easily extracted from (79) by taking the trace, which leads to

$$\frac{\partial(\bar{\rho}K)}{\partial t} + \frac{\partial}{\partial x_j}\left(\tilde{u}_j\bar{\rho}K\right) = \bar{\rho}\widetilde{\mathcal{P}} + \bar{\rho}\Pi^{dl} + \mathcal{M} + \frac{\partial}{\partial x_j}\widetilde{\mathcal{D}}_j^t - \bar{\rho}\epsilon + \frac{\partial}{\partial x_j}\left(\bar{\mu}\frac{\partial K}{\partial x_j}\right) \tag{83}$$

where the right-hand-side represents the turbulent kinetic energy transport produced by the turbulent production $\bar{\rho}\widetilde{\mathcal{P}} \equiv \bar{\rho}\widetilde{\mathcal{P}}_{ii}/2$, the pressure dilatation, $\bar{\rho}\Pi^{dl} \equiv \bar{\rho}\Pi_{ii}^{dl}/2$, the mass flux variation $\mathcal{M} \equiv \mathcal{M}_{ii}/2$, the turbulent diffusion $\widetilde{\mathcal{D}}_k^t \equiv \widetilde{\mathcal{D}}_{iik}^t/2$, the isotropic turbulent dissipation rate, $\epsilon \equiv \epsilon_{kk}/2$, and the viscous diffusion. In the case of the viscous diffusion, the term is not an exact contraction of the model given in (81); however, it is consistent with the incompressible and standard form given in (30). Similarly, the turbulent diffusion term is modelled as a variable density extension to the incompressible form given in (32). The other terms in the compressible turbulent kinetic energy equation that are unique to the compressible case will be discussed in the next section. Coupled with the compressible turbulent kinetic energy equation is the need for a variable density extension to the incompressible Boussinesq formula (34) for $\bar{\rho}\tau_{ij}$

$$\bar{\rho}\tau_{ij} = \tfrac{2}{3}\bar{\rho}K\delta_{ij} - 2\bar{\mu}_t\left(\widetilde{S}_{ij} - \tfrac{1}{3}\widetilde{S}_{kk}\delta_{ij}\right) \tag{84}$$

Note that the trace of the mean strain-rate tensor needs to be subtracted out to ensure that τ_{kk} is twice the kinetic energy. Analogous to (31), the eddy viscosity $\bar{\mu}_t$ is then given by the relation

$$\bar{\mu}_t = \bar{\rho} C_\mu \frac{K^2}{\varepsilon} \tag{85}$$

where $C_\mu = 0.09$ is a closure coefficient. As in the incompressible case, the coefficient C_μ can be made dependent on the strain-rate and rotation-rate tensors through a low-order truncation of an algebraic stress model.

The common high-Reynolds-number form of the isotropic solenoidal dissipation rate equation that is used is given by

$$\frac{\partial \bar{\rho}\varepsilon}{\partial t} + \frac{\partial}{\partial x_j}\left(\tilde{u}_j \bar{\rho}\varepsilon\right) = \bar{\rho}\tilde{\mathcal{P}}_\varepsilon - \frac{4}{3}\bar{\rho}\varepsilon \frac{\partial \tilde{u}_k}{\partial x_k} - \bar{\rho}\tilde{\mathcal{D}}_\varepsilon + \frac{\partial}{\partial x_j}\bar{\rho}\tilde{\mathcal{D}}^t_{\varepsilon j} + \frac{\partial}{\partial x_j}\left(\bar{\mu}\frac{\partial \varepsilon}{\partial x_j}\right) \tag{86}$$

with

$$\tilde{\mathcal{P}}_\varepsilon = -C_{\varepsilon 1}\frac{\varepsilon}{K}\tau_{ij}\left(\frac{\partial \tilde{u}_i}{\partial x_j} - \frac{1}{3}\frac{\partial \tilde{u}_k}{\partial x_k}\delta_{ij}\right) \tag{86a}$$

$$\tilde{\mathcal{D}}_\varepsilon = C_{\varepsilon 2}\frac{\varepsilon^2}{K} \tag{86b}$$

where the first and third terms on the right-hand-side are the production and destruction of dissipation, respectively, and the last term is the viscous diffusion. The closure constants $C_{\varepsilon 1}$ and $C_{\varepsilon 2}$ are assumed to take on their incompressible values. The form of (86) differs slightly from the incompressible form with the addition of the mean dilatation term and the traceless production-of-dissipation term. Speziale and Sarkar (1991) have added these terms in order to account properly for the behaviour of compressed isotropic turbulence. Omission of the mean dilatation term causes the model to predict incorrectly the decrease of the integral length scale for isotropic expansion and the increase for isotropic compression.

The turbulent diffusion term is given by

$$\bar{\rho}\tilde{\mathcal{D}}^t_{\varepsilon j} = \bar{\rho}C_\varepsilon\left(\tau_{ij}\frac{K}{\varepsilon}\frac{\partial \varepsilon}{\partial x_i}\right) \tag{87}$$

for Reynolds stress models and by

$$\bar{\rho}\tilde{\mathcal{D}}^t_{\varepsilon j} = \frac{\bar{\mu}_t}{\sigma_\varepsilon}\frac{\partial \varepsilon}{\partial x_j} \tag{88}$$

for two-equation models. Both C_ε and σ_ε are also assumed to take on their incompressible values. Note that the basic form of the dissipation rate equation (86) is simply a variable density extension of the incompressible form.

Another contribution to the right-hand-side, which is also generally neglected, arises from accounting for the variation of mean kinematic viscosity $\bar{\nu}$ in the development of the dissipation rate transport equation (Coleman and Mansour, 1991). This accounting leads to the additional term that is also proportional to the mean dilatation. For the most part, however, all these mean dilatation effects can be neglected in flows without shocks or in the absence of strong pressure gradients.

As mentioned previously, El Baz and Launder (1993) have taken an alternative approach by simply using a solenoidal dissipation rate equation that has been sensitized to compressibility effects through the turbulent Mach number. These effects were accounted for through a modification of the decay coefficient $C_{\varepsilon2}$ rather than the solenoidal and dilatational partitioning that has been discussed. They argue that this approach is more consistent with the single-scale framework in which the dissipation equation provides a model for the energy transfer to the small scales. In their formulation, a modified coefficient $C_{\varepsilon2c}$ was defined to account for the compressibility effects

$$C_{\varepsilon2c} = \frac{C_{\varepsilon2}}{1 + \beta_2 M_t^2} \tag{89}$$

where $\beta_2 = 3.2$ was obtained from calibration with the DNS of decaying isotropic turbulence.

Both Huang et al. (1994) and Wilcox (1992) have pointed out that the dissipation equation (86) is inappropriate in the log layer of a wall-bounded flow because it incorrectly accounts for the mean density variation near the wall. This causes the slope of the Van Driest velocity in the log-layer to be different than κ^{-1} (where κ is the Von Karman constant). However, Wilcox (1992) has shown that the K–ω formulation yields good log-law predictions in the absence of any dilatational dissipation corrections.

C Scalar flux and dilatation closure models

The closure schemes presented in the last section clearly show that the compressible formulation is complicated by the addition of second-order correlations. These include the mass flux, the turbulent heat flux, the pressure dilatation, and the compressible (dilatation) dissipation. Turbulent heat flux effects are present in the incompressible formulation but are essentially decoupled from the mean momentum field. The models to be presented and discussed reflect relatively recent contributions to the modelling of these terms. Although, this discussion may not be all inclusive, it does present the latest trends in the field. The references provided can lead the interested reader to a more complete historical record of the research in a particular topical area.

1 Mass Flux

The average fluctuating velocity $\overline{u_i''}$ is related to the mass flux $\overline{\rho' u_i'}$ through the defining relation

$$\overline{u_i''} \equiv -\frac{\overline{\rho' u_i'}}{\bar{\rho}} \tag{90}$$

Several recent attempts have been made to model this correlation, although none of the models have been extensively tested in a variety of compressible flows. Although earlier models have been proposed (e.g. see Rubesin (1990) for a brief historical survey), no dominant form has emerged. For this reason, the main attention in this survey will be on recent attempts at such modelling in conjunction with the latest advances in closure modelling and simulation data for verification.

Rubesin (1990) assumed that the density and pressure fluctuations were related through a polytropic gas law. In addition, by assuming constant specific heats to relate the fluctuating density and enthalpy, and a simple gradient transfer hypothesis between the fluctuating and mean enthalpy, he was able to relate the fluctuating velocity $\overline{u_i''}$ to moments of the turbulent Mach number

$$\overline{u_i''} \simeq \frac{1}{n-1} c_e \tau_{ij} \frac{K}{\varepsilon} \frac{\tilde{h}_{,j}}{\tilde{h}} = \frac{\gamma - 1}{n-1} c_e \frac{K}{\varepsilon} \tilde{h}_{,j} \left[\frac{\tau_{ij}}{\gamma R \tilde{T}} \right] \tag{91}$$

where $\tilde{h}$ is the static enthalpy ($\tilde{h} = \tilde{H} - \tilde{u}_i \tilde{u}_i / 2 - \widetilde{u_i'' u_i''}/2$), $c_e = 0.35$ is a closure coefficient, and n is the polytropic coefficient.

A more complicated formulation proposed by Taulbee and VanOsdol (1991) requires the solution of a modelled transport equation for the mass fluctuating velocity. In addition, because the density variance appears as a source term in the mass fluctuating velocity transport equation, an additional modelled transport equation for the density variance is derived. The general form of their mass flux equation is given by

$$\frac{D\overline{\rho' u_i'}}{Dt} = -C_{u_2} \frac{\varepsilon}{K} \overline{\rho' u_i'} + \mathcal{P}_{mf} + \mathcal{D}_{mf}^t + \mathcal{D}_{mf}^v \tag{92}$$

where $C_{u_2} = 5.3$ is a model parameter determined from comparison with experimental data. The contributions to the production $\mathcal{P}_{mf}$ are from both mean velocity and mean density gradients, and the diffusion terms $\mathcal{D}_{mf}^t$ and $\mathcal{D}_{mf}^v$ are closed by gradient diffusion hypotheses. The first term on the right-hand-side is a modelled term that represents the relaxation of $\overline{u_i''}$. The reader is referred to Taulbee and VanOsdol (1991) for details of the formulation and the full form of the modelled transport equations. Validation tests were made on an adiabatic flat-plate boundary layer flow and a compressible free shear layer flow. In both cases, the results compared favourably with experiments.

Zeman and Coleman (1991) also derived a transport equation for the mass flux in order to study the response of turbulence to a shock. Their transport equation consisted of a relaxation term needed to represent the decay of the turbulence after the shock, and two production-type terms based on mean density and mean velocity gradients. To lowest order, their transport equation was given by

$$\frac{D\overline{\rho' u_i'}}{Dt} \simeq -\frac{\overline{\rho' u_i'}}{\tau_a} - \tau_{ij}\bar{\rho}_{,j} - \overline{\rho' u_j'}\tilde{u}_{i,\,j} \tag{93}$$

where $\tau_a = 0.4 M_t K/\varepsilon$ is an acoustic time-scale and $M_t = \sqrt{2K/\gamma R\widetilde{T}}$ is the turbulent Mach number with γ as the ratio of specific heats (c_p/c_v).

In the most recent attempt at developing a mass flux model, Ristorcelli (1993) has proposed an algebraic relationship extracted from an exact evolution equation for the mass fluctuating velocity that keeps only the lowest-order terms in the fluctuating density. A linear relaxation model analogous to that used by Zeman and Coleman (1991) is used to model a fluctuating divergence term. The resulting model is given by

$$\overline{u_i''} \simeq \tau\left[\nu_0\delta_{ij} + \nu_1\tau\tilde{u}_{i,j} + \nu_2\tau^2\tilde{u}_{i,k}\tilde{u}_{k,j}\right]\tau_{jl}\frac{\bar{\rho}_{,l}}{\bar{\rho}} \tag{94}$$

where $\tau_a = 0.4 M_t K/\varepsilon$ is an acoustic time-scale and $M_t = \sqrt{2K/\gamma R\widetilde{T}}$ is the turbulent Mach number with γ as the ratio of specific heats (c_p/c_v).
kinetic energy production $\tilde{\mathcal{P}}$ as defined previously. The coefficients ν_0, ν_1, and ν_2 are functions of the invariants of $\delta_{ij} + \tilde{u}_{i,j}$. Preliminary validation tests have been made that utilize DNS data from the early stages of a compressible boundary layer flow (Dinavahi et al., 1994). These tests suggest that the mass flux is important in the vicinity of a wall and that the proposed model displays the necessary qualitative features. Further validation tests are necessary that utilize DNS data of more fully developed turbulence to substantiate these preliminary results.

2 Turbulent heat flux

Unlike the incompressible case, where the mean momentum equation is decoupled from the mean energy equation with the velocity field independent of the temperature field, a direct coupling exists in the compressible case which requires a more careful specification of the turbulent heat flux modelling.

Gradient-diffusion models have been the most popular and simplest closures used for the turbulent heat fluxes, although they have not been rigorously justified in the general situation. Although several models have been derived with the assumption of a constant Prandtl number, in the presence of walls and, in particular, with non-adiabatic wall conditions, a variable Prandtl number formulation is needed. Sommer et al. (1993) have proposed a gradient-diffusion model that displays the correct asymptotic consistency in the near-wall region.

Their model is given by

$$\widetilde{u_i''T''} = -\kappa_T \frac{\partial \widetilde{T}}{\partial x_i} \tag{95}$$

with turbulent heat diffusivity

$$\kappa_T = C_\lambda \, f_\lambda K \left[\frac{K\widetilde{T''^2}}{\varepsilon \varepsilon_T}\right]^{\frac{1}{2}} \tag{95a}$$

where $C_\lambda = 0.11$ is a model constant, f_λ is a near-wall damping function that is a function of distance from the wall and turbulent Reynolds number (see Sommer *et al.* (1992) and Sommer *et al.* (1993) for complete details of the derivation), and ε_T is the dissipation rate of the temperature variance. This variable turbulent Prandtl number formulation $Pr_t = \bar{\nu}_t/\bar{\alpha}_t$ ($\bar{\nu}_t = \bar{\mu}_t/\bar{\rho}$) required transport equations for the temperature variance and temperature variance dissipation rate (Sommer *et al.*, 1993), which were solved in conjunction with a two-equation K–ε turbulence model. Their tests on incompressible channel and pipe flows and compressible flat-plate boundary layers with adiabatic and cooled-wall boundary conditions showed that the assumption of dynamic similarity between incompressible and compressible flows is valid for high-Mach-number flows. In addition, they also verified that in highly cooled-wall, compressible boundary layer flows the Prandtl number was not constant and the assumption of dynamic similarity between momentum and heat transport was not applicable.

Another approach to determining the turbulent heat flux is to utilize a modelled transport equation for the correlation itself rather than to assume a gradient transport process at the outset and determine the turbulent heat diffusivity. As shown below, the former approach requires a knowledge of the individual turbulent stress components either through a Boussinesq approximation or a second-moment closure, and the latter approach requires the determination of the temperature variance and associated dissipation rate. Huang and Coakley (1993a) have extracted an algebraic relationship from a transport equation for the heat flux by using the assumptions associated with the development of algebraic stress models. The algebraic relation takes the form

$$-\widetilde{u_k''T''} \frac{\partial \tilde{u}_i}{\partial x_k} - c_{1T}\widetilde{u_i''T''} \frac{\varepsilon}{K} + c_{2T}\widetilde{u_k''T''} \frac{\partial \tilde{u}_i}{\partial x_k} = \tau_{ki} \frac{\partial \widetilde{T}}{\partial x_k} \tag{96}$$

where $c_{1T} = 3$ and $c_{2T} = 0.5$ are model coefficients. This equation can be solved by direct inversion to obtain the component heat flux correlations. Huang and Coakley (1993a) also tested a simple gradient transport model proposed by Ha Minh *et al.* (1985) given by

$$\widetilde{u_i''T''} = -c_T \frac{K}{\varepsilon} \tau_{ik} \frac{\partial \widetilde{T}}{\partial x_k} \tag{97}$$

where $c_T = 0.313$ was obtained from optimization of free shear flow predictions. Implicit in both approaches is the assumption of a variable Prandtl number. These heat flux models were used in conjunction with two Reynolds stress closure models for the solution of flat-plate flows. Tests were run over a range of Mach numbers and wall conditions, including a case of hypersonic flow with a shock–boundary layer interaction. In these cases, wall functions were used, which negated the need to incorporate near-wall corrections into the closures. The results of the study showed somewhat limited predictive capability of the combined stress closure and heat flux models tested, with little difference between the algebraic model (96) and the gradient diffusion model (97). A limited number of other, older proposals have been put forth for turbulent heat flux modelling; these are briefly discussed in the review by Cousteix and Aupoix (1990).

Finally, El Baz and Launder (1993) have used transport equations for the heat-flux, temperature variance and variance dissipation rate in conjunction with a full second-moment closure to compute the compressible mixing layer. Their pressure–temperature gradient term (analogous to the pressure–strain term in the turbulent stress transport equation) is similar in structure to their model for the pressure–strain correlation. No indication is given in the study whether such complexity is necessary for the heat flux term because the primary motivation was focused on an alternative formulation for the pressure–strain and turbulent dissipation rate models. Nevertheless, predictions of both single-stream and two-stream mixing layers were satisfactory, which indicates that the effects of pressure dilatation and dilatation dissipation are important and should be included in the closure.

3 Pressure dilatation

Because the velocity gradient tensor is not traceless in compressible flow, the structuring of the compressible stress transport equation (79) explicitly accounted for the pressure dilatation term. This correlation survives in the turbulent kinetic energy equation (83). In the Reynolds stress transport equation, this term also appears in the normal stress component equations. As mentioned previously, in this form of the equation the compressibility effects due to the pressure–strain correlation are attributed to the pressure dilatation term, and the deviatoric part can be modelled in much the same way as its incompressible counterpart with simple variable mean density extensions. With the exception of recent work of El Baz and Launder (1993), this is the approach that is generally followed when compressibility effects are explicitly taken into account.

Sarkar (1992) has proposed a model for the pressure dilatation obtained from a formal solution for homogeneous turbulence that is applicable to free shear flows. The pressure field is split into an incompressible and compressible part. The DNS data were utilized to show that only the contribution from the incompressible pressure field to the pressure dilatation correlation was significant and needed to be modelled. The model was validated against DNS solutions of compressible homogeneous shear and isotropic turbulence; it takes the relatively simple form of

$$\Pi^{dl} = 0.30\bar{\rho}KM_t b_{mn}\tilde{u}_{mn} + 0.20\bar{\rho}\varepsilon M_t^2 \tag{98}$$

Zeman (1993) has extracted a model for Π^{dl} from a transport equation for the rate of change of the pressure variance. The formulation was directed toward turbulent boundary layers in quasi-equilibrium. In the analysis, the pressure dilatation was shown to be proportional to the mass flux. The previous work of Zeman and Coleman (1991) for the mass flux was used to close the model. The resulting form was given by

$$\Pi^{dl} = 2f_\rho(M_t)\bar{a}^2 \tau_{ij} \frac{K}{\varepsilon}\left[\frac{\bar{\rho}_{,i}\bar{\rho}_{,j}}{\bar{\rho}}\right] \tag{99}$$

where $f_\rho(M_t) = 0.2[1 - \exp(-M_t^2/0.02)]$, and $\bar{a} = \sqrt{\gamma R\tilde{T}}$ is the sound speed. An important result of the study was the fact that the resulting model was able to preserve the Van Driest compressible law of the wall. This was attributed to the fact that the effect of the density gradient was taken into account in the pressure–strain model.

For completeness, note that Taulbee and VanOsdol (1991) proposed a composite model for the pressure dilatation and the dilatation dissipation. The resulting model required the density variance to represent turbulent–turbulent interactions and the mean dilatation to represent the influence of the mean flow. By using this approach, the remaining dissipation contribution only involved the solenoidal or incompressible contribution. The results of this study were discussed previously in the section on the mass flux. This partitioning of the turbulent dissipation rate will be discussed in the next section, and, as noted previously in the discussion of the mass flux and heat flux, the dissipation rate implied in these models is the solenoidal component.

4 Dilatation dissipation

In Section III.B, the isotropic dissipation rate was partitioned into a solenoidal part and a dilatational part. The solenoidal part was modelled through a transport equation in a manner analogous to the incompressible case. In this subsection, the closure of the dilatation dissipation is presented. In both the Zeman (1990) and Sarkar et al. (1991) formulations, the compressible component takes the form

$$\varepsilon^d = \alpha^* \mathcal{F}(M_t)\epsilon \tag{100}$$

where α^* is a closure coefficient, and $\mathcal{F}(M_t)$ is some function of the turbulent

Mach number. In the Zeman (1990) model $\alpha^* = 0.75$, and in the Sarkar *et al.* (1991) model, $\alpha^* = 0.6$. For the function $\mathcal{F}(M_t)$, Zeman (1990) prescribes the form

$$\mathcal{F}(M_t) = 1 - \exp\{-[0.5(\gamma + 1)(M_t - M_{t0})^2/\Lambda^2]\}\mathcal{H}(M_t - M_{t0}) \qquad (101)$$

where $\mathcal{H}(M_t)$ is the Heaviside function, with $M_{t0} = 0.10[2/(\gamma + 1)]^{\frac{1}{2}}$ and $\Lambda = 0.6$ for mixing layers, and $M_{t0} = 0.25[2/(\gamma + 1)]^{\frac{1}{2}}$ and $\Lambda = 0.66$ for boundary layers. Zeman's (1990) values change for the boundary-layer flows because he accounts for the change in kurtosis between the mixing-layer and boundary layer flows. Sarkar *et al.* (1991) uses the simpler form

$$\mathcal{F}(M_t) = M_t^2 \qquad (102)$$

These models have been tested on simple free shear and wall-bounded flows (Sarkar *et al.* 1991; Wilcox, 1992; Zeman, 1993). In the compressible mixing layer flow, for example, this compressibility correction to the dissipation rate has proved essential in correctly predicting the spreading rate. In simple flat-plate boundary layer flows, the results are less encouraging because the log-law is adversely affected by these models, particularly in the cold-wall case. Wilcox (1992) has developed a composite dilatation dissipation model that utilizes both the Zeman and Sarkar *et al.* models for use with his $K-\omega$ formulation. The model for ε^d follows the same functional form as (100), but with $\alpha^* = 1.5, M_{t0} = 0.25$, and

$$\mathcal{F}(M_t) = [M_t^2 - M_{t0}^2]\mathcal{H}(M_t - M_{t0}) \qquad (103)$$

If the $K-\omega$ model is used in conjunction with this dilatation dissipation model then relatively good predictions for both mixing layer and boundary layer flows are obtained. The mixing layer calculations clearly show the dramatic effect the compressibility corrections have on predicting the correct spreading rate. Figure 6 shows the unmodified $K-\omega$ model as well as the $K-\omega$ model with the Wilcox, Sarkar and Zeman compressibility corrections. Similar results are obtained from a second-moment closure formulation (Zeman, 1990; Sarkar and Lakshmanan, 1991).

As mentioned previously, Zeman (1993) has also alleviated the problem with the Van Driest scaling by accounting for density effects in the specification of his pressure dilatation model

D Wall functions for compressible flows

The discussion in the previous compressible flow sections has avoided the complicated aspects of near-wall effects. As in the incompressible case, these near-wall effects are most commonly handled through the wall function approach. The underlying assumptions used in Section II.C carry over to the

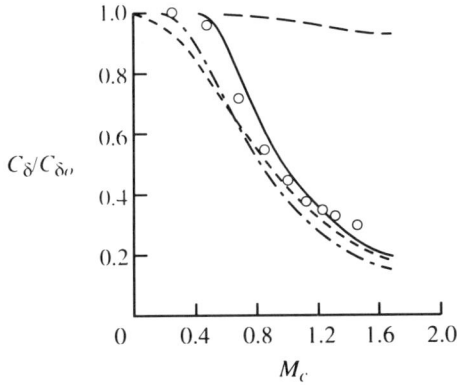

Figure 6 Comparison of computed measured spreading rate for compressible mixing layer (Wilcox, 1993). Key: $--$, unmodified $K - \omega$; $—$, Wilcox model, $- - -$, Sarkar model; $- - -$, Zeman model; $\bigcirc$, Langley curve (Kline *et al.*, 1982).

compressible case; however, the compressible formulation also includes the energy equation, which must have a temperature or heat flux specified at the wall in order to be solved. In this section, the results in Section II.C will be extended to account for this added equation. If the simple cartesian coordinate representation is used, as in the incompressible case, and, again pressure gradient effects are neglected, the momentum equation (75) and total energy equation (77) are to lowest order

$$\bar{\sigma}_{xy} - \bar{\rho}\tau_{xy} = \bar{\sigma}_{xy}\big|_w \tag{104}$$

$$\bar{q}_y + c_p\bar{\rho}\widetilde{v''T''} = \bar{q}_y\big|_w + \bar{\sigma}_{xy}\tilde{u} - \bar{\rho}\tilde{u}\tau_{xy} \tag{105}$$

where v'' is the y component of the Favre-averaged fluctuating velocity. In addition, both the mass flux and turbulent transport are neglected as well. Analogous to the incompressible case, (104) simply states that the total stress $\bar{\sigma}_{tot} = \bar{\sigma}_{xy} - \bar{\rho}\tau_{xy}$ is equivalent to the wall shear stress $\bar{\sigma}_w$. Equation (105) then equates the total heat flux $\bar{q}_{tot} = \bar{q}_y + c_p\bar{\rho}\widetilde{v''T''}$ to the wall heat flux and total stress transport:

$$\bar{q}_{tot} = \bar{q}_w + \tilde{u}\bar{\sigma}_t \tag{106}$$

To obtain explicitly the compressible law of the wall, a relation between the velocity and temperature is needed. By using either dimensional analysis or mixing length arguments (Cebeci and Bradshaw, 1984), relations for the mean velocity and mean temperature gradients are found

$$\frac{\partial \tilde{u}}{\partial y} = \frac{(\bar{\sigma}_w/\bar{\rho})^{\frac{1}{2}}}{\kappa y} \tag{107}$$

$$\frac{\partial \widetilde{T}}{\partial y} = -Pr_t \frac{(\bar{q}_{tot}/\bar{\rho} c_p)}{(\bar{\sigma}_w/\bar{\rho})^{\frac{1}{2}} \kappa y} \tag{108}$$

By equating (107) and (108), the relationship

$$\frac{\bar{q}_{tot}}{c_p} \frac{\partial \tilde{u}}{\partial y} = -\frac{\bar{\sigma}_w}{Pr_t} \frac{\partial \widetilde{T}}{\partial y} \tag{109}$$

is obtained, which leads to the heat flux at the wall $\bar{q}_w$ by using (106), (107), and (109):

$$\bar{q}_w = -c_p \frac{\bar{\mu}_t}{Pr_t} \frac{\partial \widetilde{T}}{\partial y} - \tilde{u} \bar{\sigma}_w \tag{110}$$

where $\bar{\mu}_t$ is given by

$$\bar{\mu}_t = \bar{\rho} C_\mu \frac{K^2}{\varepsilon} \tag{111}$$

In order to obtain the Van Driest velocity from (107), it is necessary also to know the temperature distribution since the density is inversely proportional to the temperature. The temperature distribution (Cebeci and Bradshaw, 1984) in the inner layer can be obtained by integrating (109) and by using (106):

$$\widetilde{T} \simeq \widetilde{T}_w - \frac{Pr_t}{c_p} \left(\frac{\bar{q}_w \tilde{u}}{\bar{\sigma}_w} + \frac{\tilde{u}^2}{2} \right) \tag{112}$$

This equation is then used to determine either the wall temperature or wall heat flux. Note that (112) is only an approximate relationship because the integration limit extends directly to the wall. Because (109) is only valid in the log region of the flow, any integration to the wall through the viscous sublayer is precluded. Nevertheless, in the context of a single layer wall-function approach, there is no alternative but to perform the integration to the wall. More complex multilayer wall-function formulations (Viegas et al., 1985) can be used, but these are cumbersome to apply to a Reynolds stress formulation. Equations (107) and (112) can be combined to obtain the Van Driest velocity as

$$\tilde{u}_c \simeq \left(\frac{2 c_p \widetilde{T}_w}{Pr_t} \right)^{\frac{1}{2}} \left[\arcsin \left(\frac{\bar{q}_w + \tilde{u} \bar{\sigma}_w}{\bar{\sigma}_w D} \right) - \arcsin \left(\frac{\bar{q}_w}{\bar{\sigma}_w D} \right) \right] \tag{113}$$

where

$$D = \left(\frac{\bar{q}_w^2}{\bar{\sigma}_w^2} + \frac{2 c_p \widetilde{T}_w}{Pr_t} \right)^{\frac{1}{2}} \tag{113a}$$

In addition, the Van Driest velocity satisfies the compressible law of the wall

$$\tilde{u}_c^+ \equiv \frac{\tilde{u}_c}{\tilde{u}_\tau} = \frac{1}{\kappa_c} \ln y^+ + B_c \qquad (114)$$

where $\tilde{u}_\tau = \sqrt{(\bar{\sigma}_w/\bar{\rho}_w)}$ and $y^+ = \bar{\rho}_w \tilde{u}_\tau y/\bar{\mu}_w$. In the compressible formulation, both κ_c and B_c are no longer constants but are functions of various compressibility parameters in the flow. Huang *et al.* (1994) and So *et al.* (1994) (see also Wilcox, 1993) have shown that $\kappa_c = \kappa_c(M_\tau, \mathcal{B}_q, Pr_t)$ and $B_c = B_c(M_\tau, \mathcal{B}_q, Pr_t)$, where $\mathcal{B}_q = \bar{q}_w/\bar{\rho}_w c_p \tilde{u}_\tau \tilde{T}_w$ is a dimensionless heat flux parameter. Although these dependencies generally hold, their effect is usually minimal in the low supersonic regime. Nevertheless, caution should be exercised whenever compressible wall functions are used because the usual log-law behaviour may not be properly represented.

The same procedure outlined in Section II.C for the wall shear stress and surface pressure is followed for the compressible case. In addition, the wall temperature or surface heat flux $\bar{q}_w$ is also specified for the mean energy equation. In addition, the turbulent kinetic energy or the turbulent stresses, and the turbulent solenoidal dissipation rate, are obtained in a manner similar to the incompressible case.

The main steps for implementing the wall functions for compressible flows can be outlined as follows:

(a) Set an initial guess for the wall shear stress, with either $\bar{\sigma}_w = 0.3\bar{\rho}_w K_{(2)}$ or $\bar{\sigma}_w$ from the previous iteration.

(b) *Isothermal condition:* Compute the heat flux at the wall $\bar{q}_w$ from (112) evaluated at the first interior point (2).
Adiabatic or specified $\bar{q}_w$ condition: Compute the temperature at the wall $\tilde{T}_w$ from (112) evaluated at the first interior point (2)

(c) Given the temperature and velocity profiles from the previous iteration (time) step, calculate $\tilde{u}_{c(2)}$ from (113).

(d) Update the wall shear stress $\bar{\sigma}_w$ from (114).

(e) Repeat steps (b), (c), and (d) until convergence.

(f) Update the turbulent kinetic energy $K_{(2)}$ from $\tilde{u}_\tau^2/\sqrt{C_\mu}$ or the turbulent stresses $\tau_{ij(2)}$ from their respective transport equations.

(g) Update the dissipation rate $\epsilon_{(2)}$ from (61).

(h) Return to (a) for the next iteration (time) step.

As noted in the discussion of incompressible wall functions, grid spacing considerations need to be taken into account in the placement of the first grid point and the next interior one. In addition, care should be exercised when applying the compressible wall functions in the hypersonic regime. As the Mach number increases, the viscous sublayer becomes thicker. To keep the first interior point inside the turbulent region ($y^+ \approx 30$), the physical distance from the first interior point to the wall should be increased as the Mach number increases. As a result, computations of hypersonic flows with wall functions have difficulty capturing flow separation (Huang and Coakley, 1993b) because the reverse flow in the viscous layer is not resolved.

The wall function implementation is straightforward and accurate when applied to flows that meet the conditions under which the various functions are derived. As in the incompressible case, more complex flows require more general conditions upon which to base the matching behaviour in the near-wall region with that of the high-Reynolds-number models valid away from the wall. The development of such near-wall models is the subject of the next section.

E Compressible near-wall models

As can be expected, considerably less research has concentrated on the development of compressible near-wall models than on their incompressible counterparts. In addition, the various compressible correlations discussed previously, such as the mass flux, heat flux, and pressure dilatation, have, in general, not been suitably developed for near-wall applications and have, therefore, been neglected in the near-wall models. Thus, the work to date has focused on modifying the variable density extensions of the pressure–strain correlation and/or the (solenoidal) dissipation rate equation.

Zhang et al. (1993b) have developed a second-moment near-wall closure model based on a variable density extension of the Lai and So (1990) model. This simply involves multiplying the near-wall forms given in (62) and (63) by the mean density $\bar{\rho}$. They invoke Morkovin's hypothesis (Morkovin, 1964) in order to eliminate the need to use explicitly any of the mass flux, pressure dilatation, or dilatation dissipation models. Unlike the Lai and So model, they include the pressure-echo term originally proposed in the Launder et al. (1975) study. The form used is given by

$$\Pi_{ij}^R = 2C^w \bar{\rho} K (\tilde{S}_{ij} - \tfrac{1}{3}\tilde{S}_{kk}\delta_{ij})\left(\frac{K^{\frac{3}{2}}}{\varepsilon x_2}\right) \tag{115}$$

with

$$C^w = \begin{cases} C^w|_{in} - (5.8 \times 10^{-4}) M_\infty & M_\infty > 2.5 \\ C^w|_{in} & M_\infty \le 2.5 \end{cases} \tag{115a}$$

with $C^w|_{in} = 4.14 \times 10^{-3} + 3 \times 10^{-3} Re_\theta$ for momentum thickness Reynolds number $Re_\theta \le 5500$ and $C^w|_{in} = 0.0153$, otherwise. Their near-wall correction function ξ^w used in the dissipation rate equation also differs from the incompressible form and is given by

$$\xi^w = f_2^w \left(-2\frac{\varepsilon \varepsilon_{w1}}{K} + 1.5\frac{\varepsilon_{w2}^2}{K} - 0.75 C_{\varepsilon 1}\frac{\varepsilon}{K} P_{ii}\right) \tag{116}$$

where the damping function $f_2^w = \exp\left[-(Re_t/40)^2\right]$ and the modelling constants

$C_{\varepsilon 1}$, $C_{\varepsilon 2}$, and C_{ε} take the values 1.5, 1.83 and 0.1, respectively (see (64)). This model has been tested by Zhang *et al.* (1993b) on flat-plate boundary layer flows with adiabatic and cooled conditions up to a free-stream Mach number of 10. In addition to good agreement with experiments throughout the boundary layer, the model was shown to predict the Van Driest compressible law of the wall. Morrison *et al.* (1993) have also tested the model in compressible ramp flows with generally good qualitative agreement with experimental results.

Two-equation compressible models have also been derived. Zhang *et al.* (1993b) have extracted a two-equation model directly from the Reynolds stress model just discussed. In addition, Zhang *et al.* (1993a) have also derived a two-equation model by using a variable density extension of the two-equation model developed by So *et al.* (1991). These models were tested on flat-plate boundary layer flows with adiabatic- and cooled-wall conditions up to free-stream Mach numbers of 10. For the adiabatic-wall cases both the log-law and skin-friction were well predicted. For the cooled-wall conditions, the log-law slope varied from model to model, and the skin-friction predictions were much lower than the measurements.

IV NUMERICAL SOLUTION OF TURBULENT MODEL EQUATIONS

The previous sections in this chapter have presented the wide spectrum of turbulent closure models that are at the disposal of today's scientists and engineers for both incompressible and compressible flows. In this section, the focus will be on the application of the appropriate numerical techniques to the solution of the system of mean and turbulent equations outlined throughout the chapter. Although this task may seem rather extensive at first, the numerical algorithms presented in Chapters 2, 3 and 4 for laminar incompressible and compressible flows are readily extendable to the turbulent flow cases. For example, for the zero- and half-equation models, the only difference between the laminar and turbulent case is the introduction of a variable turbulent eddy viscosity into the diffusion term of the mean-flow equations.

The one-equation models of Baldwin–Barth and Spalart–Allmaras treat the eddy-viscosity equation independently of the mean flow. For the eddy-viscosity equation, they use an implicit algorithm with upwind differencing for the advection terms and central differencing for the diffusive terms. The source terms are treated implicitly; they satisfy various positivity constraints. Baldwin and Barth (1991) use an implicit factored ADI solver for the turbulent model equation, and Spalart and Allmaras (1994) use an approximate factorization scheme (see Section II.B.3) for their turbulent model equation.

Higher-order models such as the two-equation and Reynolds stress models pose additional challenges because of the increase in the number of equations to be solved and the form of the source terms that arise from the production, destruction, and redistribution (Reynolds stress models) terms. Nevertheless, the extensions of the laminar solution techniques to the turbulent equation sets are relatively straightforward, so this section will focus on various unique aspects of these extensions and refer the reader to the previous chapters that deal with the respective incompressible and compressible solution methodologies.

A Pressure–velocity-based solution methodologies

The solution of the incompressible, constant–density turbulence equations is based on extensions of techniques for laminar incompressible equations that were outlined in Chapter 2. The most common approach for the incompressible equations is a pressure–velocity coupling, where the solution of the mean momentum equations is coupled with a Poisson equation for the pressure, which ensures a divergence-free velocity field. The solution procedure chosen for the mean momentum and pressure correction are only slightly modified to include the contribution due to either the turbulent Reynolds stresses or the turbulent kinetic energy. The basic solution procedure is then extended to solve additional transport equations by using the underlying discretization procedure adopted for the mean equations.

In the numerical solution of the incompressible Reynolds-averaged Navier–Stokes equations and turbulent transport equations presented in Section II, the equations can be written as

$$\frac{\partial \bar{u}}{\partial x} + \frac{\partial \bar{v}}{\partial y} + \frac{\partial \bar{w}}{\partial z} = 0 \tag{117}$$

$$\frac{\partial \mathbf{Q}}{\partial t} + \frac{\partial (\mathbf{F}^I - \mathbf{F}^v)}{\partial x} + \frac{\partial (\mathbf{G}^I - \mathbf{G}^v)}{\partial y} + \frac{\partial (\mathbf{H}^I - \mathbf{H}^v)}{\partial z} = \mathbf{S} \tag{118}$$

where $(\bar{u}, \bar{v}, \bar{w})$ are the respective (x, y, z) cartesian components of the mean velocity, $\mathbf{Q}$ is the vector of dependent (primitive) variables, $\mathbf{F}^I, \mathbf{G}^I, \mathbf{H}^I$ are the inviscid (convective) fluxes, $\mathbf{F}^v, \mathbf{G}^v, \mathbf{H}^v$ are the viscous (diffusive) fluxes, and $\mathbf{S}$ represents the source terms due to production, destruction, and redistribution. For the present purposes, only the equation in cartesian (x, y, z) coordinates will be displayed. Extensions to generalized coordinates are the same as in the incompressible Navier–Stokes formulation of Chapter 2, and as discussed in standard numerical texts (e.g., Anderson *et al.*, 1984; Fletcher, 1988). Note that the vector formulation described here is an atypical notation for the incompressible numerical formulation. However, in keeping with the spirit of this chapter to present a unified picture of turbulent flow modelling for both incompressible and compressible flow, it is appropriate to present the numerical formulations in as consistent and common a framework as possible. To aid in the discussion of the various solution algorithms used in solving incompressible turbulent flows, it is advantageous to partition the vector $\mathbf{Q}$ into mean momentum and turbulent parts given by

$$\mathbf{Q} = \left\{ \begin{array}{c} \mathbf{Q}_{\text{momentum}} \\ \mathbf{Q}_{\text{turb}} \end{array} \right\} \tag{119}$$

The mean part is composed of the dependent variables from the momentum equation (5). The turbulent part is composed of the dependent variables from the Reynolds stress or turbulent kinetic energy equations (equation (8) or (30) respectively) and the dissipation rate equation (12). The resulting forms are given by

$$\mathbf{Q}_{\text{momentum}} = \left\{ \begin{array}{c} \bar{u} \\ \bar{v} \\ \bar{w} \end{array} \right\}, \quad \mathbf{Q}_{\text{turb}}|_{\text{RS}} = \left\{ \begin{array}{c} \tau_{xx} \\ \tau_{yy} \\ \tau_{zz} \\ \tau_{xy} \\ \tau_{xz} \\ \tau_{yz} \\ \varepsilon \end{array} \right\}, \quad \mathbf{Q}_{\text{turb}}|_{K-\varepsilon} = \left\{ \begin{array}{c} K \\ \varepsilon \end{array} \right\} \tag{120}$$

These vector expressions, plus the continuity equation, show that the two-equation turbulence model yields six partial differential equations, and the second-moment model yields 11 partial differential equations required for the solution of an incompressible turbulent flow.

The inviscid fluxes are given by

$$\mathbf{F}^I_{\text{momentum}} = \left\{ \begin{array}{c} \bar{u}^2 + \bar{p} \\ \overline{uv} \\ \overline{uw} \end{array} \right\}, \quad \mathbf{F}^I_{\text{turb}}|_{\text{RS}} = \left\{ \begin{array}{c} \bar{u}\tau_{xx} \\ \bar{u}\tau_{yy} \\ \bar{u}\tau_{zz} \\ \bar{u}\tau_{xy} \\ \bar{u}\tau_{xz} \\ \bar{u}\tau_{yz} \\ \bar{u}\varepsilon \end{array} \right\}, \quad \mathbf{F}^I_{\text{turb}}|_{K-\varepsilon} = \left\{ \begin{array}{c} \bar{u}K \\ \bar{u}\varepsilon \end{array} \right\} \tag{121}$$

and the diffusive fluxes by

$$\mathbf{F}^v_{\text{momentum}} = \left\{ \begin{array}{c} \bar{\sigma}_{xx}/\bar{\rho} - \tau_{xx} \\ \bar{\sigma}_{xy}/\bar{\rho} - \tau_{xy} \\ \bar{\sigma}_{xz}/\bar{\rho} - \tau_{xz} \end{array} \right\} \tag{122}$$

$$
\mathbf{F}^v_{\text{turb}}\big|_{\text{RS}} =
\begin{Bmatrix}
\bar{\nu}\dfrac{\partial \tau_{xx}}{\partial x} + \bar{D}^t_{xxx} \\[2mm]
\bar{\nu}\dfrac{\partial \tau_{yy}}{\partial x} + \bar{D}^t_{yyx} \\[2mm]
\bar{\nu}\dfrac{\partial \tau_{zz}}{\partial x} + \bar{D}^t_{zzx} \\[2mm]
\bar{\nu}\dfrac{\partial \tau_{xy}}{\partial x} + \bar{D}^t_{xyx} \\[2mm]
\bar{\nu}\dfrac{\partial \tau_{xz}}{\partial x} + \bar{D}^t_{xzx} \\[2mm]
\bar{\nu}\dfrac{\partial \tau_{yz}}{\partial x} + \bar{D}^t_{yzx} \\[2mm]
\bar{\nu}\dfrac{\partial \varepsilon}{\partial x} + \bar{\mathcal{D}}^t_{\varepsilon x}
\end{Bmatrix}, \quad
\mathbf{F}^v_{\text{turb}}\big|_{K-\varepsilon} =
\begin{Bmatrix}
\left(\bar{\nu} + \dfrac{\bar{\nu}_t}{\sigma_K}\right)\dfrac{\partial K}{\partial x} \\[3mm]
\left(\bar{\nu} + \dfrac{\bar{\nu}_t}{\sigma_\varepsilon}\right)\dfrac{\partial \varepsilon}{\partial x}
\end{Bmatrix}
$$

In the absence of applied body forces, no source terms associated with the mean momentum equations occur on the right-hand-side of (118), so that $\mathbf{S}_{\text{momentum}} = \mathbf{0}$; however, the turbulent equations do have source terms, and these are given by

$$
\mathbf{S}_{\text{turb}}\big|_{\text{RS}} =
\begin{Bmatrix}
\bar{P}_{xx} + \Pi_{xx} - \varepsilon_{xx} \\
\bar{P}_{yy} + \Pi_{yy} - \varepsilon_{yy} \\
\bar{P}_{zz} + \Pi_{zz} - \varepsilon_{zz} \\
\bar{P}_{xy} + \Pi_{xy} - \varepsilon_{xy} \\
\bar{P}_{xz} + \Pi_{xz} - \varepsilon_{xz} \\
\bar{P}_{yz} + \Pi_{yz} - \varepsilon_{yz} \\
\bar{\mathcal{P}}_\varepsilon - \bar{\mathcal{D}}_\varepsilon
\end{Bmatrix}, \quad
\mathbf{S}_{\text{turb}}\big|_{K-\varepsilon} =
\begin{Bmatrix}
\bar{P} - \varepsilon \\
\bar{\mathcal{P}}_\varepsilon - \bar{\mathcal{D}}_\varepsilon
\end{Bmatrix}
\tag{123}
$$

The casting of the governing equation set in the form presented clearly suggests that the pressure–velocity-based techniques used for the solution of the laminar incompressible flow problems are equally applicable to the turbulent problem. Additional transport equations to account for other effects such as heat transport or reacting and non-reacting gases, can be easily added to these vector forms in a straightforward manner with the same process outlined for the turbulent transport equations.

1 Pressure–velocity discretization techniques

Many popular techniques are available to discretize the governing equations, several of which are outlined in Chapter 2. One of the most popular of the early techniques used a staggered grid approach, where the different variables to be solved were stored at alternate locations; the pressure was stored at a cell centre, and the velocities were stored on the cell faces on which they acted. The solution procedure usually used the SIMPLE (or variants thereof) procedure to couple the solution of the pressure to the solution of the velocities. This solution technique was extended to turbulent calculations with second-moment closures by Pope and Whitelaw (1976), Huang and Leschziner (1985) and Fu *et al.* (1988).

A major disadvantage of the staggered grid approach is that the different variables have different control volumes used in their solution. For a three-dimensional staggered grid, four different control volumes are required; four different sets of metrics are required for general, non-orthogonal coordinate systems. This limitation has hampered the extension of the staggered grid approach to the general, non-orthogonal coordinate systems required for complex configurations of interest to fluid dynamics engineers of today.

An alternative to the staggered grid approach is the collocated arrangement in which all solution variables are stored at the same location, usually the cell centre in a finite-volume scheme. The advantage of this arrangement is obvious for the complex grid systems because only a single control volume is used for all equations, and only a single set of grid metrics is required. The disadvantage of the collocated arrangement, historically, was the even–odd chequerboard oscillations that frequently arose in the solution but did not arise from the staggered grid arrangement. This obstacle was overcome by Rhie and Chow (1983) when they developed a collocated arrangement that included an artificial pressure dissipation to mimic the behaviour of the staggered grid. Comparisons of collocated grid with staggered grid results for laminar flows (Perić *et al.*, 1988) have shown that the collocated approach may converge faster, and has implementation advantages when non-orthogonal grids or multigrid techniques are introduced. Although no similar tests have been run for the turbulent case, the same improvements should carry over because the collocated structure is straightforwardly applied to the turbulent transport equations.

The semidiscrete second-order-accurate finite-volume form of (118) for a collocated grid arrangement is written as

$$\left[\frac{\mathcal{V}}{\Delta t}(\mathbf{Q}^{n+1} - \mathbf{Q}^n)\right]_{i,j,k} + \Big\{[\mathbf{F}\mathcal{A}_x]_{i+\frac{1}{2},j,k} - [\mathbf{F}\mathcal{A}_x]_{i-\frac{1}{2},j,k} + [\mathbf{G}\mathcal{A}_y]_{i,j+\frac{1}{2},k} - [\mathbf{G}\mathcal{A}_y]_{i,j-\frac{1}{2},k}$$

$$+[\mathbf{H}\mathcal{A}_z]_{i,j,k+\frac{1}{2}} - [\mathbf{H}\mathcal{A}_z]_{i,j,k-\frac{1}{2}} - (\mathcal{V}\mathbf{S})_{i,j,k}\Big\} = 0 \qquad (124)$$

where

$$\mathbf{F} = \mathbf{F}^I - \mathbf{F}^v, \quad \mathbf{G} = \mathbf{G}^I - \mathbf{G}^v, \quad \mathbf{H} = \mathbf{H}^I - \mathbf{H}^v \qquad (124a)$$

are the fluxes defined at the interfaces of the computational cell that bounds the cell average value $\mathbf{Q}_{i,j,k}$; $\mathcal{A}_x$, $\mathcal{A}_y$, $\mathcal{A}_z$ represent the cell-face areas in the x, y, z directions, respectively (e.g. $\mathcal{A}_x = \Delta y \Delta z$); and $\mathcal{V}$ is the cell volume ($= \Delta x \Delta y \Delta z$).

The interpolation procedure adopted in interpolating the calculated cell-centre data to the cell interface locations determines the accuracy and numerical characteristics of the scheme. The use of linear interpolation for all quantities leads to a second-order-accurate central-difference scheme that suffers from the even–odd decoupling and chequerboard oscillations. The interpolation developed by Rhie and Chow (1983) for the interface velocity in the convection term adds an additional term to the linear interpolation that is a third-order artificial pressure dissipation term. This additional term adds a fourth-order smoothing to each equation that does not affect the formal second-order accuracy of the differencing scheme. The value of the flow variables at the interfaces can be obtained from a variety of interpolation procedures including QUICK, SHARP, LODA and MUSCL. The QUICK procedure (see Chapter 2) has been commonly used for the mean momentum equations; however, QUICK can suffer from oscillations in regions of high gradients, and MUSCL (see Chapter 4) with a limiting function has been adapted to the turbulence transport equations to avoid oscillations. Both SHARP (Leonard, 1988) and LODA (Zhu and Leschziner, 1988) have also been formulated to limit oscillations, thus permitting their extension to turbulence transport equations as well (see also Lien and Leschziner, 1994).

The discretization of the diffusion terms is accomplished in an accurate and stable manner by using linear interpolation for the interface quantities, such as $\bar{\nu}_t$, and central differencing for the first derivative terms at the cell interfaces. This procedure has been satisfactory for laminar, algebraic models, and two-equation models because the components of τ_{ij} are evaluated with the eddy viscosity hypothesis, which results in a second-order derivative in the momentum equations. However, the Reynolds stress gradients are first derivative terms and are evaluated as such when a second-moment closure is used. This first-derivative evaluation can lead to a decoupling of the stresses from the momentum equations. Huang and Leschziner (1985), Obi *et al.* (1989) and Lien and Leschziner (1994) detailed interpolation procedures for the cell interface values of the τ_{ij} in the momentum equations. Huang and Leschziner (1985), in a staggered grid, split the stresses into an 'apparent' viscosity associated with one of the corresponding strain rates and a correction term to include the remainder of the stress term. The apparent viscosity term more strongly couples the Reynolds stress term to the velocity in the momentum equation and results in increased diagonal dominance and enhanced iterative stability. The approach of Obi *et al.* (1989) and Lien and Leschziner (1994) was to introduce a third-order artificial smoothing term based on the mean velocity similar to the artificial smoothing term introduced by Rhie and Chow (1983) for the convection terms. Additionally, the apparent viscosity term of Huang and Leschziner (1985) was used to enhance coupling and diagonal dominance.

2 Pressure–velocity solution algorithm

The solution of the system of equations just described is best understood by examining the discrete form of a representative turbulence transport equation for an element of $\mathbf{Q}_{\text{turb}} = \{Q_1, Q_2, \cdots Q_p\}^{\text{T}}$ and $\mathbf{S}_{\text{turb}} = \{S_1, S_2, \cdots S_p\}^{\text{T}}$, where $p = 2$ for two-equation models, and $p = 7$ for second-moment closures. The discrete component equation for Q_p can be written after the terms are collected as:

$$\left[\left(\frac{V}{\Delta t} + \alpha \right) Q_p^{n+1} \right]_{i,j,k} = \sum_m \alpha_m Q_{p_m}^{n+1} + \frac{V}{\Delta t} Q_p^n + VS^+ \tag{125}$$

where $\sum_m \alpha_m Q_{p_m}^{n+1}$ represents the convection and diffusion contributions from the neighbouring cells used in the formulation of the interface fluxes. The coefficients α, which contain metric and flow variables, are evaluated at time level n to obtain a linear relationship for the solution for flow variable Q_p. The source terms in the turbulent transport equations are split into positive and negative portions $\mathbf{S}_{\text{turb}} = \mathbf{S}_{\text{turb}}^- + \mathbf{S}_{\text{turb}}^+$. The negative part is included in the coefficient $\alpha_{i,j,k}$ to increase diagonal dominance and iterative stability. The positive portion is included as a source term to include the complete source contribution. Because the negative portion of the source term is included as $\alpha_{i,j,k} Q_p^{n+1}$, negative contributions to the source terms that do not have a factor Q_p can be included by dividing them by the previous iterate of the solution variable to arrange them in this form.

The common solution approach is based on the SIMPLE procedure of Patankar (1980) or variants such as SIMPLER or SIMPLEC (see Fletcher, 1988). The momentum equations, discretized similarly to (125), give an update procedure for each of the velocities. The discretized equations for the turbulence quantities (125) give an updated procedure for the turbulence field. Given an estimate of the pressure field, (125) can be solved for the momentum equations to give an estimate $\bar{u}^*$, $\bar{v}^*$, and $\bar{w}^*$ of the field $\bar{u}^{n+1}$, $\bar{v}^{n+1}$, and $\bar{w}^{n+1}$. In general, this intermediate value of the velocity field will not satisfy the divergence-free condition. The solution procedure then finds a correction to the pressure field $\delta \bar{p}$ that updates the pressure to the new time level $\bar{p}^{n+1} = \bar{p}^n + \delta \bar{p}$, and provides a velocity correction such that $\bar{u}^{n+1} = \bar{u}^* + \bar{u}^c$, and, similarly for $\bar{v}$ and $\bar{w}$, satisfies continuity. Complete details of this process are outlined in Chapter 2.

The complete algorithm for updating the solution of the incompressible equations from time level n to time level $n + 1$ with the SIMPLE algorithm can be summarized as follows:

(i) Calculate $\bar{u}^*$, $\bar{v}^*$, and $\bar{w}^*$ from the discretized momentum equations with the interpolation of Lien and Leschziner (1994) for the Reynolds stress gradients from the solution field at time n.

(ii) Obtain $\delta \bar{p}$ from the solution of the discrete Poisson equation for the pressure correction.

(iii) Calculate the velocity field at time level $n + 1$ as $\bar{u}^{n+1} = \bar{u}^* + \bar{u}^c$ (similarly for $\bar{v}$ and $\bar{w}$).

(iv) Calculate the pressure field at time level $n + 1$ as $\bar{p}^{n+1} = \bar{p}^n + \beta_r \delta \bar{p}$, where β_r is an underrelaxation parameter to improve stability and convergence.

(v) Update each turbulence variable from its discrete transport equation (125).

For steady-state calculations, this procedure is repeated until each of the solution variables converges to a predetermined level. Underrelaxation is usually employed in all of the equations to improve the convergence, but with different relaxation parameters for the momentum equations, the pressure correction equation, and the turbulence transport equations (see Chapter 2 for additional details). The steps of the SIMPLE procedure outlined above are very similar to the procedures for the laminar flow case; the differences are the oscillation-free (Lien and Leschziner, 1994) inclusion of τ_{ij} in the momentum equations and the additional steps to update the transport equations for the turbulent quantities.

The pressure–velocity-based scheme presented here was developed for the incompressible constant-density flow. This pressure correction scheme can be extended to work for variable density and compressible flows, as indeed the velocity interpolation scheme for the collocated grid was developed by Rhie and Chow (1983) for compressible flows with two-equation turbulence models. Recently, the work of Lien and Leschziner (1994) reports the compressible formulation with a second-moment closure. An alternative to the pressure–velocity coupling approach is the density–velocity approach, where the compressible equations are integrated in time to a steady state.

B Density–velocity-based solution methodologies

In this section, the attention is once again focused on the numerical solution of the Reynolds stress and two-equation closure models. For the numerical solution of the compressible mean conservation and turbulent transport equations, presented in Section III, the system of equations is customarily written as

$$\frac{\partial \mathbf{Q}}{\partial t} + \frac{\partial (\mathbf{F}^I - \mathbf{F}^v)}{\partial x} + \frac{\partial (\mathbf{G}^I - \mathbf{G}^v)}{\partial y} + \frac{\partial (\mathbf{H}^I - \mathbf{H}^v)}{\partial z} = \mathbf{S} \qquad (126)$$

where, once again, $\mathbf{Q}$ is the vector of dependent variables; $\mathbf{F}^I, \mathbf{G}^I, \mathbf{H}^I$ are the inviscid (convective) fluxes; $\mathbf{F}^v, \mathbf{G}^v, \mathbf{H}^v$ are the viscous (diffusive) fluxes, and $\mathbf{S}$ represents the source terms due to production, destruction and redistribution. As in the incompressible discussion, the vector $\mathbf{Q}$ can be partitioned into mean and turbulent parts given by

$$\mathbf{Q} = \left\{ \begin{array}{c} \mathbf{Q}_{mean} \\ \mathbf{Q}_{turb} \end{array} \right\} \tag{127}$$

The mean part is composed of the dependent variables from the mean conservation of mass, momentum, and energy equations, (74), (75) and (77), respectively. The turbulent part is composed of the dependent variables from the Reynolds stress or turbulent kinetic energy equations, (79) or (83), respectively, and the dissipation rate equation (86). The resulting forms are given by

$$\mathbf{Q}_{mean} = \left\{ \begin{array}{c} \bar{\rho} \\ \bar{\rho}\tilde{u} \\ \bar{\rho}\tilde{v} \\ \bar{\rho}\tilde{w} \\ \bar{\rho}\tilde{E} \end{array} \right\}, \qquad \mathbf{Q}_{turb}|_{RS} = \left\{ \begin{array}{c} \bar{\rho}\tau_{xx} \\ \bar{\rho}\tau_{yy} \\ \bar{\rho}\tau_{zz} \\ \bar{\rho}\tau_{xy} \\ \bar{\rho}\tau_{xz} \\ \bar{\rho}\tau_{yz} \\ \bar{\rho}\varepsilon \end{array} \right\}, \qquad \mathbf{Q}_{turb}|_{K-\varepsilon} = \left\{ \begin{array}{c} \bar{\rho}K \\ \bar{\rho}\varepsilon \end{array} \right\} \tag{128}$$

where $(\tilde{u}, \tilde{v}, \tilde{w})$ are the respective (x, y, z) cartesian components of the mean velocity. These vector expressions show that the two-equation turbulence model yields seven partial differential equations, and the Reynolds stress formulation yields 12 partial differential equations required for the solution of a compressible turbulent flow. These numbers are compared with the six equations for the two-equation incompressible formulation, and the 11 equations for the Reynolds stress incompressible formulations.

The inviscid fluxes are given by

$$\mathbf{F}^I_{mean} = \left\{ \begin{array}{c} \bar{\rho}\tilde{u} \\ \bar{\rho}\tilde{u}^2 + \bar{p} \\ \bar{\rho}\tilde{u}\tilde{v} \\ \bar{\rho}\tilde{u}\tilde{w} \\ \bar{\rho}\tilde{u}\tilde{H} \end{array} \right\}, \qquad \mathbf{F}^I_{turb}|_{RS} = \left\{ \begin{array}{c} \tilde{u}\bar{\rho}\tau_{xx} \\ \tilde{u}\bar{\rho}\tau_{yy} \\ \tilde{u}\bar{\rho}\tau_{zz} \\ \tilde{u}\bar{\rho}\tau_{xy} \\ \tilde{u}\bar{\rho}\tau_{xz} \\ \tilde{u}\bar{\rho}\tau_{yz} \\ \tilde{u}\bar{\rho}\varepsilon \end{array} \right\}, \qquad \mathbf{F}^I_{turb}|_{K-\varepsilon} = \left\{ \begin{array}{c} \tilde{u}\bar{\rho}K \\ \tilde{u}\bar{\rho}\varepsilon \end{array} \right\} \tag{129}$$

and the diffusive fluxes by

$$\mathbf{F}^v_{mean} = \left\{ \begin{array}{c} 0 \\ \bar{\sigma}_{xx} - \bar{\rho}\tau_{xx} \\ \bar{\sigma}_{xy} - \bar{\rho}\tau_{xy} \\ \bar{\sigma}_{xz} - \bar{\rho}\tau_{xz} \\ \bar{\Sigma}_x - \bar{q}_x - \widetilde{\bar{\rho}E''u''} \end{array} \right\} \tag{130}$$

$$
\mathbf{F}^v_{\text{turb}}\big|_{\text{RS}} = \left\{\begin{array}{c} \bar{\mu}\dfrac{\partial \tau_{xx}}{\partial x} + \bar{\rho}\widetilde{D}^t_{xxx} \\[2mm] \bar{\mu}\dfrac{\partial \tau_{yy}}{\partial x} + \bar{\rho}\widetilde{D}^t_{yyx} \\[2mm] \bar{\mu}\dfrac{\partial \tau_{zz}}{\partial x} + \bar{\rho}\widetilde{D}^t_{zzx} \\[2mm] \bar{\mu}\dfrac{\partial \tau_{xy}}{\partial x} + \bar{\rho}\widetilde{D}^t_{xyx} \\[2mm] \bar{\mu}\dfrac{\partial \tau_{xz}}{\partial x} + \bar{\rho}\widetilde{D}^t_{xzx} \\[2mm] \bar{\mu}\dfrac{\partial \tau_{yz}}{\partial x} + \bar{\rho}\widetilde{D}^t_{yzx} \\[2mm] \bar{\mu}\dfrac{\partial \varepsilon}{\partial x} + \widetilde{D}^t_{\varepsilon x} \end{array}\right\}, \quad \mathbf{F}^v_{\text{turb}}\big|_{K-\varepsilon} = \left\{\begin{array}{c} \left(\bar{\mu} + \dfrac{\bar{\mu}_t}{\sigma_K}\right)\dfrac{\partial K}{\partial x} \\[3mm] \left(\bar{\mu} + \dfrac{\bar{\mu}_t}{\sigma_\varepsilon}\right)\dfrac{\partial \varepsilon}{\partial x} \end{array}\right\}
$$

As in the incompressible case, no source terms occur on the right-hand-side of (126) associated with the mean conservation equations in the absence of applied body forces, so that $\mathbf{S}_{\text{mean}} = \mathbf{0}$; however, the turbulent equations do have source terms; these are given by

$$
\mathbf{S}_{\text{turb}}\big|_{\text{RS}} = \left\{\begin{array}{c} \bar{\rho}\widetilde{P}_{xx} + \bar{\rho}\Pi^d_{xx} + \bar{\rho}\Pi^{dl}_{xx} + M_{xx} - \bar{\rho}\varepsilon_{xx} \\[1mm] \bar{\rho}\widetilde{P}_{yy} + \bar{\rho}\Pi^d_{yy} + \bar{\rho}\Pi^{dl}_{yy} + M_{yy} - \bar{\rho}\varepsilon_{yy} \\[1mm] \bar{\rho}\widetilde{P}_{zz} + \bar{\rho}\Pi^d_{zz} + \bar{\rho}\Pi^{dl}_{zz} + M_{zz} - \bar{\rho}\varepsilon_{zz} \\[1mm] \bar{\rho}\widetilde{P}_{xy} + \bar{\rho}\Pi^d_{xy} + \bar{\rho}\Pi^{dl}_{xy} + M_{xy} - \bar{\rho}\varepsilon_{xy} \\[1mm] \bar{\rho}\widetilde{P}_{xz} + \bar{\rho}\Pi^d_{xz} + \bar{\rho}\Pi^{dl}_{xz} + M_{xz} - \bar{\rho}\varepsilon_{xz} \\[1mm] \bar{\rho}\widetilde{P}_{yz} + \bar{\rho}\Pi^d_{yz} + \bar{\rho}\Pi^{dl}_{yz} + M_{yz} - \bar{\rho}\varepsilon_{yz} \\[1mm] \bar{\rho}\widetilde{P}_\varepsilon - \bar{\rho}\widetilde{D}_\varepsilon \end{array}\right\}
$$

$$\tag{131}$$

$$
\mathbf{S}_{\text{turb}}\big|_{K-\varepsilon} = \left\{\begin{array}{c} \bar{\rho}\widetilde{P} + \bar{\rho}\Pi^{dl} + \mathcal{M} - \bar{\rho}\varepsilon \\[1mm] \bar{\rho}\widetilde{P}_\varepsilon - \bar{\rho}\widetilde{D}_\varepsilon \end{array}\right\}
$$

The vector formulation just presented for the compressible equations clearly parallels the incompressible formulation presented in Section IV.A. In both cases, the discretization and solution algorithm techniques are easily extendable to the turbulent case. This extension has already been shown for incompressible flows; as will be shown next, it is also the case for compressible flows.

1 Density–velocity discretization techniques

As expected, a variety of techniques are available for the discretization of the equation set just described. In this subsection, one such technique will be described that is commonly used in the solution of the laminar equation set and that is also used in solving turbulent flow problems. Other methodologies are also available and can be extended in a similar manner.

A popular technique that has been successfully used in the solution of the mean and turbulent transport equations is based on a control volume approach. In this approach, the semidiscrete second-order-accurate finite-volume form of (126) is written as

$$\left(\mathcal{V}\frac{\partial \mathbf{Q}}{\partial t}\right)_{i,j,k} + \Big\{ [\mathbf{F}\mathcal{A}_x]_{i+\frac{1}{2},j,k} - [\mathbf{F}\mathcal{A}_x]_{i-\frac{1}{2},j,k} + [\mathbf{G}\mathcal{A}_y]_{i,j+\frac{1}{2},k} - [\mathbf{G}\mathcal{A}_y]_{i,j-\frac{1}{2},k}$$
$$+[\mathbf{H}\mathcal{A}_z]_{i,j,k+\frac{1}{2}} - [\mathbf{H}\mathcal{A}_z]_{i,j,k-\frac{1}{2}} - (\mathcal{V}\mathbf{S})_{i,j,k} \Big\} = 0$$

$$(132)$$

where $\mathbf{F}, \mathbf{G}$, and $\mathbf{H}$ are the differences between the inviscid and viscous fluxes given in (124a) and are defined at the interfaces of the computational cell that bounds the cell-average value. Similarly, $\mathcal{A}_x, \mathcal{A}_y, \mathcal{A}_z$ represent the cell-face areas in the x, y, z directions, respectively, and $\mathcal{V}$ is the cell volume. The definition of the numerical flux function that approximates the interface flux determines the characteristics of the numerical scheme. For compressible, transonic and supersonic flows, embedded regions with shocks are always a concern, because the high Reynolds numbers of flows away from boundaries results in very thin shock structures where the physical dissipation is quite small. Therefore, the numerical flux function chosen to evaluate the interface flux must combine good shock-capturing capabilities with good accuracy to resolve the turbulent regions.

Chapter 4 outlines several schemes designed to capture shocks and maintain accuracy in smooth regions of the flow. To illustrate the extension of the numerical methods to turbulent flows, this section outlines the extension of one of these techniques (the MUSCL scheme coupled with Roe's (1981) approximate Riemann solver) to the system of mean equations coupled with the turbulent transport equations. The interface flux for the finite-volume formulation is calculated in each of the three coordinate directions as the solution of a locally one-dimensional Riemann problem normal to the cell interface (the so-called operator splitting approach). The solution of the Riemann problem results in a shock wave, a contact discontinuity, and a rarefaction that evolves in time at the interface. The interface flux can then be determined as the flux at the left or right state incremented by the flux differences crossed from that state to the interface. The exact solution of the

Riemann problem requires an iterative procedure and is quite expensive. A cheaper alternative developed by Roe is to construct the solution to an approximate, linearized problem, where the jacobian matrix $\mathbf{A}^I = \partial \mathbf{F}^I / \partial \mathbf{Q}$ is evaluated at an average state such that it satisfies the jump condition between the flux states at the right and left. The interface flux is written as the average of the interface flux calculated from the left state crossing negative running waves. This approach has been followed by Morrison (1990, 1992) and Huang and Coakley (1992) in solving both the two-equation and Reynolds stress equations.

The diffusive fluxes require the flow variables and derivatives of the flow variables to be evaluated at the cell interface. The flow variables required at this interface are obtained by averaging the neighbouring cell averages. The derivatives required are approximated with Gauss's divergence theorem by integrating around an auxiliary cell that is centred at the interface. This approximation is equivalent to a finite-volume representation of a second-order-accurate central-difference operator (Chakravarthy et al., 1985; Swanson and Turkel, 1985), which is consistent with the elliptic nature of the diffusive fluxes.

The finite-volume form gives an integral over the control volume of the source terms. This form is approximated to second-order accuracy as the value of the cell centre times the volume of the cell. The source terms require the flow variables and the velocity derivatives at the cell centre. The velocity derivatives are calculated with Gauss's divergence theorem by integrating around the computational cell with the cell interface values of the velocity approximated as averages of neighbouring values.

Finite-difference approaches have also been used in the solution of compressible flow problems. The most popular approach is the MacCormack (1969) predictor-corrector scheme mentioned in Chapters 1 and 4, and described in detail in Peyret and Taylor (1983). Vandromme and Ha Minh (1986) have applied a two-dimensional implicit version of this scheme to both the two-equation and Reynolds stress models. The full equation set is solved fully coupled between the mean and turbulent equations. Knight (1984) (see also Narayanswami et al., 1993) also uses the MacCormack method but employs a hybrid implicit–explicit approach. The implicit formulation is used in the viscous sublayer and transition wall regions in order to remove the restrictive CFL conditions necessitated by fine grid spacing in these regions. More recently, Sarkar and Lakshmanan (1991) have used an explicit MacCormack scheme to solve the coupled two-dimensional set of mean flow and turbulent Reynolds stress equations. Another finite-difference approach that has been followed is the Beam–Warming scheme (Beam and Warming, 1978; see also Peyret and Taylor, 1983). Sahu and Danberg (1986) have used a two-equation model to solve the compressible decoupled mean and turbulent equations. They restricted their study to axisymmetric flow and invoked the thin-layer approximation. In all of these approaches, the source term discretization is consistent with the spatial discretizations used on the remaining terms of the equations.

2 Density-velocity solution algorithm

The conservation and transport equations given by (126) must be integrated in time to determine the unsteady evolution or to calculate the steady-state solution. An implicit time integration is often used to integrate the equations to accelerate convergence to steady state and to avoid any stiffness problems that may occur. Although the equations can be integrated in a time accurate manner, this has not been extensively done. The limitation lies mainly with the modelled turbulent transport equations. As alluded to in the previous sections, the model equations are formed for equilibrium flows rather than for non-equilibrium flows. Although temporally evolving flows can be computed, it is not clear that such flows can be accurately predicted with the models developed to date. A common Euler implicit time integration scheme can be applied to a general discretization, either finite-volume or finite-difference, and written in delta form as

$$\left[\frac{\mathbf{I}}{\Delta t} + \delta_x \mathbf{A} + \delta_y \mathbf{B} + \delta_z \mathbf{C} - \mathbf{D}\right]\Delta \mathbf{Q} = -\mathbf{R}^n \tag{133}$$

where $\Delta \mathbf{Q} = \mathbf{Q}^{n+1} - \mathbf{Q}^n$; $\mathbf{A} = \partial \mathbf{F}/\partial \mathbf{Q}, \mathbf{B} = \partial \mathbf{G}/\partial \mathbf{Q}, \mathbf{C} = \partial \mathbf{H}/\partial \mathbf{Q}$, and $\mathbf{D} = \partial \mathbf{S}/\partial \mathbf{Q}$ are the jacobian matrices; and $\mathbf{R}^n$ is the discretized residual at time level n. In (133), δ_x, δ_y, δ_z are operators that represent the particular type of spatial differencing chosen (e.g. $[\delta_x \mathbf{B}]\Delta \mathbf{Q}$ implies $\delta_x(\mathbf{B}\Delta \mathbf{Q})$). Direct solution of (133) requires the inversion of a large banded system of block matrices at each time step to update the solution. As an alternative, Huang and Coakley (1992) have successfully used a line-by-line Gauss–Seidel approach to solve (133) for a wide variety of two-dimensional and axisymmetric flows. In three-dimensional calculations, an approximate procedure is usually used to invert the system given by (133).

   The usual approach is to factor the implicit equation into simpler systems to invert. A common approach is the spatially split, approximate factorization scheme that can be written as

$$[\mathbf{I} - \Delta t\mathbf{D}][\mathbf{I} + \Delta t\delta_x \mathbf{A}][\mathbf{I} + \Delta t\delta_y \mathbf{B}][\mathbf{I} + \Delta t\delta_z \mathbf{C}]\Delta \mathbf{Q} = -\Delta t\mathbf{R}^n \tag{134}$$

The first factor represents the implicit source term treatment. For a computational stencil in which the inviscid fluxes span five cell centres for a second-order upwind discretization or second-order central differencing with the artificial dissipation included, the three spatial factors require the inversion of a block pentadiagonal system (with a block size of 12 for Reynolds stress models or a block size of 7 for two-equation models) in each of the three coordinate directions. For central differencing of the diffusive fluxes, a three-point stencil results. Since the rate of convergence is relatively insensitive to the order of differencing (Walters and Thomas, 1989), a computational savings can be gained by treating the inviscid fluxes as first-order accurate in the implicit sweeps, which reduces the inversion work because these are tridiagonal rather than pentadiagonal in form. No reduction in accuracy occurs at steady state because of the implementation in delta form. A second benefit of first-order implicit differencing of the inviscid fluxes is gained from the unconditional diagonal dominance of the first-order differencing.

The source term is a function of both the mean and turbulent solution variables and their derivatives. As such, the implicit operator for the source terms remains a large banded system as written. Solution procedures that decouple the mean flow solution from the turbulence equation integration inherently neglect the implicit contribution due to the velocity derivatives. Gorski *et al.* (1985) and Gorski (1986) in a K–ε two-equation adaptation solve the decoupled set. Because the turbulent equations lag behind the mean equations, the source terms require no special treatment and are handled explicitly. Huang and Coakley (1992) also decouple the solution of the turbulence equations from the mean flow equations. They split the source terms, on a term-by-term basis, into positive and negative parts with the positive source treated explicitly and the negative source treated implicitly. Sahu and Danberg (1986), who used the Beam–Warming discretizations, solved the mean and turbulent equations decoupled with the source terms treated implicitly but as part of one of the spatial factors. Fourth-order dissipation terms are added explicitly to control numerical instability.

The approach adopted by Morrison (1990, 1992) is to solve the equation set fully coupled; however, the implicit contribution due to the velocity derivatives is neglected so that the implicit source term operator is now a point implicit treatment that involves only the solution at point i,j,k. The derivation of the jacobian of the source terms can be a very time-consuming project because of the complex form of the source terms, especially for near-wall models where damping functions are introduced that are functions of the solution variables. Additional approximations can be made to the implicit treatment of the source terms to reduce the work from a block-diagonal inversion and to reduce the effort required in deriving the jacobian. Morrison (1992), following the work of Vandromme and Ha Minh (1986) and Huang and Coakley (1992), replaced the jacobian matrix with a diagonal matrix that had $min\ (S_p/Q_p, 0)$ for the (p,p) element. This step results in a scalar division for each equation rather than a block inversion. Additionally, the minimum function avoids the effect of the source term when it would result in an exponentially growing solution. Morrison and Gatski (1993) have showed, using an approximate factorization (AF) scheme, that it is possible to achieve convergence rates for two-equation and Reynolds stress models that match the convergence rate of a simple algebraic turbulence model (Baldwin–Lomax). The additional cost of calculating a turbulent flow-field with either a two-equation or second-moment turbulence model then corresponds only to the additional computational time required to solve the additional equations. Table 1 shows the relative cost in time per iteration per grid point for the different types of models, and

Table 1
Comparison of CPU cost for diagonalized AF scheme[a]

Model	CPU (μsec/grid point/iteration)	Relative cost	Theoretical cost
Laminar	11.57	1.00	1.0
Two-equation	20.28	1.75	1.4
Reynolds stress	33.28	2.88	2.4

[a]CPU time for Navier–Stokes equations on NAS Cray C90 for $65 \times 65 \times 65$ grid (Morrison and Gatski, 1993).

demonstrates that it is possible to approach the 'theoretical' speed of 7:5 for two-equation models, and 12:5 for second-moment models.

Recent work on approximate factorization schemes with source terms (Shih and Chyu, 1991) recommends the inclusion of the source terms in the spatial factors rather than as a separate inversion to reduce factorization error. Including the source terms in the spatial factors would require either storing the source jacobian over the flow field to include it in all of the factors or recomputing it for each of the factors. The above approach has proven efficient enough that the additional storage or computational penalties are usually avoided. Additionally, the separation of the implicit source term into a separate factor facilitates the further approximations of the operators to reduce computational effort. An alternative to the approximate factorization method has been proposed by Lin *et al.* (1993). They use a preconditioned iterative method based on the Bi-CGSTAB method developed by Van der Vorst (1992) and the TFQMR method developed by Freund and Nachtigal (1991). The coupled set of mean and two-equation turbulence models is solved, and the convergent characteristics are compared with more traditional approaches.

Grasso and Speziale (1989) and Grasso and Falconi (1993), who use the finite-volume discretization approach described in the previous section, use a three-stage Runge-Kutta algorithm instead of the approximate factorization method just described. Because the equations are solved decoupled, they determine the source terms in a straightforward manner by using the known mean flow variables and the initial-stage turbulent flow variables in the computations.

In all of the techniques just discussed, a range of validity issues arise whenever the various turbulence models are used in transient or non-equilibrium turbulent flow computations. Initial transients sometimes have a significant effect on the robustness of the solution procedure. Improper initial conditions, coupled with poorly posed boundary conditions, and turbulence models that do not sufficiently account for violations of realizability constraints, can lead to regions of the flow where normal stress components and/or isotropic dissipation rates become negative. If these effects are truely transient, then they can usually be remedied by imposing non-negative limiters on both the normal stress components (kinetic energy) and the dissipation rate. However, these techniques must be viewed with caution, and their effect on the solution must be closely monitored.

Practical constraints associated with the solution of such a complex system of equations always arise. In addition to the validity issues just mentioned, another difficulty that may arise in solving the turbulence model equations in the vicinity of shocks is often due to the source terms of the turbulence models. The production terms in the turbulent transport equations are proportional to the velocity derivative. Because of the definition of the eddy viscosity, for two-equation turbulence models, the production terms are proportional to the velocity derivatives *squared*. At a shock, the velocity derivative is becoming unbounded – a problem that is exacerbated as the inviscid scheme is improved to provide thinner shock profiles. The result is that the production term can become unbounded. An approach that has been used to minimize this problem (J. H. Morrison, private communication) is to limit the production term to a constant multiple of the destruction term. For two-equation models, in both the kinetic energy and dissipation rate equations, this is implemented as

$$\widetilde{\mathcal{P}} = \min\left(\widetilde{\mathcal{P}}, C\bar{\rho}\epsilon\right) \qquad (135)$$

where $C \gg 1$ is a constant taken to be of the $O(10^2)$. The large value of the constant C ensures that the limit is not enforced in any region where the turbulence dynamics are important. Nevertheless, if condition (135) is invoked in some dynamically important region, then the validity of the results is seriously in question due to the fact that the turbulence models are at best validated for flows where $\widetilde{\mathcal{P}}/\bar{\rho}\epsilon$ are $O(1)$. Other such obstacles, unique to the particular flow cases that are examined and/or the numerical algorithms employed, can always arise. The difficulty lies in identifying whether the problem is a numerical or differential formulation one. A thorough knowledge of both the physical and numerical aspects of the problem is required in order to deal with the problem efficiently and effectively.

ACKNOWLEDGEMENTS

The author is indebted to a number of colleagues for their helpful comments in reading earlier versions of the manuscript, in particular, to Mr J.H. Morrison for his perceptive comments regarding the numerical aspects. Many thanks to Professors C.G. Speziale, R.M.C. So and P.A. Libby, and to Drs J.M. Cimbala, J.R. Ristorcelli, R. Abid, G. Huang and B. Younis for their comments and suggestions.

REFERENCES

Abid, R. and Speziale, C.G. (1993). *Phys. Fluids A* **5**, 1776–1782.

Anderson, D.A., Tannehill, J.C. and Pletcher, R.H. (1984). *Computational Fluid Mechanics and Heat Transfer.* McGraw Hill Book Company, New York.

Aris, R. (1962). *Vectors, Tensors, and the Basic Equations of Fluid Mechanics.* Prentice-Hall, Inc., Englewood Cliffs, NJ.

Baldwin, B.S. and Barth, T.J. (1991). *AIAA Paper 91-0610,* AIAA 29th Aerospace Sciences Meeting, Reno, NV.

Baldwin, B.S. and Lomax, H. (1978). *AIAA Paper 78-257,* AIAA 16th Aerospace Sciences Meeting, Huntsville, AL.

Beam, R.M. and Warming, R.F. (1978). *AIAA J.* **16**, 393–402.

Bradshaw, P., Launder, B.E. and Lumley, J.L. (1994). In *Advances in Computational Methods in Fluid Dynamics* (K.N. Ghia, U. Ghia and D. Goldstein, eds), FED vol. 196, pp. 77–82. ASME, New York.

Cebeci, T. and Bradshaw, P. (1984). *Physical and Computational Aspects of Convective Heat Transfer,* pp. 333–371. Springer-Verlag, New York.

Cebeci, T. and Smith, A.M.O. (1974). *Analysis of Turbulent Boundary Layers.* Academic Press, New York.

Chakravarthy, S.R., Szema, K-Y., Goldberg, U.C. and Gorski, J.J. (1985). *AIAA Paper 85-0165,* AIAA 23rd Aerospace Sciences Meeting, Reno, NV.

Chieng, C.C. and Launder, B.E. (1980). *Numer. Heat Transfer* **3**, 189–207.

Choi, K.S. and Lumley, J.L. (1984). In *Turbulence and Chaotic Phenomena in Fluids* (T. Tatsumi, ed.), pp. 267–272. North Holland, Amsterdam.

Coleman, G.N. and Mansour, N.N. (1991). *Phys. Fluids A* **3**, 2255–2259.

Cousteix, J. and Aupoix, B. (1990). *AGARD Report No. 764.* AGARD, Neuilly-sur-Seine.

Daly, B.J. and Harlow, F.H. (1970). *Phys. Fluids* **13**, 2634–2649.

Demuren, A.O. and Sarkar, S. (1993). *ASME J. Fluids Engrg.* **115**, 5–12.

Dinavahi, S.P.G. Ristorcelli, J.R. and Erlebacher, G. (1994). In *Transition, Turbulence and Combustion* (M.Y. Hussaini, T.B. Gatski and T.L. Jackson, eds), vol. 2, pp. 73–82. Kluwer, Dordrecht.

Durbin, P.A. (1991). *Theor. Comput. Fluid Dyn.* **3**, 1–13.

Durbin, P.A. (1993). *J. Fluid Mech.* **249**, 465–498.

El Baz, A.M. and Launder, B.E. (1993). In *Engineering Turbulence Modelling and Experiments 2* (W. Rodi and F. Martelli, eds), pp. 63–72. Elsevier Science Publishers, Amsterdam.

Favre, A. (1965). *J. Mecanique* **4**, 361–390.

Favre, A. (1991). In *Studies in Turbulence* (T. B. Gatski, S. Sarkar, C. G. Speziale, eds), pp. 324–341. Springer-Verlag, New York.

Fletcher, C.A.J. (1988). *Computational Techniques for Fluid Dynamics,* vol. II. Springer-Verlag, New York.

Freund, R.W. and Nachtigal, N. M. (1991). *Numer. Math.* **60**, 315–339.

Fu, S., Launder, B.E., and Tselepidakis, D.P. (1987). *UMIST Technical Report TFD/87/5.*

Fu, S., Huang, P.G., Launder, B.E. and Leschziner, M. A. (1988). *ASME J. Fluids Engrg.* **110**, 216–221.

Gatski, T.B. and Speziale, C.G. (1993). *J. Fluid Mech.* **254**, 59–78.

Gibson, M.M. and Launder, B.E. (1978). *J. Fluid Mech.* **86**, 491–511.

Gorski, J.J. (1986). *AIAA Paper 86-0556,* AIAA 24th Aerospace Sciences Meeting, Reno, NV.

Gorski, J.J., Chakravarthy, S.R. and Goldberg, U.C. (1985). *AIAA Paper 85-1665,* AIAA 18th Fluid Dynamics and Plasmadynamics and Lasers Conference, Cincinnati, OH.

Grasso, F. and Falconi, D. (1993). *AIAA Paper 93-0778,* AIAA 31st Aerospace Sciences Meeting, Reno, NV.

Grasso, F. and Speziale, C.G. (1989). *AIAA Paper 89-1951,* AIAA 9th Computational Fluid Dynamics Conference, Buffalo, NY.

Gulyaev, A.N., Kozlov, V.E. and Sekundov, A.N. (1993). *Fluid Dynamics* **28**, 485–494.

Hallbäck, M., Groth, J. and Johansson, A.V. (1990). *Phys. Fluids A* **2**, 1859–1866.

Ha Minh, H., Rubesin, M.W., Vandromme, D. and Viegas, J.R. (1985). *International Symposium on Computational Fluid Dynamics,* 12–16 September, Tokyo, Japan.

Hanjalić, K. (1994). *Int. J. Heat Fluid Flow* **15**, 178–203.

Hanjalić, K. and Launder, B.E. (1972). *J. Fluid Mech.* **52**, 609–638.

Hanjalić, K. and Launder, B.E. (1976). *J. Fluid Mech.* **74**, 593–610.

Hanjalić, K., Launder, B.E. and Schiestel, R. (1979). *Proc. Second Symposium on Turbulent Shear Flows,* pp. 10.31–10.35. Imperial College, London.

Huang, P.G. and Coakley, T.J. (1992). *AIAA Paper 92-0547,* AIAA 30th Aerospace Sciences Meeting, Reno, NV.

Huang, P.G. and Coakley, T.J. (1993a). In *Near-Wall Turbulent Flows* (R.M.C. So, C.G. Speziale and B.E. Launder, eds.), pp. 199–208. Elsevier Science Publishers, Amsterdam.

Huang, P.G. and Coakley, T.J. (1993b). In *Engineering Turbulence Modelling and Experiments 2* (W. Rodi and F. Martelli, eds), pp. 731–739. Elsevier Science Publishers, Amsterdam.

Huang, P.G. and Leschziner, M.A. (1985). In *Proc. Fifth Symposium on Turbulent Shear Flows,* pp. 20.7–20.12. Cornell University, Ithaca, NY.

Huang, P.G. Bradshaw, P. and Coakley, T.J. (1994). *AIAA J.* **32**, 735–740.

Johnson, D.A. (1987) *AIAA J.* **25** (2), 252–259.

Johnson, D.A. and Coakley, T.J. (1990) *AIAA J.* **28**, 2000–2003.

Johnson, D.A. and King, L.S. (1985). *AIAA J.* **23**, 1684–1692.

Johnson, R.W. and Launder, B.E. (1982). *Numer. Heat Transfer* **5**, 493–496.

Kebede, W., Launder, B.E. and Younis, B.A. (1985). In *Proc. Fifth Symposium on Turbulent Shear Flows,* pp. 16.23–16.29. Cornell University, Ithaca, NY.

Kim, J., Moin, P. and Moser, R.D. (1987). *J. Fluid Mech.* **177**, 133–186.

Kline, S.J., Morkovin, M.V., Sovran, G. and Cockrell, D.J. (eds) (1969). *Computation of Turbulent Boundary Layers – 1968 AFOSR-IFP-STANFORD Conference,* vol. 1. Thermosciences Division, Mechanical Engineering Department Stanford Univ., Stanford, CA.

Kline, S.J., Cantwell, B.J. and Lilley, G.M. (eds) (1982). *The 1980-81 AFOSR-HTTM-STANFORD on Complex Turbulent Flows: Comparison of Computation and Experiment.* Thermosciences Division, Mechanical Engineering Department Stanford Univ., Stanford, CA.

Knight, D.D. (1984). *AIAA J.* **2**, 1056–1063.

Kolmogorov, A.N. (1942). *Izv. Akad. Nauk., SSSR; Ser. Fiz.* **6**, 56–58.

Kundu, P.K. (1990). *Fluid Mechanics.* Academic Press, San Diego, CA.

Lai, Y.G. and So, R.M.C. (1990). *J. Fluid Mech.* **221**, 641–673.

Laufer, J. (1951). *NACA Report No. 1053.*

Launder, B.E. and Reynolds, W.C. (1983). *Phys. Fluids* **26**, 1157–1158.

Launder, B.E. and Shima, N. (1989). *AIAA J.* **27**, 1319–1325.

Launder, B.E. and Spalding, D.B. (1974). *Comput. Methods Appl. Mech. Engrg.* **3**, 269–289.

Launder, B.E. and Tselepidakis, D.P. (1991). *AIAA Paper 91-0219,* AIAA 29th Aerospace Sciences Meeting, Reno, NV.

Launder, B.E., Reece, G.J. and Rodi, W. (1975). *J. Fluid Mech.* **68**, 537–566.

Leonard, B.P. (1988). *Int. J. Numer. Methods Fluids* **8**, 1291–1318.

Lien, F.-S., and Leschziner, M.A. (1994). *Comput. Methods Appl. Mech. Engrg.* **114**, 123–148.

Lin, H., Yang, D.Y. and Chieng C-C (1993). *AIAA Paper 93-3316,* AIAA 11th Computational Fluid Dynamics Conference, Orlando, FL.

Lumley, J.L. (1978). In *Advances in Applied Mechanics,* vol. 18, pp. 123–176. Academic Press, New York.

MacCormack, R. W. (1969). *AIAA Paper 69-0354,* AIAA Hypervelocity Impact Conference, Cincinnati, OH.

Magnaudet, J. (1993). *Appl. Sci. Res.* **51**, 525–531.

Mellor, G.L. and Herring, H.J. (1973). *AIAA J.* **11**, 590–599.

Mohammadi, B. and Pironneau, O. (1994). *Analysis of the K–Epsilon Turbulence Model.* John Wiley & Sons, Chichester.

Morkovin, M.V. (1964). In *The Mechanics of Turbulence* (A. Favre, ed.), pp. 367–380. Gordon and Breach, New York.

Morrison, J.H. (1990). *AIAA Paper 90-5251,* AIAA Second International Aerospace Planes Conference., Orlando, FL.

Morrison, J.H. (1992). *NASA Contractor Report 4440.*

Morrison, J.H. and Gatski, T.B. (1993). *APS 46th Annual Meeting of the Division of Fluid Dynamics,* 21–23 November, Albuquerque, NM.

Morrison, J.H., Gatski, T.B., Sommer, T.P., Zhang, H.S. and So, R.M.C. (1993). In *Near-Wall Turbulent Flows* (R.M.C. So, C.G. Speziale and B.E. Launder, eds.), pp. 239–250. Elsevier Science Publishers, Amsterdam.

Myong, H.K. and Kasagi, N. (1990). *JSME Int. J., Ser. II* **33**, 63–72.

Naot, D., Shavit, A. and Wolfshtein, M. (1970). *Isr. J. Technol.* **8**, 259–269.

Narayanswami, N., Horstman, C.C. and Knight, D.D. (1993). *AIAA Paper 93-0779,* AIAA 31st Aerospace Sciences Meeting, Reno, NV.

Obi, S., Perić, M. and Scheuerer, G. (1989). *Proc. Seventh Symposium on Turbulent Shear Flows,* pp. 17.4–17.9. Stanford Univ., Stanford, CA.

Perić, M., Kessler, R. and Schuerer, G. (1988). *Computers and Fluids* **16**, 389–403.

Peyret, R. and Taylor, T.D. (1983). *Computational Methods for Fluid Flow.* Springer-Verlag, New York.

Pope, S.B. (1975). *J. Fluid Mech.* **72**, 331–340.

Pope, S.B. and Whitelaw, J.H. (1976). *J. Fluid Mech.* **73**, 9–32.

Prandtl, L. (1925). *ZAMM* **5**, 136–139.

Reynolds, W.C. (1976). *Annual. Review of Fluid Mechanics,* vol 8, pp. 183–208. Annual Reviews Inc., Palo Alto, CA.

Reynolds, W.C. (1987). *Lecture Notes for Von Karman Institute,* AGARD Lecture Series 86.

Rhie, C.M. and Chow, W. L. (1983). *AIAA J.* **21**, 1525–1532.

Ristorcelli, J.R. (1993). *ICASE Report 93-88.*

Ristorcelli, J.R., Lumley, J.L. and Abid, R. (1995). *J. Fluid Mech.* **292**, 111–152.

Rodi, W. (1976). *ZAMM* **56**, T219–T221.

Roe, P.L. (1981). *J. Comput. Phys.* **43**, 357–372.

Rubesin, M.W. (1990). *NASA Contractor Report 177556.*

Rubesin, M.W. and Rose, W. C. (1973). *NASA Technical Memorandum X-62248.*

Rubinstein, R. and Barton, J.M. (1990). *Phys. Fluids A* **2**, 1472–1476.

Saffman, P.G. (1970). *Proc. Roy. Soc., Lond.* **A317**, 417–433.

Sahu, J. and Danberg, J.E. (1986). *AIAA J.* **24**, 1744–1751.

Sarkar, S. (1992). *Phys. Fluids A* **4**, 2674–2682.

Sarkar, S, and Lakshmanan, B. (1991). *AIAA J.* **29** (5), 743–749.

Sarkar, S., Erlebacher, G., Hussaini, M.Y. and Kreiss, H.O. (1991). *J. Fluid Mech.* **227**, 473–493.

Schumann, U. (1977). *Phys. Fluids* **20**, 721–725.

Schwarz, W.R. and Bradshaw, P. (1994). *Phys. Fluids* **6**, 986–998.

Sekundov, A.N. (1971). *Fluid Dynamics* **6**, 828–840.

Shih, T.H. and Lumley, J.L. (1985). *Cörnell University Technical Report No. FDA-85-3.*

Shih, T.I.-P. and Chyu, W.J. (1991). *AIAA J.* **29**, 1759–1760.

Shima, N. (1988). *ASME J. Fluids Engrg.* **110**, 38–44.

Shur, M., Strelets, M., Zaikov, L., Gulyaev, A., Kozlov, V. and Secundov, A. (1995). *AIAA Paper 95-0863,* AIAA 33rd Aerospace Sciences Meeting, Reno, NV.

So, R.M.C., Zhang, H.S. and Speziale, C.G. (1991). *AIAA J.* **29**, 2069–2076.

So, R.M.C., Zhang, H.S., Gatski, T.B. and Speziale, C.G. (1994). *AIAA J.* **32**, 2162–2168

Sommer, T.P., So, R.M.C. and Lai, Y.G. (1992). *Int. J. Heat Mass Transfer* **35**, 3375–3387.

Sommer, T.P., So, R.M.C. and Zhang, H.S. (1993). *AIAA J.* **31**, 27–35.

Spalart, P.R. and Allmaras, S.R. (1994). *La Recherche Aérospatiale* **1994–1**, 5–21.

Speziale, C.G. (1987). *J. Fluid Mech.* **178**, 459–475.

Speziale, C.G. (1989). *Theor. Comput. Fluid Dyn.* **1**, 3–19.

Speziale, C.G. (1991). *Annual Review of Fluid Mechanics,* vol. 23, pp. 107–157, Annual Reviews Inc, Palo Alto, CA.

Speziale, C.G. and Sarkar, S. (1991). *AIAA Paper 91-0217,* AIAA 29th Aerospace Sciences Meeting, Reno, NV.

Speziale, C.G., Gatski, T.B. and Mac Giolla Mhuiris, N. (1990). *Phys. Fluids A* **2**, 1678–1684.

Speziale, C.G., Sarkar, S. and Gatski, T. B. (1991). *J. Fluid Mech.* **227**, 245–272.

Speziale, C.G., Abid, R. and Anderson, E. C. (1992a). *AIAA J.* **30**, 324–331.

Speziale, C.G., Gatski, T. B. and Sarkar, S. (1992b). *Phys. Fluids A* **4**, 2887–2899.

Speziale, C.G., Abid, R. and Durbin, P.A. (1994) *J. Sci. Comput.* **9**, 369–403.

Speziale, C.G., Abid, R. and Mansour, N. N. (1995). *ZAMP* (to be published).

Swanson, R.C. and Turkel, E. (1985). *AIAA Paper 85-0035* AIAA 24th Aerospace Sciences Meeting, Reno, NV.

Taulbee, D.B. and VanOsdol, J. (1991). *AIAA Paper 91-0524,* AIAA 29th Aerospace Sciences Meeting, Reno, NV.

Taulbee, D.B., Sonnenmeier, J.R. and Wall, K.M. (1994). *Phys. Fluids A* **6**, 1399–1401.

Tavoularis, S. and Karnik, U. (1989). *J. Fluid Mech.* **204**, 457–478.

between the boundaries. Furthermore, variations in the grid-line spacing can be used advantageously to concentrate points in regions of expected high gradients, as shown in Figure 3. In a structured mesh, the physical location of the mesh points must be stored, but the identity of neighbouring mesh points is still known implicitly. Structured meshes are the method of choice for relatively-simple but not trivial geometries, such as airfoil sections in two dimensions, or wing configurations in three dimensions. For more complex geometries, it often becomes impossible to create a mapping in which the grid simultaneously conforms to all geometry boundaries. Therefore, an overset grid or a block-structured approach is generally adopted. In the overset grid approach, multiple overlapping grids are used to discretize the domain, and the flow solver is required to interpolate values between the various grids in the regions of overlap. In block-structured techniques, the domain is decomposed into a small number of topologically simpler domains, and each domain is meshed

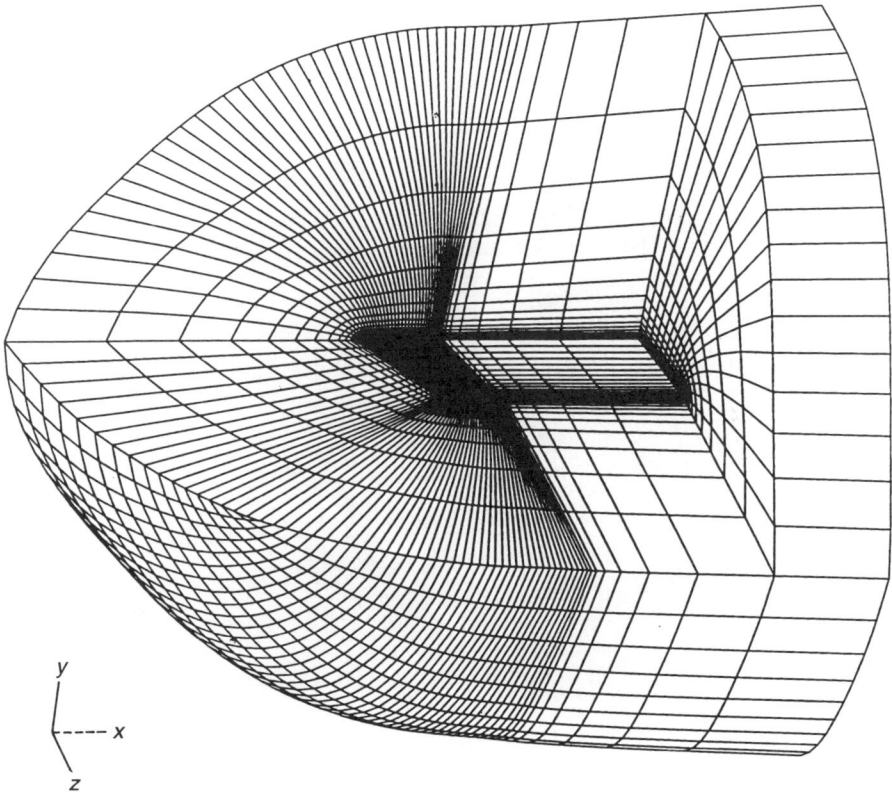

Figure 3 Example of a single block structured mesh about a wing configuration. The C–O topology of the mesh permits concentration of points at the wing leading edge and tip. (Reproduced from Vatsa and Wedan (1989).)

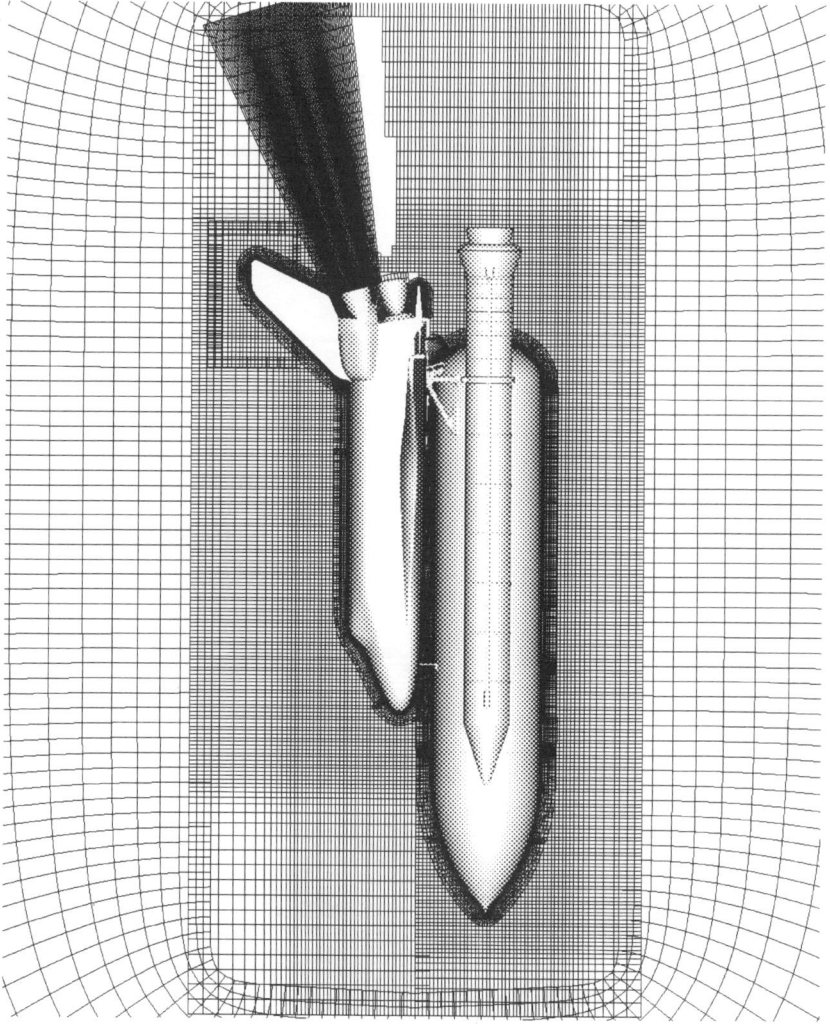

Figure 4 Example of overset grid method for a complex geometry. (Reproduced from Gomez and Ma (1994).)

independently with a structured grid. The block-structured grid technique is probably the most commonly employed technique for computational fluid dynamics problems. Examples of overset and block-structured grids are given in Figures 4 and 5.

An alternative approach which has gained in popularity over the last decade is the unstructured mesh approach. This approach consists of dividing the entire domain into a large set of geometrically simple elements such as tetrahedra, prisms or hexahedra. In contrast to the structured mesh approach, these elements are not ordered in a regular fashion, but they do conform exactly to

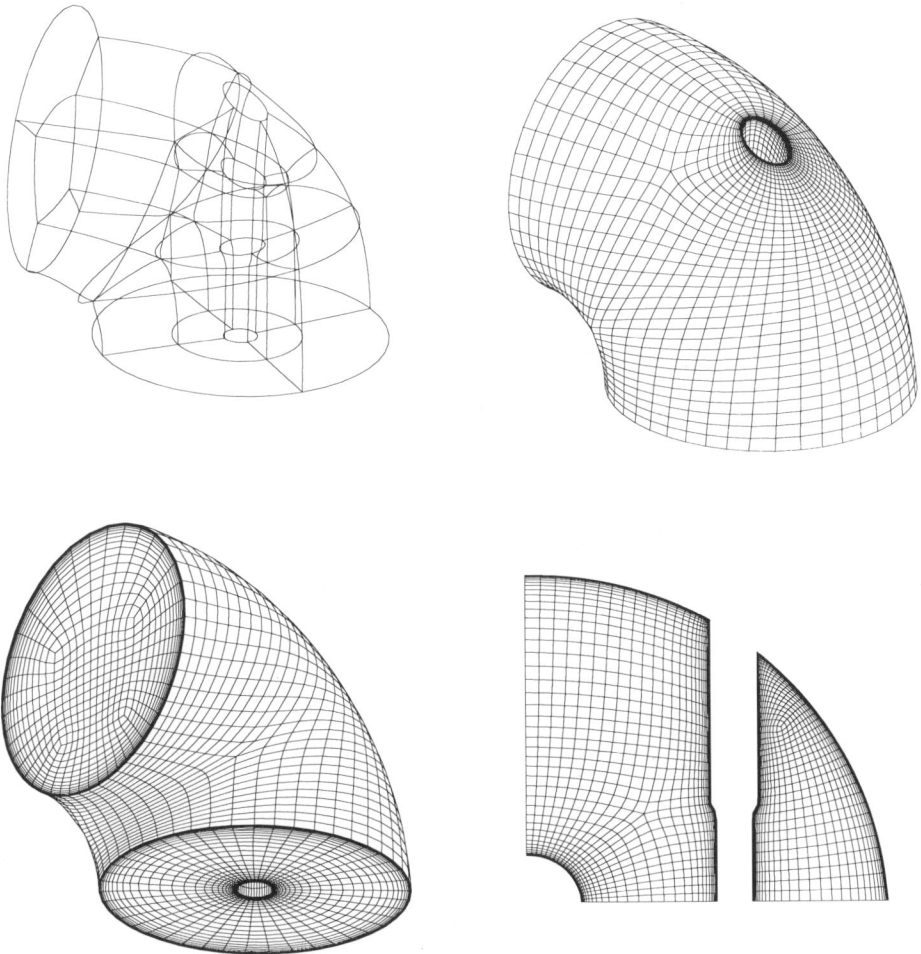

Figure 5 Example of a block structured grid for a complex geometry. (Reproduced from Eiseman *et al.* (1994).)

the boundaries of the domain. This may also be regarded as an extreme case of the block structuring approach, where the blocks become small enough that the local structured meshes are no longer required (i.e. micro-unstructuring versus macro-unstructuring). For unstructured meshes, the physical locations of the grid points must be stored, as well as the identity of the neighbouring grid points for each vertex, or the connectivity of the mesh, since there is no regular distribution pattern to which the grid points conform. Unstructured grid techniques have proven very flexible for discretizing complex geometries. Also, since no implicit connectivity is assumed, adaptive meshing techniques can be implemented in a straightforward manner by inserting and deleting mesh points and locally recomputing the mesh connectivity. An example of an unstructured mesh about a complex three-dimensional configuration is given in Figure 6. The drawbacks of unstructured meshes centre mostly around the increased computational resources required due to the explicit nature of the mesh connectivity.

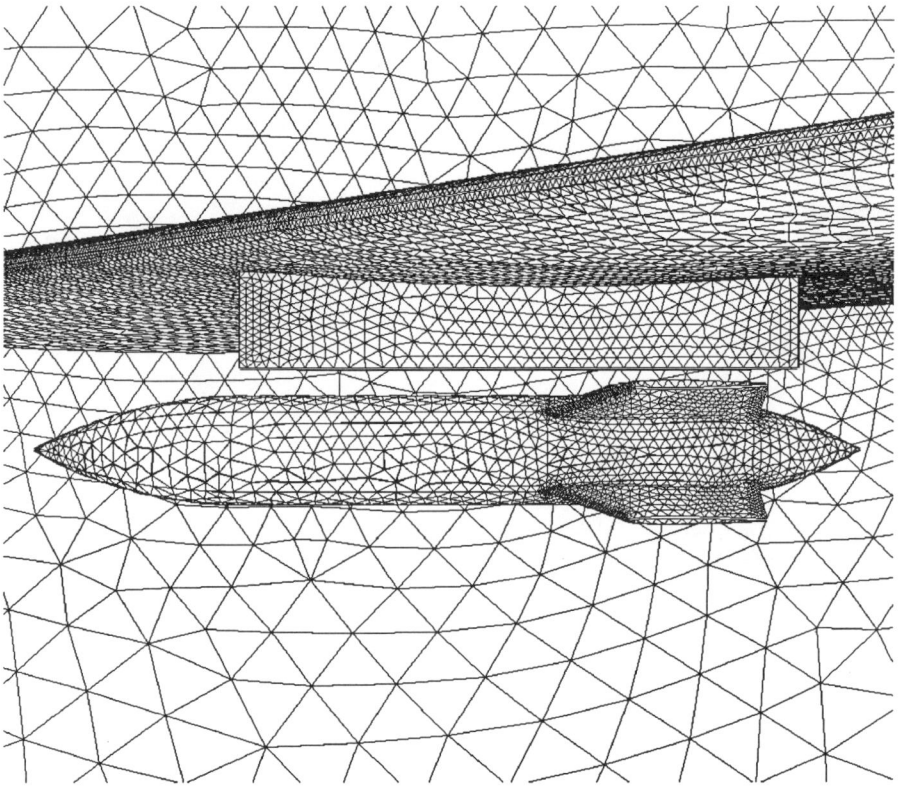

Figure 6 Example of a tetrahedral unstructured grid for a complex three-dimensional geometry. (Reproduced from Parikh *et al.* (1994).)

Finally, the so-called 'gridless' approach goes one step further by dropping any required mesh connectivity altogether (Batina, 1993). The 'grid' now consists of a set of disconnected points distributed throughout the flow field. As previously, higher concentrations of vertices in regions of large flow gradients are desirable, but the point distributions are no longer constrained by the need to form non-overlapping tetrahedra or hexahedra. This represents the ultimate in flexibility for space discretization. However, the burden of determining the stencil for the discrete equations is shifted entirely to the numerical scheme. Numerical techniques such as multivariate interpolation are capable of operating on randomly scattered vertices. However, the efficiency of such methods is inadequate for computational fluid dynamics problems. In order to construct a feasible numerical scheme, a set of neighbouring points for each vertex must be defined. Least-squares type techniques can then be employed, for example, to estimate gradients at these vertices, using the defined set of neighbouring points. Thus, although the mesh does not contain any connectivity information, the numerical algorithm is required to construct neighbour subset information for efficiency reasons.

As can be seen from the above discussion, these various different strategies achieve simplification or increased flexibility in one aspect of the mesh generation–flow solution strategy at the expense of another aspect in the solution procedure. The structured and unstructured mesh approaches outlined above are by far the most commonly employed strategies for computational fluid dynamics problems, and will be considered almost exclusively in the remainder of this chapter.

Unfortunately, structured and unstructured meshes are often classified by element-type (i.e. quadrilateral and triangular meshes in two dimensions or hexahedral and tetrahedral meshes in three dimensions). In fact, the type of element is not related to the structure of the grid. It is possible to have hexahedral unstructured grids, or tetrahedral structured grids. Other types of elements such as prisms, pyramids, and even dodecahedra, are also possible. Combinations of various types of elements, particularly for unstructured grids, have often been advocated.

Similarly, finite-difference or finite-volume methods are often associated with structured grids, while finite-element methods are associated with unstructured meshes. While the unknowns in a finite-volume method typically represent cell averages, the solution of a finite-element method is represented as a piecewise varying function which, for second-order methods, varies linearly over the elements of the mesh. In practice, the particular type of mesh employed does not dictate the use of a particular discretization scheme. Finite-difference, finite-volume and finite-element methods may be devised for use on structured or unstructured grids. In fact, such schemes merely refer to the mathematical theory employed for deriving the discrete equations from the continuous system on a given grid.

For finite-volume formulations, a distinction between vertex-based and cell-centred schemes can be made. In a vertex-based scheme, the flow variables are stored at the vertices of the mesh. The discrete equations usually involve stencils which are defined by the nearest or next to nearest neighbours of each mesh point. In a cell-centred scheme, the flow variables are stored at the centres of the cells or elements of the mesh. The stencils of the discrete equations usually involve the nearest or next to nearest neighbouring cells. The equivalence between vertex and cell-centred schemes can be demonstrated with the concept of a dual mesh. If a dual mesh point is created at each cell centre, and dual mesh edges are drawn by joining neighbouring cells, the cell-centred scheme can be seen to be equivalent to a vertex-based scheme operating on the dual mesh. Such dual meshes can be constructed for structured and unstructured grids, using various types of elements.

Figure 7 illustrates the dual mesh for a mixed quadrilateral and triangular mesh. For a purely quadrilateral mesh, the dual mesh also contains quadrilateral elements, and the number of vertices and edges in the original and dual meshes is identical. Thus, the cell-centred and vertex-based schemes are nearly identical for quadrilateral and hexahedral meshes. The main differences occur at boundaries. In the vertex scheme, flow variables on the

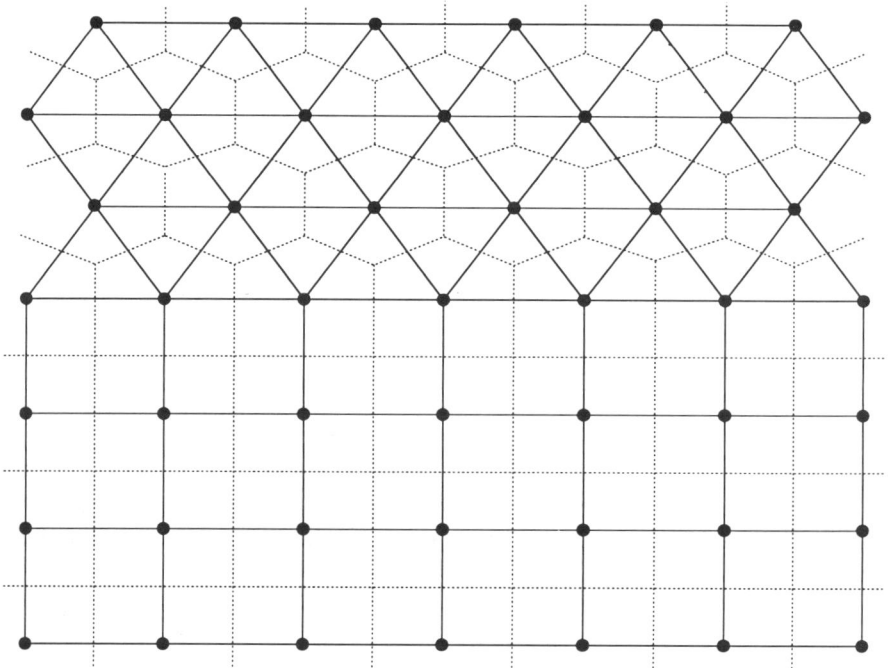

Figure 7 Original and dual mesh for a mixed triangle-quadrilateral unstructured mesh in two-dimensions.

boundaries must be computed, whereas in the cell-centred scheme, only boundary fluxes are required. Depending on the type of boundary condition to be implemented (e.g. Dirichlet or Neumann), one scheme may be more suitable than the other.

For triangular or tetrahedral meshes, the number of dual mesh vertices is larger than the number of original mesh vertices. This is due to the fact that a triangular mesh contains twice as many triangles as vertices (neglecting boundary effects), and a tetrahedral mesh five to six times more elements than vertices. On the dual mesh, the degree of a vertex (number of incoming edges) is fixed and equal to 3 for triangular elements, or 4 for tetrahedral elements, whereas on the original mesh, the degree of each vertex is variable. Similar relationships hold for other elements such as prisms and pyramids. Thus, cell-centred and vertex-based schemes operating on the same grid result in vastly different discretizations when non-quadrilateral or non-hexahedral elements are present.

In particular, a cell-centred scheme can be expected to result in a much larger number of unknowns while generating relatively simple stencils of fixed size. The vertex scheme on the other hand, will result in a smaller number of unknowns with larger variable-size stencils. The choice of a particular scheme depends on the relative balance between delivered accuracy and computational cost.

A Numerical implications

It is useful to examine the numerical implications of the choice of a particular meshing strategy. In the case of the regular cartesian mesh, the vertices are all equidistant and the identity of neighbouring vertices is implicitly known. Thus, the matrix of the discrete operator which arises from the discretization of a partial differential equation on such a mesh is a banded matrix where, for a regular discretization of a constant coefficient equation, all off-diagonal entries are equal. For a structured body-fitted grid, the resulting matrix is also banded, but off-diagonal terms now have variable magnitude which depends on the local mesh spacing. For an unstructured mesh, the resulting operator matrix is no longer banded. A true sparse matrix with an arbitrary pattern of non-zeros is obtained. This pattern depends on the numbering scheme used to identify the vertices, which can in principle be random. In the gridless approach, the resulting matrix is similar to the unstructured mesh case, although the pattern of the sparse matrix is not determined by the grid, but by the phase of the algorithm where neighbouring point-sets are determined.

For the cartesian and structured mesh cases, since the matrix structure is known at the outset, simple storage schemes for the non-zero matrix entries are adequate. Furthermore, banded matrix solution techniques such as tridiagonal, pentadiagonal or alternating-direction-implicit schemes may be employed to solve efficiently the discrete equations.

For unstructured meshes, the pattern of non-zeros as well as their values must be stored. The pattern of non-zeros is related to the connectivity of the mesh. For example, it can be shown that for a Galerkin finite-element discretization of Laplace's equation on tetrahedral elements, each off-diagonal non-zero element in the sparse matrix corresponds to an edge of the mesh. In sparse matrix terminology, the graph of a matrix is defined as the graph obtained by drawing a line joining the column and row number corresponding to each non-zero entry in the matrix (Duff *et al.*, 1986). Thus, for the Galerkin approximation of a laplacian, the graph of the matrix is identical to the graph of the grid. Banded matrix solution techniques can no longer be used to solve the discrete equations arising from unstructured mesh discretizations. Instead, sparse matrix techniques must be implemented either for direct inversion of the resulting matrices (Duff *et al.*, 1986), or iterative sparse matrix techniques such as incomplete LU factorization (Venkatakrishnan and Mavriplis, 1993; Whitaker, 1993).

An alternative to the above matrix inversion techniques is a grid-based solution strategy such as multigrid (McCormick, 1987). In multigrid methods, multiple coarse grid levels are employed to accelerate the convergence of the solution on the finest grid. This strategy can be applied equally well to structured and unstructured grids. Although a matrix interpretation of any multigrid strategy can be developed, these methods are often best formulated and understood as grid-based approaches. Therefore, this represents a situation in which the mesh generation operation contributes not only to the discretization of the continuous equations, but also to the solution process.

A similar expanded role of the mesh generation process can be found in adaptive meshing strategies. In this case, the mesh is modified, by moving points, adding points, deleting points, and/or changing connectivity, as the solution evolves. Adaptive meshing can be employed in conjunction with any type of mesh, cartesian, structured, unstructured, or gridless. For steady-state problems, several isolated adaptation phases are usually sufficient to provide the desired level of final accuracy, whereas for transient problems, the mesh must be continuously modified every few time steps.

Adaptive meshing and multigrid techniques are examples of comprehensive strategies for simulating physical problems in which the grid generation and numerical solution phases are closely coupled and interdependent. This philosophy is taken even further by methods such as the full adaptive composite scheme (FAC) (McCormick and Thomas, 1986), which combines adaptive meshing, multigridding, and numerical solutions into a unified approach.

The remainder of this chapter describes the various approaches to mesh generation in more detail. Section II begins with a look at three-dimensional geometry definition techniques. Geometry definition invariably constitutes the initial phase of any three-dimensional grid generation technique. In Sections III and IV, we examine the various volume grid generation techniques. Only structured and unstructured mesh techniques are considered, since they are by far the most commonly employed techniques. In each case, mesh adaptivity is also discussed. In Section V, we take a closer look at the interplay between grid generation and solution techniques, and finally, draw overall conclusions.

This chapter is not meant to be an all encompassing survey of all available mesh generation techniques. Rather, after having surveyed and classified the major types of approaches to the problem, the chapter concentrates in somewhat more detail on the most common techniques in use today, usually providing a few representative references for each major approach.

II GEOMETRY MODELLING AND SURFACE DEFINITION

Prior to generating a mesh and simulating the flow in a given domain, the geometry of the configuration must be defined in an unambiguous and recognizable manner. For many two-dimensional and simple three-dimensional geometries, this can easily be performed by a number of simple techniques (e.g. analytic description of simple geometries). However, for more complex configurations, particularly in three dimensions, the geometry definition phase can prove to be complicated, time-consuming and error-prone. The fact that geometry modelling can represent the largest bottleneck in many present-day numerical simulations has perhaps not been fully appreciated in the research community. While many of the difficulties are related to a lack of standard representation formats for data interchange, there exist fundamental approaches to the problem that can greatly alleviate ambiguity and improve fidelity and robustness.

Many early numerical simulations employed geometry definitions based on point sets. Multiple two-dimensional cross-sections of the geometry could easily be obtained either from blueprints or computer aided design (CAD) software packages. These cross-sections could then be discretized as a set of ordered points in two dimensions. The union of all the points thus generated forms a discretization of the surface. These points could be used directly as the surface grid points for generating a volume grid. A more practical alternative is to fit a spline or piecewise spline curves or surfaces to these points, and obtain the actual grid points by interpolation along the splined curves and surfaces. In today's era of complex geometry, high-resolution numerical simulations, point-set based geometry definition is no longer acceptable. Such techniques can be both inaccurate and ambiguous. Preservation of geometry details such as small gaps and slope or curvature continuity are necessary.

The most commonly employed approach for geometry definition in numerical simulations is known as a boundary representation or b-rep. In this method, the surface of the geometry is defined by a set of curved surface patches which cover the entire geometry surface. These patches are most often three- or four-sided (topologically triangular or quadrilateral). Their shapes are mathematically defined, and represent the two-dimensional equivalents of one-dimensional spline curves. Many types of surface patches exist, mostly differentiated by the mathematical equations which define them. A comprehensive description of these can be found in the book by Farin (1990). The non-uniform rational Bezier surface (NURB) is increasingly becoming the surface type of choice for many numerical simulations. As the name implies, these surfaces are defined by piecewise rational polynomials of high order. They are capable of accurately describing surface features such as curvature and slope. In addition, their rational construction permits exact representation of simple but common types of geometries, such as spheres and conics (Farin, 1990).

In many instances, a complete description of a complex geometry using only three- or four-sided patches becomes overly burdensome. This is often due to the fact that many geometry features are created by the intersection of basic components (such as the wing and fuselage of an aircraft), and as such are irregular and difficult to model accurately. This leads to the notion of 'trimmed'

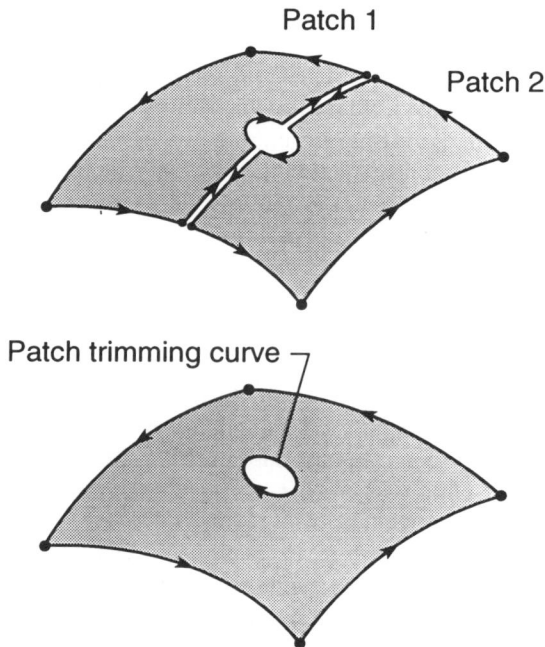

Figure 8 Example of the approximation of a surface with hole by two simple surfaces (top) and by a single trimmed surface (bottom).

surfaces. A trimmed surface consists of a regular (possibly NURB) surface patch with superimposed curves which delimit regions of the surface which are not part of the geometry. Consider the simple case of a surface with a hole drilled in its centre (cf. Figure 8). Using regular three- or four-sided patches, it is not possible to represent this topology without using multiple patches, or making a cut through the domain. Along any such cuts or boundaries of neighbouring patches, surface quantities such as slope and curvature must remain continuous. With a trimmed surface, on the other hand, it is a simple matter to represent such a geometry. First the 'undrilled' surface is modelled with a single patch, and then the hole is modelled by placing a circular trimming curve in the centre of the surface.

There are two possible approaches for generating surface grids on the geometry patches. The grid may either be generated in physical space, or in parametric space (more common). If the physical space approach is chosen, grid point coordinates which coincide with the actual surface must be determined from the defining equations for the surface. If the parametric space approach is adopted, the surface grid generation problem reduces to a two-dimensional meshing problem in a parametric space, and is thus greatly simplified. Once the parametric surface mesh is generated, it must be mapped back to physical space. If an irregular parametrization has been employed, or the surface is extremely warped, invalid physical surface meshes may result. Since each patch is generally parametrized individually, this approach preserves patch boundaries as surface grid lines. In cases where many patches are employed to model a given surface, this may restrict the flexibility of the surface grid. Thus, a common practice is to form global patches, or 'quilted' patches with new parametrizations, solely for the grid generation process.

Often, in the surface grid generation phase, it is required to modify or 'repair' the surfaces. Boundary representations obtained from CAD software packages often contain gaps, discontinuities, or small overlaps between neighbouring surface patches. Most CAD packages were designed with other end uses in mind (such as manufacturing), and the delivered accuracy is not always sufficient for surface grid generation. Thus, many grid generation codes contain facilities for modifying the input geometry. An alternative solution to this problem is afforded by the use of a projection surface (Akdag and Wulf, 1992). In this approach, the surface grid is constructed on a projection surface which is then placed over the collection of surface patches, which defines the actual geometry. Another benefit of this approach is the ability to perform small changes to the surface geometry without the need to regenerate the grid.

The increasing use of adaptive meshing techniques in numerical simulations serves to emphasize the importance of geometrical modelling. When a body fitted mesh is adaptively refined during the flow solution phase, the new adaptively generated boundary points must be placed on the original geometry surface patch. Similarly, when mesh redistribution is employed, surface mesh points must be constrained to slide along the geometric surface. This requires that the parametric coordinates of all boundary points be preserved throughout the flow solution phase, as well as the mathematical description of each and every surface patch.

A completely different approach to geometry definition can be found in solid modelling techniques. Solid modelling actually provides a description of an entire volumetric region, rather than just the surface of a configuration, as in the b-rep case. Solid modelling is based on boolean operations of primitive objects. For example, in order to describe a cube with a hole drilled through it, we would instance a cube and a cylinder, and take the boolean sum of the cube and the negative of the cylinder. The boundary representation of this same object would require the use of six flat surfaces, of which two must be trimmed, and at least one curved surface. Solid modelling results in a hierarchical data-structure description of the geometry, which often parallels the manufacturing process (but may be non-unique). In principle, a solid model can be converted to a boundary representation (b-rep), however this operation can be complex and expensive. Solid modelling techniques are not often used in computational fluid dynamics using body-fitted grids. However, they have been demonstrated successfully for cartesian grid or voxel methods (Tsuboi et al., 1991).

III STRUCTURED GRIDS

The use of structured grids constitutes the most commonly employed mesh generation strategy for computational fluid dynamics problems. The implicit grid structure alleviates the need to store the mesh connectivity, and permits the construction of rapid solution algorithms, which may make use of banded matrix solvers. However, single block-structured grids are generally only applicable to relatively simple geometries, although occasional successes with complex geometries have been reported (Baker, 1991). For complex configurations, multiple overset or block-structured grids must be employed. Block-structured or overset-structured grid generation has developed into a relatively sophisticated discipline, and its applications now lie firmly in the domain of large public or commercially supported software packages (Benek et al., 1985; Thompson, 1988; Steinbrenner et al., 1990; Boerstoel and Spekreyse, 1991; Fujitani and Himeno, 1991; Dener and Hirsch, 1992; Sorensen and McCann, 1992). The aim of this section is not to examine the various available codes for complex geometry grid generation, but rather to discuss the fundamental approaches to structured grid generation and to illustrate the advantages and drawbacks of the various methods.

First the basic methods for generating single-block structured meshes are considered. Next adaptive meshing techniques are examined, and finally the issues involved in block-structured and overset grids for complex geometries are discussed.

A Algebraic methods

The mathematical formulation of structured grid generation consists of determining the value of the functions

$$x = x(\xi, \eta)$$
$$y = y(\xi, \eta)$$

$$(1)$$

at the discrete locations $\xi = 1, 2, \ldots, N$ and $\eta = 1, 2, \ldots, M$. The discrete values of the couple (ξ, η) represent individual grid points, and the curvilinear lines defined by $\xi =$ constant and $\eta =$ constant correspond to structured gridlines as shown in Figure 9, using the two-dimensional case for simplicity. For body-fitted grids, by definition, the extremum values $\xi = 1$, $\xi = N$ and $\eta = 1$, $\eta = M$ coincide with the physical boundaries. The solution of equations (1) amounts to the specification of the physical coordinates for each mesh point. Prescribing a distribution of mesh points along the boundary amounts to specifying the functions at locations

$$x = x(1, \eta), \quad x = x(N, \eta)$$
$$y = y(1, \eta), \quad y = y(N, \eta)$$

$$(2)$$

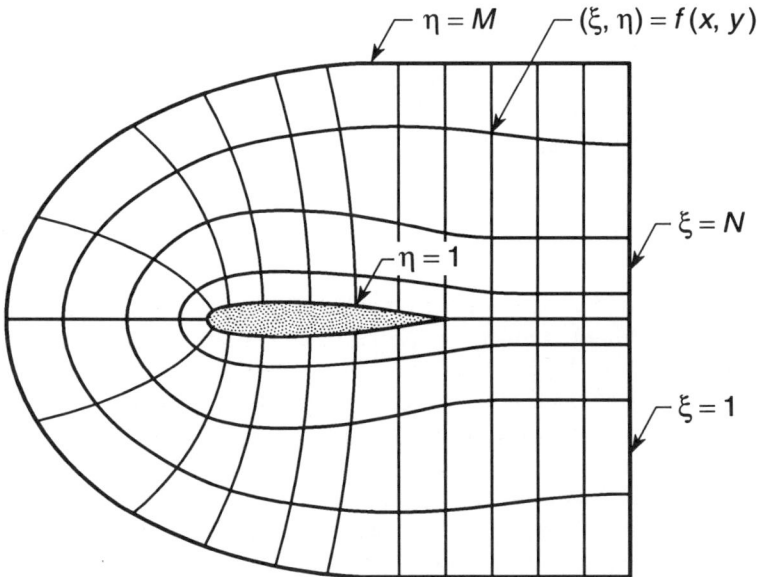

Figure 9 Parametrization for two dimensional structured grid.

with similar expressions for the $\eta = 1$, M boundaries. A simple and effective method for determining functional values at interior points is to interpolate the values given on the boundaries into the interior. Since blending of the two coordinate directions must be achieved while still observing the prescribed boundary values, transfinite interpolation is required (Thompson *et al.*, 1985). Much work has been devoted to using different polynomial basis functions or spline representations for the interpolation, as well as alternate parametric representations for (ξ, η) (normalized boundary curve arc length, for example; Soni, 1992). While these methods are simple and inexpensive, methods based on partial differential equations generally yield better control over grid smoothness, especially for more complicated topologies.

B Elliptic mesh generation

Because certain elliptic systems such as Laplace's equation satisfy a maximum principle, the set of equations

$$\xi_{xx} + \xi_{yy} = 0$$
$$\eta_{xx} + \eta_{yy} = 0 \tag{3}$$

can be used to generate a valid structured mesh. Since the $\xi = 1$, $\xi = N$ and $\eta = 1$, $\eta = M$ lines represent curvilinear grid lines which coincide with the physical boundaries of the domain, a monotonic specification of $\xi(x, y)$ and $\eta(x, y)$ along these boundaries (which is required to define a unique boundary point distribution) results in a monotonic variation of $\xi(x, y)$ and $\eta(x, y)$ in the interior of the domain, upon solving equations (3), owing to the maximum principle. A monotonic variation of ξ and η is sufficient to guarantee a unique and valid structured grid with no cross-overs. Numerically, we wish to determine the values of the functions $x = x(\xi, \eta)$ and $y = y(\xi, \eta)$, thus we must solve (3) in computational space rather than physical space. In computational space, equations (3) can be shown to be equivalent to the set of equations (4):

$$\alpha x_{\xi\xi} - 2\beta x_{\xi\eta} + \gamma x_{\eta\eta} = 0$$
$$\alpha y_{\xi\xi} - 2\beta y_{\xi\eta} + \gamma y_{\eta\eta} = 0 \tag{4}$$

with

$$\alpha = x_\eta^2 + y_\eta^2 \quad \beta = x_\xi x_\eta + y_\xi y_\eta \quad \gamma = x_\xi^2 + y_\xi^2$$

Although these equations are non-linear, they are still elliptic and readily solved by standard iterative techniques. The use of Laplace's equation for elliptic mesh generation tends to equidistribute the mesh points in the flow field and provides little control over the mesh point distribution in the interior of the domain. This situation may be rectified by the addition of source terms, thus yielding the following equations

$$\xi_{xx} + \xi_{yy} = P(\xi, \eta)$$
$$\eta_{xx} + \eta_{yy} = Q(\xi, \eta)$$

(5)

The source terms can be used to cluster grid points near boundaries and in regions of interest in the interior of the domain. Much effort has been devoted to devising different source term formulations (Steger and Sorensen, 1979; Thomas and Middlecoff, 1979; Hilgenstock, 1988). Most of those currently in use are evaluated from an initial algebraically generated grid or interpolated from boundary point distributions. Another approach is to solve a set of Poisson equations for the source terms themselves, thus effectively using a biharmonic formulation for generation of the mesh (Sparis, 1985).

Elliptic methods have been arguably the most successful mesh generation algorithms, and are certainly the most prevalent in large industrial codes. Their main drawback is perhaps the amount of computational effort required to solve the elliptic equations.

C Hyperbolic mesh generation

Hyperbolic methods offer a much less expensive alternative for generating structured meshes for problems where the outer boundary need not be fully specified, such as external aerodynamics problems.

In two dimensions (for simplicity) the governing grid generation equations can be chosen as partial differential constraints of orthogonality and mesh cell volume (Steger and Chaussee, 1980):

$$x_\xi x_\eta + y_\xi y_\eta = 0$$
$$x_\xi x_\eta - y_\xi x_\eta = \Delta V$$

(6)

where ΔV is the area of a given mesh cell and must be specified throughout the flow field. Alternatively, the constraints of orthogonality and arc length Δs may be employed:

$$x_\xi x_\eta + y_\xi y_\eta = 0$$
$$x_\xi^2 + y_\xi^2 + x_\eta^2 + y_\eta^2 = (\Delta s)^2$$

(7)

where Δs must now be specified throughout the domain. It can be shown that both of the above formulations are hyperbolic in nature. Hence, if initial data is specified (i.e. a boundary point distribution along $\eta = 0$) the equations may be marched out in the η direction, thus generating a structured mesh. Hyperbolic mesh generators are one to two orders of magnitude faster than elliptic generators, and produce orthogonal or nearly orthogonal grids. Their main drawback is of course the inability to match prescribed point distributions on all boundaries.

Finally, it should be mentioned that hybrid elliptic–hyperbolic methods are possible (Spardling *et al.*, 1991) as well as parabolic methods (Nakamura and Suzuki, 1987), although those are not discussed further in this chapter.

D Adaptive meshing

Mesh-point movement or mesh redistribution has been the most commonly used form of mesh adaptivity for structured meshes. Such methods are relatively easily implemented, since they preserve the structure and connectivity of the mesh, and simply involve recomputing the grid metrics in the flow solver at each adaptation cycle. Almost all mesh-redistribution schemes are based on a variational principle, although often these principles are derived by intuitive reasoning.

One such approach is the physically based idea of linking mesh points with springs whose stiffnesses depend on the quantity (or error) to be minimized (Gnoffo, 1983; Nakahashi and Deiwert, 1985). Other approaches consider variational formulations directly to ensure equidistribution (of the product of the target function with the mesh spacing) throughout the mesh (Jacquotte and Cabello, 1990; Warsi and Thompson, 1990). Alternatively, the source terms or control functions of the elliptic mesh generation equations (5) may be modified to produce grid clustering based on solution gradients or truncation errors (Kim and Thompson, 1990). Since all of these approaches are variational in nature, minimization functionals may be constructed which contain various solution-based criteria (e.g. truncation error, gradients) as well as grid-quality criteria (orthogonality, mesh smoothness, etc.) simultaneously.

There is, however, a limit to the amount of improvement that can be obtained by mesh redistribution, and at some point it becomes necessary to refine the mesh by adding extra vertices, thus upsetting the regular structure of the mesh. This approach to adaptivity involves considerably more complications and has therefore not been employed as often. There are two main approaches to structured grid refinement. The first approach attempts to define regular shaped regions which are then uniformly refined. This results in a block-structured type data-structure (Berger and Jameson, 1984; Quirk, 1994). The other approach only refines where necessary, and treats the resulting meshes as unstructured datasets (Davis and Dannehoffer, 1991; Aftosmis, 1993). The trade-offs between the two methods are evident: locally structured datasets for simple and efficient solver implementations but suboptimal mesh enrichment in the first case, and near optimal mesh enrichment at the expense of unstructured-mesh type overheads in the second case.

Finally, several studies have demonstrated the advantages of performing both mesh enrichment and mesh redistribution simultaneously, particularly for viscous flows (Dannehoffer, 1991).

E Overset grids and blocking techniques

For complex three-dimensional geometries, it becomes increasingly difficult to discretize the domain with a single structured grid. Thus, more sophisticated techniques must be pursued. One possible approach is known as the 'Chimera' or overset grid scheme (Benek *et al.*, 1985). In this method, the region of the domain in the proximity of each geometry component is discretized with a local structured grid. This results in multiple local grids which overlap each other, the union of which covers the entire domain. The flow equations are discretized on each individual mesh, and communication between the various grids in the flow solver is performed by inter-grid interpolation in the regions of overlap. The advantages of this approach are the ability to grid any complex geometry easily, while the drawbacks include complexities in the flow solver, as well as issues concerning numerical conservation (cf. Figure 4).

By far the most prevalent technique for discretizing complex configurations is the block-structured technique. As the name implies, this technique consists of partitioning a geometrically complex domain into a number of (topologically) simpler regions which can then be individually discretized with a structured mesh. The basic properties of structured mesh solvers are retained, although extra logic must be added to treat the inter-block boundaries. Block-structured mesh generation forms the basis for many large commercial software packages (Thompson, 1988; Steinbrenner *et al.*, 1990; Boerstoel and Spekreyse, 1991; Dener and Hirsch, 1992; Sorensen and McCann, 1992) .Such codes incorporate many options, such as interfaces with different CAD and geometry definition packages, various techniques for structured mesh generation, different blocking techniques, and varied user interfaces. Some of the issues in block-structured mesh generation involve the type of inter-block communication allowed (i.e. neighbouring blocks may be required to share the same point distribution on their common face, to have nested point distributions of differing resolution, or to have arbitrary point distributions).

In most applications, the blocking of the domain is performed manually through a graphical user interface (Steinbrenner *et al.*, 1990). Although this can be extremely time consuming, many industrial applications involve repeated simulations of nearly similar geometries. Thus, new configurations can often be generated quickly by locally modifying one or several of the blocks.

A more recent and interesting development from a theoretical standpoint has been the development of automatic blocking techniques. Most automatic blocking techniques require some degree of user input, but simplify the task by representing the physical geometry in simplified form in a topology plane (Allwright, 1988; Cordova, 1992; Eiseman *et al.*, 1994). Automatic blocking generally results in a larger number of smaller blocks, but this does not appear to pose a problem for current flow solvers.

IV UNSTRUCTURED MESH GENERATION

An alternative to the complications inherent in the overset mesh and block-structuring techniques is the use of unstructured meshes. While block-structured techniques attempt to preserve the local mesh structure using a large-scale partitioning of the domain, unstructured mesh techniques obtain increased flexibility by removing any requirements of structure in the mesh. Thus, mesh points may be numbered in any order, even randomly, and any vertex may have an arbitrary number of neighbours. Since the mesh connectivity is not implicitly known, it must be maintained explicitly in storage. While this connectivity may be arbitrary, it is however homogeneous. This means that a single data-structure can be employed to describe the connectivity over the entire mesh. This is not the case for block-structured techniques, where differentiation between intra- and inter-block connectivity is required. This homogeneity greatly facilitates the implementation of adaptive meshing techniques, as well as certain aspects of vectorization and parallelization. Unstructured meshes and their associated algorithms are much more closely tied to computational geometry and computer science principles, whereas structured grids are more closely related to the field of partial differential equations. This may offer a partial explanation for the slow acceptance of unstructured grid methods into the field of computational fluid dynamics.

In this section, the two most prevalent unstructured mesh generation approaches are examined: the advancing front method and the Delaunay triangulation method. The former is somewhat heuristic in nature, while the latter is firmly based in computational geometry principles. In practice, however, both methods have resulted in successful three-dimensional grid generation tools for truly arbitrary geometries. Both approaches have also demonstrated robustness problems and grid-quality issues which have led to the development of more sophisticated methods. These alternative triangulation approaches are examined briefly, followed by a discussion of adaptive meshing techniques for unstructured meshes.

A Advancing front methods

Advancing front techniques (Lo, 1985; Peraire *et al.*, 1987; Gumbert *et al.*, 1989) begin with a discretization of the geometry boundaries as a set of edges in two dimensions. These edges form the initial front which is to be advanced out into the field. A particular edge of this front is selected, and a new triangle is formed with this edge as its base, by joining the two ends of the current edge either to a newly created point, or to an existing point on the front. The current edge is then removed from the front, since it is now obscured by the new triangle. Similarly, the remaining two edges of the new triangle are either assigned to the front or removed from the front, depending on their visibility, as shown in Figure 10. The front thus constitutes a stack (or priority queue), and edges are continuously added to or removed from the stack. The process terminates when the stack is empty, i.e. when all fronts have merged upon each other and the domain is entirely covered. One of the critical features of such methods is the placement of new points. Upon generating a new triangle, a new point is first placed at a position which is determined to result in a triangle of optimal size and shape. The parameters which define this optimum triangle as a function of field position are obtained by a prescribed field function (which may be interpolated from a background grid). The triangle generated with this new point may result in a cross-over with other front edges, and thus may be rejected. This is determined by computing possible intersections with 'nearby' front edges. Alternatively, an existing point on the front may coincidentally be located very close to the new point, and thus should be employed as the forming point for the new triangle, to avoid the appearance of a triangle with a very small edge at some later stage. Existing candidate points are thus also searched by locating all 'nearby' front points.

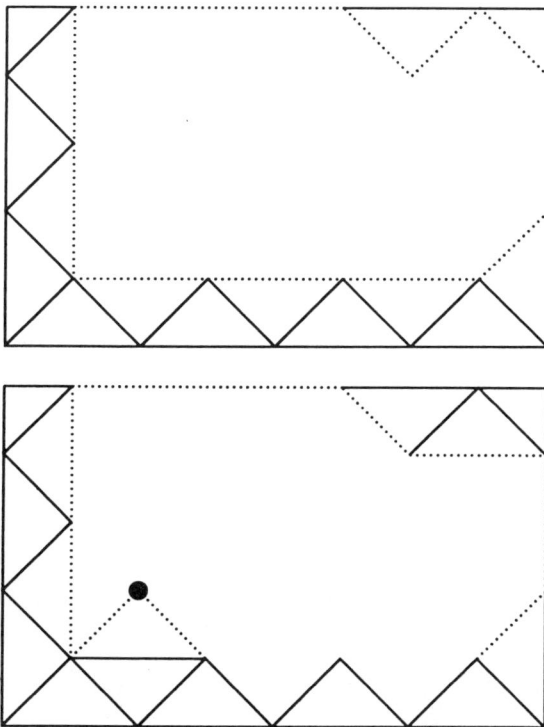

Figure 10 Illustration of advancing front mesh generation concept in two-dimensions. Dotted line represents current front. New triangles are generated by joining two end points of a front edge either to a newly created point, or an existing front point.

For three-dimensional grid generation, a surface grid is first constructed by generating a two-dimensional triangular mesh on the surface boundaries of the domain. This triangular mesh forms the initial front, which is then advanced into the flow-field by placing new points ahead of the front and forming tetrahedral elements. The required intersection checking now involves triangular front faces, rather than edges as in the two-dimensional case.

One of the advantages of such an approach is the automatic point placement strategy, which generally results in high-quality elements throughout most of the flow field because of the optimum positioning of these new points. Additionally, all real operations performed (such as intersection checking) are of a local nature; i.e. intersection checks are performed with neighbouring edges or faces of similar length, thus reducing the chances for round-off error induced failure. Finally, boundary integrity is guaranteed, since the boundary discretization constitutes the initial condition.

The disadvantages of advancing front techniques mainly relate to their efficiency and complexity of implementation. The intersection checking phase is a rather brute-force technique for ensuring the acceptability of a new triangle or tetrahedron, which is relatively expensive. Sophisticated spatial searching data-structures such as quad-trees or octrees (Samet, 1990) are required in order to reduce the number of possible intersections which must be evaluated, and to reduce the overall asymptotic algorithmic complexity from $O(N^2)$ to $O(N \log N)$, where N represents the total number of grid points. Since only front edges and faces must be checked for intersections, the quad or octrees can be based on the front alone. However, as the front evolves, these data-structures must be dynamically updated, which can lead to a considerable amount of code in the final implementation.

Finally, even though advancing front techniques rely only on local operations, they may still suffer from robustness problems. Central to the issue of determining acceptable triangles and 'best' points, is the determination of a local length scale which defines the region of 'nearby' points and edges. This length scale is generally obtained from the field function (which may be constructed on a background grid). If this field function varies rapidly over the region between two merging fronts, the relative sizes of the edges/faces on one front may be much larger than those on the other front. If a search is initiated from the front with the smaller length scale, the region of 'nearby' edges/faces may not be sufficiently large, resulting in failure of the algorithm. Thus, the advancing front technique can only be guaranteed to produce a valid triangulation if certain non-heuristic constraints are derived and imposed on the variation of the field function.

B Delaunay triangulation methods

Given a set of points in the plane, there exists many possible triangulations of these points. A Delaunay construction represents a unique triangulation of these points which exhibits a large class of well defined properties (Preparata and Shamos, 1985). Particular properties can be employed to construct algorithms for generating the Delaunay triangulation of a given set of points.

The empty circumcircle property forms the basis of the Bowyer–Watson algorithm (Bowyer, 1981; Watson, 1981). This property states that no triangle in a Delaunay triangulation can contain a point other than its three forming vertices within its circumcircle. Thus, given an initial triangulation, a new point may be inserted into the triangulation by first locating and deleting all existing triangles whose circumcircles contain the newly inserted point. A new triangulation is then formed by joining the new point to all boundary vertices of the cavity created by the previous removal of intersected triangles, as shown in Figure 11.

Point insertion algorithms can be employed as the basis for a mesh generation strategy where the mesh points have been predetermined. The mesh points are put in a list, and an initial triangulation is artificially constructed (with auxiliary points) which completely covers the entire domain to be gridded. The mesh points in the list are then inserted sequentially into the existing triangulation

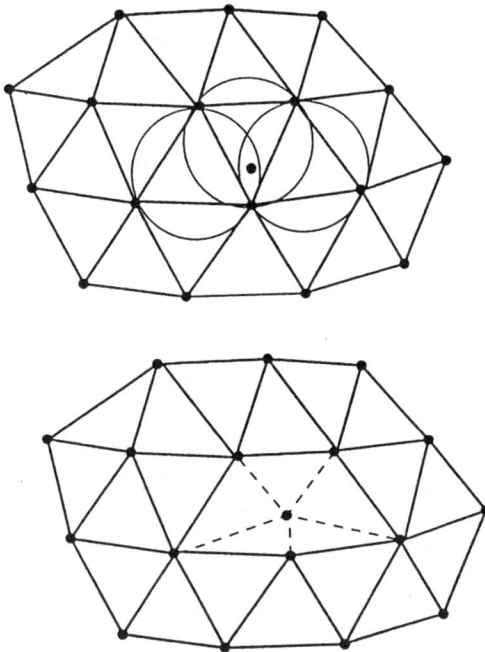

Figure 11 Illustration of Bowyer's point insertion algorithm for Delaunay triangulation.

using the Bowyer–Watson algorithm. The final mesh is obtained when all points from the list have been inserted.

Most often, automatic point placement strategies are desired in the context of mesh generation. A very simple method (Holmes and Snyder, 1988; Weatherill and Hassan, 1992) can be employed in conjunction with the Bowyer–Watson algorithm described above. Starting with an initial coarse triangulation which covers the entire domain, a priority queue is constructed based on some parameter of the individual triangles (e.g. circumradius). A field value is assumed to exist which determines the local maximum permissible value for the circumradius of the triangles (or other parameter). The first triangle in the queue is examined, and a point is added at its circumcentre if the triangle circumradius is larger than the locally prescribed maximum. This new point is inserted into the triangulation using Bowyer's algorithm, and the newly formed triangles are inserted into the queue if their circumcircles are too large, otherwise they are labelled as acceptable, and do not appear in the queue. The final grid is obtained when the priority queue empties out.

The main problem of the Bowyer–Watson algorithm relates to its worst case complexity which is of order $O(N^2)$. There are many other Delaunay triangulation algorithms (i.e. the divide and conquer method (Lee and Schachter, 1980; Preparata and Shamos, 1985) and the sweepline method (Fortune, 1987)), most of which are rooted in computational geometry, and are designed to provide optimal $O(N \log N)$ worst case complexity. In practice, however, the Bowyer–Watson algorithm exhibits near optimal complexity the worst case scenario being somewhat pathological in nature (Guibas *et al.*, 1990). Furthermore, this algorithm extends in a straightforward manner to three-dimensions, by considering the circumsphere associated with each tetrahedron.

Delaunay triangulation techniques based on the point insertion approach are much more efficient than advancing front techniques. The absence of a sophisticated spatial data-structure for locating neighbouring points, and the lack of an intersection checking routine, make these very simple and efficient algorithms. Furthermore, the mesh is generated point by point, rather than one triangle or tetrahedron at a time. Each time a point is inserted, all triangles/ tetrahedra neighbouring that point are formed simultaneously, which results in increased efficiency due to the larger number of elements than points in an unstructured mesh.

The main disadvantages of Delaunay triangulation techniques relate to their inability to guarantee boundary integrity. In two dimensions, the set of edges which define the domain boundaries must form a subset of the triangulation edges. If this is not the case, edges of the triangulation exist which penetrate the boundary, as shown in Figure 12. In general, boundary integrity is not guaranteed unless the domain is convex. In three dimensions, the equivalent problem is that of constructing a tetrahedralization of the domain which contains, as a subset of its faces, a given surface boundary triangulation. A somewhat less restrictive problem is that of constructing a tetrahedralization of

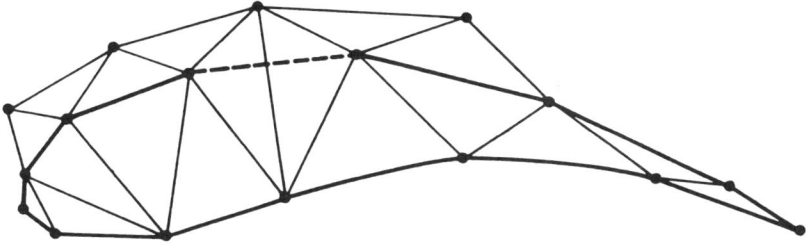

Figure 12 Illustration of breakthrough of boundary surface for a two-dimensional airfoil in the context of Delaunay triangulation.

the boundary and interior points for which no edges or faces intersect the boundary surfaces. One way of achieving this is by modifying the boundary point resolution (by judiciously adding surface points) in regions where breakthroughs occur (Baker, 1988). The problem of matching a prescribed boundary triangulation is more difficult. The use of local transformations (face- and edge-swapping) has been developed (George *et al.*, 1991; Weatherill *et al.*, 1993) for locally modifying the domain tetrahedralization such that the surface triangulation can be recovered as a subset of the domain faces. In practice, the addition of extra surface points is often required in order to permit the recovery of the boundary triangulation. Thus, a Delaunay triangulation of the domain which conforms exactly to a given surface triangulation may not always be possible. In two dimensions, the existence of constrained Delaunay triangula- tions (Chew, 1989) guarantees the possibility of constructing triangulations which contain a prescribed subset of edges, and thus which respect a given boundary discretization.

C Edge- and face-swapping techniques

The Delaunay triangulation represents but one of many possible triangulations of a given point set. An algorithm for transforming one triangulation to another triangulation of the same point set is given by Lawson (1972). Since all two- dimensional planar graphs obey Euler's formula (Mortenson, 1985), all possible triangulations of a given set of points contain the same number of edges and triangles. Thus, any one triangulation may be obtained by simply rearranging the edges of another triangulation of the same set of points. The edge-swapping algorithm of Lawson (1972) consists of successively examining pairs of neighbouring triangles in the mesh. For each pair of triangles, there exist two possible configurations of the diagonal edge, as shown in Figure 13. The algorithm chooses the diagonal configuration which optimizes some given criterion. The algorithm terminates when no further optimizations are possible.

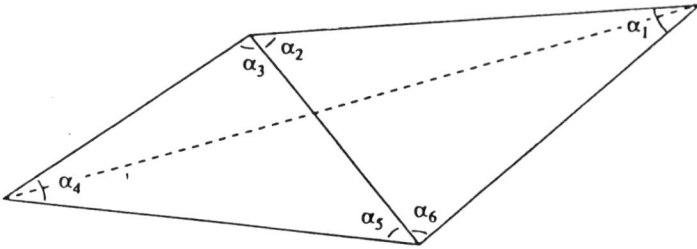

Figure 13 Two possible configurations for the diagonal edge delimiting two neighbouring triangles in the edge-swapping algorithm.

The edge-swapping algorithm can be used to transform an arbitrary triangulation into a Delaunay triangulation. In this case, the edge-swapping criterion to be used is the edge configuration which maximizes the minimum angle within the two neighbouring triangles. (Delaunay triangulations are also known as max–min triangulations: they maximize the smallest angles in the triangulation.)

Edge-swapping can be used to construct triangulations other than the Delaunay triangulation, such as the minimum–maximum angle (min–max) triangulation (Barth, 1991), or the minimum total edge length triangulation (Preparata and Shamos, 1985). However, the edge-swapping procedure is not guaranteed to result in the global optimum triangulation for these cases: the algorithm may terminate within local optima. A more sophisticated procedure known as the edge insertion algorithm (Edelsbrunner *et al.*, 1992) has been shown to be capable of transforming a given triangulation into the global optimum triangulation for these cases.

In three dimensions, analogous local transformations can be achieved through face- and edge-swapping. By considering pairs of neighbouring tetrahedra, the common triangular face may be removed and the edge joining the two opposing end points of the tetrahedra can be drawn, as shown in Figure 14. Note that the original configuration which contained two tetrahedra is transformed into a configuration containing three tetrahedra. The main difficulty in three-dimensional applications is that of achieving a global optimum, even for the case of a Delaunay triangulation. It has been shown that under certain conditions, the Delaunay triangulation can be recovered from an initial triangulation (Joe, 1989).

Edge-swapping provides an alternative to Bowyer's point insertion algorithm for constructing a Delaunay triangulation. Each time a new point is inserted, rather than performing a search for intersected circumcircles, the triangle which contains the new point can simply be subdivided. The Delaunay triangulation can then be recovered by locally swapping the edges in the vicinity of the new point. By modifying the edge-swapping criterion, this procedure can be used

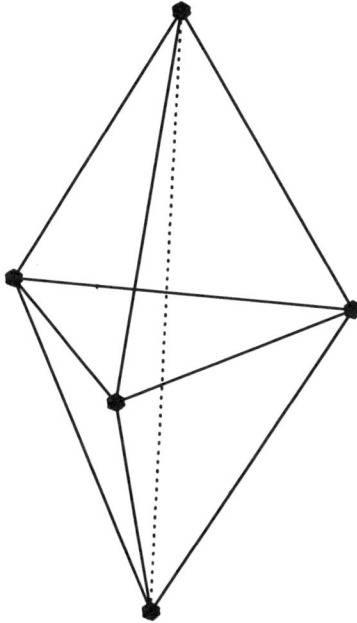

Figure 14 Two possible configurations for the edge-face swapping algorithm in three dimensions.

incrementally to construct other triangulations such as the min–max triangulation (Marcum and Weatherill, 1994). Finally, face- and edge-swapping constitutes the basis for the boundary recovery techniques described above for three-dimensional Delaunay triangulations.

D Other triangulation methods

Various techniques have been developed which combine the advantages of advancing front methods with the theoretical rigorousness of Delaunay triangulations. Rebay (1993) and Muller et al. (1992) have developed similar algorithms, where the triangulation method is based on Bowyer's Delaunay triangulation algorithm, but the point insertion process is designed to mimic that achieved by the advancing-front method. The result is the very smooth point distribution and high-quality mesh usually associated with advancing front techniques, coupled with the efficiency of the Delaunay point insertion algorithm. A method for producing the Delaunay triangulation of a given set of points by successively adding triangles to an advancing front has been discovered and rediscovered several times throughout the literature (Nelson,

1978; Tanemura *et al.*, 1983; Maus, 1984; Merriam, 1991). While the point insertion algorithms maintain a triangulation over the entire domain at all stages of the construction process, this algorithm only maintains triangles behind the front, and the area ahead of the front is untriangulated, just as in the advancing front method. Although the algorithm becomes somewhat more complex, robustness is increased since the highly skewed triangles which appear at intermediate stages of the point insertion triangulation algorithm are never formed. Also, boundary integrity can easily be guaranteed by preserving the initial front. An extension of this algorithm to simultaneous point generation is given in Mavriplis (1993).

A completely different approach to unstructured mesh generation is provided by the quad or octree method (Yerry and Shephard, 1984; Cheng *et al.*, 1988). In this method, a recursive subdivision of the domain is first performed down to a prescribed boundary resolution level, thus generating a quad-tree (2D) or octree (3D) data-structure. The nodes of the quad-tree provide the vertices for the mesh. At boundaries the quad-tree nodes are displaced or warped to coincide with the boundary, and the triangular elements are formed by subdivision of the quad-tree cells. The method is relatively simple, inexpensive, and produces good-quality meshes in interior regions of the domain. The main drawback is an irregular cell distribution near boundaries.

E Mixed element meshes

While Delaunay constructions are only relevant for triangular or tetrahedral meshes, advancing front methods can operate using a variety of element types. Thus advancing front quadrilateral or hexahedral meshing techniques have been devised, as well as methods which use combinations of quadrilaterals and triangles (Blacker and Stephenson, 1991; Zhu *et al.*, 1991). Triangular meshes can also be converted to mixed quadrilateral–triangle grids by identifying and removing appropriate diagonal edges (Merriam, 1991; Tembulkar and Hanks, 1992). Since a grid of quadrilaterals represents a sparser connectivity graph than an equivalent grid of triangles, the goal in using mixed elements is a reduction in flow solution time for comparable accuracy. Another technique consists of employing structured grids in regions or subregions of simple geometries, and filling the remainder of the domain with unstructured triangular elements (Nakahashi and Obayashi, 1987; Weatherill, 1990; Shaw, 1991). The resulting mesh may either be treated as a hybrid structured–unstructured mesh, or as a mixed element unstructured mesh.

The use of prismatic elements has been advocated for the computation of high-Reynolds-number viscous flows (Nakahashi, 1991; Ward and Kallinderis, 1993). For such flows, the presence of thin boundary layers requires high grid resolution in the direction normal to the boundary surface. One method of easily obtaining high normal resolution is by maintaining grid structure in the normal direction. Thus, if the boundary surface has been triangulated, the construction of a structured grid in the normal direction results in prismatic elements. A more comprehensive discussion of the issues involved in grid generation for viscous flows is given in the next section.

F Adaptivity

One of the main advantages of the use of unstructured meshes is the ease with which adaptive meshing techniques may be incorporated. Since there is no inherent structure to be preserved, new mesh points may be added or deleted and the connectivity may be locally updated.

One of the simplest methods for adaptively refining a mesh is element subdivision. Elements in regions of high local truncation errors are first flagged for refinement. The refinement operation is then performed by placing new points midway along the edges of the flagged cells, and subdividing the cells into multiple smaller elements. A set of rules can be constructed for guiding the subdivision of cells into the desired number (usually two, four or eight) of finer cells. This method can be applied to tetrahedral meshes as easily as hexahedral, prismatic or mixed element meshes. The method is simple and very efficient, and has thus been used often for transient flow simulations, where very frequent mesh adaptation cycles are required (Löhner and Baum, 1992). However, when multiple levels of refinement are employed, the mesh connectivity and element shapes may degrade, resulting in skewed elements and vertices of high degree. In order to prevent this behaviour, strict rules for allowable subdivision patterns need to be enforced.

A more elegant method for adaptive mesh refinement makes use of Bowyer's Delaunay point insertion algorithm. This method is only valid for triangular or tetrahedral elements, and for initial meshes which are Delaunay in construction. The mesh is refined by adding new points in regions of high gradients or truncation error (Berger and Collela, 1989; Danenhoffer, 1991). These points may either be placed midway along existing mesh edges, or at element centres. Each point is inserted into the mesh by recomputing the local connectivity of the mesh using Bowyer's algorithm. The final mesh is also a Delaunay triangulation, thus preserving element shape and connectivity characteristics. Element subdivision followed by edge-swapping can also be used to recover a Delaunay triangulation and avoid the degeneration of element shape and mesh connectivity. However, this can only be applied to tetrahedral element meshes. Adaptive meshing, therefore, provides strong incentives for employing uniquely tetrahedral unstructured meshes.

Remeshing has also been proposed as an adaptive meshing strategy (Peraire et al., 1987). In this technique, an entire new mesh is generated where the local mesh spacing is governed entirely by some measure of error in the current solution. Remeshing can obviously be used for any types of elements with any mesh generator (Delaunay or not). While this approach offers the most flexibility for optimizing the adapted mesh, it is expensive and is thus limited to steady-state calculations where only occasional adaptive phases are employed. Local remeshing represents an alternative to global remeshing, which can lessen the expense of each adaptive cycle. This approach is especially attractive for the advancing front method, where local regions of the mesh may be cut out, thus creating a new front, and then remeshed (Pirzadeh, 1992).

Finally, mesh point movement can also be applied in the context of unstructured meshes. The strategies for computing new mesh point locations are similar to those discussed for structured grids (Palmerio, 1994). When the elements become too distorted, the mesh can be improved by recomputing the connectivity through an edge-swapping procedure, or simply remeshing. Moving meshes of this type have been used for transient problems with moving boundaries (Trepanier *et al.*, 1991).

V NUMERICAL ISSUES

In this section we discuss the issues which arise in numerical simulations from the interaction between the grid generation and flow solution phases. These include issues of solution quality and accuracy (grid quality and adaptivity) as well as overall numerical efficiency (adaptivity and grid-based strategies). The central theme of this section is the need for a tight coupling between these two operations in order to produce a reliable and economical numerical solver.

A Grid quality

The issue of grid quality is concerned with the ability of a particular discretization scheme to represent accurately the continuous governing equations on a given grid. While most studies of grid quality immediately launch into grid optimization strategies for improving the discrete approximation, it is important to realize that optimizations of the accuracy and stability of a given discretization represents a burden which is shared by both the numerical scheme as well as the grid generation scheme. For example, it is possible to construct a numerical scheme which takes no account of the local grid metrics. Such a scheme will be inexpensive and stable, since the elements of the matrix of the associated discrete operator will all have the same magnitude. However, the accuracy achieved by such a scheme is highly dependent on the smoothness of the grid on which it operates. The formal accuracy of the scheme will only be recovered on a cartesian or smoothly varying structured grid. By including more grid metric information in the numerical scheme, the dependence of the accuracy on the quality of the grid can be reduced. However, the stability of the scheme may be more difficult to guarantee since the entries in the matrix of the discrete operator now depend on the grid cell shapes. As an example, vertex-based schemes for structured meshes have often been advocated as being less sensitive in terms of accuracy to grid skewness than cell-centred schemes, due to the larger stencils upon which they operate (Swanson and Radespiel, 1991).

Clearly, the goal of any numerical simulation should be the optimization of both the discretization scheme as well as the grid generation scheme. A simple Taylor series analysis of truncation error reveals terms which depend on the relative changes of grid spacing, the grid skewness, as well as higher-order derivatives in the flow solution. This indicates that grid spacing and element shape need to be optimized, and that the final accuracy also depends on the solution itself.

Most grid optimization methods are based upon adaptive meshing techniques. For structured meshes, mesh-point movement is most often employed where the optimization targets are element shape and size variations. This can be performed either automatically (Luong et al., 1993), manually (Eiseman, 1991), or using a general variational approach (Jacquotte, 1991).

For unstructured meshes, grid quality is somewhat more difficult to quantify. In finite-element theory, accuracy has been shown to degrade for triangular elements with obtuse angles (close to 180°) (Babuska and Aziz, 1976). This has led to considerable research into the field of non-obtuse ($\alpha < 90° + \varepsilon$) triangular mesh generation (Baker et al., 1988). One method of quantifying the quality of a triangular mesh is by observing its effect on the discretization of a simple equation such as a laplacian. Since the continuous Laplace's equation exhibits a maximum principle, an equivalent discrete maximum principle is often desirable. A Galerkin finite-element discretization of a laplacian on a non-obtuse triangulation naturally satisfies a discrete maximum principle. However, it has also been shown that any Delaunay triangulation (even those containing obtuse triangles) satisfies a discrete maximum principle for Laplace's equation (Barth, 1991). This provides a strong motivation for the use of Delaunay-triangulations (although a similar result in three dimensions does not hold).

However, since any point set may be triangulated, it is possible to generate Delaunay triangulations with extreme jumps in cell sizes, simply by distributing the points unevenly, which results in reduced approximation accuracy. Thus, as in the structured grid case, mesh movement (or smoothing) is often employed to optimize the mesh. Various measures of grid quality can be (heuristically) defined (Babuska and Aziz, 1976; Formaggia, 1991; Marcum and Weatherill, 1994), and these can then be optimized through adaptive meshing techniques which include mesh movement, edge-swapping/reconnection and point insertion/deletion, or local remeshing.

B Grid stretching for viscous flows

The drive towards full Navier–Stokes solvers has necessitated the development of stretched grid generation techniques in order to resolve the thin boundary layers, wakes, and other viscous regions which are characteristic of high-

Reynolds-number viscous flows. Proper boundary layer resolution usually requires mesh spacing several orders of magnitude smaller in the direction normal to the boundaries than in the streamwise direction, resulting in large cell aspect ratios in these regions.

For structured meshes, the source terms in the elliptic solvers may be adjusted to increase the normal resolution (Steger and Sorenson, 1979; Hilgenstock, 1988). However, for extreme grid stretching these source terms can become very large, resulting in stiff sets of elliptic equations which can be difficult to solve. Hyperbolic techniques appear to be better suited for highly stretched mesh generation (Steger and Chaussee, 1980). These methods are capable of generating meshes of any normal resolution without difficulty. Their main drawback is their inability to match prescribed inner and outer boundary point distributions.

For unstructured meshes, the situation is again more complex. The practitioner is first confronted with the choice of elements to be employed in the stretched regions. Methods employing highly stretched triangular elements have been developed using mapping techniques (Mavriplis, 1990, 1991; Vallet et al., 1991). Since obtuse triangles must be avoided for accuracy reasons (Babuska and Aziz, 1976), the resulting grids are designed to generate mostly elongated nearly right-angled triangles. The use of quadrilateral elements may appear more natural for problems with a preferred direction, and thus many hybrid methods have been proposed (Nakahashi and Obayashi, 1987; Holmes and Connell, 1989). The main drawbacks here are increased solver and adaptive meshing complexity. In three dimensions, the element choices are between hexahedra, prisms and tetrahedra. Hybrid methods using tetrahedra and prisms appear promising (Nakahashi, 1991; Ward and Kallinderis, 1993), since this allows the use of triangular surface meshes.

One approach to prismatic mesh generation attempts to preserve the mesh structure in the normal direction, either throughout the entire domain (Nakahashi, 1991), or up to a specified distance away from the boundary, after which fully unstructured meshing techniques are employed (Ward and Kallinderis, 1993). Alternative strategies can be devised to generate fully unstructured prismatic and/or tetrahedral meshes by capitalizing on the similarities between advancing front unstructured and hyperbolic structured mesh generation (Lohner, 1993; Hassan et al., 1994; Pirzadeh, 1994). It should finally be noted that the more complex elements such as hexahedra and prisms can easily be subdivided into tetrahedra to simplify subsequent flow solution and mesh adaptivity operations.

C Grid-based strategies

While the grid generation and flow solution phases of a numerical simulation are often thought of as separate and distinct steps, there is a growing body of powerful techniques which take a more comprehensive approach to the general problem of numerically simulating a physical phenomenon by closely coupling the grid generation and numerical solution aspects.

The most obvious of these approaches is the multigrid strategy (McCormick, 1987). The idea of a multigrid algorithm is to accelerate the convergence of a set of fine-grid discrete equations by computing corrections to these equations on a coarser grid, where the computation can be performed more economically. This process is applied recursively to an entire set of coarse-grid levels. Each multigrid cycle begins on the finest grid level and cycles through the various levels up to the coarsest mesh. At this stage the computed corrections are successively interpolated back to the finest level, and the cycle is repeated. The accuracy of the final discretization is solely determined by the fine-grid discretization, and the coarser levels may be viewed simply as artifacts employed to accelerate convergence.

For structured grid simulations, multigrid can be implemented in a relatively straightforward manner. Coarser levels are constructed by removing every 2^n-th point in each coordinate direction from the fine meshes, where n represents the coarse mesh level. This requires the original fine mesh to be generated with grid point distributions where the number of mesh points is a power of 2 in each coordinate direction. For block-structured meshes, multigrid can easily be applied within each block of the grid, but inter-block multigrid becomes more complex and is seldom attempted.

For unstructured meshes, various approaches may be taken. If the fine mesh has been generated by recursive subdivision of a coarser mesh level, the nested parent meshes are natural candidates for coarse multigrid levels (Perez, 1985; Connell and Holmes, 1993). However, since this situation is most often not present, multigrid methods using non-nested coarse levels are often employed (Mavriplis and Jameson, 1988; Leclercq, 1990; Mavriplis, 1992; Peraire et al., 1992). In this approach, a grid generator is employed to construct various coarse-level meshes which need not be nested or even contain common points with the fine grid. The patterns for interpolating flow variables, residuals, and corrections between the various meshes of the sequence are determined in a preprocessing phase. Once the interpolation weights and addresses have been computed and stored, multigrid cycling can proceed as in the structured or nested unstructured grid cases.

A third alternative consists of agglomerating or fusing fine-grid cells or control volumes together to form a smaller set of larger polyhedral coarse-grid cells or control volumes (Lallemand et al., 1992; Mavriplis and Venkatakrishnan, 1994; Venkatakrishnan and Mavriplis, 1994) (Figure 15). The main advantage of this approach is that coarse levels can be generated in a fully automatic manner on any type of mesh (block-structured, unstructured, arbitrary element types) using graph-based algorithms. On the other hand, the flow solver must be modified to enable discretization on polyhedral coarse meshes.

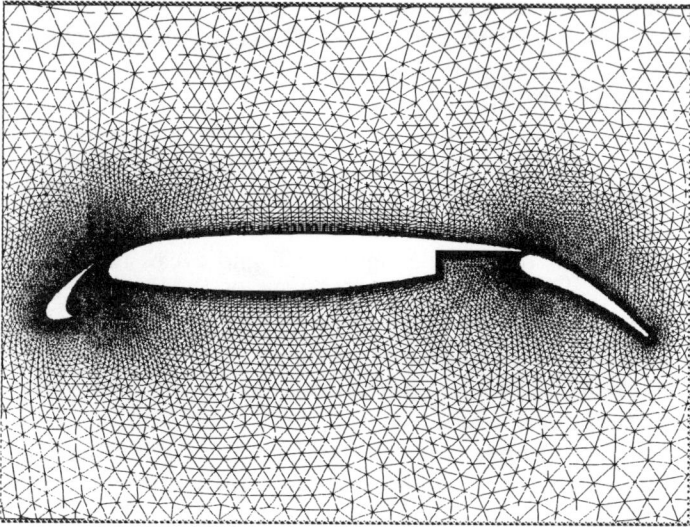

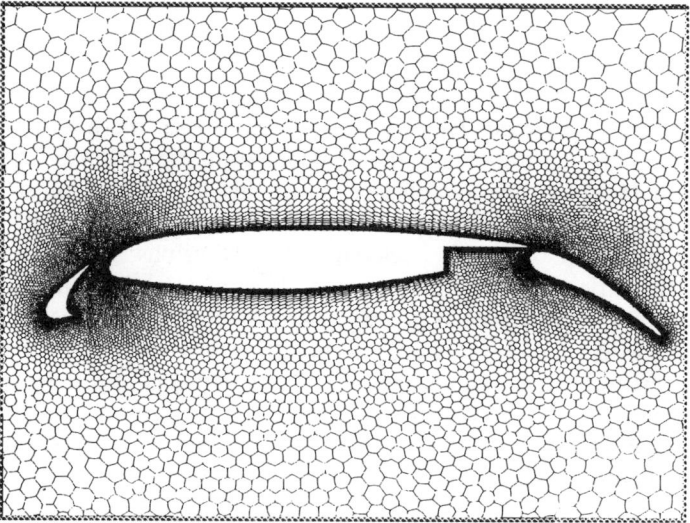

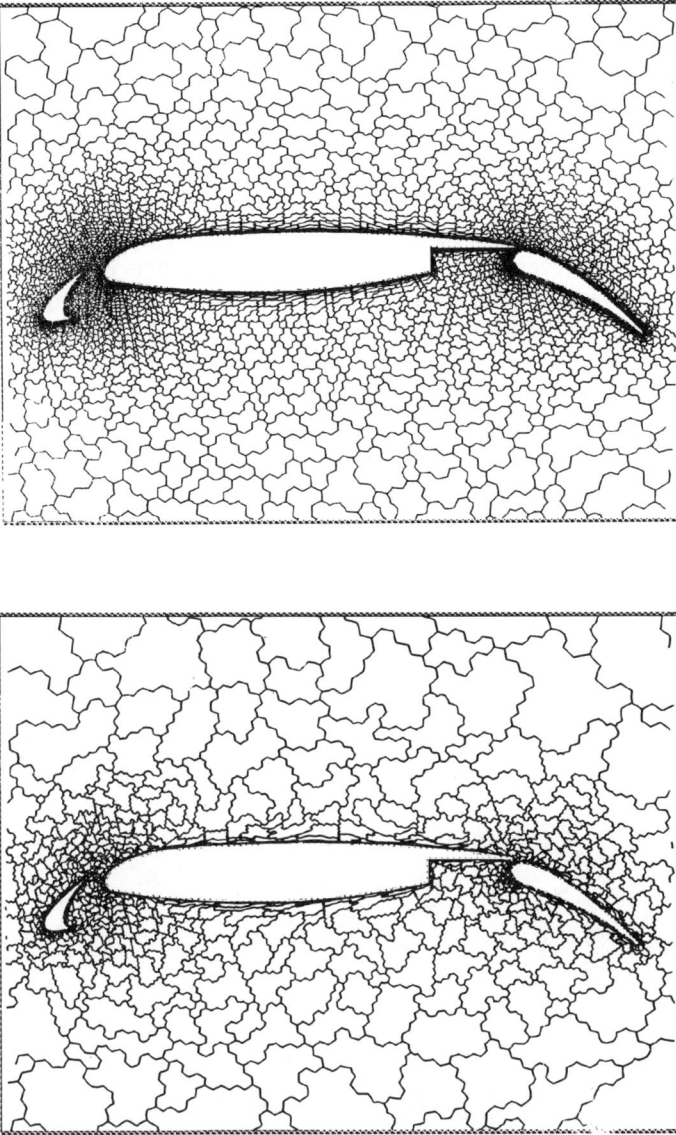

Figure 15 Fine mesh, dual control volume mesh and coarse agglomerated meshes employed for multigrid acceleration in the agglomeration multigrid method.

Finally, another alternative known as algebraic multigrid does not make use of any coarse-level meshes, but rather attempts to solve coarser or smaller representations of the sparse matrix of the discrete fine-grid operator (Ruge and Stuben, 1987). While algebraic multigrid methods are potentially more general than grid-based methods, their development has lagged that of grid-based multigrid methods, perhaps a tribute to the fact that useful physical information can be obtained by looking at the problem in a geometric sense, rather than purely from a mathematical standpoint.

There is also a reverse trend in the literature, known as the de-algebraization of multigrid (Brandt, 1984). This philosophy consists of viewing multigrid not simply as a convergence acceleration technique, but as a broad multilevel approach to simulating continuous partial differential equations, which may also involve adaptive iteration schemes and adaptive meshing techniques. A good illustration of this approach is the full adaptive composite scheme (FAC) (McCormick and Thomas, 1986). This method can be applied to structured as well as unstructured meshes (McCormick and Thomas, 1986; Mavriplis, 1989). The method is initiated by discretizing and solving the governing equations on a coarse mesh which covers the entire domain. The mesh is then adaptively refined in regions of high local truncation error. Rather than adding points to the existing mesh, new finer local meshes are created in regions where high error is detected. By recursive application, this procedure results in a multilevel technique which combines the efficiency of multigrid with the accuracy of

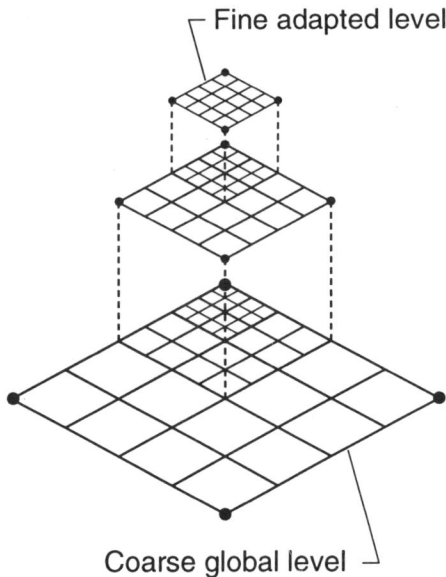

Figure 16 Illustration of the multiple composite levels in the FAC strategy.

adaptive meshing. The final solution is obtained by assembling the finest mesh patches from all levels which cover the entire domain, as illustrated in Figure 16. At each level, a constant resolution is afforded for approximating the solution. Thus, the function of the various levels in the FAC scheme is to isolate coarse and fine features of the solution. This information can also be utilized beneficially in other aspects of the simulation, such as in the visualization process where unwanted details can be filtered out by changing levels.

Another example of a comprehensive grid-based strategy can be found in the h-p refinement approach (Oden *et al.*, 1990). Rather than operating with a fixed discretization scheme and refining the mesh to obtain higher spatial accuracy, h-p refinement schemes achieve higher accuracy either by mesh refinement (h-refinement), or by increasing the approximation order of the discretization (p-refinement). By examining the character of the truncation error and judiciously choosing the type and amount of refinement to be applied, exponential convergence to the continuous solution can be obtained. This represents a significant gain over the typically low-order polynomial convergence which is obtained by most adaptive meshing techniques.

In conclusion, significant gains in efficiency and accuracy can be achieved by designing comprehensive strategies which tightly couple the grid generation and flow solution phases. As the state of the art of numerical simulations progresses, increased coupling of these two aspects of the simulation strategy can be expected in the future.

REFERENCES

Aftosmis, M.J. (1993). *AIAA Paper 93-0772*, 31st Aerospace Sciences Meeting, Reno, NV.

Akdag, V. and Wulf, A. (1992). In *Proc. Software Systems for Surface Modeling and Grid Generation Workshop* (R.E. Smith, ed.). NASA Langley Research Center, Hampton, VA.

Allwright, S.E. (1988). *Proc. 2nd International Conference on Numerical Grid Generation in CFD* (S. Sengupta, J. Hauser, P.R. Eisman and J.F. Thompson, eds), pp. 559–568. Pineridge Press Ltd., Swansea.

Babuska, I. and Aziz, A.K. (1976). *SIAM J. Numer. Anal.* **13**, 214–226.

Baker, T.J. (1988). *Proc. 2nd International Conference on Numerical Grid Generation in Computational Fluid Dynamics* (S. Sengupta, J. Hauser, P.R. Eisman and J.F. Thompson, eds), pp. 675–686. Pineridge Press, Swansea.

Baker, T.J. (1991). *Proc. 3rd International Conference on Numerical Grid Generation in CFD and Related Fields* (A.S. Arcilla, J. Hauser, P. R. Eiseman and J.F. Thompson, eds), pp. 261–272. North Holland, New York.

Baker, B.S., Grosse, E. and Rafferty, C.S. (1988). *Discrete Comput. Geom.* **3**, 147–168.

Barth, T.J. (1991). *AIAA Paper 91-0721*, 29th Aerospace Sciences Meeting, Reno, NV.

Batina, J.T. (1993). *AIAA Paper 93-0333*, 31st Aerospace Sciences Meeting, Reno, NV.

Benek, J.A., Buning, P.G. and Steger, J.L. (1985). *AIAA Paper 85-1523-CP*, 7th Computational Fluid Dynamics Conference, Cincinnati, OH.

Berger, M. and Collela, P. (1989). *J. Comput. Phys.* **82**, 64–84.

Berger, M. and Jameson, A. (1984). *9th International Conference on Numerical Methods in Fluid Dynamics*, Lecture Notes in Physics, vol. 218, pp. 92-97. Springer-Verlag, Berlin.

Blacker, T.D. and Stephenson, M.B. (1991). *Int. J. Numer. Methods in Engrg.* **32**, 811–847.

Boerstoel, J.W. and Spekreyse, S.P. (1991). *Comput. Methods Appl. Mech. Engrg.* **89**, 237–257.

Bowyer, A. (1981). *Computer J.* **24**, 162–166.

Brandt, A. (1984). *Multigrid Techniques with Applications to Fluid Dynamics: 1984 Guide*, Lecture Notes for the Computational Fluid Dynamics Lecture Series, Von Karman Institute for Fluid Dynamics, Belgium.

Cheng, J.H., Finnigan, P.M., Hathaway, A.F., Kela, A. and Schroeder, W.J. (1988). *Proc. 2nd International Conference on Numerical Grid Generation in CFD* (S. Sengupta, J. Hauser, P.R. Eisman and J.F. Thompson, eds), pp. 633–642. Pineridge Press, Swansea.

Chew, L.P. (1989). *Algorithmica* **4**, 97–108.

Connell, S. and Holmes, G. (1993). *AIAA Paper 93-3339*, 11th Computational Fluid Dynamics Conference, Orlando, FL.

Cordova, J.Q. (1992). In *Proceedings of Software Systems for Surface Modeling and Grid Generation Workshop* (R. E. Smith, ed.). NASA Langley Research Center, Hampton, VA.

Dannehoffer, J.F. (1991). *Int. J. Numer. Methods Engrg.* **32**, 653–663.

Davis, R.L. and Dannehoffer, J.F. (1991). *J. Propulsion* **7**, 792–799.

Dener, C. and Hirsch, C. (1992). *AIAA Paper 92-0073*, 30th Aerospace Sciences Meeting, Reno, NV.

Duff, I.S., Erisman, A.M. and Reid, J.K. (1986). *Direct Methods for Sparse Matrices*. Oxford Science Publications, Clarendon Press, Oxford.

Edelsbrunner, H., Tan, T.S. and Waupotitsch, R. (1992). *SIAM J. Sci. Stat. Comput.* **13**, 994–1008.

Eiseman, P.R. (1991). *Comput. Methods Appl. Mech. Engrg.* **91**, 1151–1156.

Eiseman, P.R., Cheng, Z. and Hauser, J. (1994). *Proc. 4th International Conference on Numerical Grid Generation in CFD and Related Fields* (N.P. Weatherill, P.R. Eiseman, J. Hauser and J.F. Thompson, eds), pp. 123–134. Pineridge Press, Swansea.

Farin, G.E. (1990). *Curves and Surfaces for Computer Aided Geometric Design: A Practical Guide*, 2nd edn. Academic Press, London.

Formaggia, L. (1991). *Proc. 3rd International Conference on Numerical Grid Generation in CFD and Related Fields* (A.S. Arcilla, J. Hauser, P.R. Eiseman and J.F. Thompson, eds), pp. 249–260. North Holland, New York.

Fortune, S. (1987). *Algorithmica* **2**, 153–174.

Fujitani, K. and Himeno, R. (1991). *Proc. 3rd International Conference on Numerical Grid Generation in CFD and Related Fields* (A.S. Arcilla, J. Hauser, P.R. Eiseman and J.F. Thompson, eds), pp. 755–768. North Holland, New York.

George, P.L., Hecht, F. and Saltel, E. (1991). *Comput. Methods. Appl. Mech. Engrg.* **33**, 975–995.

Gnoffo, P.A. (1983). *AIAA J.* **21**, 1249–1254.

Gomez, R.J. and Ma, E.C (1994). *AIAA Paper 94-1859-CP*, 12th Applied Aerodynamics Conference, Colorado Springs, CO.

Guibas, L.J., Knuth, D.E. and Sharir, M. (1990). *Stanford Univ. Computer Science Rep. No. STAN-CS-90-1300.*

Gumbert, C., Lohner, R., Parikh, P. and Pirzadeh, S. (1989). *AIAA Paper 89-2175*, 20th Fluid Dynamics Conference, Buffalo, NY.

Hassan, O., Probert, E.J., Weatherill, N. P., Marchant, M. J., Morgan, K. and Marcum, D.L. (1994). *AIAA Paper 94-2346*, 25th Fluid Dynamics Conference, Colorado Springs, CO.

Hilgenstock, A. (1988). *Proc. 2nd International Conference on Numerical Grid Generation in CFD* (S. Sengupta, J. Hauser, P.R. Eisman and J.F. Thompson, eds), pp. 17–146. Pineridge Press, Swansea.

Holmes, D.G. and Connell, S.D. (1989). *AIAA Paper 89-1932*, 9th Computational Fluid Dynamics Conference.

Holmes, D.G. and Snyder, D.D. (1988). *Proc. 2nd International Conference on Numerical Grid Generation in CFD* (S. Sengupta, J. Hauser, P.R. Eisman and J.F. Thompson, eds). Pineridge Press, Swansea.

Jacquotte, O.P. (1991). *Proc. 3rd International Conference on Numerical Grid Generation in CFD and Related Fields* (A.S. Arcilla, J. Hauser, P.R. Eiseman and J.F. Thompson, eds), pp. 581–596. North Holland, New York.

Jacquotte, O.-P. and Cabello, J. (1990). *La Recherche Aerospatiale* 1990-**4**, 7–19.

Joe, B. (1989). *SIAM J. Sci. Stat. Comput.* **10**, 718–741.

Kim, H.J. and Thompson, J.F. (1990).*AIAA J.* **28**, pp. 470–477.

Lallemand, M., Steve, H. and Dervieux, A. (1992) *Computers and Fluids* **21**, 397–433.

Lawson, C.L. (1972). *Discrete Math.* **3**, 365–372.

Leclercq, M.P. (1990). PhD Thesis, Applied Mathematics, Université de Saint-Etienne.

Lee, D.T. and Schachter, B. (1980). *Int. J. Comput. Inform. Sci.* **9**, 219–242.

Lo, S.H. (1985). *Int. J. Numer. Methods Engrg.* **21**, 1403–1426.

Löhner, R. (1993). *AIAA Paper 93-3348*, 11th Computational Fluid Dynamics Conference, Orlando, FL.

Löhner, R. and Baum, D. (1992). *Int. J. Numer. Methods Fluids* **14**, 1407-1419.

Luong, P.V., Thompson, J.F. and Gatlin, B. (1993). *J. Aircraft* **30**, 227–234.

Marcum, D.L. and Weatherill, N.P. (1994). *AIAA Paper 94-1926*, 12th Applied Aerodynamics Conference, Colorado Springs, CO.

Maus, A. (1984). *BIT* **24**, 151–163.

Mavriplis, D.J. (1989). *ICASE Report No. 89-35, NASA CR 181848.*

Mavriplis, D.J. (1990). *J. Comput. Phys.* **90**, 271–291.

Mavriplis, D.J. (1991). *Proc. 3rd International Conference on Numerical Grid Generation in CFD and Related Fields* (A.S. Arcilla, J. Hauser, P.R. Eiseman and J.F. Thompson, eds), pp. 79–92. North Holland, New York.

Mavriplis, D.J. (1992). *AIAA J.* **30**, 1753–1761, July 1992.

Mavriplis, D.J. (1993). *AIAA Paper 93-0671*, 31st Aerospace Sciences Meeting, Reno, NV.

Mavriplis, D.J. and Jameson, A. (1988). *AIAA J.* **26**, 824–831, July 1988.

Mavriplis, D.J. and Venkatakrishnan, V. (1994). *AIAA Paper 94-2332*, 25th Fluid Dynamics Conference, Colorado Springs, CO.

McCormick, S., ed. (1987). *Multigrid Methods*, SIAM Frontiers in Applied Mathematics, SIAM, Philadelphia.

McCormick, S. and Thomas, J. (1986). *Math. Comput.* **46**, 439–456.

Merriam, M.L. (1991). *AIAA Paper 91-0792*, 29th Aerospace Sciences Meeting, Reno, NV.

Mortenson, M.E. (1985). *Geometric Modeling*, John Wiley and Sons.
Muller, J.D., Roe, P.L. and Deconinck, H. (1992). *Unstructured Grid Methods for Advection Dominated Flows*, VKI Lecture Notes, pp. 9-1, 9-7, AGARD, Neuilly-sur-Seine.
Nakahashi, K. (1991).*AIAA Paper 91-0103*, 29th Aerospace Sciences Meeting, Reno, NV.
Nakahashi, K. and Deiwert, G.S. (1985). *AIAA Paper 85-1525.*
Nakahashi, K. and Obayashi, S. (1987). *AIAA Paper 87-0604*, 25th Aerospace Sciences Meeting, Reno, NV.
Nakamura, S. and Suzuki, M. (1987). *AIAA Paper 87-0277*, 25th Aerospace Sciences Meeting, Reno, NV.
Nelson, J.M. (1978). *Appl. Math Modeling* 2, 151–159.
Oden, J.T., Demkowicz, L., Liszka, T. and Rachowicz, W. (1990). *Proc. Symp. on Comp. Technology on Flight Vehicles* (A.K. Noor, S.L. Venneri, eds), pp. 523–534. Permagon Press, Washington, DC.
Palmerio, B. (1994). *Computers and Fluids* 23, 487–506.
Parikh, P., Pirzadeh, S. and Frink, N.T. (1994). *J. Aircraft* 31, 1291–1296.
Peraire, J., Vahdati, M., Morgan, K. and Zienkiewicz, O.C. (1987). Adaptive Remeshing for Compressible Flow Computations *J. Comput. Phys.* **72**,.
Peraire, J., Peiro, J. and Morgan, K. (1992). *A Three Dimensional Finite Element Multigrid Solver for the Euler Equations. AIAA Paper 92-0449*, 30th Aerospace Sciences Meeting, Reno, NV.
Perez, E. (1985). *INRIA Report No. 442.*
Pirzadeh, S. (1992). *AIAA Paper 92-0445*, 30th Aerospace Sciences Meeting, Reno, NV.
Pirzadeh, S. (1994). *AIAA Paper 94-0417*, 32nd Aerospace Sciences Meeting, Reno, NV.
Preparata, F.P. and Shamos, M.I. (1985). *Computational Geometry, An Introduction.* Springer-Verlag, Berlin.
Quirk, J.J. (1994). *Computers and Fluids* 23, 125–142.
Rebay, S. (1993). *J. Comput. Phys.* **106**, 125–138.
Ruge, J.W. and Stuben, K. (1987). In *Multigrid Methods* (S.F. McCormick, ed.), SIAM Frontiers in Applied Mathematics, pp. 73–131. SIAM, Philadelphia.
Samant, S.S., Bussoletti, J.E., Johnson, F.T., Burkhart, R.H., Everson, B.L., Melvin, R.G., Yound, D.P., Erickson, L.L., Madson, M.D. and Woo, A.C. (1987). *AIAA Paper 87-0034*, 25th Aerospace Sciences Meeting, Reno, NV.
Samet, H. (1990). *The Design and Analysis of Spatial Data Structures*, Addison-Wesley, Reading, MA.
Shaw, J.A. (1991). *Proc. 3rd International Conference on Numerical. Grid Generation in CFD and Fields* (A.S. Arcilla, J. Hauser, P.R. Eiseman and J.F. Thompson, eds), pp. 887–898. North Holland, New York.
Soni, B. (1992). *Comput. Math. Appl.* **24**, 191–201.
Sorensen, R.L. and McCann, K. (1992). *Proc. of Software Systems for Surface Modeling and Grid Generation Workshop* (R.E. Smith, ed.), NASA Langley Research Center, Hampton, VA.
Spardling, M.L., Nakamura, S. and Kuwahara, K. (1991). *Proc. 3rd International Conference on Numerical Grid Generation in CFD and Related Fields* (A.S. Arcilla, J. Hauser, P.R. Eiseman and J.F. Thompson, eds), pp. 237–245. North Holland, New York.
Sparis, P.D. (1985). *J. Comput. Phys.* **61**, 445–462.

Steger, J.L. and Sorenson, R.L. (1979). *J. Comput. Phys.* **33**, 405–410.

Steger, J.L. and Chaussee, D.S. (1980). *SIAM J. Sci. Stat. Comput.* **1**, 431–437.

Steinbrenner, J.P., Chawner, J.R. and Fouts, L.F. (1990). *AIAA Paper 90-1602*, 21st Fluid Dynamics Conference, Seattle, WA.

Swanson, R.C and Radespiel, R. (1991). *AIAA J.* **29**, 697–703.

Tanemura, M., Ogawa, T. and Ogita, N. (1983). *J. Comput. Phys.* **51**, 191–207.

Tembulkar, J.M. and Hanks, B.W. (1992). *Computers and Structures* **42**, 665–667.

Thomas, P.D. and Middlecoff, J.F. (1979). *AIAA J.* **18**, 652–656.

Thompson, F.T, Warsi, Z.U.A. and Mastin, C.W. (1985). *Numerical Grid Generation, Foundations and Applications*, North Holland, New York.

Thompson, J.F. (1988). *AIAA J.* **26** (3), 271–272.

Trepanier, J.Y., Zhang, H., Reggio, M. and Camarero, R. (1991). *Proc. 3rd International Conference on Numerical Grid Generation in CFD and Related Fields* (A.S. Arcilla, J. Hauser, P.R. Eiseman and J.F. Thompson, eds), pp. 43–45. North Holland, New York.

Tsuboi, H., Miyakoshi, K. and Kuwahara, K. (1991). *Proc. 3rd International Conference on Numerical Grid Generation in CFD and Related Fields* (A.S. Arcilla, J. Hauser, P.R. Eiseman and J.F. Thompson, eds), pp. 379–390. North Holland, New York.

Vallet, M.G., Hecht, F. and Mantel, B. (1991). *Proc. 3rd International Conference on Numerical Grid Generation in CFD and Related Fields* (A.S. Arcilla, J. Hauser, P.R. Eiseman and J.F. Thompson, eds), pp. 93–103. North Holland, New York.

Vatsa, V.N. and Wedan, B.W. (1989). *AIAA Paper 89-1791*, 20th Fluid Dynamics Conference, Buffalo, NY.

Venkatakrishnan, V. and Mavriplis, D.J. (1993). *J. Comput. Phys.* **105**, 83–91.

Venkatakrishnan, V. and Mavriplis, D.J. (1994).*AIAA Paper 94-0069*, 32nd Aerospace Sciences Meeting, Reno, NV.

Ward, S. and Kallinderis, Y. (1993). *AIAA Paper 93-0669*, 31st Aerospace Sciences Meeting, Reno, NV..

Warsi, Z.U.A. and Thompson, J.L. (1990). *Comput. Math. Appl.* **19**, 31–41.

Watson, D.F. (1981). *Computer J.* **24**, 167–172.

Weatherill, N.P. (1990). *Aeronautical J.* 94.934, pp. 111–123.

Weatherill, N.P. and Hassan, O. (1992).*Proc. 1st European CFD Conference* (C. Hirsch, J. Periaux and W. Kordulla). Elsevier, Brussels.

Weatherill, N.P. Hassan, O. and Marcum, D.L. (1993). *AIAA Paper 93-0341*, 31st Aerospace Sciences Meeting, Reno, NV.

Whitaker, D.L. (July 1993). *AIAA Paper 93-3337-CP*, 11th Computational Fluid Dynamics Conference, Orlando, FL.

Yerry, M.A. and Shephard, M.S. (1984).*Int. J. Numer. Methods Engrg.* **20**, 1965–1990.

Zhu, J.Z., Zienkiewicz, O.C., Hinton, E. and Wu, J. (1991). *Int. J. Numer. Methods Engrg.* **32**, 849–866.

Index